The Illustrated Wavelet Transform Handbook

Introductory Theory and Applications in Science, Engineering, Medicine and Finance

SECOND EDITION

The Illustrated Wavelet Transform Handbook

Introductory Theory and Applications in Science, Engineering, Medicine and Finance

SECOND EDITION

Paul S. Addison
Edinburgh, Scotland

CRC Press is an imprint of the
Taylor & Francis Group, an **informa** business

CRC Press
Taylor & Francis Group
6000 Broken Sound Parkway NW, Suite 300
Boca Raton, FL 33487-2742

First issued in paperback 2020

© 2017 by Taylor & Francis Group, LLC
CRC Press is an imprint of Taylor & Francis Group, an Informa business

No claim to original U.S. Government works

ISBN 13: 978-0-367-57400-0 (pbk)
ISBN 13: 978-1-4822-5132-6 (hbk)

This book contains information obtained from authentic and highly regarded sources. Reasonable efforts have been made to publish reliable data and information, but the author and publisher cannot assume responsibility for the validity of all materials or the consequences of their use. The authors and publishers have attempted to trace the copyright holders of all material reproduced in this publication and apologize to copyright holders if permission to publish in this form has not been obtained. If any copyright material has not been acknowledged please write and let us know so we may rectify in any future reprint.

Except as permitted under U.S. Copyright Law, no part of this book may be reprinted, reproduced, transmitted, or utilized in any form by any electronic, mechanical, or other means, now known or hereafter invented, including photocopying, microfilming, and recording, or in any information storage or retrieval system, without written permission from the publishers.

For permission to photocopy or use material electronically from this work, please access www.copyright.com (http://www.copyright.com/) or contact the Copyright Clearance Center, Inc. (CCC), 222 Rosewood Drive, Danvers, MA 01923, 978-750-8400. CCC is a not-for-profit organization that provides licenses and registration for a variety of users. For organizations that have been granted a photocopy license by the CCC, a separate system of payment has been arranged.

Trademark Notice: Product or corporate names may be trademarks or registered trademarks, and are used only for identification and explanation without intent to infringe.

Library of Congress Cataloging-in-Publication Data

Names: Addison, Paul S., author.
Title: The illustrated wavelet transform handbook : introductory theory and applications in science, engineering, medicine and finance / Paul S. Addison.
Description: Second edition. | Boca Raton, FL : CRC Press, Taylor & Francis Group, [2016] | Includes bibliographical references and index.
Identifiers: LCCN 2016033578| ISBN 9781482251326 (hardback ; alk. paper) | ISBN 1482251329 (hardback ; alk. paper) | ISBN 9781482251333 (e-book) | ISBN 1482251337 (e-book)
Subjects: LCSH: Wavelets (Mathematics) | Fourier analysis. | Biomedical engineering.
Classification: LCC QA403.3 .A35 2016 | DDC 515/.2433--dc23
LC record available at https://lccn.loc.gov/2016033578

Visit the Taylor & Francis Web site at
http://www.taylorandfrancis.com

and the CRC Press Web site at
http://www.crcpress.com

To Hannah, Stephen, Anthony, Michael and, this time, David; to my dear parents, Josephine and Stanley, gone but still with us; and to my loving wife, Stephanie.

Thank you.

Contents

Preface to Second Edition, xiii

Preface to First Edition, xv

CHAPTER 1 ■ Getting Started		1
1.1	INTRODUCTION	1
1.2	WAVELET TRANSFORM	2
1.3	READING THE BOOK	4
CHAPTER 2 ■ The Continuous Wavelet Transform		7
2.1	INTRODUCTION	7
2.2	THE WAVELET	7
2.3	REQUIREMENTS FOR THE WAVELET	10
2.4	THE ENERGY SPECTRUM OF THE WAVELET	10
2.5	WAVELET TRANSFORM	12
2.6	IDENTIFICATION OF COHERENT STRUCTURES	14
2.7	EDGE DETECTION	22
2.8	INVERSE WAVELET TRANSFORM	24
2.9	SIGNAL ENERGY: WAVELET-BASED ENERGY AND POWER SPECTRA	28
2.10	WAVELET TRANSFORM IN TERMS OF THE FOURIER TRANSFORM	33
2.11	COMPLEX WAVELETS: THE MORLET WAVELET	34
2.12	WAVELET TRANSFORM, SHORT-TIME FOURIER TRANSFORM AND HEISENBERG BOXES	44
2.13	ADAPTIVE TRANSFORMS: MATCHING PURSUITS	51
2.14	WAVELETS IN TWO OR MORE DIMENSIONS	55
2.15	THE CWT: COMPUTATION, BOUNDARY EFFECTS AND VIEWING	56
2.16	RIDGE FOLLOWING AND SECONDARY WAVELET FEATURE DECOUPLING	61
2.17	RIDGE HEIGHTS	66
2.18	RUNNING WAVELET ARCHETYPING	67

2.19	WAVELET TRANSFORM REPHASING		69
2.20	REASSIGNMENT AND SYNCHROSQUEEZING		73
2.21	COMPARING TWO SIGNALS USING WAVELET TRANSFORMS		75
	2.21.1	Transform Differences and Ratios	76
	2.21.2	Cross-Wavelet Transform	78
	2.21.3	Wavelet Cross-Correlation	81
	2.21.4	Phase Comparison Measures	83
	2.21.5	Wavelet Coherence	85
2.22	BICOHERENCE AND CROSS-BICOHERENCE		86
2.23	ENDNOTES		88
	2.23.1	Chapter Key Words and Phrases	88
	2.23.2	Additional Notes and Resources	88
	2.23.3	Things to Try	89
	2.23.4	Final Note: The CWT as a 'Soft Tool' for Algorithm Development	91

CHAPTER 3 ■ The Discrete Wavelet Transform   93

3.1	INTRODUCTION		93
3.2	FRAMES AND ORTHOGONAL WAVELET BASES		93
	3.2.1	Frames	93
	3.2.2	Dyadic Grid Scaling and Orthonormal Wavelet Transforms	95
	3.2.3	Scaling Function and Multiresolution Representation	98
	3.2.4	Scaling Equation, Scaling Coefficients and Associated Wavelet Equation	101
	3.2.5	Haar Wavelet	102
	3.2.6	Coefficients from Coefficients: Fast Wavelet Transform	105
3.3	DISCRETE INPUT SIGNALS OF FINITE LENGTH		107
	3.3.1	Approximations and Details	107
	3.3.2	Multiresolution Algorithm: An Example	111
	3.3.3	Wavelet Energy	114
	3.3.4	Alternative Indexing of Dyadic Grid Coefficients	115
	3.3.5	A Simple Worked Example: The Haar Wavelet Transform	117
3.4	EVERYTHING DISCRETE		122
	3.4.1	Discrete Experimental Input Signals	122
	3.4.2	Smoothing, Thresholding and Denoising	127
3.5	DAUBECHIES WAVELETS		135
	3.5.1	Filtering	143

		3.5.2 Symmlets and Coiflets	148
3.6	TRANSLATION INVARIANCE		149
3.7	BIORTHOGONAL WAVELETS		150
3.8	TWO-DIMENSIONAL WAVELET TRANSFORMS		151
3.9	ADAPTIVE TRANSFORMS: WAVELET PACKETS		163
3.10	'X-LETS': CONTOURLETS, RIDGELETS, CURVELETS, SHEARLETS AND SO ON		168
3.11	ENDNOTES		173
	3.11.1	Chapter Key Words and Phrases	173
	3.11.2	Further Resources	173

CHAPTER 4 ■ Fluids   177

4.1	INTRODUCTION		177
4.2	STATISTICAL MEASURES FOR FLUID TURBULENCE		178
	4.2.1	Moments, Energy and Power Spectra	178
	4.2.2	Intermittency and Correlation	185
	4.2.3	Wavelet Thresholding	186
	4.2.4	Wavelet Selection Using Entropy Measures	191
4.3	ENGINEERING FLOWS		194
	4.3.1	Experimental Flows: Jets, Wakes, Turbulence and Coherent Structures	194
	4.3.2	Computational Fluid Dynamics: Simulation and Analysis	200
	4.3.3	Fluid–Structure Interaction	206
4.4	GEOPHYSICAL FLOWS		210
	4.4.1	Atmospheric Processes: Wind, Boundary Layers and Turbulence	211
	4.4.2	Ocean Processes: Waves, Large-Scale Oscillations, Ocean–Atmosphere Interactions and Biological Processes	215
	4.4.3	Rainfall and River Flows	221
4.5	TWO-PHASE FLOWS		224
4.6	OTHER APPLICATIONS IN FLUIDS		227

CHAPTER 5 ■ Engineering Testing, Monitoring and Characterization   231

5.1	INTRODUCTION		231
5.2	DYNAMICS		232
	5.2.1	Fundamental Behaviour	232
	5.2.2	Chaos	236

5.3	NON-DESTRUCTIVE TESTING OF STRUCTURAL ELEMENTS		238
5.4	CONDITION MONITORING OF ROTATING MACHINERY		249
	5.4.1	Gears	249
	5.4.2	Shafts, Bearings and Blades	255
5.5	MACHINING PROCESSES		259
5.6	CHARACTERIZATION OF SURFACES AND FIBROUS MATERIALS		261
5.7	OTHER APPLICATIONS IN ENGINEERING		263
	5.7.1	Compression	263
	5.7.2	Control	264
	5.7.3	Electrical Systems and Circuits	265
	5.7.4	Miscellaneous	265

CHAPTER 6 ■ Medicine 267

6.1	INTRODUCTION		267
6.2	ELECTROCARDIOGRAM		267
	6.2.1	ECG Beat Detection and Timings	268
	6.2.2	Detection of Abnormalities	271
	6.2.3	Heart Rate Variability	275
	6.2.4	Cardiac Arrhythmias	278
	6.2.5	ECG Data Compression	289
	6.2.6	Hardware Implementation	289
6.3	NEUROELECTRIC WAVEFORMS		290
	6.3.1	Evoked Potentials and Event-Related Potentials	291
	6.3.2	Epileptic Seizures and Epileptogenic Foci	297
	6.3.3	Sleep Studies	299
	6.3.4	Other Areas	302
6.4	PHOTOPLETHYSMOGRAM		302
	6.4.1	Respiratory Modulations, Respiratory Rate and Respiratory Effort	303
	6.4.2	Oxygen Saturation	307
	6.4.3	The Video Photoplethysmogram (Video-PPG)	309
	6.4.4	Other Areas	312
6.5	PATHOLOGICAL SOUNDS, ULTRASOUNDS AND VIBRATIONS		315
	6.5.1	Cardiovascular System	315
	6.5.2	Lung Sounds, Swallowing, Snoring and Speech	318
	6.5.3	Acoustic Response	321

	6.6	BLOOD FLOW AND BLOOD PRESSURE	323
	6.7	MEDICAL IMAGING	328
		6.7.1 Optical Imaging	328
		6.7.2 Ultrasonic Images	330
		6.7.3 Computed Tomography, Magnetic Resonance Imaging and Other Radiographic Images	330
	6.8	OTHER APPLICATIONS IN MEDICINE	332
		6.8.1 Electromyographic Signals	332
		6.8.2 Posture, Gait and Activity	333
		6.8.3 Analysis of Multiple Biosignals	334
		6.8.4 Miscellaneous	334

CHAPTER 7 ■ Fractals, Finance, Geophysics, Astronomy and Other Areas 337

	7.1	INTRODUCTION	337
	7.2	FRACTALS	337
		7.2.1 Exactly Self-Similar Fractals	338
		7.2.2 Stochastic Fractals	341
		7.2.3 Multifractals	350
	7.3	FINANCE	355
	7.4	GEOPHYSICS	361
		7.4.1 Properties of Subsurface Media: Well Logging, Cores and Seismic Methods	361
		7.4.2 Remote Sensing	366
	7.5	ASTRONOMY: SIGNALS AND IMAGES	371
	7.6	OTHER AREAS	378

REFERENCES, 381

APPENDIX: USEFUL BOOKS, PAPERS AND WEBSITES, 441

INDEX, 443

Preface to Second Edition

SINCE THE PUBLICATION OF the first edition in 2002, there has been an explosion of interest in wavelet transform methods. In the intervening period, I have moved on from academia to a start-up company, and then to a global medical device company; and a significant part of this time has been spent working with wavelet methods. Working in a small start-up requires rolling up your sleeves and get involved in all aspects of the business. I was lucky enough to get the opportunity work closely with many interesting individuals from a wide range of diverse backgrounds across disciplines and across the world: amazing engineers and scientists from many sub-specialities; dedicated clinicians working at the forefront of medical science in a variety of areas of care; skilful lawyers practicing corporate law and the management of intellectual property; and some quite remarkably astute businessmen and women. This has given me a deep appreciation of what constitutes real professionalism in other disciplines; many of which were previously a complete mystery to me. I have made many friends and met many good folk along the way: smart, dedicated professionals with integrity. None more so, nor more enjoyable company, than Jamie Watson, with whom I began the rollercoaster ride that was our start-up company CardioDigital Ltd, and Peter Galen, who joined us en route with still quite a few peaks and troughs to ride out. Thanks guys.

I have been fortunate enough to be inventor on around 100 US patents to date, with around 200 US patent applications in the system stemming from approximately 400 submitted ideas (dyadic scaling!) – and many of these concerning wavelet methods. I find that grasping the concepts before tackling the mathematics is my best way forward when it comes to understanding and innovation, and I know no better way of achieving the former than through the use of an illustration: a quickly scribbled diagram, flow chart or cartoon drawing. I often write instructions for colleagues in terms of diagrams rather than text and I like drawing mind maps as I explain things. Hence, *The Illustrated Wavelet Transform Handbook*. Wavelet transform methods produce distinct morphologies in 'wavelet space'- a concept that is so much easier to grasp, and more importantly feel comfortable 'within', when viewing a well-constructed plot: whether it be a modulus or phase plot or something that combines both. Once familiar with these mathematical 'landscapes', we may begin to explore unchartered territory. That's when the real fun begins!

I must thank a large number of people for their considerable time and effort in providing comment on various sections of the draft text, without all of whom I really could not have got to the end of this task. These are, in roughly the order of the text: Dr. Ge, CFD group lead, American Bureau of Shipping, Houston, Texas, USA; Dr Christopher Torrence, Harris Geospatial Solutions, Boulder, Colorado; and Professor Stephen Payne of the University of Oxford for reading over various parts of the new theory sections in Chapter 2; Dr André Atunes, a colleague and multi-talented research scientist, for providing detailed comment on the mathematics and explanatory text for the whole of Chapter 2; Professor James Walker, Department of Mathematics at the University of Wisconsin-Eau Claire for his comments on various sections in the theory chapters, 2 and 3; Associate Professor Paul Brandner of the Cavitation Research Laboratory within the Australian Maritime College at the University of Tasmania for his assiduous reading and most useful comments with respect to all aspects of fluid phenomena in Chapter 4; Professor Wieslaw Staszewski of the AGH University of Science & Technology in Krakow for his comments on various engineering sections in Chapter 5; James Ochs, an erstwhile colleague and excellent biomedical engineer, for reading over the whole of Chapter 6; Dr Paul Mannheimer, another former colleague and a recognised world expert in pulse oximetry, for his comments on the section concerning the analysis of the photoplethysmogram in Chapter 6; Dr Dean Montgomery, a colleague and outstanding scientist and engineer, for reading through the medical images section of Chapter 6; Professor Leontios Hadjileontiadis of the Aristotle University of Thessaloniki, Greece and the Khalifa University, Abu Dhabi who provided comment on the wavelet coherence theory in Chapter 2 and the section on physiological sounds in Chapter 6. Finally, a number of people were extremely helpful in checking Chapter 7, with its very wide ranging selection of topics. These included Dr Maria Haase of the Institut für Höchstleistungsrechnen, Universität Stuttgart for reading over the fractal and multifractal sections of Chapter 7, as she did for the first edition with such enthusiasm and rigor; Dr Ginanjar Dewandaru of INCEIF, The Global University of Islamic Finance, Lorong University A, Kuala Lumpur, Malaysia and Claudiu Albulescu, Associate Professor in the Management Department at the Politehnica University of Timisoara, Romania for comments regarding the finance section of Chapter 7; Mathieu J. Duchesne, Research Scientist, Geological Survey of Canada, Quebec City for his thoughts on the geophysics section, and Matthew S. Tiscareno, Senior Research Scientist at the SETI Institute and Dr. Florent Mertens, Kapteyn Astronomical Institute, Groningen, The Netherlands for proof reading the astronomy section in Chapter 7 in meticulous detail.

I'd also like to take the opportunity to express my gratitude to the Wellcome Trust for their unwavering support for our university spin-off company, CardioDigital Ltd, over the years. In the same vein, thank you to the Scottish Government and Scottish Enterprise for their invaluable grant funding.

Finally, thanks to my daughter Hannah for her unbelievable attention to detail while proof reading over the whole manuscript from a non-technical vantage point; my sons, Michael, Anthony, Stephen and David for their interest in what I do (and help picking the new cover); and my wife Stephanie, for everything.

Paul S. Addison

Preface to First Edition

OVER THE PAST DECADE or so, wavelet transform analysis has emerged as a major new time–frequency decomposition tool for data analysis. This book is intended to provide the reader with an overview of the theory and practical applications of wavelet transform methods. It is designed specifically for the 'applied' reader, whether he or she is a scientist, engineer, medic, financier or other.

The book is split into two parts: theory and application. After a brief first chapter which introduces the main text, the book tackles the theory of the continuous wavelet transform in Chapter 2 and the discrete wavelet transform in Chapter 3. The rest of the book provides an overview of a variety of applications. Chapter 4 covers fluid flows. Chapter 5 tackles engineering testing, monitoring and characterization. Chapter 6 deals with a wide variety of medical research topics. The final chapter, Chapter 7, covers a number of subject areas. In this chapter, three main topics are considered initially – fractals, finance and geophysics – and then these are followed by a general discussion which includes many other areas that are not covered in the rest of the book.

The chapters which address theory (2 and 3) are written at an advanced undergraduate level. In these chapters, I have used *italics* for both mathematical symbols and key words and phrases. The key words and phrases are listed at the end of each chapter and the reader who is new to the subject might find it useful to jot down the meaning of each key word or phrase to test his or her understanding of them. The chapters which address the various applications of the theory (Chapters 4 through 7) are at the same level, although a considerable amount of useful information can be gained without an in-depth knowledge of the theory in Chapters 2 and 3, especially in providing an overview of these applications.

It is envisaged that the book will be of use both to those new to the subject, who want somewhere to begin learning about the topic, and also to those who have been working in a particular area for some time and would like to broaden their perspective. It can be used as a handbook, or 'handy book', which can be referred to when appropriate for information. The book is very much 'figure driven' as I believe that figures are extremely useful for illustrating the mathematics and conveying the concepts. The application chapters of the book aim to make the reader aware of the similarities that exist in the uses of wavelet transform analysis across disciplines. In addition, and perhaps more importantly, it is intended to make the reader aware of wavelet-based methods in use in unfamiliar disciplines which may be transferred to his or her own area – thus promoting an interchange of ideas across discipline boundaries.

The application chapters are essentially a whistle-stop tour of work by a large number of researchers around the globe. Some examples of this work are discussed in more detail than others and, in addition, a large number of illustrations have been used which have been taken (with permission) from a variety of published material. The examples and illustrations used have been chosen to provide an appropriate range to best illustrate the wavelet-based work being carried out in each subject area. It is not intended to delve deeply into each subject but rather to provide a brief overview. It is then left to the reader to follow up the relevant references cited in the text for himself or herself in order to delve more deeply into each particular topic as he or she requires.

I refer to over 700 scientific papers in this book which I have collected and read over the past three or so years. I have made every effort to describe the work of others as concisely and accurately as possible. However, if I have misquoted, misrepresented, misinterpreted or simply missed out something I apologize in advance. Of course, all comments are welcome – my e-mail address can be found below.

The book stems from my own interest in wavelet transform analysis over the past few years. This interest has led to a number of research projects concerning the wavelet-based analysis of both engineering and medical signals; including non-destructive testing signals, vortex-shedding signals in turbulent fluid flows, digitized spatial profiles of structural cracks, river bed sediment surface data sets, phonocardiographic signals, pulse oximetry traces (photoplethysmograms) and the electrocardiogram (ECG), the latter leading to patent applications and a university spin-off company, Cardiodigital Ltd.

Quite a mixed bag, at first appearance, but with a common thread of wavelet analysis running throughout. I have featured some of this work in the appropriate chapters. However, I have tried not to swamp the application chapters with my own work – although the temptation was high for a number of reasons, including knowledge of the work, ease of reproduction, etc. I hope that I have struck the correct balance.

All books reflect, to some extent, the interests and opinions of the author and, although I have tried to cover as broad a range of examples as possible, this one is no exception. Coverage weighs more heavily towards those areas in which I have more interest: fluids, engineering, medicine and fractal geometry. Geophysics and finance are given less space and other areas (e.g. astronomy, chemistry, physics, non-medical biology, power systems analysis) are detailed briefly in the final chapter.

There are some idiosyncrasies in the text which are worth pointing out. I am an f person not an ω person: I prefer hertz to radians per second. I can tap my fingers at approximately 5 Hz, or 1 Hz, I know what 50 Hz means (mains hum in the UK) and so on; however, ω I have to convert. Hence the frequencies in the text are in the form of 1/time either in hertz or non-dimensionalized. The small downside is that the mathematics, in general, contains a few more terms – mostly 2s and πs. I have devoted a whole chapter to the continuous wavelet transform. It is noticeable that many current wavelet texts on the market deal only with the discrete wavelet transform, or give the continuous wavelet transform a brief mention en route to the theory of the discrete wavelet transform. I believe that the continuous wavelet transform has a wide variety of data analysis tasks to offer, and I attempt, through this text, to redress the balance somewhat. (Actually, the proportion of published papers which concern the continuous wavelet transform as opposed to the discrete wavelet transform is much higher than that represented by the currently available wavelet textbooks.) The book is focused on the wavelet transform and makes only passing reference in the application chapters to some of the other time–frequency methods now available. However, I have added sections on the short-time Fourier transform and matching pursuits towards the end of Chapter 2 and on wavelet packets at the end of Chapter 3, respectively. Finally, note that I have developed the discrete wavelet transform theory in Chapter 3 in terms of scale rather than resolution, although the relationship between the alternative notations is explained.

I would like to thank the following people for taking the time to comment on various drafts of the manuscript: Andrew Chan of Birmingham University, Gareth Clegg of Edinburgh University (formerly at the Royal Infirmary of Edinburgh), Maria Haase of Stuttgart University and Alexander Droujinine of Heriot-Watt University. I would like to thank Jamie Watson of CardioDigital Ltd. for his comments on the draft manuscript and for his close collaboration over the years (and various bits of computer code!). I would also like to thank all of the authors and publishers who gave their consent to reproduce their figures within this text. I am grateful to those funding bodies who have supported my research in wavelet analysis and other areas over the years, including the Engineering and Physical Science Research Council (EPSRC), the Medical Research Council (MRC) and the Leverhulme Trust. And to the other colleagues and collaborators with whom my wavelet research is conducted and who make it so interesting, thanks.

Special thanks to my wife, Stephanie, who has supported and encouraged me during the writing of this book. Special thanks also to my parents for their support and great interest in what I do.

Although it has been a long hard task, I have enjoyed putting this book together. I have certainly got a lot out of it. I hope you find it useful.

Paul S. Addison

CHAPTER **1**

Getting Started

1.1 INTRODUCTION

The wavelet transform (WT) has been found to be particularly useful for analyzing signals which can best be described as aperiodic, noisy, intermittent, transient and so on. Its ability to examine the signal simultaneously in both time and frequency in a distinctly different way from the traditional short-time Fourier transform (STFT) has spawned an ever-increasing number of sophisticated wavelet-based methods for signal manipulation and interrogation. Wavelet transform analysis has now been applied in the investigation of a multitude of diverse physical phenomena, from climate analysis to the analysis of financial indices, from heart monitoring to the condition monitoring of rotating machinery, from seismic signal denoising to the denoising of astronomical signals and images, from surface characterization to the characterization of turbulent intermittency, from video image compression to the compression of medical signal records, and so on.

Many of the ideas behind wavelet transforms have been in existence for a long time. However, wavelet transform analysis as we now know it really began in the mid-1980s, when it was developed to interrogate seismic signals. Interest in wavelet analysis remained within a small, mainly mathematical community during the rest of the 1980s, with only a handful of scientific papers coming out each year. The application of wavelet transform analysis in science and engineering really began to take off at the beginning of the 1990s, with a rapid growth in the numbers of researchers turning their attention to wavelet analysis during that decade. There are now thousands of refereed journal papers concerning the wavelet transform, and these cover all numerate disciplines. The wavelet transform is a mathematical tool which is now an indispensible part of many data analysts' toolboxes. This book aims to provide the reader with both an introduction to the theory of wavelet transforms and a wide-ranging overview of their use in practice. The two remaining sections of this short introductory chapter contain, respectively, a brief non-mathematical description of the wavelet transform and a guide to subsequent chapters of the book.

1.2 WAVELET TRANSFORM

Wavelet transform analysis uses little wavelike functions known as *wavelets*. Actually, *localized* wavelike function is a more accurate description of a wavelet. Figure 1.1a shows a few examples of wavelets commonly encountered in practice. Wavelets are used to transform the signal under investigation into another representation which presents the signal information in a more useful form. This transformation of the signal is known as the *wavelet transform*. Mathematically speaking, the wavelet transform may be interpreted as a convolution of the signal with a wavelet function, and we will see exactly how this is done in Chapters 2 and 3. Here, we stick to schematics.

The wavelet can be manipulated in two ways: it can be moved to various locations on the signal (Figure 1.1b) and it can be stretched or squeezed (Figure 1.1c). Figure 1.2 shows a schematic of the wavelet transform which basically quantifies the local matching

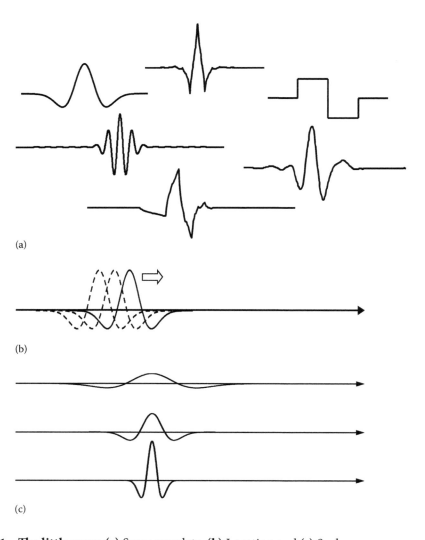

FIGURE 1.1 **The little wave: (a)** Some wavelets, **(b)** Location and **(c)** Scale.

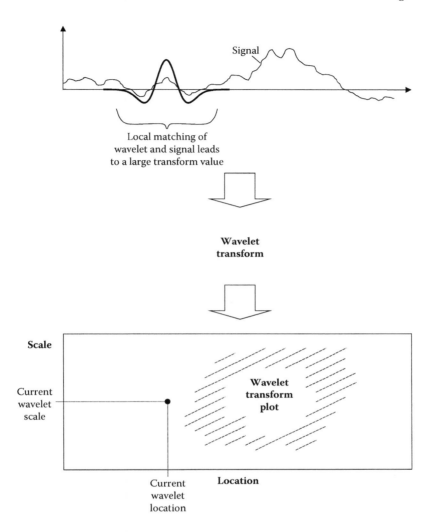

FIGURE 1.2 The wavelet, the signal and the transform.

of the wavelet with the signal. If the wavelet matches the shape of the signal well at a specific scale and location, as it happens to do in the top plot of Figure 1.2, then a large transform value is obtained. If, however, the wavelet and the signal do not correlate well, a low transform value is obtained. The transform value is then located in the two-dimensional transform plane shown at the bottom of Figure 1.2 (indicated by the black dot). The transform is computed at various locations of the signal and for various scales of the wavelet, thus filling up the transform plane: this is done in a smooth continuous fashion for the *continuous wavelet transform* (CWT) or in discrete steps for the *discrete wavelet transform* (DWT). Plotting the wavelet transform allows a picture to be built up of the correlation between the wavelet – at various scales and locations – and the signal. In subsequent chapters, we will cover the wavelet transform in more mathematical detail.

1.3 READING THE BOOK

The purpose of the book is both to introduce the wavelet transform and to convey its multidisciplinary nature. This is achieved in subsequent chapters by first providing an elementary introduction to wavelet transform theory and then presenting a wide range of examples of its application. It will quickly become apparent that very often the same wavelet methods are used to interrogate signals from very different subject areas, where quite unrelated phenomena are under investigation.

The book is split into two distinct parts: the first – comprising Chapters 2 and 3 – deals, respectively, with continuous wavelet transform theory and discrete transform theory, while the second – comprising Chapters 4 through 7 – presents examples of their application in science, engineering, medicine and finance. There are a number of ways to read this book, from the linear (beginning to end) via the targeted (employing the index) to the random (flicking through) approach. The reader unfamiliar with wavelet theory should read Chapters 2 and 3 before moving on to the sections of particular relevance to his or her own area of interest. The reader is, however, also advised to look outwith his or her own area to see how wavelets are being employed elsewhere. (The author cannot emphasize this enough!) Details of further resources concerning the theory and applications of wavelet analysis are provided at the end of each chapter. The appendix lists a selection of useful books, papers and websites. The book contents are outlined in more detail as follows:

Chapter 2: This chapter presents the basic theory of the continuous wavelet transform. It outlines what constitutes a wavelet and how it is used in the transformation of a signal. The continuous wavelet transform is compared with both the short-time Fourier transform and the matching pursuit method. Transform features such as ridges and local phase cycling are then considered before examining various useful manipulations of the transform such as rephasing, the production of a transform archetype, synchrosqueezing and reassignment. Finally, the relationship between two transformed signals is considered via transform differences and ratios, the cross-wavelet transform, wavelet cross-correlation, phase comparison measures, wavelet coherence and bicoherence.

Chapter 3: The discrete wavelet transform is described in this chapter. Orthonormal discrete wavelet transforms are considered in detail, in particular those of Haar and Daubechies. These wavelets fit into a multiresolution analysis framework where a discrete input signal can be represented at successive approximations by a combination of a smoothed signal component plus a sum of detailed wavelet components. The chapter also briefly covers wavelet packets – a generalization of the discrete wavelet transform which allows for adaptive partitioning of the time–frequency plane – prior to ending with a section on time–frequency functions that stem from the wavelet method and allow efficient representation of directionally dependent two-dimensional data.

Chapter 4: This chapter deals with fluid mechanics, a subject that is always open to new mathematical techniques. The time-frequency localization properties of the wavelet transform have been employed extensively in the study of a wide variety of fluid phenomena, including the intermittent nature of fluid turbulence, the characteristics of turbulent jets, the nature of fluid–structure interactions and the behaviour of large-scale geophysical flows. Chapter 4 also contains the mathematics for discrete wavelet statistics and power spectra following on from some of the basic theory given in Chapter 3. Thus, it is worthwhile reading Section 4.2 of this chapter even if fluids is not your area of interest.

Chapter 5: In this chapter, a close look is taken at the application of wavelet transforms to a variety of pertinent problems in engineering. These applications include the analysis of fundamental dynamical behaviour and chaotic motions, non-destructive testing of structural elements and the condition monitoring of machinery, machining processes, and the characterization of surfaces and fibrous materials.

Chapter 6: Medical applications of wavelet transform analysis are covered in this chapter. Wavelet transform methods have been used to characterize a wide variety of medical signals. Many of these are reviewed in this chapter, including the electrocardiogram (ECG), electroencephalogram (EEG), electromyogram (EMG) photoplethysmogram (PPG), pathological sounds (heart murmurs, lung sounds, swallowing sounds, snoring, speech, otoacoustic emissions), blood flows, blood pressures, medical images (optical, x-ray, NMR, ultrasound, etc.), and posture, gait and activity signals.

Chapter 7: This final chapter covers a variety of areas of application. Most of the chapter is devoted to four main subjects – fractal geometry, finance, geophysics and astronomy – with a separate section given over to each one of them. The final part of the chapter provides a brief account of the role that wavelet transform analysis has played in a number of other areas, including quantum mechanics, chemistry, ecological processes and patterns, and more.

Appendix 1: The appendix contains a brief list of useful books and websites concerning wavelet transform theory and its application. These have been chosen by the author for their extensive content and/or clarity of presentation.

CHAPTER 2

The Continuous Wavelet Transform

2.1 INTRODUCTION

This chapter covers the basic theory of the continuous wavelet transform (CWT). We will first determine what constitutes a wavelet, how it is used in the transformation of a signal and what it can tell us about the signal. The inverse wavelet transform and the reconstruction of the original signal are then considered prior to an investigation of the energy-preserving features of the wavelet transform (WT) and how it may be used to produce wavelet power spectra. We will compare the wavelet transform to both the short-time Fourier transform (STFT) and matching pursuit (MP) method. Properties of transform features are then dealt with, such as ridges and local phase cycling, and their relationship to the original signal. Various useful manipulations of the transform are also studied, including rephasing, wavelet archetyping, synchrosqueezing and reassignment. Towards the end of the chapter, we will delve into the use of transform techniques to compare two related signals, including transform differences and ratios, the cross-wavelet transform (CrWT), wavelet cross-correlation (WCC), phase comparison measures, wavelet coherence (WCH) and wavelet bicoherence (WBC).

2.2 THE WAVELET

The wavelet transform is a method of converting a function (or signal) into another form which either makes certain features of the original signal more amenable to study or enables the original data set to be described more succinctly. To perform a wavelet transform, we need a *wavelet* which, as the name suggests, is a localized waveform. In fact, a wavelet is a function, $\psi(t)$, which satisfies certain mathematical criteria. As we saw briefly in the previous chapter, these functions are manipulated through a process of translation (i.e. movements along the time axis) and dilation (i.e. the spreading out of the wavelet) to transform the signal into another form which 'unfolds' in time and scale. Note that in this chapter and the next, we assume that the signal to be analyzed is a temporal signal, that is, some function of time such as a velocity trace from a fluid, vibration data from an engine casing, an electrocardiogram (ECG), a financial signal and so on. However, many of the applications discussed later in the book concern wavelet analysis of spatial signals, such as well logged geophysical data, crack surface profiles and so on. In these cases, the independent variable is space rather than time; however, the wavelet analysis is performed in exactly the same way.

A selection of wavelets that are commonly used are shown in Figure 2.1; we will consider some of them in more detail as we proceed through the text. As we can see from the figure, they have the form of a small wave, localized on the time axis. There are, in fact, a large number of wavelets to choose from for use in the analysis of our data. The best one for a particular application depends on both the nature of the signal and what we require from the analysis (i.e. what physical phenomena or process we are looking to interrogate, or how we are trying to manipulate the signal). We will begin this chapter by concentrating on a specific wavelet, the Mexican hat, which is very good at illustrating many of the properties of continuous wavelet transform analysis. The Mexican hat wavelet is shown in Figure 2.1b and in more detail in Figure 2.2a. The *Mexican hat wavelet* is defined as

$$\psi(t) = (1-t^2)e^{-t^2/2} \tag{2.1}$$

The wavelet described by Equation 2.1 is known as the *mother wavelet* or *analyzing wavelet*. This is the basic form of the wavelet from which dilated and translated versions are derived and used in the wavelet transform. The Mexican hat is, in fact, the second derivative of the Gaussian distribution function $e^{-(t^2/2)}$, that is, with unit variance but without the usual

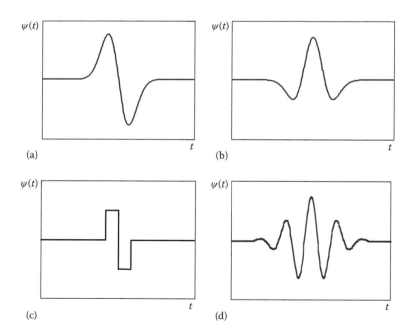

FIGURE 2.1 Four wavelets: **(a)** Gaussian wave (first derivative of a Gaussian). **(b)** Mexican hat (second derivative of a Gaussian). **(c)** Haar. **(d)** Morlet (real part).

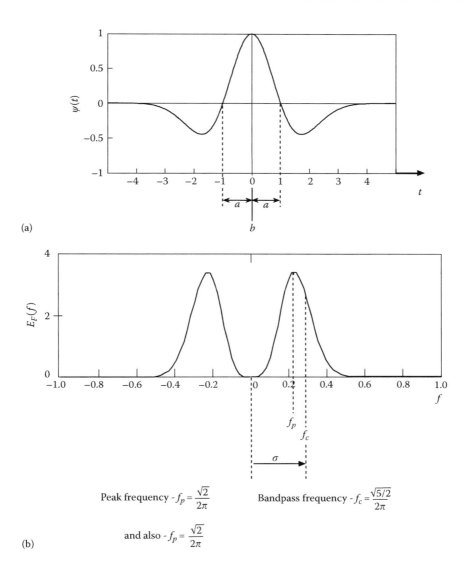

FIGURE 2.2 **The Mexican hat mother wavelet and its associated energy spectrum:** (a) The Mexican hat mother wavelet (named for an obvious reason!). Notice that, for the Mexican hat, the dilation parameter a is the distance from the centre of the wavelet to where it crosses the horizontal axis. (b) The energy spectrum of the Mexican hat shown in (a). Note that as it is a real wavelet, its Fourier spectrum is mirrored around the zero axis. (σ is the standard deviation of the spectrum around the vertical axis.)

$1/\sqrt{2\pi}$ normalization factor. The Mexican hat normally used in practice (i.e. that given by Equation 2.1 and shown in Figure 2.2a) is actually the negative of the second derivative of the Gaussian function. All derivatives of the Gaussian function may be employed as a wavelet; which is the most appropriate to use depends on the application. Both the first and second derivatives of the Gaussian function are shown in Figure 2.1a,b. These are the two that are most often used in practice. Higher-order derivatives are less commonplace.

2.3 REQUIREMENTS FOR THE WAVELET

In order to be classified as a wavelet, a function must satisfy certain mathematical criteria. These are

1. A wavelet must have finite energy:

$$E = \int_{-\infty}^{\infty} |\psi(t)|^2 \, dt < \infty \tag{2.2}$$

   where E is the energy of a function equal to the integral of its squared magnitude, the vertical brackets '| |' represent the *modulus operator* which gives the magnitude of $\psi(t)$. If $\psi(t)$ is a complex function, the magnitude must be found using both its real and complex parts.

2. If $\hat{\psi}(f)$ is the *Fourier transform* of $\psi(t)$, that is

$$\hat{\psi}(f) = \int_{-\infty}^{\infty} \psi(t) e^{-i(2\pi f)t} \, dt \tag{2.3}$$

   then the following condition must hold:

$$C_g = \int_{0}^{\infty} \frac{|\hat{\psi}(f)|^2}{f} \, df < \infty \tag{2.4}$$

   This implies that the wavelet has no zero-frequency component, $\hat{\psi}(0) = 0$, or to put it another way, the wavelet $\psi(t)$ must have a zero mean. Equation 2.4 is known as the *admissibility condition* and C_g is called the *admissibility constant*. The value of C_g depends on the chosen wavelet and is equal to π for the Mexican hat wavelet given in Equation 2.1.

3. An additional criterion that must hold for complex wavelets is that the Fourier transform must both be real and yet also vanish for negative frequencies. We will consider complex wavelets towards the end of this chapter when we take a close look at the Morlet wavelet.

2.4 THE ENERGY SPECTRUM OF THE WAVELET

Wavelets satisfying the admissibility condition (Equation 2.4) are in fact *bandpass filters*. This means, in simple terms, that they let through only those signal components within a finite range of frequencies (the *passband*) and in proportions characterized by the energy spectrum of the wavelet. A plot of the squared magnitude of the Fourier transform against the frequency of the wavelet gives its *energy spectrum*. For example, the Fourier energy spectrum of the Mexican hat wavelet is given by

$$E_F(f) = |\hat{\psi}(f)|^2 = 32\pi^5 f^4 e^{-4\pi^2 f^2} \tag{2.5}$$

where subscript F is used to denote the Fourier spectrum as distinct from the wavelet-based spectrum defined in Section 2.9. A plot of the energy spectrum of the Mexican hat wavelet is shown in Figure 2.2b. Note that as the Mexican hat wavelet is a real function, its Fourier spectrum is symmetric about zero. We will see later, in Section 2.11, that complex wavelets do not have negative frequency components (requirement 3). The peak of the energy spectrum occurs at a dominant frequency of $f_p = \pm\sqrt{2}/2\pi$. The second moment of area of the energy spectrum is used to define the *passband centre* of the energy spectrum, f_c, as follows:

$$f_c = \sqrt{\frac{\int_0^\infty f^2 |\hat{\psi}(f)|^2 \, df}{\int_0^\infty |\hat{\psi}(f)|^2 \, df}} \tag{2.6}$$

f_c is simply the standard deviation of the energy spectrum about the vertical axis. For the Mexican hat mother wavelet, f_c is equal to $\sqrt{5/2}/2\pi$ or 0.251 Hz. In practice, we require a characteristic frequency of the mother wavelet, such as f_p, f_c or some other, in order to relate the frequency spectra obtained using Fourier transforms to those obtained using wavelet transforms. In the next section we will see how these characteristic frequencies change as the mother wavelet is stretched and squeezed through its dilation parameter. When performing wavelet transform analysis, it is important that the energy spectrum of the wavelet is considered as it indicates the range and character of the frequencies making up the wavelet.

From Equations 2.1 and 2.2, we see that the total energy of the Mexican hat wavelet is finite and given by

$$E = \int_{-\infty}^{\infty} \psi(t)^2 \, dt = \int_{-\infty}^{\infty} \left((1-t^2) e^{-\frac{t^2}{2}} \right)^2 dt = \frac{3}{4}\sqrt{\pi} \tag{2.7}$$

The energy of a function is also given by the area under its energy spectrum. For the Mexican hat wavelet, this gives us:

$$E = \int_{-\infty}^{\infty} |\hat{\psi}(f)|^2 \, df = \int_{-\infty}^{\infty} 32\pi^5 f^4 e^{-4\pi^2 f^2} \, df = \frac{3}{4}\sqrt{\pi} \tag{2.8}$$

Hence,

$$\int_{-\infty}^{\infty} |\psi(t)|^2 \, dt = \int_{-\infty}^{\infty} |\hat{\psi}(f)|^2 \, df \tag{2.9}$$

This is a result that we would expect for any function from Parseval's theorem.

Often in practice, the wavelet function is normalized so that it has unit energy. To do this for the Mexican hat, we modify its definition given in Equation 2.1. From Equation 2.7, we see that it is normalized to have unit energy by dividing it by $(3\sqrt{\pi}/4)^{1/2}$. This gives

$$\psi(t) = \frac{2}{\sqrt{3}\sqrt[4]{\pi}}(1-t^2)e^{-t^2/2} \qquad (2.10)$$

Both Equations 2.1 and 2.10 are commonplace in the literature. The only alteration necessary when employing the normalized Mexican hat of Equation 2.10 rather than that defined in Equation 2.1, is in the value of the admissibility constant, C_g, which must be changed from π to $4\sqrt{\pi}/3$. In the rest of this chapter, we will stick to our original definition of the Mexican hat as given by Equation 2.1.

2.5 WAVELET TRANSFORM

Now that we have chosen a mother wavelet, how do we put it to good use in a signal analysis capacity? First, we require our wavelet to be more flexible than that defined earlier; that is, $\psi(t)$. We can perform two basic manipulations to make our wavelet more flexible: we can stretch and squeeze it (dilation) or we can move it (translation). Figure 2.3a shows the Mexican hat wavelet stretched and squeezed to, respectively, double and half its original width on the time axis. This dilation and contraction of the wavelet is governed by the dilation parameter a which, for the Mexican hat wavelet, is (helpfully) the distance between the centre of the wavelet and its crossing of the time axis. The movement of the wavelet along the time axis is governed by the translation parameter b. Figure 2.3b shows the movement of a wavelet along the time axis from b_1 via b_2 to b_3. We can include the *dilation parameter*, a, and the *location parameter*, b, within our definition of a wavelet given by Equation 2.1. These shifted and dilated versions of the mother wavelet are denoted $\psi(t-b/a)$. For example, in this form the Mexican hat wavelet becomes

$$\psi\left(\frac{t-b}{a}\right) = \left(1-\left(\frac{t-b}{a}\right)^2\right)e^{-\frac{1}{2}[(t-b)/a]^2} \qquad (2.11)$$

The original mother wavelet $\psi(t)$, given by Equation 2.1, simply had $a=1$ and $b=0$. In the form of Equation 2.11, we can now transform a signal, $x(t)$, using a range of a's and b's. The *wavelet transform* of a continuous signal with respect to the wavelet function is defined as

$$T(a,b) = w(a)\int_{-\infty}^{\infty} x(t)\psi^*\left(\frac{t-b}{a}\right)dt \qquad (2.12)$$

where $w(a)$ is a weighting function. The '*' indicates that the complex conjugate of the wavelet function is used in the transform. We need not consider this when using the Mexican hat wavelet as it is a real function, but we do need to take this into account when using complex wavelets later in the chapter. The wavelet transform can be thought of as the cross-correlation of a signal with a set of wavelets of various 'widths'. Typically, $w(a)$ is set to $1/\sqrt{a}$ for reasons of energy conservation (i.e. it ensures that the wavelets at every scale

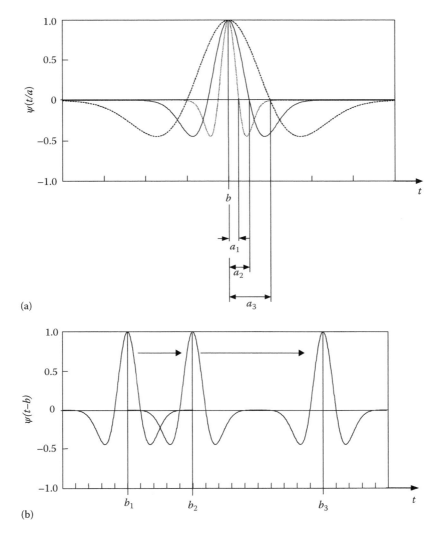

FIGURE 2.3 Dilation and translation of a wavelet: (a) Stretching and squeezing a wavelet: Dilation ($a_1 = a_2/2$; $a_3 = a_4 \times 2$). (b) Moving a wavelet: Translation.

all have the same energy) and we will use this value for the rest of this chapter. However, $w(a) = 1/a$ is sometimes used and there is nothing to stop the user defining a function more appropriate to the application; for example, for the visual enhancement of the transform plot (see Section 2.15).

From here on, we will use $w(a) = 1/\sqrt{a}$. Thus, the wavelet transform is written

$$T(a,b) = \frac{1}{\sqrt{a}} \int_{-\infty}^{\infty} x(t)\psi^*\left(\frac{t-b}{a}\right) dt \qquad (2.13)$$

This is the continuous wavelet transform. Take a closer look at this equation. It contains both the conjugate of the dilated and translated wavelet, $\psi((t-b)/a)$, and the signal, $x(t)$, where $x(t)$ could be a beating heart, an audio signal, a gearbox vibration, a financial index or perhaps even a spatial signal such as a crack profile or land surface heights. The normalized wavelet function is often written more compactly as

$$\psi_{a,b}(t) = \frac{1}{\sqrt{a}}\psi\left(\frac{t-b}{a}\right) \tag{2.14}$$

where the normalization is in the sense of wavelet energy. Hence, the transform integral may be written as

$$T(a,b) = \int_{-\infty}^{\infty} x(t)\psi_{a,b}^{*}(t)\ dt \tag{2.15}$$

This is the nomenclature we will use in this chapter and we will refer to $\psi_{a,b}(t)$ simply as 'the wavelet'. We can express the wavelet transform in an even more compact form as an inner product:

$$T(a,b) = \langle x, \psi_{a,b} \rangle \tag{2.16}$$

Figure 2.4 shows the effect that the dilation of a Mexican hat wavelet $\psi_{a,b}(t)$ has on its corresponding energy spectrum. As the wavelet expands, its corresponding energy spectrum contracts. This is an obvious consequence, as expansion in the time domain must involve the lengthening of time periods and a corresponding lowering of the associated frequencies. The wavelet a scale is therefore inversely proportional to all of its characteristic frequencies, including its passband centre frequency, peak frequency, central frequency (for complex wavelets) and so on. We will come back to this relationship in more detail later in this chapter when we consider wavelet power spectra in Section 2.9.

The wavelet transform has been called a 'mathematical microscope', where b is the location on the time series being 'viewed' and a is associated with the magnification at location b. Now that we have defined the wavelet and its transform, we are ready to see how the transform is used as a signal analysis tool.

2.6 IDENTIFICATION OF COHERENT STRUCTURES

Figure 2.5a, attempts to visualize the mechanics of the wavelet transform given by Equation 2.15. In the figure, a wavelet of scale a centred at location b on the time axis is shown superimposed on top of an arbitrary signal. The time segments where the wavelet and the signal are both positive result in a positive contribution to the integral of Equation 2.15 – for example, region A in the figure. Similarly, the time segments where the wavelet and the signal are both negative result in a positive contribution to the integral (region B). Regions where the signal and wavelet are of opposite sign result in negative contributions to the integral – for example, regions C, D and E in the figure.

Figure 2.5b shows a wavelet of fixed dilation at four locations on a signal. At the first location (b_1), the wavelet is located on a segment of the signal in which the positive and

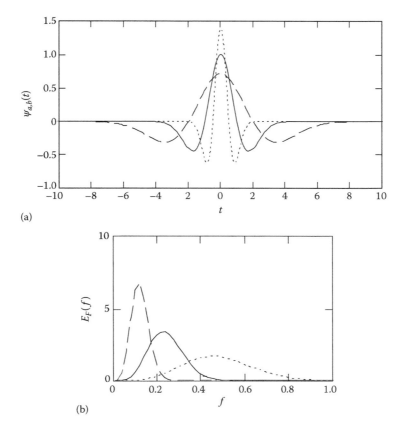

FIGURE 2.4 The Fourier energy spectra of dilating wavelets: (a) Three Mexican hat wavelets $\psi_{a,b}(t) = (1/\sqrt{a})\psi[(t-b)/a]$ at three dilations, $a = 0.5, 1.0, 2.0$ and all located at $b = 0$. (b) Energy spectra corresponding to the wavelets in (a) all have identical energy and hence the same area under their curves. Note that only the positive part of the energy spectrum is shown.

negative parts of the signal are reasonably coincidental with those of the wavelet. This results in a relatively large positive value of $T(a,b)$ given by Equation 2.15. At location b_2, the positive and negative contributions to the integral act to cancel each other out, resulting in a value near zero returned from Equation 2.15. At location b_3, the signal and the wavelet are essentially out of phase which results in a large negative value returned for $T(a,b)$. At location b_4, the wavelet and the signal are again out of phase, similar to location b_3. This time, however, the signal portion in the vicinity of the wavelet contains a large local mean component. It is easy to see that the mean component contributes equal positive and negative values to $T(a,b)$. Thus, only the local signal feature is highlighted by the wavelet at this location and the mean is disregarded. It is through this process that the wavelet transform picks out 'coherent structures' in a time signal at various scales. By moving the wavelet along the signal (increasing b), coherent structures relating to a specific a scale in the wave are identified. This process is repeated over a range of a scales until all the coherent structures within the signal, from the largest to the smallest, can be distinguished.

16 ■ The Illustrated Wavelet Transform Handbook

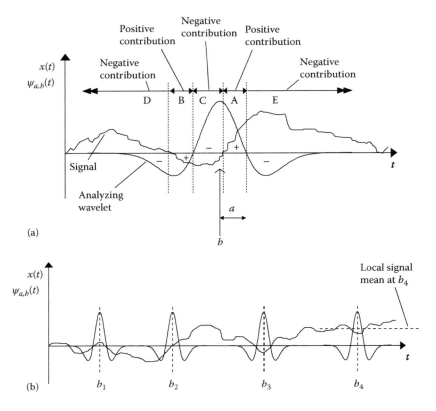

FIGURE 2.5 **The wavelet interrogation of the signal:** (a) The wavelet of specific dilation and location on the signal. The regions which give positive and negative contributions to the integral are delineated in the sketch and marked with '+' and '−', respectively. (b) A wavelet of fixed dilation at four distinct locations on the signal. A large positive value of $T(a,b)$ is returned at location b_1. A near zero value of $T(a,b)$ is returned at b_2 and a large negative value of $T(a,b)$ is returned at b_3. At b_4 a local minimum in the signal corresponds with the positive part of the wavelet and relatively higher parts of the signal correspond with the negative parts of the wavelet. This combines to give a large negative value of $T(a,b)$.

Let us look at another simple example. Figure 2.6 shows a simple sinusoidal waveform 'interrogated' at various locations by Mexican hat wavelets of various dilations. The value of the wavelet transform integral (Equation 2.13) depends on both the location and the dilation of the wavelet. Figure 2.6a shows a wavelet of similar 'periodicity' to the signal waveform superimposed on the signal at location b, which produces a reasonable local matching of the wavelet and signal. From the figure, it can be seen that there is obviously a high correlation between the signal and wavelet at this a scale and b location. The integral of the product of the signal with the wavelet here produces a large positive value of $T(a,b)$. Figure 2.6b shows the wavelet in a new location where the wavelet and signal appear to be out of phase. In this case, the wavelet transform integral produces a large negative value of $T(a,b)$. In between these two extremes, the value of the transform reduces from a maximum (Figure 2.6a) to a minimum (Figure 2.6b). Figure 2.6c shows the point at which

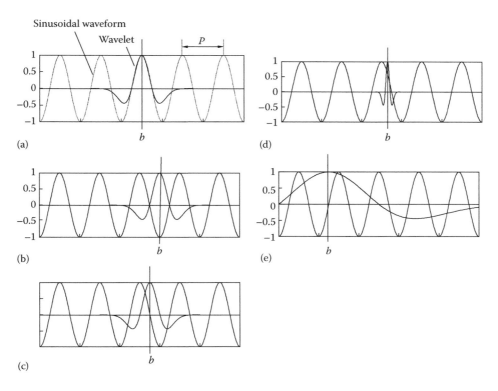

FIGURE 2.6 **A wavelet interrogating a sinusoidal waveform:** (a) The wavelet is in phase with the waveform giving good positive correlation. (b) The wavelet is now out of phase with the waveform giving good negative correlation. (c) The wavelet is again out of phase with the waveform, this time giving zero correlation. (d) The squeezed wavelet does not match the waveform locally. (e) The stretched wavelet does not match the waveform locally.

the wavelet and signal produce a zero value of $T(a,b)$. In Figure 2.6a–c consider a wavelet which matches the signal locally, that is, it has approximately the same 'shape' and 'size' as the signal in the vicinity of b. Figure 2.6d shows the effect that using a smaller a scale has on the transform. From the plot, we see that the positive and negative parts of the wavelet are all coincident with roughly the same part of the signal, producing a value of $T(a,b)$ near zero. Hence, $T(a,b)$ tends to zero as the dilation a tends to zero width. $T(a,b)$ also tends to zero as a becomes very large (Figure 2.6e) as now the wavelet covers many positively and negatively repeating parts of the signal, again producing a near zero value of $T(a,b)$ in the integral of Equation 2.15. Thus, when the wavelet function is either very small or very large compared with the signal features the transform gives near zero values.

Continuous wavelet transforms are not usually computed at arbitrary dilations and isolated locations but rather over a continuous range of a and b. A plot of $T(a,b)$ versus a and b for a sinusoidal signal is shown in Figure 2.7 where the Mexican hat wavelet has been used. This plot of $T(a,b)$ against a and b is known as a wavelet transform plot. Two methods are employed to present the resulting transformed signal in Figure 2.7: a contour plot (Figure 2.7b) and a surface plot (Figure 2.7c). The contour plot is most commonly used

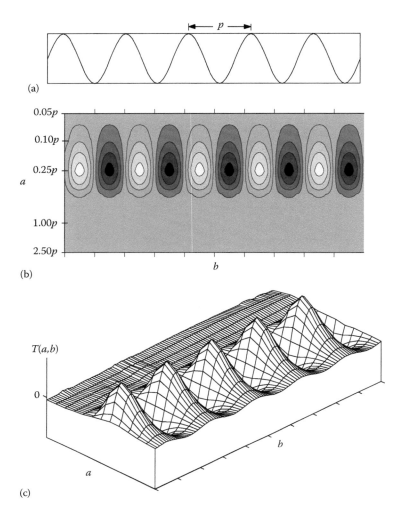

FIGURE 2.7 Wavelet transform plots of a sinusoidal waveform: (a) Five cycles of a sinusoid of period p. (b) Contour plot of $T(a,b)$ for the sinusoid in (a). (Note the logarithmic scaling of the a axis. Note also that a greyscale is used where white corresponds to transform maxima and black to minima. This is the format used in all subsequent transform plots unless otherwise stated.) (c) Isometric surface plot of $T(a,b)$. Viewed with the smallest a scales to the fore.

in practice. The near zero values of $T(a,b)$ are evident in the plot at both large and small values of a. However, at intermediate values of a, we can see large undulations in $T(a,b)$ corresponding to the sinusoidal form of the signal. We can explain these large undulations by referring back to Figure 2.6a–c, where wavelets of a 'size' comparable with the waveform move in and out of phase with the signal. For the Mexican hat wavelet, the a scale is required to be roughly one-quarter of the period, p, of the sine wave for this to occur. (This is covered in more detail in Section 2.9). In Figure 2.7b, we can see that the maxima and minima of the transform plot actually do occur at an a scale of approximately $0.25p$, indicating maximum correlation between the wavelet and signal at this scale. In the figure, the

a axis has logarithmic scaling. This is the most common form seen in practice; however, note that linear scales are sometimes employed. All wavelet transform plots in this chapter have logarithmic a scales, with the exception of Figure 2.12. In addition, in Chapters 2 and 3, we stick to the convention of showing transform plots with the smallest a scales at the top, although there is nothing to stop the reader from plotting them the other way round.

The signal shown in Figure 2.8 is composed of two sinusoidal waveforms, one with a period (p_1) five times the other (p_2). The transform plot shows up very well the two periodic waveforms in the signal at a scales of one-quarter of each of the periods. This figure clearly shows the ability of the transform to decompose the signal into its separate components. The transform has unfolded the signal to show its two constituent waveforms. Figure 2.9

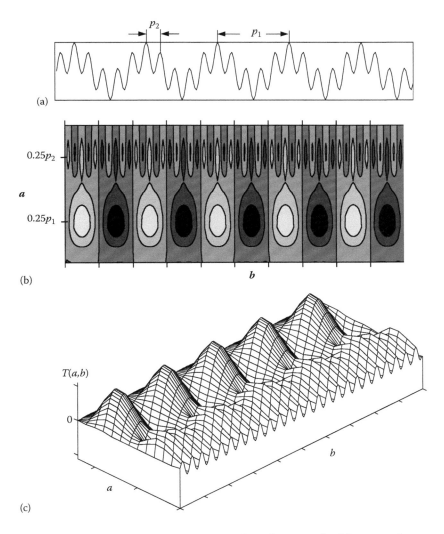

FIGURE 2.8 Wavelet transform plots of two combined sinusoids: (a) A signal composed of a combination of two sinusoids of period p_1 and p_2, where $p_2 = p_1/5$. (b) Contour plot of $T(a,b)$ for the waveform in (a). (c) Isometric surface plot of $T(a,b)$. Viewed with the smallest a scales to the fore.

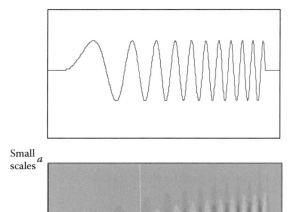

FIGURE 2.9 Segment of a chirp signal with an associated transform plot – Mexican hat wavelet.

contains a segment of a chirp signal which has the form $x(t) = \sin x^2$. The chirp begins just into the signal window and finishes just before the end of the window. The increase in the frequency of the oscillation can be seen in the signal. The transform plot of the transformed chirp is shown below the signal. The transform plot shows the oscillation as peaks at decreasing a scales from left to right.

Figure 2.10a contains a signal with a number of isolated features: four identical wavegroups containing three sinusoidal oscillations of the same periodicity, a group of three bumps, a small negative constant region (a block pulse) and a further sinusoidal wavegroup at a higher frequency. Three representations of the wavelet transform are plotted directly below the signal in Figure 2.10b–d: a filled plot, a contour plot using 2 equally spaced contours and a contour plot using 12 equally spaced contours, respectively. The three different representations are shown to illustrate their advantages and disadvantages as a method for viewing the signal in the timescale plane ('wavelet domain' or 'wavelet space'). We can see from all three transform plots that the four identical wavegroups all have the same morphology in the wavelet domain. In addition, the periodicity of the wavegroup can be easily differentiated from the periodicity of the waves within the group, found by using the associated wavelet scale indicated in Figure 2.10c. Wavelet transforms are particularly good at picking out recognizable signal features in this way, where the features occur intermittently. The wavegroup with the higher frequency oscillation appears at a smaller a scale towards the top of the plot as we would expect. The group of three bumps appears to have a similar form to the oscillations in the filled plot, the difference being more apparent in the two-contour plot. The edges of the block pulse are more apparent

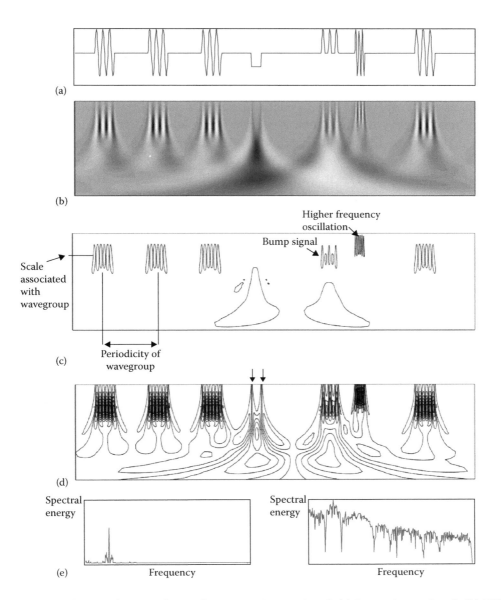

FIGURE 2.10 **The wavelet transform of an intermittent signal: (a)** Intermittent signal. **(b)** Filled transform plot. **(c)** The unfilled transform plot using two contours. **(d)** The unfilled transform plot using 12 contours. (The arrows point to the edges of the constant discontinuity.) **(e)** Fourier power spectrum of the signal. (Arbitrary axis units. The left-hand plot has both axes with linear scales, the right-hand plot has a linear horizontal scale and a logarithmic vertical scale.)

in the 12-contour plot which points to these discontinuities, these are located by the two arrows at the top of the plot. Figure 2.10e contains the Fourier energy spectrum of the signal. The energy spectrum on the left-hand side is plotted with linear scaling axes. The same energy spectrum is plotted on the right-hand side with a logarithmic vertical axis, which shows up the rich structure in the Fourier domain. However, we can see that the Fourier representation does not provide us with any useful information regarding the coherent (obvious even) nature of the localized features within the signal.

2.7 EDGE DETECTION

Another useful property of the wavelet transform is its ability to identify abrupt discontinuities ('edges') in the signal. A simple example of a discontinuity is shown in Figure 2.11a, where a constant signal, $x(t) = 1$, suddenly drops to a constant negative value, $x(t) = -1$. To see how the wavelet picks out such a discontinuity, we follow a wavelet of arbitrary dilation a as it traverses the signal discontinuity. The effect of wavelet

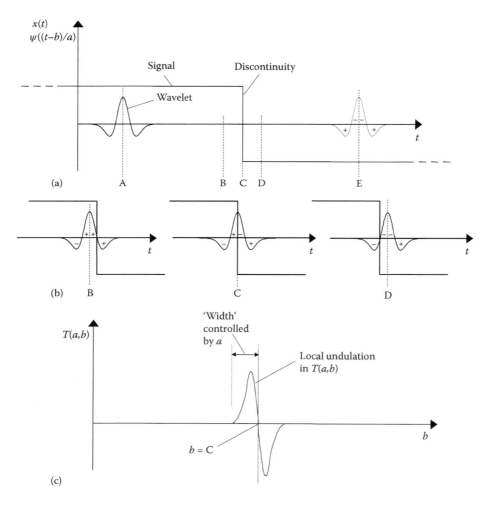

FIGURE 2.11 A schematic illustration of the wavelet interrogation of a signal discontinuity: **(a)** A schematic diagram of the wavelet interrogation of a signal discontinuity. The regions which give positive and negative contributions to the wavelet transform integral are indicated. **(b)** A blow-up of the wavelet as it traverses the discontinuity. **(c)** A plot of $T(a,b)$ against location b at a specific a scale.

location b on the transform $T(a,b)$ is discussed for each of five locations on the signal, A, B, C, D and E:

Location A: At locations much earlier than the discontinuity, for example at location A, the wavelet and the (constant) signal combine to give near zero values of the integral for $T(a,b)$. As it is a localized function, the wavelet becomes approximately zero at relatively short distances from its centre. Hence, the wavelet transform (Equation 2.15) effectively becomes an integral of the wavelet with a constant valued signal producing a zero value.

Location B: At this location, the wavelet is just beginning to traverse the discontinuity. The left-hand lobe of the wavelet produces a negative contribution to the integral, the right-hand lobe of the wavelet produces an equal positive contribution, leaving the central bump of the wavelet to produce a significant positive value for the integral at this location.

Location C: When the signal discontinuity coincides with the wavelet centre, b, the right and left halves of the wavelet contribute to a zero value of the integral. Note that, as the wavelet has zero mean by definition, we can see that the four regions of the wavelet in the figure all have the same area.

Location D: This is similar to location B. As the wavelet traverses the discontinuity further, the left-hand lobe of the signal produces a negative contribution to the integral and the right-hand portion of the wavelet produces an equal positive contribution, as with location B. This time however, the central portion of the wavelet coincides with the negative constant signal and hence the integral produces a significant negative value at this location.

Location E: At locations far greater than C, the wavelet and signal combine to give near zero values of the integral in a manner similar to location A.

Hence, as the wavelet traverses the discontinuity there are first positive then negative values returned by the transform integral. These values are localized in the vicinity of the discontinuity. This is illustrated in Figure 2.11c, where a schematic diagram is given of $T(a,b)$ plotted against the b location on the time axis. From the figure, we see an undulation in $T(a,b)$ centred at the signal discontinuity. The width of this ripple in $T(a,b)$ is controlled by the width of the wavelet, a. In fact, it is directly proportional to it. This is illustrated in Figure 2.12, which plots the wavelet transform of a signal discontinuity for various a scales. Below these, the wavelet transform plot is shown. Notice how the ripple in the transform plot becomes more localized as the dilation parameter reduces. This has the effect of making the transform plot 'point' to the location of discontinuity in the signal. Think for yourself what the result would be of an antisymmetric wavelet (e.g. first derivative of the Gaussian shown in Figure 2.1a) passing across the edge shown in Figures 2.11 and 2.12. Can you see that $T(a,b)$ plotted against b would have the shape of a single bump with a width proportional to the wavelet a scale? Actually, if we were using the first derivative of the Gaussian function as a wavelet, the bump would be Gaussian in shape. Figure 2.13 contains another example of a signal discontinuity: a sudden spike in the signal half way along its length followed by a smooth exponential decay. As the transform plot has been orientated with the smallest a scales at the top, it 'points' to the signal discontinuity in the signal above.

24 ■ The Illustrated Wavelet Transform Handbook

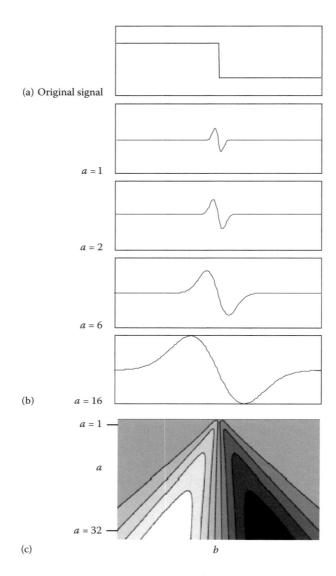

FIGURE 2.12 **The wavelet decomposition of a signal discontinuity: (a)** A signal with a step discontinuity. **(b)** Plots of $T(a,b)$ against b at four arbitrary a scales for the discontinuity: $a = 1, 2, 6$ and 16. (Total window length = 64 units.) **(c)** The transform plot for the discontinuity in (a). Light greys correspond to large positive values of $T(a,b)$ and dark greys to large negative values of $T(a,b)$. Note that here a linear a scale axis is used. The plots in (b) represent vertical slices taken through the transform surface $T(a,b)$ plotted in (c).

2.8 INVERSE WAVELET TRANSFORM

As with its Fourier counterpart, there is an *inverse wavelet transform*, defined as

$$x(t) = \frac{1}{C_g} \int_{-\infty}^{\infty} \int_{0}^{\infty} T(a,b)\, \psi_{a,b}(t)\, \frac{da\,db}{a^2} \tag{2.17}$$

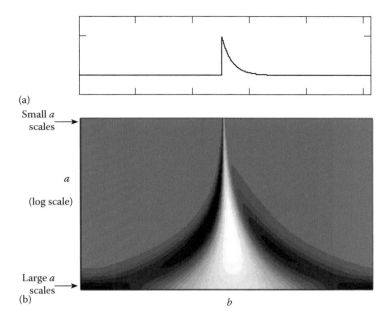

FIGURE 2.13 Pointing to an exponential discontinuity: (a) A sudden spike with an exponential tail. (b) The transform plot for the discontinuity. (Note that in the transform plot, a logarithmic *a* scale is used and small *a* scales are located at the top of the plot. This is the format adopted for all of the transform plots in Chapters 2 and 3 with the exception of Figure 2.12.)

This allows the original signal to be recovered from its wavelet transform by integrating over all scales and location, *a* and *b*. Note that for the inverse transform, the original wavelet function is used, rather than its conjugate which is employed in the forward transformation. If we limit the integration over a range of *a* scales rather than all *a* scales, we can perform a basic filtering of the original signal. Figures 2.14 and 2.15 illustrate this on a segment of signal constructed from two sinusoidal waveforms, one with a period one-quarter the size of the other, plus a local burst of noise. Figure 2.14a–c shows the three component waveforms which are added together to make the composite signal shown in Figure 2.14d. The transform plot of the composite signal (Figure 2.14e) shows up the two constituent waveforms at scales a_1 and a_2. In addition, the high-frequency (i.e. small *a* scale) burst of noise is shown as a patch within the top left-hand quadrant of the transform plot. Figure 2.15 shows two reconstructions of the signal where the components in the transform plot, $T(a,b)$, are set to zero above the white line indicated. In effect, we are reconstructing the signal using

$$x(t) = \frac{1}{C_g} \int_{-\infty}^{\infty} \int_{a^*}^{\infty} T(a,b) \psi_{a,b}(t) \frac{da\,db}{a^2} \qquad (2.18)$$

that is, over a range of scales, $a^* < a < \infty$. The lower integral limit, a^*, is the cutoff scale indicated by the white lines in the figures. The reduction in the high-frequency noise

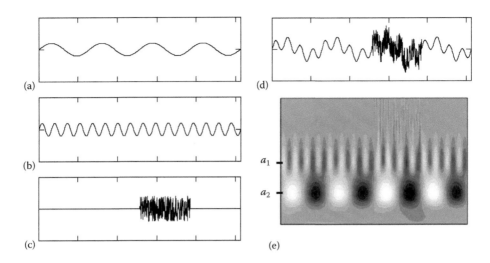

FIGURE 2.14 **A composite signal and its transform plot: (a)** Sinusoidal waveform. **(b)** Sinusoidal waveform with a period one-quarter of that in (a). **(c)** A burst of high-frequency noise. **(d)** Composite signal obtained by combining (a), (b) and (c). **(e)** The transform plot of the composite signal (d).

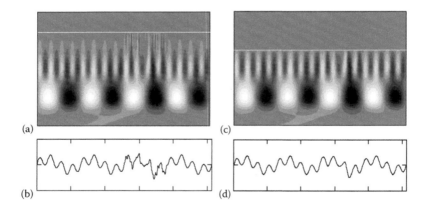

FIGURE 2.15 **Wavelet filtering of the composite signal: (a)** Small-scale (i.e. high-frequency) components above the line indicated are removed from the transform plot in Figure 2.14e. **(b)** Reconstructed signal using the transform plot in (a). **(c)** Small-scale components above the line indicated are removed from the transform plot in Figure 2.14e. **(d)** Reconstructed signal using the transform plot in (c).

components in the reconstructed signal is evident as the cutoff a scale value increases. This simple noise reduction method is known as scale-dependent thresholding.

Figure 2.16a,b show, in a very simple fashion, the ability of the wavelet transform to perform a manipulation of the signal which is localized in both time and scale. Only those $T(a,b)$ values in the region contained within the box in the transform plot are set to zero. In this way, the burst of noise can be dealt with locally in the signal and thus, the denoising does not affect other parts of the signal. Figure 2.16c–e show the effect of using a global cutoff at a much higher a scale. Here, the transform plot components are restricted to those

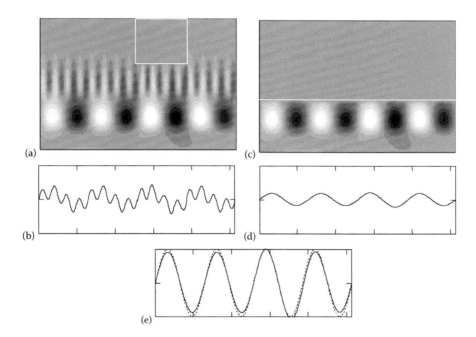

FIGURE 2.16 Further wavelet filtering: (a) Small-scale (i.e. high-frequency) components are removed from the transform plot in Figure 2.14e in the location within the box shown. (b) Reconstructed signal using the transform plot in (a). (c) Components above the line indicated are removed from the transform plot in Figure 2.14e. (d) Reconstructed signal using transform plot in (c). (e) Blow-up of the reconstructed signal (solid line) in (d), together with the original low-frequency sinusoidal waveform (dotted line).

associated mainly with the low-frequency waveform. The inverse transform Figure 2.16d shows a sinusoidal-like waveform. Figure 2.16e plots the reconstructed signal of Figure 2.16d at a greater vertical scale and compares it with the original waveform component. We can see that a reasonably good match is obtained. However, as the spectral information in the transform components is smeared across scales, perfect reconstruction of the individual sinusoidal components is not achievable (it would be for the Fourier transform, in the specific case of a signal composed of sinusoidal components). The denoising strategies shown in Figures 2.15 and 2.16 are very simple in nature and are shown here as an illustration of the inverse transform. A better way to separate pertinent signal features from unwanted noise, or other larger-scale artefacts, using the continuous wavelet transform is by using a *wavelet transform modulus maxima* method. Figure 2.17 shows a composite signal together with its transform plot and corresponding modulus maxima lines. The modulus maxima lines are the loci of the local maxima and minima of the transform plot, with respect to b, traced over wavelet scales. Various signal features are identified within the modulus maxima plot. Modulus maxima plots allow the salient information within the transform plot to be expressed in a much more compact form. Following maxima lines from large to small a scales allows the high-frequency information, corresponding to large features within the signal, to be differentiated from high-frequency noise components. This lends

28 ■ The Illustrated Wavelet Transform Handbook

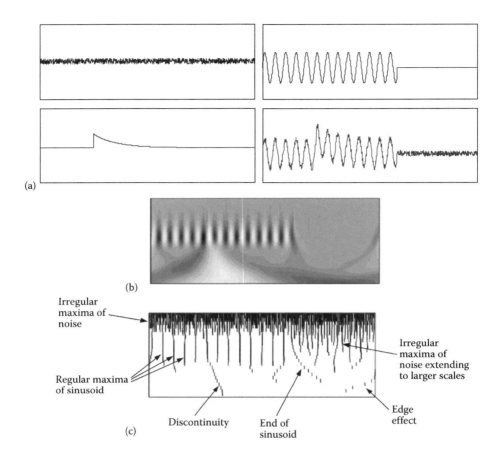

FIGURE 2.17 Modulus maxima of a composite signal: (a) A composite signal (bottom right) constructed from the noise, curtailed sinusoid and exponential decay signals shown. (b) Wavelet transform plot of the composite signal (Mexican hat wavelet). (c) Modulus maxima plot corresponding to the transform plot in (b).

itself to novel approaches in filtering out noise from coherent signal features. We will come across examples of the use of modulus maxima methods as we proceed through the rest of the book, in the context of the analysis of turbulent fluid measurements in Chapter 4 (Figure 4.20); the filtering of non-destructive testing data in Chapter 5 (Figures 5.8 through 5.10); the interrogation of the ECG (Figures 6.3, 6.17 and 6.18) in Chapter 6; and the analysis of multifractal signals (Figures 7.10 and 7.11), geological strata, remote sensing data and astronomical images in Chapter 7.

2.9 SIGNAL ENERGY: WAVELET-BASED ENERGY AND POWER SPECTRA

The total energy contained in a signal, $x(t)$, is defined as its integrated squared magnitude:

$$E = \int_{-\infty}^{\infty} |x(t)|^2 dt = \|x(t)\|^2 \tag{2.19}$$

For this equation to be useful, the signal must contain finite energy. We have already come across this expression in Equation 2.7, where we found the energy in the Mexican hat function (i.e. substitute $x(t)$ for $\psi(t)$). The relative contribution of the signal energy contained at a specific a scale and b location is given by the two-dimensional wavelet energy density function:

$$E(a,b) = |T(a,b)|^2 \qquad (2.20)$$

A plot of $E(a,b)$ is known as a *scalogram* (analogous to the *spectrogram* – the energy density surface of the short-time Fourier transform – see Section 2.12.) In practice, all functions which differ from $|T(a,b)|^2$ by only a constant multiplicative factor are also called 'scalograms', for example $|T(a,b)|^2/C_g$, $|T(a,b)|^2/C_g f_c$ and so on. The scalogram can be integrated across a and b to recover the total energy in the signal using the admissibility constant, C_g, as follows:

$$E = \frac{1}{C_g} \int_{-\infty}^{\infty} \int_{0}^{\infty} |T(a,b)|^2 \frac{da}{a^2} db \qquad (2.21)$$

Figure 2.18a–c show an experimental signal, $x(t)$, with an associated wavelet transform plot, $T(a,b)$, and scalogram, $E(a,b)$. A Mexican hat wavelet was used in the signal transformation. The scalogram (Figure 2.18c) is very similar in form to the wavelet transform plot. This is to be expected when using real wavelets as the scalogram is simply the squared magnitude of the wavelet transform values. For complex wavelets (see Section 2.11), we can view the modulus, phase, real and complex parts separately. The scalogram's surface highlights the location and scale of dominant energetic features within the signal.

The relative contribution to the total energy contained within the signal at a specific a scale is given by the scale-dependent energy distribution:

$$E(a) = \frac{1}{C_g} \int_{-\infty}^{\infty} |T(a,b)|^2 db \qquad (2.22)$$

Peaks in $E(a)$ highlight the dominant energetic scales within the signal. Figure 2.18d plots $E(a)$ against a for the signal segment in Figure 2.18a. The plot shows that two dominant scales exist within the signal which are linked to the dominant oscillatory regime of the original experimental signal.

We may convert the scale-dependent wavelet energy spectrum of the signal $E(a)$ to a frequency-dependent wavelet energy spectrum, $E_W(f)$, in order to compare it directly with the Fourier energy spectrum of the signal $E_F(f)$. To do this, we must convert from the wavelet a scale (which can be interpreted as a representative temporal, or spatial, period for physical data) to a characteristic frequency of the wavelet. One of the most commonly used characteristic frequencies in practice is the passband centre of the wavelet's power spectrum. We will use this here, but note that another representative frequency of the mother wavelet, such as either the spectral peak frequency, f_p, or the central frequency, f_0, could

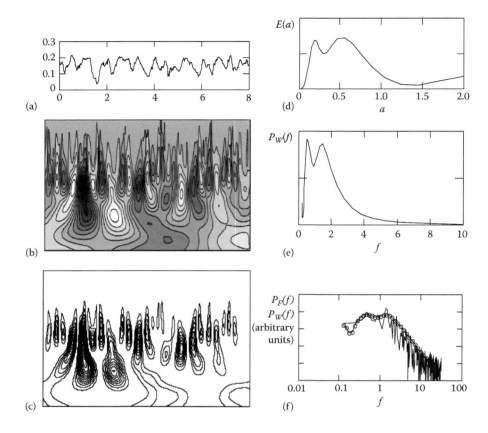

FIGURE 2.18 **Wavelet energy density and power spectra:** (a) Experimental signal $x(t)$: A velocity trace taken within a vortex-shedding regime in a fluid. (b) Transform plot $T(a,b)$ using a Mexican hat wavelet. (Negative values are shown dark grey to black. Positive values are shown light grey to white.) (c) The wavelet scalogram $|T(a,b)|^2$. (Note that all values are positive as the square of the modulus is plotted. Hence, all contours enclose peaks.) (d) Wavelet energy distribution $E(a)$. (e) Wavelet power spectral density $PW(f)$. The horizontal axis is related to that in (d) through $f = 0.251/a$ (Mexican hat wavelet). (f) Power spectral densities (logarithmic plot). Fourier spectrum $P_F(f)$, continuous line. Wavelet spectrum $P_W(f)$, circles. (Note that times are in seconds, frequencies are in hertz and the scale $a = 1$ corresponds to 1 second.)

be chosen and would be equally valid in the following discussion. We saw in Section 2.5, Figure 2.4, that the spectral components are inversely proportional to the dilation, that is, $f \propto 1/a$, and in Section 2.4 we defined the passband centre frequency of the mother wavelet (i.e. $a = 1$) as f_c. Hence, using this passband frequency, the characteristic frequency associated with a wavelet of arbitrary a scale is given by

$$f = \frac{f_c}{a} \quad (2.23)$$

where the passband centre of the mother wavelet, f_c, becomes a scaling constant and f is the representative or characteristic frequency for the wavelet at scale a. We saw in Section 2.4

that f_c is equal to $\sqrt{5/2}/2\pi$ or 0.251 for the Mexican hat mother wavelet. Hence, for this specific wavelet, we have $f = 0.251/a$ and this is why the peaks in the transform plot of the sinusoid in Figure 2.7 occurred at around 0.25 of its period, $p(=1/f)$. Using Equation 2.23, we can now associate the scale-dependent energy, $E(a)$, with the passband frequency of our wavelet. We can also see from Equations 2.21 and 2.22 that the total energy in the signal is given by

$$E = \int_0^\infty E(a) \frac{da}{a^2} \qquad (2.24)$$

We can rewrite this equation in terms of passband frequency by making the change of variable $f = f_c/a$. The relationship between the derivatives is $da/a^2 = -df/f_c$ and, after modifying and then swapping the integral limits to get rid of the negative sign, we get

$$E = \int_0^\infty E_W(f) \, df \qquad (2.25)$$

where we define $E_W(f) = E(a)/f_c$ for $f = f_c/a$, and the subscripted 'W' corresponds to 'wavelet' to differentiate it from its Fourier counterpart. A plot of the wavelet energy – $E_W(f)$ against f, the wavelet energy spectrum – has an area underneath it equal to the total signal energy and may be compared directly with the Fourier energy spectrum, $E_F(f)$, of the signal. (Remember that $E_F(f)$ is defined as the squared magnitude of the Fourier transform of the signal. We have already come across it in Equation 2.5 where the energy spectrum of the Mexican hat wavelet function was given.) From Equation 2.22, we see that the total energy in the signal is given by

$$E = \frac{1}{C_g f_c} \int_{-\infty}^\infty \int_0^\infty |T(f,b)|^2 \, df \, db \qquad (2.26)$$

where we define $T(f,b) = T(a,b)$ for $f = f_c/a$. We can see also that the energy density surface in the time–frequency plane, defined by $E(f,b) = (|T(f,b)|^2)/(C_g f_c)$, contains a volume equal to the total energy of the signal, that is

$$E = \int_{-\infty}^\infty \int_0^\infty E(f,b) \, df \, db \qquad (2.27)$$

This energy density surface can be compared directly with the energy density surface of the short-time Fourier transform (the spectrogram). Note that the timescale representations of the scalogram, $E(a,b)$, and scale-dependent energy distribution, $E(a)$, do not, respectively, enclose volumes and areas proportional to the energy of the signal, whereas their time–frequency counterparts, $E(f,b)$ and $E_W(f)$, do. In fact, the way we have defined $E(f,b)$ and $E_W(f)$ means that they enclose a volume and an area equal to the energy of the signal, respectively. However, the peaks in $E(a,b)$ and $E(a)$ do correspond to the most energetic parts of the signal, as do the peaks in $E(f,b)$ and $E(f)$. We can, therefore, use both the scalogram and scale-dependent energy distribution to determine the energy distribution relative to wavelet scale. Scalograms are normally plotted with a logarithmic a scale axis. As $f = f_c/a$ and hence $\log(f) = \log(f_c) - \log(a)$, the plot of $|T(f,b)|^2$ using a logarithmic frequency scale is simply a shifted, inverted plot of $|T(a,b)|^2$ using a logarithmic a scale. For example, Figure 2.18c containing $|T(a,b)|^2$ with logarithmically decreasing a scales towards the top of the plot can also be interpreted as a plot of $|T(f,b)|^2$ with logarithmically increasing frequencies towards the top of the plot. In the literature both representations are commonplace.

If the signal in Figure 2.18a was infinitely long, we can see that its energy would be infinitely large. However, in practice, experimental signals (see Chapters 4 through 7) are of finite length – usually long enough for the pertinent statistics of the signal to settle down sufficiently for analysis. Hence, in practice, power spectra are more often used to characterize experimental signals of finite length. The power spectrum is simply the energy spectrum divided by the time period of the signal under investigation. Hence, the area under the power spectrum gives the average energy per unit time (i.e. the power) of the signal. For example, for a signal of length τ, the Fourier and wavelet power spectra are, respectively,

$$P_F(f) = \frac{1}{\tau} E_F(f) \qquad (2.28)$$

$$P_W(f) = \frac{1}{\tau} E_W(f) = \frac{1}{\tau f_c C_g} \int_0^\tau |T(f,b)|^2 db \qquad (2.29)$$

Figure 2.18e plots the wavelet power spectrum, $P_W(f)$, for the experimental signal shown in Figure 2.18a. The wavelet power spectral density plot contains the same two peaks as those of the scale energy distribution plot of Figure 2.18d but in reverse order, as the horizontal frequency axis is the rescaled inverse of the scale axis. The area underneath the $P_W(f)$ plot is equal to the power of the signal. Figure 2.18f again contains the wavelet power spectrum; this time logarithmic axes are employed and the corresponding Fourier power spectrum is also drawn for comparison. Such logarithmic power spectral plots are commonly used in practice (e.g. see the fluid turbulence spectra of Chapter 4) where, for example, some form of power–law scaling is expected or when the pertinent spectral components span quite different orders of magnitude. Due to the frequency distribution within each wavelet, the resulting wavelet power spectrum is smeared compared with the Fourier spectrum. However, the wavelet spectrum is more than simply a smeared version of the Fourier spectrum as the shape of the wavelet itself is an important parameter in the analysis of the signal. Some wavelets will correlate better with specific signal features than others, accentuating these features in the resulting spectra. Note also that we are using the passband centre of the wavelet as its representative frequency in our discussion, if we had used another characteristic frequency of the wavelet then this would affect the resulting wavelet power spectrum of the signal (and energy density plots), either squashing it or stretching it while retaining the same overall shape and of course, the same power (and, respectively, energy).

Finally, it is worth noting that the *wavelet variance*, defined for the continuous wavelet transform as

$$\sigma^2(a) = \frac{1}{\tau}\int_0^\tau |T(a,b)|^2 db \qquad (2.30)$$

is often used in practice to determine dominant scales in the signal. Again we assume that τ is of sufficient length to gain a reasonable estimate of $\sigma^2(a)$. We can see that this expression is very similar to both the scale-dependent energy distribution of Equation 2.22 and the power spectral density function of Equation 2.29, differing from both equations only by constant multiplicative factors. We will come across many examples of the use of wavelet and Fourier spectra (and wavelet variance) as we proceed through the application chapters of this book.

2.10 WAVELET TRANSFORM IN TERMS OF THE FOURIER TRANSFORM

Referring back to the definition of the wavelet transform given by Equation 2.15, it can be seen that we may employ the convolution theorem to express the wavelet transform in terms of products of the Fourier transforms of the signal, $\hat{x}(f)$, and wavelet, $\hat{\psi}_{a,b}(f)$, as follows:

$$T(a,b) = \int_{-\infty}^{\infty} \hat{x}(f)\hat{\psi}^*_{a,b}(f) df \qquad (2.31)$$

where, notably, the conjugate of the wavelet function is used. The Fourier transform of the dilated and translated wavelet is

$$\hat{\psi}_{a,b}(f) = \int_{-\infty}^{\infty} \frac{1}{\sqrt{a}} \psi\left(\frac{t-b}{a}\right) e^{-i(2\pi f)t} dt \qquad (2.32a)$$

Making the substitution $t' = (t-b)/a$ (hence $dt = adt'$), we obtain

$$\hat{\psi}_{a,b}(f) = \frac{1}{\sqrt{a}}\int_{-\infty}^{\infty} \psi(t') e^{-i(2\pi f)(at'+b)} adt' \qquad (2.32b)$$

Separating out the constant part of the exponential function and dropping the superscripted dash from t', we get

$$\hat{\psi}_{a,b}(f) = \sqrt{a} e^{-i(2\pi f)(b)} \int_{-\infty}^{\infty} \psi(t) e^{-i(2\pi af)(t)} dt \qquad (2.32c)$$

The integral expression in the previous equation is simply the Fourier transform of the wavelet at rescaled frequency af, hence we can write Equation 2.32b as

$$\hat{\psi}_{a,b}(f) = \sqrt{a}\,\hat{\psi}(af)e^{-i(2\pi f)b} \qquad (2.33)$$

The Fourier transform of the wavelet function conjugate is then simply

$$\hat{\psi}^*_{a,b}(f) = \sqrt{a}\,\hat{\psi}^*(af)e^{i(2\pi f)b} \qquad (2.34)$$

Hence, Equation 2.31 can be written in expanded form as

$$T(a,b) = \sqrt{a}\int_{-\infty}^{\infty} \hat{x}(f)\hat{\psi}^*(af)e^{i(2\pi f)b}\,df \qquad (2.35)$$

which we can see has the form of an inverse Fourier transform. This is a particularly useful result when using discretized approximations of the continuous wavelet transform in practice with large signal data sets, as the 'fast Fourier transform' (FFT) algorithm may be employed to facilitate rapid calculation of the wavelet transform and its inverse. In addition, the Fourier transform of the wavelet function, $\hat{\psi}_{a,b}(f)$, is usually known in analytic form and hence need not be computed using an FFT. Only an FFT of the original signal, $\hat{x}(f)$, is required. Then, to get $T(a,b)$ we take the inverse FFT of the product of the signal Fourier transform and the wavelet Fourier transform for each required a scale and multiply the by $\sqrt{a}$. The equivalence between the time-based and frequency-based methods for determining $T(a,b)$ is depicted in Figure 2.19. The bandpass nature of the wavelet is evident from Figure 2.19b. The inverse transform (Equation 2.17) can similarly be written in terms of an inverse Fourier function.

2.11 COMPLEX WAVELETS: THE MORLET WAVELET

So far, we have utilized the Mexican hat wavelet to illustrate many of the features of the wavelet transform. In this section, we consider wavelets which have both a real and imaginary part. Complex or *analytic wavelets* have Fourier transforms which are zero for negative frequencies (requirement 3 in Section 2.3). By using such complex wavelets, we can separate the phase and amplitude components within the signal. Actually, we can easily make a complex version of the Mexican hat wavelet by taking its Fourier transform, setting the negative frequency components in the Fourier domain to zero and then performing an inverse Fourier transform to get the complex wavelet. This low-oscillation wavelet provides a useful temporally localized complex wavelet (Addison et al., 2002a). However, in this section, we focus on the most commonly used complex wavelet, the *Morlet wavelet*, which is defined as

$$\psi(t) = \pi^{-1/4}\left(e^{i2\pi f_0 t} - e^{-(2\pi f_0)^2/2}\right)e^{-t^2/2} \qquad (2.36)$$

where f_0 is the central frequency of the mother wavelet. The second term in brackets is known as the correction term, as it corrects for the non-zero mean of the complex sinusoid

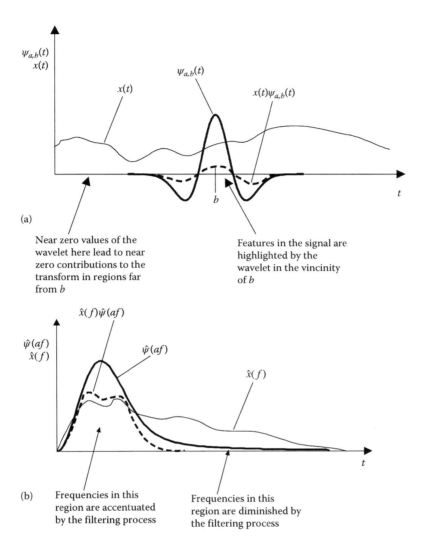

FIGURE 2.19 Schematic representation of the wavelet transform in its time and frequency representations: (a) The convolution of the wavelet with the signal. (b) The convolution in (a) expressed in the Fourier domain involves a product of the signal Fourier transform and the wavelet Fourier transform.

of the first term. In practice, it becomes negligible for values of $f_0 \gg 0$ and can be ignored, in which case the Morlet wavelet can be written in a simpler form as

$$\psi(t) = \frac{1}{\pi^{1/4}} e^{i2\pi f_0 t} e^{-t^2/2} \qquad (2.37)$$

Normalization factor — Complex sinusoid — Gaussian bell curve

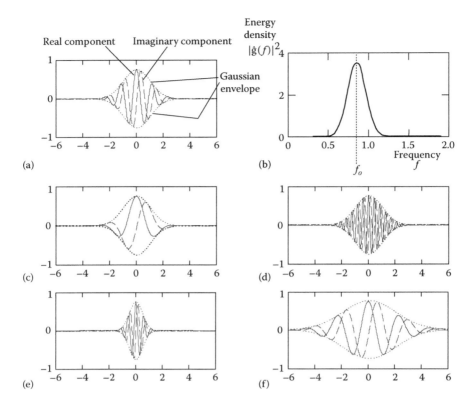

FIGURE 2.20 **Morlet wavelets:** (a) The Morlet wavelet $a=1$ and $f_0=0.894$. (b) Energy spectrum of the Morlet wavelet. (c) $f_0=0.318$ ($a=1$). (d) $f_0=1.909$ ($a=1$). (e) $a=0.5$ ($f_0=0.894$). (f) $a=2$ ($f_0=0.894$).

This wavelet is simply a complex wave within a Gaussian envelope. We can see this by looking at Equation 2.37 in conjunction with Figure 2.20a. The complex sinusoidal waveform is contained in the term $e^{i2\pi f_0 t}$ ($=\cos(2\pi f_0 t) + i\sin(2\pi f_0 t)$). The Gaussian envelope $e^{-t^2/2}$ has unit standard deviation and 'confines' the complex sinusoidal waveform. Figure 2.20a shows the real and imaginary parts of the Morlet wavelet together with its confining Gaussian envelope. We can see that the real and imaginary sinusoids differ in phase by a quarter period. The $\pi^{1/4}$ term is a normalization factor which ensures that the wavelet has unit energy. Note that the function given by Equation 2.37 is not really a wavelet as it has a non-zero mean, that is, the zero frequency term of its corresponding energy spectrum is non-zero and hence it is inadmissible according to Equation 2.4. However, it can be used in practice with $f_0 \gg 0$ with minimal error.

The Fourier transform of the Morlet wavelet is given by

$$\hat{\psi}(f) = \pi^{1/4}\sqrt{2} e^{-\frac{1}{2}(2\pi f - 2\pi f_0)^2} \qquad (2.38)$$

which has the form of a Gaussian function displaced along the frequency axis by f_0. Note that the central frequency of the Gaussian spectrum is generally chosen to be the characteristic frequency of the analytic Morlet wavelet rather than the passband frequency, which we used previously for the Mexican hat wavelet. The energy spectrum (the squared magnitude of the Fourier transform) is given by

$$\left|\hat{\psi}(f)\right|^2 = 2\pi^{1/2} e^{-(2\pi f - 2\pi f_0)^2} \tag{2.39}$$

The integral of this gives the energy of the Morlet wavelet which is equal to unity according to our definition given by Equation 2.37. The energy spectrum of the Morlet wavelet is shown in Figure 2.20b. The central frequency f_0 is the frequency of the complex sinusoid and its value determines the number of 'effective' or 'significant' sinusoidal waveforms contained within the envelope, that is, those which are not very close to 0 amplitude. The value of 0.849 ($=\sqrt{1/(2\ln 2)}$) is often used in practice. It produces a decay where the magnitudes of the two peaks in the real waveform that are adjacent to the central peak are half its amplitude. (In the literature the angular frequency $\omega_0 = 2\pi f_0$ of the wavelet is often quoted, hence $f_0 = 0.849$ becomes $\omega_0 = 5.336 = \pi\sqrt{2/\ln 2}$. Values of ω_0 equal to 5 and 6 (and in between) are also commonly employed in practice. For values of ω_0 less than 5 ($f_0 < 0.8$), the full or 'complete' Morlet wavelet of Equation 2.36 should be used as the simplified wavelet function of Equation 2.37 contains a significant non-zero mean.) Figure 2.20c and d show Morlet wavelets with f_0 equal to 0.318 (another low-oscillation complex wavelet [Addison et al., 2002a]) and 1.909, respectively. The figure shows that the number of effective oscillations contained within the Gaussian window increases with f_0.

To construct the dilated and translated Morlet wavelet, we replace t with $(t-b)/a$ as we did for the Mexican hat in Equation 2.11, its form is then

$$\psi\left(\frac{t-b}{a}\right) = \frac{1}{\pi^{1/4}} e^{i2\pi f_0[(t-b)/a]} e^{-\frac{1}{2}[(t-b)/a]^2} \tag{2.40}$$

Figure 2.20e and f show Morlet wavelets with a scales of 0.5 and 2, respectively. We can see directly from Equation 2.40 that the standard deviation of the Gaussian envelope on the time axis is, in fact, simply equal to a. Figure 2.20e and f show the stretching and squeezing of the wavelet with a scale.

Figure 2.21 illustrates the use of the Morlet wavelet ($f_0 = 0.849$) in analyzing a two-component sinusoidal waveform. Performing the wavelet transform on the signal using the complex Morlet wavelet results in complex transform values $T(a,b)$ which we may view in a number of ways. The real part of $T(a,b)$ is shown in Figure 2.21b. As expected, the two wavelet components are picked up and displayed as ripples in the transform plot at two distinct scales. Figure 2.21c also contains the real part of $T(a,b)$. This time, a coarser shading is used and contour lines are added to enhance the visualization of the periodic structure of the transform plot. Figure 2.21d contains the imaginary part of $T(a,b)$. Notice the similarity between this transform plot and that of Figure 2.21c. In fact, the imaginary plot is a phase-shifted version of the real plot. The reason for this is apparent if we consider the form of the Morlet wavelet as shown in Figure 2.20a. In the figure, we see that, for the Morlet wavelet defined previously, the imaginary part of the wavelet

38 ■ The Illustrated Wavelet Transform Handbook

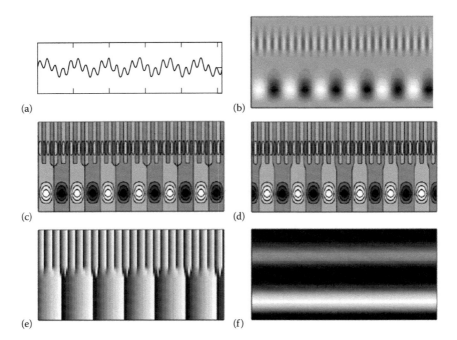

FIGURE 2.21 **Morlet wavelet analysis of a two-component sinusoidal waveform:** (a) Original signal. (b) The real part of the wavelet transform $Re(T(a,b))$ (positive maxima in white, negative minima in black). (c) The real part of the wavelet transform $Re(T(a,b))$ (same plot as (b) but with contours added and a coarser shading used). (d) The imaginary part of the wavelet transform $Im(T(a,b))$. (e) The phase of the wavelet transform. $\phi(a,b) = \tan^{-1}\{[Im(T(a,b))]/[Re(T(a,b))]\}$ ($-\pi$ phase in black, π phase in white, zero phase in mid-grey tone). (f) The modulus of the wavelet transform. $T(a,b) = \sqrt{[Re(T(a,b))]^2 + [Im(T(a,b))]^2}$ (maximum values in white, zero values in black).

comprises a sinusoidal waveform within a Gaussian envelope which leads the real part by one-quarter of a cycle. In other words, it is *phase-shifted* by one-quarter of a cycle from the real part. However, as we use the complex conjugate in the transform, the imaginary part is inverted leading to an imaginary waveform which lags behinds by one-quarter of a cycle from the real part. Hence, the imaginary part of the Morlet best matches one-quarter of a cycle later than the real part. Therefore, the maxima of the transform plot for the imaginary part is phase-shifted *forward* by one-quarter of a cycle. This we see when comparing Figure 2.21c and d.

As $T(a,b)$ is a complex number, that is $T(a,b) = Re(T(a,b)) + Im(T(A,b))$, we can write $T(a,b)$ in terms of its phase $\phi(a,b)$ and modulus $|T(a,b)|$. The phase of the Morlet transform plot is shown in Figure 2.21e. The phase varies cyclically between $-\pi$ and π over the duration of the component waveforms. Zero phase corresponds to the real part of the Morlet wavelet centred at the maximum amplitude of the sinusoidal waveforms. Hence, zero phase corresponds to the peaks on the real transform plot of Figure 2.21c. A phase of π (and $-\pi$) corresponds to the minima of the real transform plot. Figure 2.21f contains the modulus of the transform plot where we can see that the periodic sinusoidal waveforms manifest themselves as continuous bands across the modulus plot.

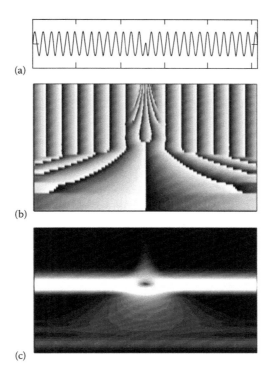

FIGURE 2.22 **Phase-shifted sinusoid:** **(a)** Original signal. **(b)** Phase. **(c)** Modulus.

Figures 2.22 through 2.28 show various simple signals together with their associated Morlet wavelet transform plots. Figure 2.22 shows a sinusoidal waveform which is shifted by half a cycle in the middle. Hence, the resulting signal contains both a discontinuity and a phase shift. Both transform plots indicated the location of the discontinuity. The phase plot shows up the location of the phase shift in the waveform. Figure 2.23 shows the Morlet wavelet decomposition of a signal which contains a change in periodicity midway along its length. The change in periodicity is clearly shown in all four transform plots. Figure 2.24 shows the effect of interrogating the same signal as Figure 2.23 using different Morlet wavelets (i.e. different values of f_0). On the left-hand side of the figure, the signal is decomposed using a Morlet wavelet with $f_0 = 0.318$. We see that the real part of this wavelet is more like a Mexican hat in form. (As mentioned, we should not really use such a low frequency f_0 for a Morlet wavelet in practice, as its power spectrum is significantly non-zero at the origin. Instead, its complete form given by Equation 2.36, should be employed. An example of the use of the complete Morlet wavelet can be found in Figure 5.11 of Chapter 5.) Comparing the transform plots with those of the previous figure, we see some differences. The phase plot for the $f_0 = 0.318$ wavelet is much smoother at the transition point. This is because this wavelet has essentially only a single positive bump which matches the signal smoothly over the transition region, whereas the $f_0 = 0.849$ wavelet has five distinct peaks within the window which correspond to the five ridges converging at small scales in the phase plot in Figure 2.23c. The right-hand side of Figure 2.24 shows the Morlet wavelet decomposition of the signal using $f_0 = 1.909$. The power spectra for the three Morlet wavelet decompositions

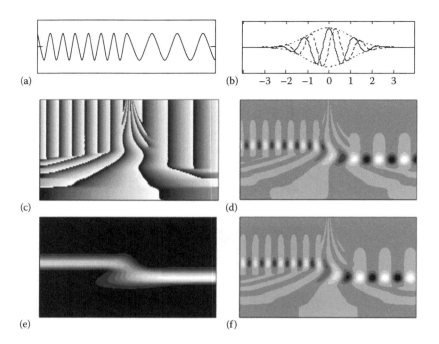

FIGURE 2.23 Morlet decomposition of signal with abrupt change in periodicity: (a) Original signal. (b) Morlet wavelet $f_0 = 0.849$ (period = $1/f_0 = 1.117$) at scale $a = 1$. (c) Phase of the transform. (d) Real part of the transform. (e) Modulus of the transform. (f) Imaginary part of the transform.

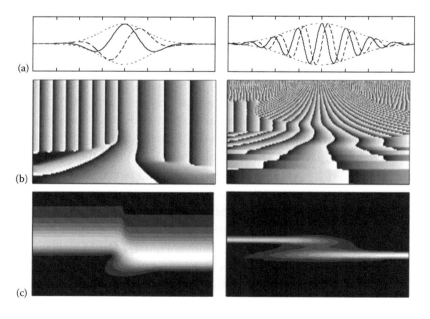

FIGURE 2.24 The effect of f_0 on the Morlet decomposition of signal with abrupt change in periodicity: (a) Morlet wavelets $f_0 = 0.318$ (left) and $\omega_0 = 1.909$ (right). (b) Phase plots corresponding to $f_0 = 0.318$ (left) and $\omega_0 = 1.909$ (right). (c) Modulus plots corresponding to $f_0 = 0.318$ (left) and $\omega_0 = 1.909$ (right).

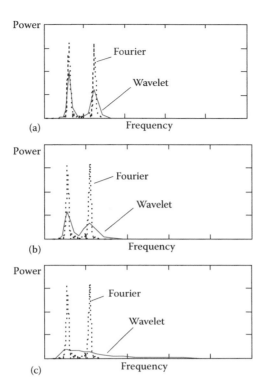

FIGURE 2.25 Comparison of Morlet power spectra for three values of f_0: (a) Power spectrum for $f_0 = 1.909$ (arbitrary axis units). (b) Power spectrum for $f_0 = 0.849$ (arbitrary axis units). (c) Power spectrum for $f_0 = 0.318$ (arbitrary axis units). (Note that the Fourier spikes are shown with a finite height and width as they were calculated numerically using a discrete Fourier algorithm. In theory, they are Dirac delta functions.)

given in Figures 2.23 and 2.24 are presented in Figure 2.25. We can see greater degree frequency localization as f_0 increases (and hence the number of cycles within the Gaussian envelope increases). This is also evident from the transform plots in Figures 2.23 and 2.24, where we can see that narrower bands in the modulus transform plot are associated with higher values of f_0. (Remember from Section 2.9 that we can interpret the logarithmic vertical axis of the transform plot in terms of characteristic wavelet frequency rather than scale.) However, this greater degree of frequency localization with increasing f_0 is associated with much poorer temporal localization as can be seen when comparing both the modulus and phase plots in Figures 2.23 and 2.24.

Figure 2.26 shows two signals containing repeating features – bumps and spikes. The transform plots are plotted below the signals. We can see that, even for these relatively simple repeating signals, the phase plots are already exhibiting a significant degree of complexity. The modulus plot of the bump and spike signal points to the discontinuous spikes in the signal (compared with the ECG signal shown in Chapter 6, Figure 6.10). Figure 2.26g shows a schematic of the wavelet at approximately the same periodicity as the bump signal of Figure 2.26a. At this scale, a maximum correlation is produced which shows up as

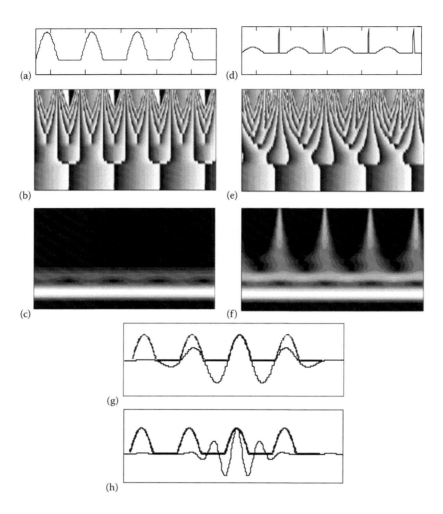

FIGURE 2.26 **Morlet decomposition of bump signals:** (a) Single bump – positive part of sinusoid. (b) Wavelet phase of signal (a). (c) Wavelet modulus of signal (a). (d) Bumps and spikes. (e) Wavelet phase of signal (d). (f) Wavelet modulus of signal (d). (g) The Morlet wavelet at a periodicity correlating best with the bump signal. This value of the Morlet wavelet results in the dominant band of (c). (h) The Morlet wavelet with half the period of that in (g). This value of the Morlet wavelet results in the second band in (c) at an a scale half that of (g).

the dominant (white) band in the modulus plot of Figure 2.26c. The next, most dominant (grey) band in the modulus plot is generated when the Morlet wavelet correlates with the signal as shown in Figure 2.26h. Figure 2.27 contains the same chirp signal as that shown earlier in Figure 2.9. The Morlet wavelet with $f_0 = 0.849$ was used to transform the signal. The real part of the transformed signal is plotted in Figure 2.27b and has similarities with the Mexican hat transform plot in Figure 2.9. The discontinuities at the beginning and end of the chirp segment are picked up well in the phase plot of Figure 2.27c. These are located using arrows at the top of the phase plot. The continuous increase in the instantaneous frequency associated with the chirp is highlighted in the modulus plot of Figure 2.27d.

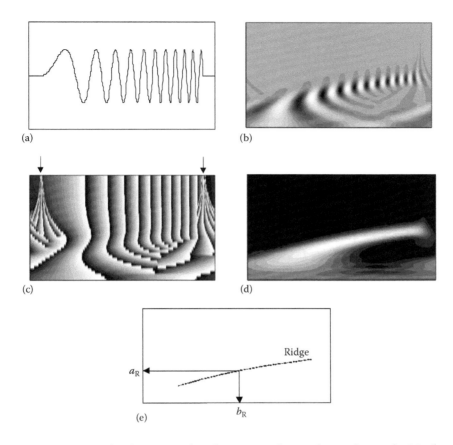

FIGURE 2.27 Segment of a chirp signal with associated transform plots – the Morlet wavelet: (a) Chirp signal segment. (b) Real part of the Morlet wavelet transform. (c) Phase. (d) Modulus. (e) A schematic of the ridge found from the maxima of the rescaled scalogram $|T(a,b)|^2/a$. The instantaneous frequency at time b_R can be found from a_R. We can see the relation between the maxima in the rescaled scalogram and the instantaneous frequency by substituting a complex sinusoid for the signal $x(t)$ in the wavelet transform integral given by Equation 2.13, and using a Morlet wavelet. Then, using a change of variable $t' = (t-b)/a$, it can be shown that maxima in the rescaled scalogram correspond to the instantaneous frequencies through their associated scales.

The instantaneous frequency associated with a signal can be found by way of its *wavelet transform ridges*. These are the maxima found in the rescaled wavelet transform scalogram, $|T(a,b)|^2/a$ associated with the instantaneous frequency of the signal. The ridge associated with the chirp signal is shown schematically in Figure 2.27e, where the instantaneous scale a_R at time b_R can be used to find the instantaneous frequency f_R ($=f_0/a_R$). The instantaneous amplitude and phase can also be found from the ridge. Further, if we plot the rescaled scalogram in terms of a characteristic wavelet frequency $|T(f,b)|^2/a$ where $f=f_0/a$ (refer back to Equations 2.26 and 2.27), then the instantaneous frequency can be read directly from this plot. The reader is referred to Sections 2.16, 2.17 and 2.23 at the end of this chapter for more information on ridges.

44 ■ The Illustrated Wavelet Transform Handbook

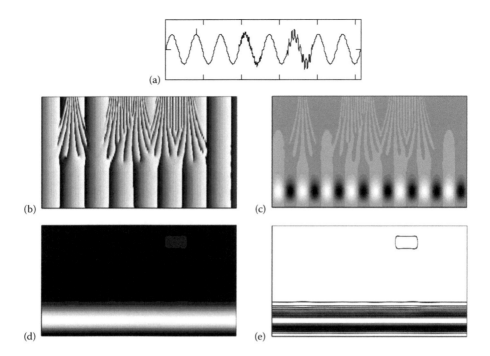

FIGURE 2.28 Wavelet decomposition of a sinusoidal waveform with added small-scale features: (a) Signal. (b) Phase. (c) Real part. (d) Modulus. (e) Unfilled contour plot of modulus.

A final test signal for the Morlet wavelet is shown in Figure 2.28. It is composed of a dominant sinusoid of unit amplitude which has three features superimposed on it. At the second peak, a small spike has been added. The fourth oscillation has noise added to it. A high-frequency oscillatory component has been added to the sixth oscillation of the dominant sinusoid. This has a frequency 10 times that of the large waveform. The location of the spike can be easily identified in the phase plot, as can the regions of random noise and the high-frequency component. The real transform plot locates the dominant peaks and troughs of the signal together with superimposed signal features. The modulus is plotted as a filled plot in Figure 2.28d and as a contour plot in Figure 2.28e. The dominant waveform is obvious in both figures. The high-frequency oscillatory feature in the signal is made much more obvious in the contour plot. In general, as the signal becomes more complex in nature, the phase and modulus information quickly becomes more difficult to interpret.

2.12 WAVELET TRANSFORM, SHORT-TIME FOURIER TRANSFORM AND HEISENBERG BOXES

In this section, we take a brief look at the time–frequency characteristics of the wavelet transform and compare it to the short-time Fourier transform. We will consider the specific cases of the Morlet wavelet transform and the Gabor STFT, both of which employ a Gaussian window.

As we will see, the Morlet wavelet has a form very similar to the analyzing function used for the short-time Fourier transform within a Gaussian window. The important difference is that, for the Morlet wavelet transform, we scale the window and enclosed sinusoid together, whereas for the STFT we keep the window length constant and scale only the enclosed sinusoid. The wavelet can, therefore, localize itself in time for short-duration, that is, high-frequency, fluctuations. There is, however, an associated spreading of the frequency distribution associated with wavelets of short duration. Conversely, there is a spreading of temporal resolution at low frequencies. This is illustrated in Figure 2.29a. The middle of the figure contains a schematic of a Morlet wavelet (real part only) shown at three a scales. The energy densities of the wavelets are then plotted in both the time and frequency domains, that is, respectively, $|\psi_{a,b}(t)|^2$ and $|\hat{\psi}_{a,b}(f)|^2$. We can see from the figure that, as the wavelet contracts in time, it becomes composed of higher frequencies with a wider spread. The spread f $|\psi_{a,b}(t)|^2$ and $|\hat{\psi}_{a,b}(f)|^2$ can be quantified using σ_t and σ_f, respectively – the standard deviations around their respective means. We can represent the spread of the wavelets in the time–frequency plane by drawing boxes of side lengths $2\sigma_t$ by $2\sigma_f$. These are shown at the top of Figure 2.29a. These boxes are known as *Heisenberg boxes* after the Heisenberg uncertainty principle which tells us the minimum area that these boxes can have. Specifically, the product $\sigma_t \sigma_f$ must be greater than or equal to $1/4\pi$, and thus, the area of the Heisenberg box is $1/\pi$. In fact, for the Gaussian windowed functions used in the Morlet wavelet transform and STFT considered here, $\sigma_t \sigma_f$ is exactly equal to $1/4\pi$ as the Gaussian distribution is the optimal window shape. (You can verify this for the Morlet wavelet given in the time domain by Equation 2.37 and the frequency domain by Equation 2.38, integrating their squared modulus leads to $\sigma_t = 1/\sqrt{2}$ and $\sigma_f = \sqrt{2}/4\pi$.) The Heisenberg uncertainty principle actually addresses the problem of the simultaneous resolution in time and frequency that can be attained when measuring a signal. To get a good idea of the frequency composition, we need to sample a long period of the signal. If instead we pinpoint a small region of the signal to measure it with accuracy, then it becomes very difficult to determine the frequency makeup of the signal in that region. That is, the more accurate the temporal measurement (smaller σ_t) the less accurate the spectral measurement (larger σ_f) and vice versa. Note that the Morlet central frequency f_0 sets the location of the Heisenberg box in the time–frequency plane for the mother wavelet and hence the relative locations for all dilated wavelets. This is shown in Figure 2.29b. Thus, when comparing Heisenberg boxes centred at the same location in the time–frequency plane, lower values of f_0 correspond to Heisenberg boxes that are wider in frequency and narrower in time than boxes corresponding to higher values of f_0. Thus, Morlet wavelets with lower values of f_0 correspond to time–frequency decompositions that are 'more temporal than spectral' as compared with their higher, central-frequency counterparts. (See the example in Chapter 5, Figure 5.11.)

The Fourier transform of a signal $x(t)$ is defined as

$$\hat{x}(f) = \int_{-\infty}^{\infty} x(t) e^{-i(2\pi f)t} \, dt \qquad (2.41)$$

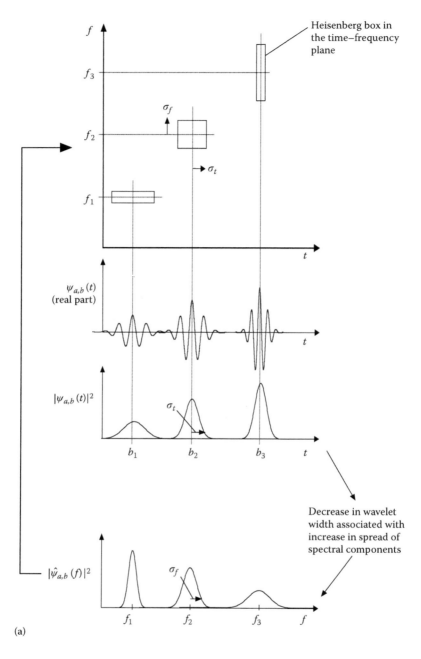

FIGURE 2.29 **Heisenberg boxes in the time–frequency plane:** (a) Heisenberg boxes in the time–frequency plane for a wavelet at various scales. Do not confuse σ_f with f_c, the passband frequency, given earlier in Section 2.4: f_c is the standard deviation of the spectrum around the origin; σ_f is the standard deviation of the spectrum around the mean spectral components f_1, f_2 and f_3 shown in the figure.

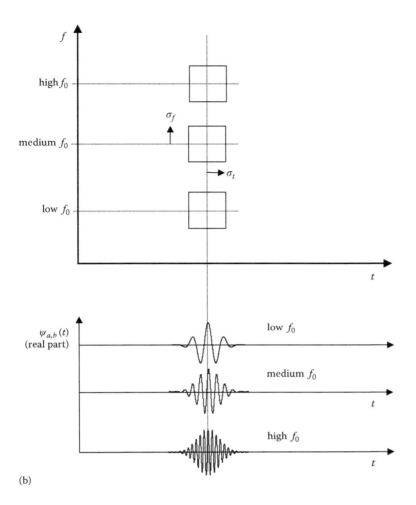

FIGURE 2.29 (CONTINUED) Heisenberg boxes in the time–frequency plane: **(b)** Heisenberg boxes in the time–frequency plane for a Morlet mother wavelet with three different central frequencies set to a low, medium and high value. The confining Gaussian windows are all of the same dimensions. Notice that altering the central frequency of the mother wavelet simply shifts the associated 'mother' Heisenberg box up and down the time–frequency plane without altering the box dimensions. This mother Heisenberg box then defines the relative shapes of all the others in the time–frequency plane associated with each wavelet, that is, the pattern shown in (a) is simply shifted up or down the plane.

We can modify the Fourier transform to allow localized features in the signal to be interrogated. This *short-time Fourier transform* employs a window function to localize the complex sinusoid. It is defined as

$$F(f,b) = \int_{-\infty}^{\infty} x(t)\, h(t-b)\, e^{-i2\pi ft}\, dt \qquad (2.42)$$

where $h(t-b)$ is the window function which confines the complex sinusoid $e^{-i2\pi ft}$. The STFT is also commonly known as the windowed Fourier transform. There are many shapes of

48 ■ The Illustrated Wavelet Transform Handbook

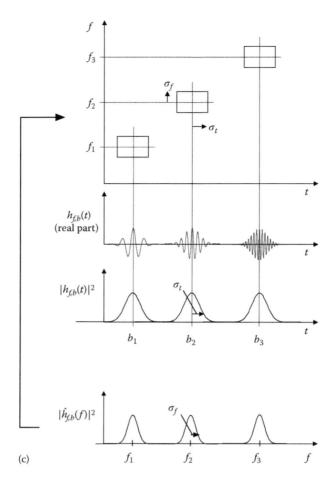

(c)

FIGURE 2.29 (CONTINUED) Heisenberg boxes in the time–frequency plane: (c) Heisenberg boxes in the time–frequency plane for the STFT. Other window lengths will produce longer and thinner or shorter and fatter boxes. However, once the window length is fixed, all Heisenberg boxes in the time–frequency plane corresponding to the STFT will have the same dimensions.

windows available, for example Hanning, Hamming, cosine, Kaiser and Gaussian. We will consider the Gaussian windowed STFT, known as the Gabor transform, which has the form:

$$h(t) = \frac{1}{\sqrt{\sigma}\,\pi^{1/4}} e^{-\frac{1}{2}\frac{t^2}{\sigma^2}} \quad (2.43a)$$

where σ is a fixed parameter (the standard deviation) which sets the width of the Gaussian window on the time axis. The combined window plus complex sinusoid is known as a *windowed Fourier atom*, or more generally as a *time-frequency atom*, denoted

$$h_{f,b}(t) = h(t-b)e^{i2\pi ft} \quad (2.43b)$$

The complex conjugate of this atom may be used to perform a time–frequency decomposition of the signal $x(t)$: the short-time Fourier transform. We can see that, when the Gaussian window is employed within the STFT integral of Equation 2.42, we obtain

$$F(f,b) = \int_{-\infty}^{\infty} x(t) h_{f,b}^*(t)\, dt$$

$$= \int_{-\infty}^{\infty} x(t)\, \frac{1}{\sqrt{\sigma}\, \pi^{1/4}}\, e^{-\frac{1}{2}\frac{(t-b)^2}{\sigma^2}}\, e^{-i2\pi ft}\, dt \qquad (2.44)$$

This has a very similar form to the Morlet wavelet transform. We can see this by combining Equations 2.13 and 2.40 and rearranging the terms as follows:

$$T(a,b) = \int_{-\infty}^{\infty} x(t)\, \psi_{a,b}^*(t)\, dt$$

$$= \int_{-\infty}^{\infty} x(t)\, \frac{1}{\sqrt{a}\, \pi^{1/4}}\, e^{-\frac{1}{2}[(t-b)^2/a^2]}\, e^{-i2\pi(f_0/a)(t-b)}\, dt \qquad (2.45)$$

The main difference between the Gabor STFT and the Morlet WT is now obvious: the internal frequency, f, is allowed to vary within a Gaussian window of *fixed* width (given by σ) in the former whereas the latter employs an internal frequency $f\ (=f_0/a)$ which is linked to the window width (given by a). There is another less significant difference, in that the wavelet's complex sinusoid is centred at b on the time axis whereas the complex sinusoid contained within the Gabor atom is 'centred' at the origin ($t=0$). To put it another way, in the STFT the sinusoid remains fixed in relation to the origin and the window slides across it, whereas for the wavelet transform, the origin of the complex sinusoid is at b and hence the sinusoid and window move together.

The atom used for the Gabor STFT is shown in Figure 2.29c for three internal frequencies within the fixed-width window. This atom also has a constant width in the frequency domain. This leads to boxes in the time–frequency plane of equal shape, regardless of the internal frequency. The dimensions of the Heisenberg boxes shown in the time–frequency plane at the top of Figure 2.29c are determined by the preselected window width σ. Longer and thinner or shorter and fatter, Heisenberg boxes can be obtained by changing the window width. However, whatever length of window is employed, once fixed, the corresponding Heisenberg boxes associated with the STFT all have exactly the same dimensions in the time–frequency plane. This is true for all window shapes used in the STFT – not just Gaussians. As with the Morlet wavelet transform, the Gaussian windowed complex sinusoids used in the Gabor STFT have the smallest areas of Heisenberg boxes in the time–frequency plane – that is, they have optimal time–frequency energy distributions. An example of the difference between the WT and STFT is shown in Figure 2.30, which contains a test signal together with corresponding Morlet-based wavelet transforms and Gabor STFTs. The signal comprises a sinusoid plus a single spike located by the arrow in

50 ■ The Illustrated Wavelet Transform Handbook

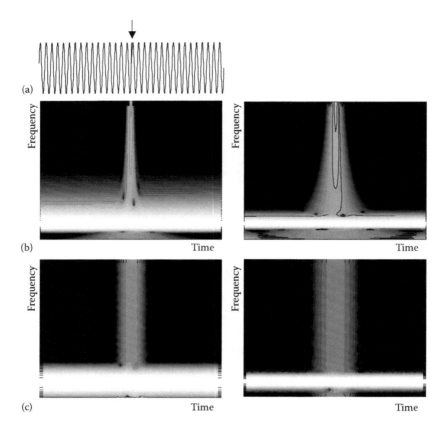

FIGURE 2.30 STFT and WT time–frequency plots: (a) Sinusoid containing a spike at the location indicated by the arrow. (b) Morlet wavelet transforms of the signal in (a). The wavelet transform on the left has been generated using a Morlet wavelet with a central frequency half that used to produce the transform plot on the right. (c) Gabor STFTs of the signal in (a). The left-hand plot corresponds to a Gaussian window which is half the width of the one used to generate the right-hand plot. Note that the horizontal axis is time and the vertical axis is frequency. In addition, high to low energies correspond with white to black in the greyscale used.

Figure 2.30a. We can see that the sinusoid manifests itself as a dominant horizontal ridge in both the WT and STFT plots. The spike, however, is localized in time at high frequencies by the wavelet transform whereas it corresponds to the vertical ridge of constant width in the STFT due to the constant width of its Heisenberg boxes. Hence, the wavelet transform can discern individual high-frequency features located close to each other in the signal whereas the STFT smears such high-frequency information occurring within its fixed-width window (a good example of this effect is shown in Figure 6.15 of Chapter 6). We can also see from Figure 2.30b that the Morlet wavelet transform corresponding to the lower central frequency (left-hand plot) produces a sharper resolution in time but correspondingly poorer resolution in frequency than the transform based on the higher central frequency (right-hand plot). Similarly, comparing the plots in Figure 2.30c, we can see that the STFT corresponding to the shorter window width produces a sharper resolution in time but a poorer resolution in frequency than that generated using the longer window width.

2.13 ADAPTIVE TRANSFORMS: MATCHING PURSUITS

In the previous section, we saw that the wavelet transform becomes more localized as it interrogates the signal at smaller scales, whereas the STFT has a fixed window length for all scales. We will now take a brief look at an adaptive transform called the *matching pursuit* which offers an alternative, more flexible way of providing time–frequency information. Although more flexible, it does not provide a regular, repeatable or even full coverage of the time–frequency plane but rather one that adapts to each signal and is hence signal dependent.

The matching pursuit method involves the decomposition of the signal piece by piece using a dictionary of analyzing functions. At each stage in the decomposition, an analyzing function is chosen which 'best' represents part or all of the remaining signal. After a number of decompositions, the original signal $x(t)$ can be represented by some arbitrary resolution using a series of expansion coefficients, M_i, $i = 1, 2, \ldots n$; where n is the number of iterations of the decomposition algorithm. The signal approximation reconstructed from these expansion coefficients is given by

$$x_n(t) = \sum_{i=0}^{n-1} M_i h_i(t) \qquad (2.46)$$

where $h_i(t)$ are the functions used in the decomposition. The differences between wavelet transform analysis and matching pursuit analysis lie both in the way that the transform coefficients are selected and also in the flexibility around choosing the analyzing function used in the signal decomposition. The signal is first examined using each of the analyzing functions contained within a preselected dictionary of functions, and the one which takes the most energy from the signal is chosen to decompose the signal. The residual signal is constructed and then examined to find the next function from the dictionary which takes the most energy from this new signal. The process is repeated until the residual signal falls below some predetermined cutoff. The procedure is illustrated in Figure 2.31. In the figure, we can see the original signal with the first analyzing function – a time–frequency atom – used to decompose it shown directly below. The bottom plot in the figure contains the first residual signal.

The example in Figure 2.31 uses a Gabor atom, one of the most commonly employed analyzing functions in matching pursuit analysis. We have already come across the Gabor atom used in the STFT. For the matching pursuit method the atom is defined as

$$h_{a,b,f}(t) = \frac{1}{\sqrt{a}} h\left(\frac{t-b}{a}\right) e^{i(2\pi f)t} \qquad (2.47)$$

where the scale, a, location, b and frequency, f, can all be varied independently. Thus, it has an increased flexibility over the Gabor STFT atom. We will define the Gaussian window as

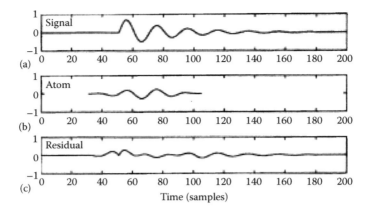

FIGURE 2.31 The matching pursuit method: (a) A damped sinusoidal signal. (b) The first Gabor atom chosen to represent the signal in (a). (c) The residual signal. (Note that this example comes from a paper by Goodwin and Vetterli, 1999, which goes on to show the difficulties in using symmetric window functions in representing transient signals.) (After Goodwin and Vetterli, *IEEE Transactions on Signal Processing*, 47(7), 1890–1902, © (1999) IEEE.)

$$h(t) = 2^{1/4} e^{-\pi t^2} \tag{2.48}$$

Note that this is different in form from that used for the STFT in Equation 2.43. Both have unit energy and are equally valid; however, Equation 2.48 is prevalent in the literature for the matching pursuit method, hence we use it here. The time–frequency atom defined by Equations 2.47 and 2.48 has a similar form to both those used in the Morlet wavelet transform and the short-time Fourier transform of Equations 2.44 and 2.45. However, the Gabor atom used in the matching pursuit method is more flexible in that its scale, location and internal frequency may all be varied independently.

In order to get a decomposition with real expansion coefficients and real residuals, which is often required in practice, real-only atoms are employed of the form:

$$h_{a_i,b_i,f_i,\phi_i}(t) = K_i \frac{2^{1/4}}{\sqrt{a_i}} e^{-\pi\left(\frac{t-b_i}{a_i}\right)^2} \cos(2\pi f_i t + \phi_i) \tag{2.49}$$

where:
a_i and b_i are the scale and location factors for the Gaussian envelope
f_i is the frequency of the real sinusoid within the Gaussian envelope and
ϕ_i is the phase of the real sinusoid within the Gaussian envelope
K_i is a normalization factor used to maintain unit energy for for $h_{a_i,b_i,f_i,\phi_i}(t)$

The subscript *i* relates to the specific set of parameters a, b, f and ϕ used for the *i*th decomposition to get M_i.

The expansion coefficients M_i are determined in turn by examining the signal with the analyzing atom and selecting the parameter set (a_i,b_i,f_i,ϕ_i) which provides the largest value of $|M_i|^2$ where

$$M_i(a_i,b_i,f_i,\phi_i) = \int_{-\infty}^{\infty} x(t) h_i^*(t) \, dt \qquad (2.50)$$

and where we have used h_i as a more compact representation of h_{a_i,b_i,f_i,ϕ_i}. In addition to a_i, b_i, f_i and ϕ_i, we need to retain K_i and, of course, M_i at each iteration of the matching pursuit method in order to perform the reconstruction. The matching pursuit method is very flexible and the dictionary of analyzing functions can include functions other than the Gabor function h_i, such as sinusoids within a differently shaped window, continuous sinusoids (i.e. a Fourier function), wavelet functions, Dirac delta functions and so on. All these functions are kept within a preselected *dictionary* and each are used in turn to determine which gives the maximum value $|M_i|^2$ and, hence, takes the most energy from the signal.

The matching pursuit method is an iterative technique. At the first iteration, the signal is decomposed into two orthogonal components:

$$x(t) = x_1(t) + R^1 x(t) \qquad (2.51)$$

which can also be written as

$$x_0(t) = M_0 h_0(t) + R^1 x(t) \qquad (2.52)$$

where $R^1 x(t)$ is the residual vector after approximating the original signal $x_0(t)$ (now subscripted with '0') in the direction of $h_0(t)$. $R^1 x(t)$ and $h_0(t)$ are orthogonal to each other, hence the energy of the signal can be expressed as

$$\|x_0(t)\|^2 = \|M_0 h_0(t)\|^2 + \|R^1 x(t)\|^2 \qquad (2.53)$$

where $\|x\|^2 = \int |x|^2 dx$. In addition, as $h_0(t)$ has unit energy by definition and M_0 is a constant, then we can write

$$\|x_0(t)\|^2 = |M_0|^2 + \|R^1 x(t)\|^2 \qquad (2.54)$$

The first coefficient M_0 is found by applying each of the dictionary functions to the signal in turn and choosing the one which maximizes the value of M_0, or conversely, the one which minimizes the energy of the residual $\|R^1 x(t)\|^2$. This residual signal is then examined in the same way to find the second coefficient M_1 and the new signal residual $R^2 x(t)$, and so on. At any stage in the decomposition, the original signal can be partitioned into two components: the reconstructed part using all the expansion coefficients (given by Equation 2.46), and the residual component after n iterations, $R^n x(t)$. This can be written as

$$x_0(t) = \sum_{i=0}^{n-1} M_i h_i(t) + R^n x(t) = x_n(t) + R^n x(t) \qquad (2.55)$$

where $x_n(t)$ is the signal approximation and $R^n x(t)$ is the residual signal left after the nth iteration of the matching pursuit algorithm. In addition, the energy at the nth decomposition is given by

$$\|x_0(t)\|^2 = \sum_{i=0}^{n-1} |M_i|^2 + \|R^n x(t)\|^2 \qquad (2.56)$$

In practice, the matching pursuit algorithm is terminated either when the residual energy is below a preset cutoff level ε, defined by

$$\|R^n x(t)\|^2 < \varepsilon^2 \left(x_0(t)^2\right) \qquad (2.57)$$

or, alternatively, after a predetermined number of iterations, n.

Figure 2.32 shows an example of a synthetic signal decomposed using the matching pursuit method selecting from a dictionary containing three basic elements: Gabor atoms, Dirac delta functions (spikes in the time domain) and sinusoidal Fourier functions (spikes in the frequency domain and continuous sinusoids in the time domain). We can see from the figure that the MP algorithm leads to an intermittent, patchy covering of the time–frequency plane as only those coefficients obtained when the time–frequency atom is matched to the signal at locations of maximum energy removal are retained. This

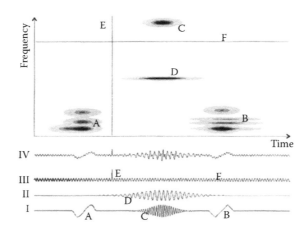

FIGURE 2.32 The time–frequency decomposition of a test signal using the matching pursuit method. The figure contains the time–frequency plane with the energy distributions associated with the MP decomposition of the signal IV. Letters mark signal components and corresponding atoms or groups of atoms: A, B = single transients; C, D = sinusoids modulated by a Gaussian function; E = single sharp transient; F = sinusoid. Signal IV is a sum of I, II and III. The MP analysis time–frequency plot for signal IV is shown above the signals. The darkness of the plot elements is proportional to the logarithm of the energy distribution in the time–frequency plane according to a Wigner–Ville distribution. With kind permission from P.J. Durka.

contrasts with both the wavelet transform and short-time Fourier transform, where the time–frequency plane is covered evenly. The time–frequency plot used to represent the energy distribution of the matching pursuit decomposition is based on a Wigner distribution. We do not go into the details of this representation herein, but instead the reader is referred elsewhere at the end of this chapter.

2.14 WAVELETS IN TWO OR MORE DIMENSIONS

The two-dimensional Mexican hat specified on a t_1, t_2 coordinate plane is given by

$$\psi(\mathbf{t}) = \left(2 - |\mathbf{t}|^2\right) e^{-\frac{|\mathbf{t}|^2}{2}} \tag{2.58}$$

where $\mathbf{t}$ is the coordinate vector (t_1, t_2) and $|\mathbf{t}| = \sqrt{t_1^2 + t_2^2}$. Figure 2.33 shows a couple of two-dimensional Mexican hats on the plane. The two-dimensional wavelet transform is given by

$$T(a, \mathbf{b}) = \frac{1}{a} \int_{-\infty}^{\infty} \psi^*\left(\frac{\mathbf{t} - \mathbf{b}}{a}\right) x(\mathbf{t}) \, d\mathbf{t} \tag{2.59}$$

where $\mathbf{b}$ is the coordinate vector (b_1, b_2). Note that the weighting function $w(a)$ has been set to $1/a$ required to conserve energy across scales for the 2-D wavelets. Note also that the coordinate vector $\mathbf{t} = (t_1, t_2)$ is likely to specify two spatial (rather than temporal) coordinates in practice, where the function $x(\mathbf{t})$ could, for example, be surface heights (e.g. fractures surfaces, topographic features) or image greyscales (e.g. medical images, fluid visualization studies). The corresponding inverse wavelet transform is

$$x(\mathbf{t}) = \frac{1}{C_g} \int_{-\infty}^{\infty} \int_0^{\infty} \frac{1}{a} g\left(\frac{\mathbf{t} - \mathbf{b}}{a}\right) T(a, \mathbf{b}) \frac{da}{a^3} d\mathbf{b} \tag{2.60}$$

Wavelet transforms in higher dimensions, D, are also possible simply by extending the length of the vectors $\mathbf{t}$ and $\mathbf{b}$ to D components. To preserve energy in the D-dimensional transformation, the weighting function becomes $1/a^{D/2}$. Hence, the D-dimensional wavelet is defined as

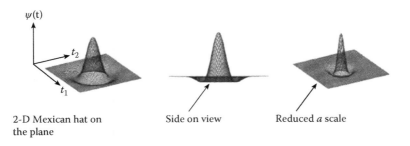

2-D Mexican hat on the plane   Side on view   Reduced a scale

FIGURE 2.33 Two-dimensional Mexican hat.

$$\psi_{a,\mathbf{b}}(t) = \frac{1}{a^{D/2}} \psi\left(\frac{\mathbf{t}-\mathbf{b}}{a}\right) \qquad (2.61)$$

The transform in D dimensions becomes

$$T(a,\mathbf{b}) = \int_{-\infty}^{\infty} \psi_{a,b}^{*}(t)\, x(\mathbf{t})\, d\mathbf{t} \qquad (2.62)$$

with inverse:

$$x(\mathbf{t}) = \frac{1}{C_g} \int_{-\infty}^{\infty} \int_{0}^{\infty} \psi_{a,b}(t)\, T(a,\mathbf{b}) \frac{da}{a^{D+1}}\, d\mathbf{b} \qquad (2.63)$$

and energy of the signal may be found from

$$E = \frac{1}{C_g} \int_{-\infty}^{\infty} \int_{0}^{\infty} |T(a,\mathbf{b})|^2 \frac{da}{a^{D+1}}\, d\mathbf{b} \qquad (2.64)$$

2.15 THE CWT: COMPUTATION, BOUNDARY EFFECTS AND VIEWING

As with all mathematical tools used to investigate physical phenomena, a number of practical issues must be taken into consideration. This is no less the case when using the wavelet transform. The results obtained by the investigator must be viewed in terms of the limitations in the data analysis method employed. These limitations stem from a number of sources, including the discrete nature of the data, the finite resolution of the data, the finite extent of the data, the wavelet used, the discretization and numerical computation of the transform and so on.

To compute the continuous wavelet transform, we could simply perform a naive discretization of the transform integral given by Equation 2.13, replacing the integral with a discrete summation involving the sampling interval of the measured time series Δt together with a suitable discretization for the a and b parameters – usually logarithmic for a and linear steps of Δt for b. This is a very cumbersome way to compute the CWT integral and a much better approach is to use the FFT method described in Section 2.10 and given by Equation 2.35. The FFT provides a much faster algorithm for the computation of the transform integral. In fact, this approach has been used in all the transform plots presented in this chapter. We will consider again the discretization of the wavelet transform in the next chapter when we tackle the discrete wavelet transform (DWT). However, note that the DWT is fundamentally different from the discretized CWT (see Chapter 3) and in the scientific literature the term 'continuous wavelet transform' generally includes all discretizations using continuous wavelets such as the Mexican hat, Morlet and so on.

In practice, experimental data sets are finite in extent and often the investigator wants to analyze the whole of the available data. However, an obvious consequence of the wavelet analysis of a finite data set is that, as the wavelet gets closer to the edge of the data, parts of

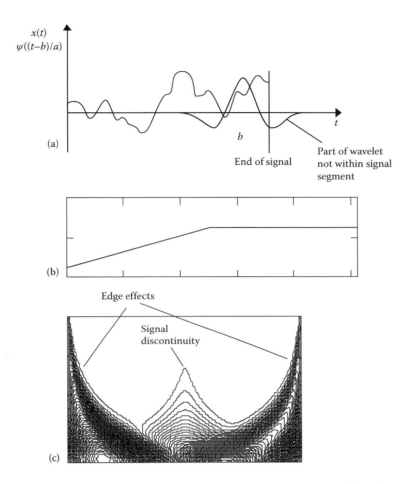

FIGURE 2.34 The manifestation of boundary effects in the scalogram: (a) Schematic diagram of the wavelet encountering the signal boundary. (b) A simple signal with a discontinuity halfway along its length. (c) The scalogram of the signal in (b).

it begin to 'spill over' the edge as illustrated in Figure 2.34a. This creates a *boundary effect*, where transform values close to the boundary of the signal are tainted by the discontinuous nature of the signal edge. The affected region increases in extent as the dilation of the analyzing wavelet increases. An example of a boundary effect is shown in the transform plot of Figure 2.34c, which is generated for the simple signal shown in Figure 2.34b using a Mexican hat wavelet. The transform plot points to the location of the signal discontinuity as expected, but also treats the edges of the signal as additional discontinuities. Hence, large $T(a,b)$ values are realized close to the edge of the transform plot, which increase in extent as a increases. This region is affected by a discontinuity and is known as the 'cone of influence'. The extent of the cone increases linearly with a, that is, it is proportional to the temporal support (or 'width') of the wavelet. (We have already seen the cone of influence corresponding to a signal discontinuity in Figure 2.12.) However, when plotting $T(a,b)$ using a logarithmic a scale (e.g. Figure 2.34c), the cone contours become curved. The cone boundaries at either end of the signal define the region, which is significantly influenced by the signal edges. It is very much up to the investigator where they take the boundary of the cone by choosing a limiting value of the distance from the wavelet centre – a multiple of the wavelet a scale – at which point pertinent information contained in the transform plot is not considered masked by edge effects.

A number of methods have been developed to cope with the boundaries of signals of finite extent. A variety of these are illustrated in Figure 2.35 and include adding a line of zero values at either end of the signal (zero padding – Figure 2.35a); adding a line of constant values at either end of the signal equal to the last value of the signal (value padding – Figure 2.35b); adding some form of decay to zero for the last value at each end of the signal (decay padding – Figure 2.35c); continuing the signal on from the last point back to the first point (periodization – Figure 2.35d); reflecting the signal at the edges (reflection – Figure 2.35e); convolving the signal with a window function which reduces the edge values of the signal to zero (windowing – Figure 2.35f); using a polynomial extrapolation of the

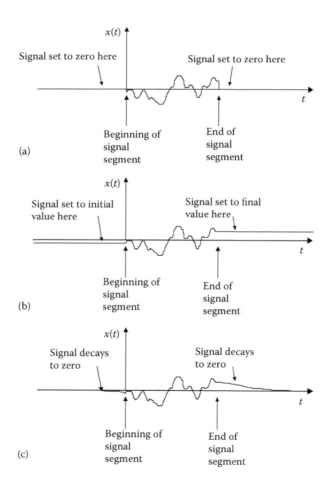

FIGURE 2.35 **Schemes to deal with signal boundaries: (a)** Zero padding. **(b)** Value padding. **(c)** Decay padding.

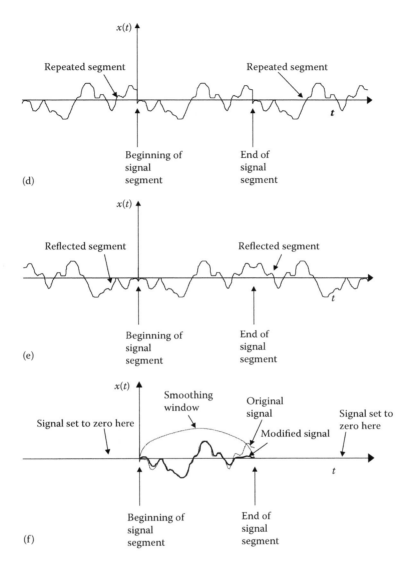

FIGURE 2.35 (CONTINUED) **Schemes to deal with signal boundaries: (d)** Periodization involves the repetition of the signal along the time axis. **(e)** Reflection. **(f)** Smoothing window.

signal at either end (polynomial fitting – Figure 2.35g); and, if we are focussing in on a small segment of a much larger signal available to us, we may simply use the known data points outwith the segment under consideration (signal following – Figure 2.35h). In addition, there is the wraparound method which uses only the length of signal available but wraps the parts of the wavelet which fall off each end of the signal back to the other end. Wraparound is similar to periodization but gives wavelet transform values only within the extent of the signal. Wraparound is used extensively in the next chapter. Whichever technique is employed to cope with the signal edges, we must be aware that features appearing near the edges of the transform plot will contain information (synthetic or real) from outwith the region of the signal segment under consideration. In addition, this effect will

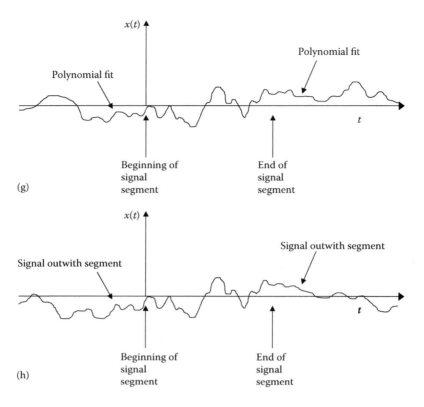

FIGURE 2.35 (CONTINUED) Schemes to deal with signal boundaries: **(g)** Polynomial fitting. **(h)** Signal following.

increase in extent as the wavelet a scale increases and the wavelet extends further beyond the edge of the signal when close to it.

Finally, when plotting the wavelet transform, we want to be able to discern features associated with the signal at specific a scales. It is often the case that features at certain scales dominate the transform plot, obscuring the detail at other scales. In order to accentuate this hidden detail, we can change the weighting parameter $w(a)$ in the transform from the usual $1/\sqrt{a}$ value to a value which is more suitable for highlighting features at a specific a scale of interest. This is shown in Figure 2.36 for a sinusoidal signal containing a burst of noise. The transform plot with $w(a)$ set to $1/\sqrt{a}$ shows up the sinusoidal waveform well, but this dominates and the noise is not highlighted particularly well. By changing the weighting to $w(a) = a^{-1.5}$, we can enhance the noise within the plot at lower a scales. When $w(a)$ is set to $a^{-2.5}$, we can see that the noise now dominates the transform plot. It is sometimes useful to vary $w(a)$ in this way to accentuate features at different scales in the transform plot which might otherwise have been missed. Another way to weight the transform plot to show up small and large amplitude features is to use a logarithmic scale for $T(a,b)$. However, two problems arise with this approach. The first is that we cannot take the logarithm of the negative values of $T(a,b)$ the second is that taking the logarithm of near zero values of $T(a,b)$ produces very large negative numbers. We can get round

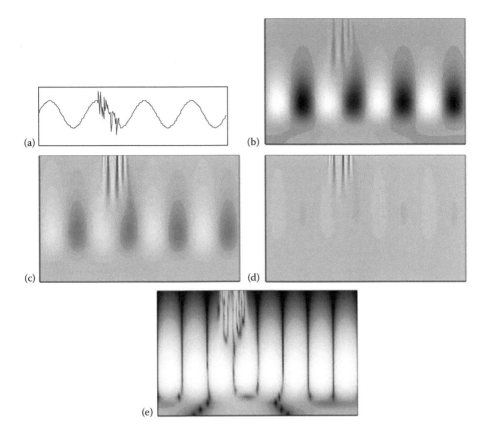

FIGURE 2.36 Accentuating features in the wavelet transform plot: **(a)** Original signal. **(b)** Transform plot with $w(a)=1/\sqrt{a}$. **(c)** Transform plot with $w(a)=a^{-1.5}$. **(d)** Transform plot with $w(a)=a^{-2.5}$. **(e)** Logarithmic plot of the modulus of the Mexican hat wavelet transform $|T(a,b)|$. Near zero values of $|T(a,b)|$ produce large negative logarithmic values. These are omitted using a minimum cutoff value, where all $|T(a,b)|$ values less than the cutoff are set to this value before computing logarithms.

these two problems, respectively, by using the modulus of the transform $|T(a,b)|$ to avoid negatives and setting a cutoff, or floor, of $|T(a,b)|$ in order to both avoid zero and near zero values and limit the extent of the logarithmic scale. This is illustrated in Figure 2.36e. Logarithmic plotting is particularly good at simultaneously highlighting features in the signal occurring at very different orders of magnitude, see, for example, the logarithmic amplitude plots of the ECG signals in Figures 6.10 and 6.11 in Chapter 6.

2.16 RIDGE FOLLOWING AND SECONDARY WAVELET FEATURE DECOUPLING

Interrogation of the wavelet transform ridges corresponding to dominant signal components may be a relatively simple and effective way of extracting signal information in terms of frequency or amplitude modulation. Figure 2.37a contains a test sinusoidal signal with both a frequency modulation and a decreasing frequency drift. Inspection

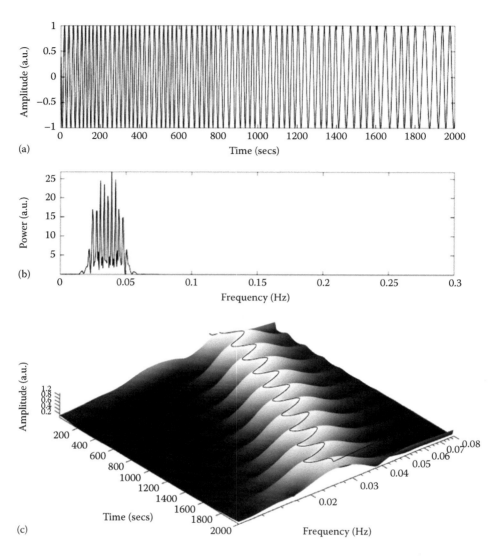

FIGURE 2.37 **Wavelet transform of a frequency drift and a modulated test signal:** (a) Test signal, sinusoidally modulated and drifting in frequency. (b) Fourier transform of the test signal. (c) Wavelet transform modulus of the test signal. (Greyscale: Black to white = lowest to highest amplitude.) (From Addison, P.S., *IEEE Reviews in Biomedical Engineering*, 8, 78–85, 2015b. © 2004 IEEE.)

of the signal in the time domain provides limited information, with perhaps a vague realization that, in general, the frequencies are drifting and modulating over time (Addison, 2015b). The corresponding Fourier transform (Figure 2.37b) also provides limited information apart from a general picture of the range of frequencies visited over the signal segment. However, if we view the corresponding wavelet transform modulus (Figure 2.37c,d), we may observe very high-resolution detail of the underlying signal characteristics: the drift is evident, decreasing from 0.05 to 0.025 Hz, and

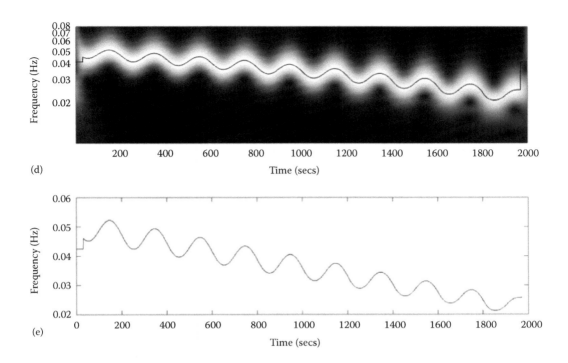

FIGURE 2.37 (CONTINUED) Wavelet transform of a frequency drift and a modulated test signal: **(d)** Plan view of (c). **(e)** Ridge extracted from the wavelet transform modulus surface. (From Addison, P.S., *IEEE Reviews in Biomedical Engineering*, 8, 78–85, 2015b. © 2004 IEEE.)

the modulation (of 0.005 Hz) is obvious from the ridge of the band (shown as a black curve superimposed on the band maxima). In fact, we may easily extract the instantaneous frequencies from the signal by projecting this ridge vertically (Figure 2.37e). Contrasting this information with that available in the time and/or frequency domains of Figure 2.37a,b illustrates the power of the wavelet method to extract useful temporally resolved spectral information.

We may further examine the nature of transform ridge modulations by performing a second wavelet transform on the ridge itself using the *secondary wavelet feature decoupling* (SWFD) method illustrated in Figures 2.38 and 2.39. The example signal shown in Figure 2.38a is composed of a sinusoid of frequency 0.04 Hz which is modulated in amplitude at 0.005 Hz. The wavelet transform modulus for the signal is shown in Figure 2.38b. A constant frequency for the main component can be observed on the ridge. However, the transform ridge is modulated in amplitude, as can just be seen in the figure by the cyclical change in shading along the ridge. This is more obvious in the 3-D plot of the transform modulus in Figure 2.38c which has the ridge locus superimposed. We may extract this amplitude modulation of the ridge by projecting the ridge horizontally and performing a second wavelet transform on it. This projection, which we call the 'ridge amplitude projection signal' (RAP), is shown in Figure 2.38d. By performing a second wavelet transform of the RAP signal, we obtain a new transform plot ($T_{RAP}(a,b)$) with a secondary band occurring at the frequency of the amplitude modulation of the original component of 0.005 Hz.

64 ■ The Illustrated Wavelet Transform Handbook

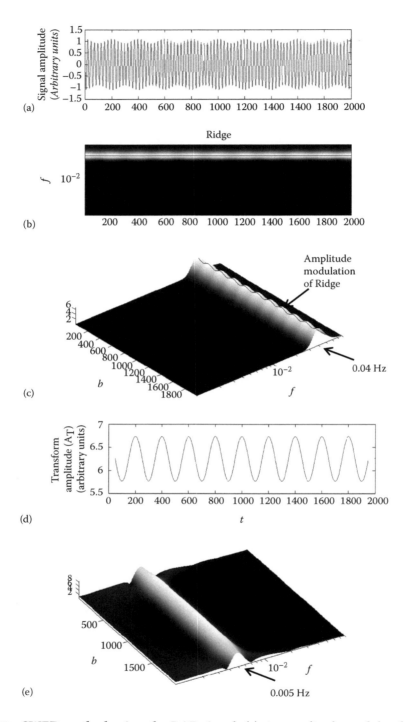

FIGURE 2.38 SWFD method using the RAP signal: (a) An amplitude-modulated test signal. **(b)** Wavelet transform modulus of the signal. **(c)** 3-D view of the wavelet transform modulus. **(d)** Amplitude modulations derived from the scalogram surface ridge – the RAP signal. **(e)** The modulus of the secondary wavelet transform of the RAP signal – $T_{RAP}(a,b)$. (Transform plot axes are in terms of f against b and signal axes are the amplitude in arbitrary units against time.)

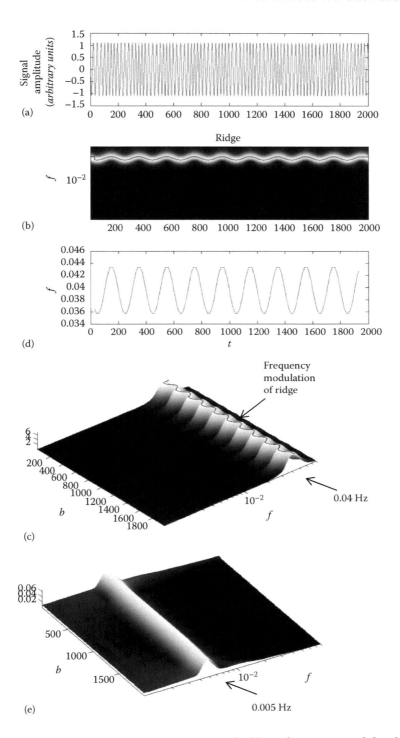

FIGURE 2.39 SWFD method using the RFP signal: (a) A frequency-modulated test signal. **(b)** Wavelet transform modulus of the signal. **(c)** 3-D view of the wavelet transform modulus. **(d)** Frequency modulations derived from the scalogram surface ridge – the RFP signal. **(e)** The modulus of the secondary wavelet transform of the RFP signal – $T_{RFP}(a,b)$.

This is shown in Figure 2.38e. In this way, the amplitude modulation of a signal component may be examined in wavelet space.

We may similarly investigate the frequency modulation of a signal component by projecting the ridge vertically to obtain a ridge frequency perturbation (RFP) signal. This is shown in Figure 2.39 for a test signal with a frequency modulation of 0.05 Hz. Note that here we can see the frequency modulation in the original 2-D plot of the transform modulus (Figure 2.39b). The vertical projection of the ridge gives the plot of Figure 2.39d. The secondary transform modulus of the RFP signal ($T_{RFP}(a,b)$) is shown in Figure 2.39e where a secondary band appears at the frequency of the frequency modulation of the signal.

Although simple test signals are used in these examples, the SWFD method described here is particularly useful for examining non-stationary modulations of non-stationary signal components (Addison and Watson, 2003, 2004a; Bozhokin and Suslova, 2013, 2014). An example of this is the use of the method for the extraction of the respiratory modulations of a heartbeat signal from a pulse oximeter medical device, which is given in Chapter 6, Figure 6.35 (Addison and Watson, 2004a).

2.17 RIDGE HEIGHTS

We know that a sinusoidal signal will create a smooth band in wavelet space when a Morlet wavelet transform is used. The ridge of this band will scale with the amplitude of the signal component; for example, if we double the amplitude of the sinusoid, the ridge of the band amplitude will double and so on. If we use the rescaled transform modulus $|T(a,b)|/\sqrt{a}$, then the ridge amplitude remains constant regardless of the frequency of the sinusoidal signal as long as its amplitude remains the same. We can show that for a Morlet wavelet (Section 2.11) the rescaled transform ridge height, A_{TR}, is related to the amplitude of the sinusoid A_S by

$$A_{TR} = \frac{\sqrt[4]{\pi}}{\sqrt{2}} A_S \quad (2.65)$$

An example of this is shown in Figure 2.40. A sinusoidal component of amplitude $A_S = 1$ is shown in Figure 2.40a. The corresponding rescaled transform modulus is given in Figure 2.40b, with the ridge superimposed on the plot. Figure 2.40c shows a 3-D view of the rescaled wavelet transform modulus with the ridge of height, A_{TR}, drawn as a black line along the top of the band. The amplitude of this ridge may be read off the transform surface as 0.941 as expected ($=\sqrt[4]{\pi}/\sqrt{2}$). By using a rescaled transform modulus plot, we can follow non-stationary ridges across the transform plane in the knowledge that any change in their amplitude is not related to the change in scale (or characteristic frequency), but rather some other factor; for example, the diminishing strength of that particular component or, in more complex signals, morphological changes to the component. Note that we may read the sinusoidal signal amplitude directly from the ridge (i.e. without the $\sqrt[4]{\pi}/\sqrt{2}$ scaling constant in Equation 2.65) by setting the absolute value of the maximum of the Fourier transform of the wavelet to 2, as described in Lilly and Gascard (2006). For more general information on wavelet ridges, see Carmona et al. (1997); Staszewski (1997); Wang et al. (2003); Tse and Lai (2007); Dien (2008); and Lilly and Olhede (2010). Additional

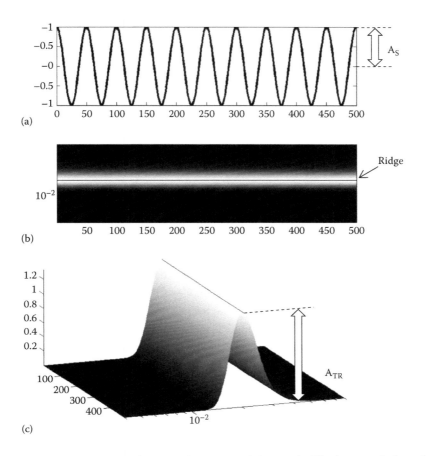

FIGURE 2.40 (a) The real part of a complex sinusoidal signal: (b) The rescaled modulus plot, $|T(a,b)|\sqrt{a}$, showing the ridge on the band. (c) The 3-D plot of (b) showing the ridge height A_{TR}. (Transform plot axes are f against b and signal axes are the amplitude in arbitrary units against time.)

information is also provided in Section 2.23.3, including reasons for the differences in formulations of wavelet transform ridge amplitude found in the literature.

2.18 RUNNING WAVELET ARCHETYPING

We may often be interested in a persistent repeating component within a signal – for example, the heartbeat in a biosignal, regular vortex shedding within a fluid flow signal, recurring beating within a condition monitoring signal from rotating machinery and so on. However, the signal may be relatively weak, intermittent or suffer from excessive noise, thus making it difficult to follow the component on a cycle-by-cycle or beat-by-beat basis as it is acquired in real time. In signal processing, we may build up a characteristic, or archetype, signal segment which is representative of this repeating waveform. In practice, this is often performed in the time domain where each beat is extracted from the signal and averaged with others in its vicinity. For example, we may simply average the latest beat with a number, N, of preceding beats in an ensemble average, or we may want to give more weight to the more recent beats used in such averaging.

One of the main problems for traditional methods in performing this kind of averaging is in the identification of the characteristic, or fiducial, points for each beat. Their location is required in order to align the averaging of the same points along each beat. Many schemes for deriving these fiducial points are available. However, an error in the temporal position of one beat relative to the others will lead to morphological errors occurring in the archetype. This is particularly so for highly non-stationary signals such as, for example, a heartbeat signal where the heart rate may vary naturally and hence the beat lengths change as well as the beat morphology. Performing the archetyping in wavelet space makes the problem simpler as the natural periodicity of the wavelet may be used at each scale to guide the cyclical averaging process. Thus, we may generate a *running wavelet archetype* (RWA) using a simple 'infinite impulse response' (IIR) weighted averaging scheme (Addison, 2015a) as follows:

$$T_{RWA}(a,b) = w.T(a,b) + (1-w).T_{RWA}(a,b-P(a)) \qquad (2.66)$$

Each time a wavelet transform value, $T(a,b)$, is computed, it is weighted by w and used with the previous archetype transform value $T_{RWA}(a,b-P(a))$, separated from the current value by a period $P(a)$, to form a new value of the archetype transform $T_{RWA}(a,b)$. This is illustrated schematically in Figure 2.41. Hence, at each step in the process, an archetype wavelet transform value is computed for each wavelet scale. The question is, what period to use for $P(a)$? The wavelet already separates out the signal information into natural scales; hence, we may use the characteristic periodicity of the wavelet at the scale considered. In this way, a running weighted average may be computed at each scale in the scalogram. As the signal component of interest moves to another scale (e.g. a pulse band moving due to a change in heart rate), more energy appears in the transform at that scale and thus, through the archetyping process, a new dominant archetype of the component evolves at the new scale.

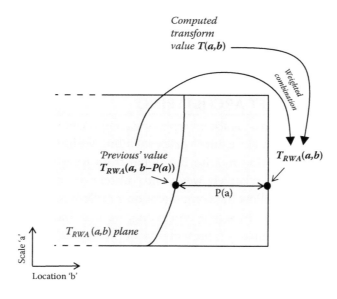

FIGURE 2.41 Schematic of running wavelet archetyping method.

Figure 2.42 contains an example signal and its running wavelet archetype computed using a complex Morlet wavelet transform. The underlying signal in Figure 2.42a is a sinusoid with a frequency of 0.04 Hz. However, it contains additional high-frequency noise, low-frequency modulations at 0.005 Hz, a large amplitude, a segment with no underlying signal (at 1250–1350 seconds) and a low-frequency feature (at around 1800 seconds). Figure 2.42b,c shows the wavelet transform modulus of the signal in 2-D and 3-D. The main band is noisy and broken up. The RWA plot is shown in Figure 2.42d,e in 2-D and 3-D. The RWA has reduced the frequency modulation of the band, reduced the high-frequency noise, filled in the missing signal information and smoothed out the large-scale feature to leave a distinct band 'emerging' at 0.04 Hz. Note that the method employed for the figure used only the modulus of the transform, $|T(a,b)|$, in Equation 2.66 which provides a more robust technique for realizing a dominant band. The reader is encouraged to try both the full transform and the modulus-only versions. Using the full complex transform allows for the computation of an inverse transform in order to obtain a modified signal. (Note that, in practice, this may be aided by increasing the temporal resolution of the original signal.) From the example, we can see that the scale-dependent periodic nature of the wavelet transform lends itself to the method which is particularly useful for identifying and following components in real time in very noisy conditions. The application of the RWA technique to a noisy biosignal in order to better determine a pulse rate has been reported by Addison (2015a). This is described in more detail in Chapter 6, Section 6.4 and Figure 6.38. In the same section, a modular CWT approach to processing noisy biosignals is detailed where the RWA method is one of a number of key components. This is summarized in Figures 6.39 through 6.41 in Chapter 6, and in more detail in Addison (2016).

2.19 WAVELET TRANSFORM REPHASING

We can use the wavelet transform to rephase a signal; that is, change the phase of some or all of its components by an amount, $\Delta\phi$, while retaining the underlying signal modulus information. As the transform is simply a map of complex numbers, which for simplicity we will denote here as $T(a,b) = x + iy$, we can rephase these to new values $T'(a,b) = x' + iy'$ by rotating the complex numbers by the rephasing angle. This may be performed using the standard matrix transformation:

$$\begin{bmatrix} x' \\ y' \end{bmatrix} = \begin{bmatrix} \cos(\Delta\phi) & -\sin(\Delta\phi) \\ \sin(\Delta\phi) & \cos(\Delta\phi) \end{bmatrix} \begin{bmatrix} x \\ y \end{bmatrix} \qquad (2.67)$$

The rotation of the complex number from (x,y) to (x',y') is shown schematically in Figure 2.43 where the magnitude of the complex number is $|T(a,b)|$. The method is illustrated in Figure 2.44. A signal comprising three sinusoidal components is shown in Figure 2.44a. The components have been arranged so that the signal is similar in morphology to the kind of heartbeat signal waveform used in pulse oximetry or blood pressure monitoring, where higher frequency 'pulses' ride on a lower-frequency baseline modulation. We could, for example, consider this a combination of an idealized heartbeat signal and a lower-frequency respiratory signal. In the method, a wavelet transform is performed

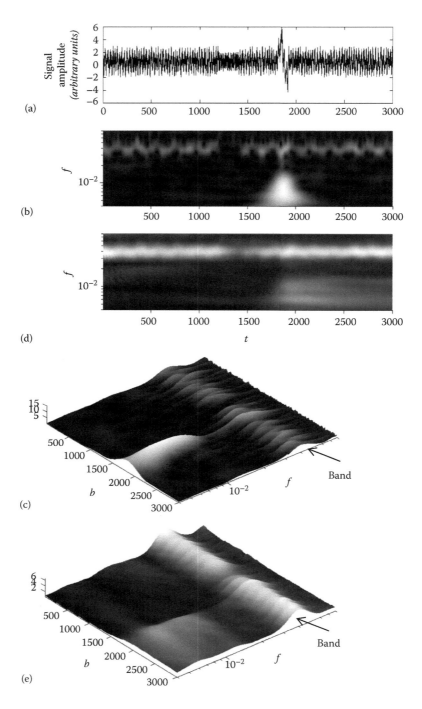

FIGURE 2.42 **Wavelet archetyping:** (a) Test signal. (b) Wavelet transform modulus of the signal. (c) 3-D view of the wavelet transform modulus. (d) Running wavelet archetype – $T_{RWA}(a,b)$ – using a weighing factor of 0.1. (e) 3-D plot of (d). (Note that wavelet scale a has been converted to characteristic frequency for the plots.) (From figure 2 of Addison, P. S., *Electronics Letters*, 51(15), 1153–1155, 2015a. Reproduced with permission from the Institution of Engineering & Technology.)

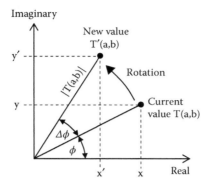

FIGURE 2.43 Rotation of a complex number.

on the signal to be rephased. The modulus of the transform is shown in Figure 2.44b. The corresponding phase and real part of the transform are shown in Figure 2.44c,d, respectively, both highlight the scale-dependent cyclical nature of the signal. The regular cycling of the constituent components is obvious in these latter two plots. In the example shown, we rephase the signal at all scales across the transform plane by $\pi/2$ using the complex rotation given by Equation 2.67. The phase and real part of the rephased transform are shown in Figure 2.44e,f, respectively. The change in phase of the components at all scales by $\pi/2$ can be observed through close inspection of these figures and by comparison with Figure 2.44c,d. It is interesting to note that the phase maps of Figure 2.44c,e take no account of the strength of signal whereas the real transform plot does. Hence, perhaps counter-intuitively, the real transform may be better in some circumstances for examining phase, especially when the relative importance of signal components in terms of their energies needs to be taken account of.

The rephased signal itself can be constructed through an inverse wavelet transform of the rephased transform $T'(a,b)$. This is shown in Figure 2.44g. We can see that the *wavelet rephasing* has changed the shape of the 'pulses'. Such morphological changes may be expected from the rephasing technique. Note that the transform modulus remains the same (and hence the energy distribution will too). This can be seen by comparing the transform modulus for the rephased signal in Figure 2.44h with that of the original signal in Figure 2.44b. (It is easy to understand from these two examples why simply comparing wavelet transform modulus [or energy] plots does not reveal a full picture of the signal content.)

We may only want to rephase certain components in a signal. This is easily performed by manipulating only those transform values associated with the components of interest. Figure 2.44i, shows a partially rephased version of the signal whereby only the low-frequency component is rephased (by $\pi/2$), with the higher frequency 'pulses' retaining their shape. Comparing this with Figure 2.44b, we can see which components have been rephased and which have not. This selective rephasing is also apparent when comparing the real parts of the original and rephased transform of Figure 2.44d,j.

Rephasing may be useful when we have a number of signals, each of which contains a component with information of interest to us but with different phases. A good example

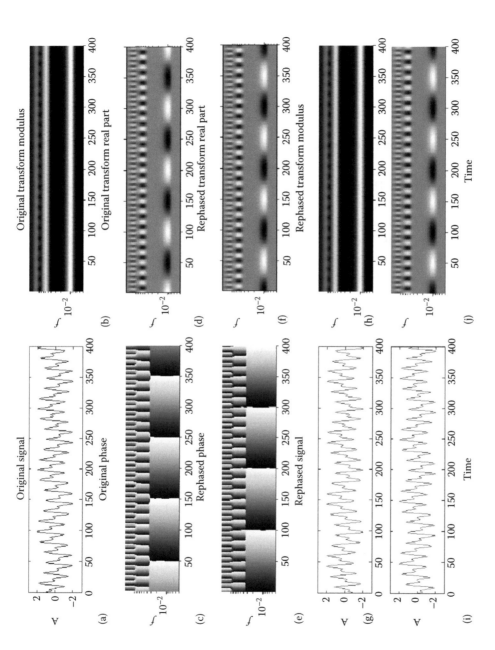

FIGURE 2.44 **Rephasing a test signal:** (a) Original signal. (b) Transform modulus of the original signal (a). (c) Phase of the transform of the original signal (a). (d) Real part of the transform. (e) Phase of the rephased transform. (f) Real part of the rephased transform. (g) The rephased signal reconstructed using the inverse transform. (h) Transform modulus of the rephased signal. (i) The partially rephased signal reconstructed using the inverse transform of the selectively rephased transform. (j) Real part of the partially rephased transform. (Transform plot axes are characteristic frequency against time and signal axes are amplitude in arbitrary units against time.)

of this would be the respiratory modulations derived from a photoplethysmogram (a pulse oximeter biosignal, see Chapter 6, Section 6.4), where by using SWFD (Section 2.16), we may obtain three respiratory signals: a baseline modulation, amplitude modulation and frequency modulation signal. However, these are usually out of phase as they originate from different physiological mechanisms and hence they may destructively interfere with each other if we want to average them to derive a representative signal. An example of rephasing a number of these respiratory biosignals is given in Addison et al. (2011), where individual rephased signals are combined to form a characteristic respiratory signal.

2.20 REASSIGNMENT AND SYNCHROSQUEEZING

We have already seen in many previous examples of sinusoidal signals that the wavelet transform produces a band with components smeared out across scales. This is unlike the Fourier transform, which for a sinusoid produces a localized component present only at the frequency of the signal. *Reassignment* and *synchrosqueezing* are two popular techniques for 'sharpening' the wavelet representation of a signal by localizing the components. Reassignment was devised by Kodera et al. (1978) and extended by Auger and Flandrin (1995) to the wavelet transform. It uses two operators. The first is the group delay time defined as

$$b_g = b - \frac{1}{2\pi}\frac{\partial \phi(a,b)}{\partial f} \tag{2.68}$$

where $\phi(a,b)$ is the transform phase in radians and the 2π term in the denominator converts from radians per second to hertz. The second component is the instantaneous frequency, defined as

$$f_i = \frac{1}{2\pi}\frac{\partial \phi(a,b)}{\partial b} \tag{2.69}$$

This is the derivative of phase with respect to time and so gives the frequency of phase cycling corresponding to the transform component at that point in the transform domain. Once we compute b_g and f_i for a given location in the transform plane, we move the transform component to the new location given by these 'coordinates'.

It is easiest to begin with an example where we simply reassign according to phase only (in a method called 'synchrosqueezing'). The signal of Figure 2.45a contains two sinusoids at a frequency of 0.2 and 0.8 Hz. The wavelet transform modulus of the signal is shown in Figure 2.45b,c. We can see the smearing in frequency of the signal components. However, if we look at the corresponding wavelet phase map of Figure 2.45d, we see two distinct phase cycling frequencies in the transform plane. We can determine the frequency of this phase cycling at each point by taking the derivative of the local phase with respect to time. We then move the corresponding transform component 'up' or 'down' the transform domain to this frequency. In effect, all transform components in the region marked 'phase cycling at 0.2 Hz' in the plot of Figure 2.45d are moved to a position at 0.2 Hz. Similarly,

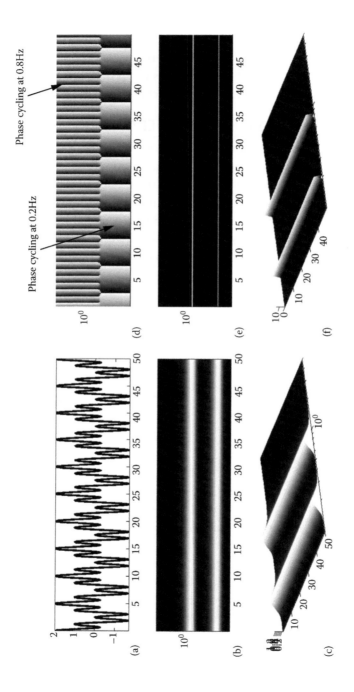

FIGURE 2.45 **Synchrosqueezing a two-sinusoid signal:** (**a**) Test signal comprising two sinusoids. (**b**) Wavelet transform of the test signal. (**c**) 3-D view of (b). (**d**) Wavelet phase. (**e**) Synchrosqueezed transform components. (**f**) 3-D view of (e).

all transform components in the region marked 'phase cycling at 0.8 Hz' are moved to a position at 0.8 Hz. We are therefore moving transform components from a characteristic frequency to a phase cycling frequency. Hence, by doing so for this two-frequency signal, all transform components are moved to localized or 'sharpened' bands. Thus, all the points in the lower region of the phase map cycling at 0.2 Hz are moved to the 0.2 Hz sharpened band and all the points at the top of the phase map cycling at 0.8 Hz are moved to the 0.8 Hz sharpened band. These sharpened bands can be seen in Figure 2.45e,f.

The initial ideas of reassignment derived by Kodera et al. (1978) were developed for the STFT. Auger and Flandrin (1995) further developed the theory to include the mathematics for the wavelet transform and the provision of computationally efficient methods. A few applications of wavelet transform reassignment can be found in the literature, for example, in the vibration signal analysis of rotating machinery (Peng et al., 2002), modal parameter estimation (Peng et al., 2011), characterization of the ECG (Clifton et al., 2003) and ultrasonic non-destructive testing signals (Addison et al., 2006) where low-oscillation complex wavelets were used. Reassignment was developed independently of synchrosqueezing, the latter being a 'special case' according to Auger et al. (2013). For more information on wavelet transform synchrosqueezing, and the underlying mathematics, the reader is referred to Daubechies and Maes (1996) and Daubechies et al. (2011). Note that the synchrosqueezed transform has the property that it offers signal reconstruction. A number of applications of synchrosqueezing can be found in the literature, for example, gearbox fault diagnosis (Li and Liang, 2012); seismic signal analysis (Herrera et al., 2014); the investigation of cardiorespiratory interactions (Franco et al., 2012; Iatsenko et al., 2013); the analysis of breathing dynamics, ventilator weaning (Wu et al. 2012) and atrial fibrillation (Wu et al., 2014); paleoclimate analysis (Thakur et al., 2013); and the examination of financial systems (Guharay et al., 2013). More recently, Addison (2015c) provided a method of synchrosqueezing the cross-wavelet transform in order to identify stable phase coupling between blood pressure and near-infrared spectroscopy (NIRS) signals. The method also employs low-oscillation complex wavelets (Addison et al., 2002a). (The cross-wavelet transform is discussed in Section 2.21.2.)

2.21 COMPARING TWO SIGNALS USING WAVELET TRANSFORMS

Often, we want to compare two signals to determine whether there is some kind of relationship between them. There are many instances of this in the literature including the analysis of cerebral dynamics, cardiorespiratory coupling, geophysical flow processes, wind–wave interactions, fluid–structure interactions, economic index time series, astronomical signals and so on – many of which are cited in later chapters. The rest of this section describes a number of methods for comparing signals in wavelet space, but first we consider some simple definitions.

As we have seen, the wavelet transform produces complex numbers for a real signal when a complex mother wavelet function is used, such as the Morlet wavelet (Equation 2.40). Thus, it may be written as

$$T(a,b) = |T(a,b)| e^{i\phi(a,b)} \tag{2.70}$$

In terms of its real and imaginary components, $Re(T(a,b))$ and $Im(T(a,b))$, the modulus of the transform may be written as

$$|T(a,b)| = \sqrt{Re(T(a,b))^2 + Im(T(a,b))^2} \quad (2.71)$$

and the phase of the transform as

$$\phi(a,b) = \tan^{-1}\left[\frac{Im(T(a,b))}{Re(T(a,b))}\right] \quad (2.72)$$

2.21.1 Transform Differences and Ratios

Consider the transforms of two signals, $g(t)$ and $h(t)$: $T_g(a,b)$ and $T_h(a,b)$. There are simple ratio and difference metrics that we can construct to examine the relationship between the transform components on the transform plane (a,b). The difference of the moduli of the two transforms is given by

$$DiffMOD_{g,h}(a,b) = |T_h(a,b)| - |T_g(a,b)| \quad (2.73)$$

This *modulus difference* and gives the difference in the strength of the local signal component regardless of phase. The *modulus ratio* is given by

$$RatioMOD_{g,h}(a,b) = \frac{|T_h(a,b)|}{|T_g(a,b)|} \quad (2.74)$$

and measures the local ratio of the strengths of components in the two transforms. This might be useful for local regions in the timescale plane, but can blow up if near zero values occur in the transform of the signal, $g(t)$ – so use with care. (Such near zero values may be due to a lack of signal components at a particular location (a,b) or could be caused by localized noise in the signal driving some transform value down to near zero. However, a good example of its use is the wavelet ratio surface method for the determination of oxygen saturation given in Chapter 6, Figure 6.36 [Addison and Watson, 2005].)

In addition to modulus differences, we may also compare the relative phases of the two transforms through their local phase differences, given by

$$\Delta\phi_{g,h}(a,b) = \phi_h(a,b) - \phi_g(a,b) \quad (2.75)$$

Figure 2.46a,b show two simple signals both composed of two sinusoids. The signals differ only in that the high-frequency sinusoid of signal $h(t)$ is half the amplitude of the corresponding high-frequency sinusoid on the signal $g(t)$ and the phase of the lower frequency component differs by $\pi/2$ between the two signals. The corresponding transform modulus

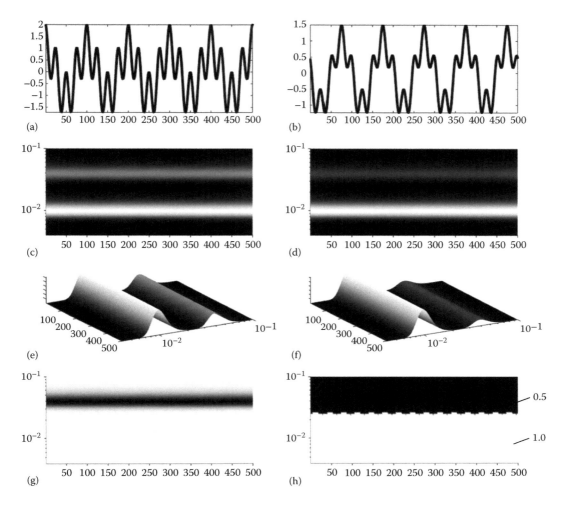

FIGURE 2.46 Comparing the wavelet transform modulus of two signals: (a) Test signal $g(t) = A_{g1} \cos(2\pi f_{g1} t) + A_{g2} \cos(2\pi f_{g2} t)$. **(b)** Test signal $h(t) = A_{h1} \cos(2\pi f_{h1} t + \phi_{h1}) + A_{h2} \cos(2\pi f_{h2} t)$. **(c)** Modulus of $T_g(a,b)$. **(d)** Modulus of $T_h(a,b)$. **(e)** 3-D plot of (c). **(f)** 3-D plot of (d). **(g)** Modulus difference. **(h)** Modulus ratio. The amplitudes of the components of these test signals are: $A_{g1} = 1$, $A_{g2} = 1$, $A_{h1} = 1$, $A_{h2} = 0.5$, their frequencies are $f_{g1} = 0.01$ Hz, $f_{g2} = 0.04$ Hz, $f_{h1} = 0.01$ Hz, $f_{h2} = 0.04$ Hz and the phase shift of the first component of the second signal from the first is $\phi_{h1} = \pi/2$.

plots are shown in Figure 2.46c,d, respectively, and in 3-D in Figure 2.46e,f where the morphology of the surface can be seen more clearly. Two bands are evident in both plots corresponding to the two sinusoidal components. Figure 2.46g shows the difference in the two modulus plots – where the difference is zero (white in the plot) except at the higher frequency band corresponding to the two high-frequency sinusoids of different amplitudes. Figure 2.46h shows the modulus ratio between the two transforms. As expected, the ratio is one (white region in the plot) in the vicinity of the two low-frequency bands and 0.5 (black region) in the vicinity of the high-frequency bands.

2.21.2 Cross-Wavelet Transform

The *cross-wavelet transform* is useful for highlighting regions of coincidental energy between signals in the transform domain as well as determining relative phase. It is defined as

$$\mathrm{CrWT}_{g,h}(a,b) = T_g^*(a,b)\, T_h(a,b) \tag{2.76}$$

The squared absolute value is given by

$$\left|\mathrm{CrWT}_{g,h}(a,b)\right|^2 = \left|T_g^*(a,b) T_h(a,b)\right|^2 = \left|T_g(a,b)\right|^2 \left|T_h(a,b)\right|^2 \tag{2.77}$$

The modulus, $\left|\mathrm{CrWT}_{g,h}(a,b)\right|$ is often plotted in the literature for visualization purposes and is analogous to the wavelet energy density function ($E(a,b)$) – the scalogram as per Equation 2.20 – for a single signal. We may call $\left|\mathrm{CrWT}_{g,h}(a,b)\right|$ the 'cross-scalogram'. Note that, if function g is identical to h, we would get a standard scalogram for function g (and h). Regions of coincidentally high energies in both transforms produce large values in the cross-scalogram. We may also plot its square root, $\sqrt{\left|\mathrm{CrWT}_{g,h}(a,b)\right|}$, which is the modulus form that we use in the examples in the figures shown here.

Figure 2.47e contains the cross-wavelet transform modulus of the two test signals in Figure 2.46a,b calculated using a complex Morlet wavelet. The end views of the individual transform surfaces and the cross-wavelet transform are given in Figure 2.47b,d,f. We can see that the amplitude of the low-frequency ridge in the cross-wavelet transform is the same as that in the two individual transforms (which are the same). However, the higher frequency ridges on the two individual transforms are different in amplitude and the cross-wavelet transform results in a ridge with an amplitude measuring somewhere between these two.

We may write the cross-wavelet transform out in complex exponential form as

$$\mathrm{CrWT}_{g,h}(a,b) = \left|T_g(a,b)\right| e^{-i\phi_g(a,b)} \left|T_h(a,b)\right| e^{i\phi_h(a,b)} = \left|T_g(a,b)\right| \left|T_h(a,b)\right| e^{i(\phi_h(a,b) - \phi_g(a,b))} \tag{2.78}$$

from which we can see that the phase angle of $\mathrm{CrWT}_{f,g}(a,b)$ is given by the expression:

$$\phi_{\mathrm{CrWT}}(a,b) = \phi_h(a,b) - \phi_g(a,b) \tag{2.79}$$

Note that this is the phase difference between the transforms of the individual signals as given in Equation 2.75, that is,

$$\phi_{\mathrm{CrWT}}(a,b) = \Delta\phi_{g,h}(a,b) \tag{2.80}$$

From this, we see that the local phase angle of the cross-wavelet transform of Equation 2.76 is equal to the local difference in phase of the individual signal transforms. Thus, by

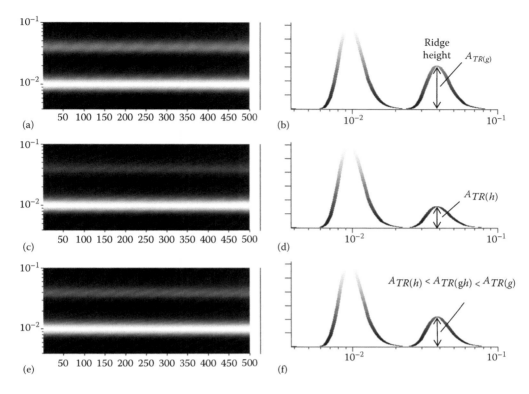

FIGURE 2.47 Plan and end views of two transforms and the corresponding cross-wavelet transform: (a) Plan view of $|T_g(a,b)|$. (b) End view of $T_g(a,b)$ against frequency. (c) Plan view of $|T_h(a,b)|$. (d) End view of $|T_h(a,b)|$ against frequency. (e) Plan view of $|\sqrt{\mathrm{CrWT}_{g,h}(a,b)}|$. (f) End view of $|\sqrt{\mathrm{CrWT}_{g,h}(a,b)}|$ against frequency. A is ridge amplitude.

computing the cross-wavelet transform, we can extract a map of the phase differences between the signal components in wavelet space. An example of this is shown in Figure 2.48, which contains the phases of the original transforms $\phi_g(a,b)$ and $\phi_h(a,b)$ together with the phase map computed using the cross-wavelet transform. We can see from the plot of Figure 2.48e,f that $\phi_{\mathrm{CrWT}}(a,b)$ is a constant value of 0 for the higher frequency components and $\pi/2$ for the lower-frequency components, which is to be expected from the phase maps for the individual transforms (Figure 2.48a–d). For more information on the cross-wavelet transform, the reader is referred to Ge (2008), who calls it the 'wavelet cross-spectrum' and uses the nomenclature we follow with the conjugate on the first transform on the right-hand side of Equation 2.76. See also Torrence and Webster (1999), who call it the 'cross-wavelet spectrum' and Grinsted et al. (2004), who call it the 'cross-wavelet transform'. The cross-wavelet transform has been exploited by researchers in many areas, including fluid flows, engineering, medicine, geophysics and astronomy. Many examples of its use are provided in subsequent chapters of this book. Recently, a method of synchrosqueezing the cross-wavelet transform in order to identify stable phase coupling between two biosignals has been described by the author (Addison, 2015c).

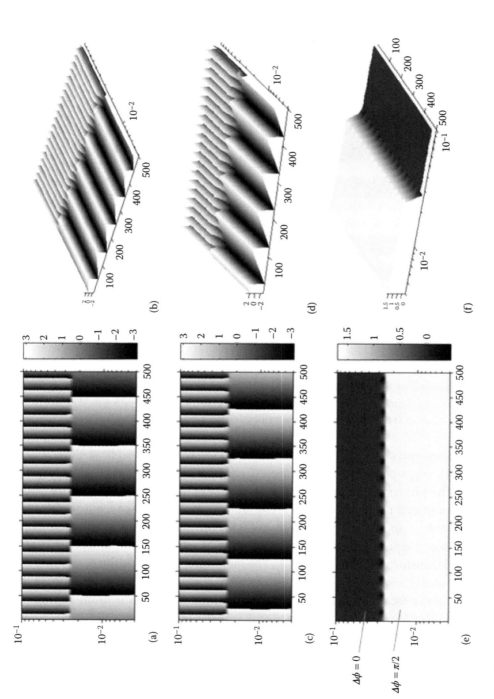

FIGURE 2.48 Phases of signal transforms and the cross-wavelet transform: (a) The transform phase $\phi_g(a,b)$, of the signal in Figure 2.46a. (b) The transform phase of the signal in Figure 2.45a in 3-D. (c) The transform phase $\phi_h(a,b)$ of the signal in Figure 2.45b. (d) The transform phase of the signal in Figure 2.46b in 3-D. (e) The phase of the cross-wavelet transform $\phi_{CrWT}(a,b) = \Delta\phi_{g,h}(a,b)$. (f) The phase of the cross-wavelet transform in 3-D. (Note that this has been rotated relative to (b) and (d) to aid viewing.)

2.21.3 Wavelet Cross-Correlation

We may integrate the cross-wavelet transform over time to get a scale-dependent *wavelet cross-correlation* measure as follows:

$$\text{WCC}_{g,h}(a) = \frac{\left|\int T_g^*(a,b) T_h(a,b) \, db\right|}{\left[\int |T_g(a,b)|^2 \, db \int |T_h(a,b)|^2 \, db\right]^{1/2}} = \frac{\left|\int \text{CrWT}_{g,h}(a,b) \, db\right|}{\left[\int |T_g(a,b)|^2 \, db \int |T_h(a,b)|^2 \, db\right]^{1/2}}$$

(2.81)

This includes a normalization using the individual transform values. This form of wavelet cross-correlation (WCC) is often seen in the literature expressed as a plot of WCC against scale, *a*, or corresponding characteristic frequency, *f*. The measure lies in the range between 0 and 1, a value of 1 indicating complete correlation at that scale. Examples of WCC plots are shown in Figure 2.49 for the test sinusoidal signals of Figure 2.46 and also for two random signals. The two sinusoidal signals are plotted in Figure 2.49a,b. Figure 2.49c plots the components of the integrals in Equation 2.81 calculated using a complex Morlet wavelet. The thick line is the numerator of Equation 2.81. The thin line and dashed thin line represent the two integrals in the denominator of Equation 2.81; these combine to produce a denominator with the same values as the numerator and hence the WCC value is equal to unity at all scales as shown in Figure 2.49d. Thus, although the two sinusoids making up each signal differ in amplitude or phase, they are completely correlated across scales. The wavelet cross-correlation plot for the sinusoidal test signals may be contrasted with that for two random signals in the example of Figure 2.49e–h. Here, the random signals produce a greater denominator (thick dashed line) than numerator (thick continuous line) in Equation 2.81 and hence low values of $\text{WCC}_{g,h}(a)$ result (Figure 2.49h). Note that the random signals here are of quite a short duration and that the WCC of the random signals becomes flatter and tends to values nearer 0 as longer signal segments are considered. (Note also that, in practice, only the thick continuous lines in Figure 2.49d,h are plotted. The other lines in Figure 2.49c,g are plotted here only to illustrate the method.)

We may extend the WCC given to consider the effect of shifting the signals relative to each other by a time delay (τ) as follows:

$$\text{WCC}_{g,h}(a,\tau) = \frac{\left|\int T_g^*(a,b) T_h(a,b-\tau) \, db\right|}{\left[\int |T_g(a,b)|^2 \, db \int |T_h(a,b)|^2 \, db\right]^{1/2}}$$

(2.82)

This produces a range of correlation values with respect to scale and time delay. (Note that either '+τ' or '−τ' may be used in the equation depending on the direction of the shift in the transforms relative to each other.) One way to think of this is as a set of $\text{WCC}_{g,h}(a)$

82 ■ The Illustrated Wavelet Transform Handbook

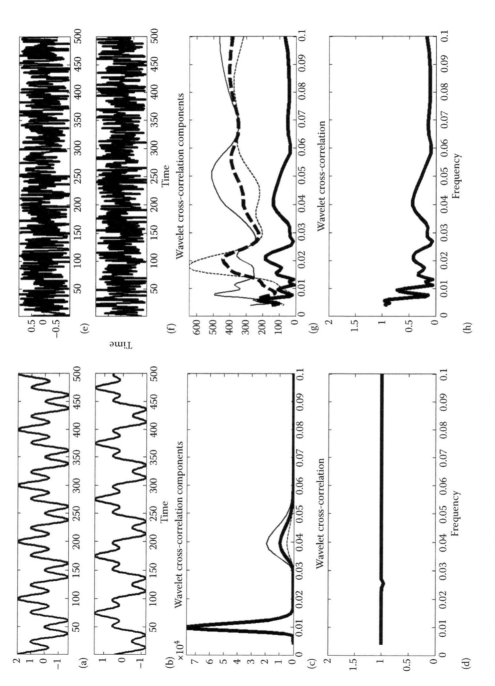

FIGURE 2.49 Wavelet cross-correlation of two test sinusoidal signals and two random signals: (a) and (b) Test sinusoidal signals. (c) Cross-correlation components. (d) $WCC_{g,h}(a)$ for the test sinusoidal signals. (e) and (f) Random signals. (g) Cross-correlation components. (h) $WCC_{g,h}(a)$ for the random signals. (Note that the WCC is plotted against characteristic frequency and not wavelet scale.)

plots of Equation 2.81 for the two signal transforms shifted successively in time relative to each other (by τ). (Note also that Equation 2.81 may be written as $WCC_{g,h}(a,0)$ using the expanded nomenclature of Equation 2.82.) A stronger WCC for a non-zero shift, τ, may indicate that the components at that scale are linked but shifted in time: an effect that $WCC_{g,h}(a)$ of Equation 2.81 would not 'see'. (Note that if functions g and h in Equation 2.82 are identical, then it becomes the scale-dependent autocorrelation of a single transformed signal.)

An example of $WCC_{g,h}(a,\tau)$ is given in Figure 2.50 for two signals containing a 'triangular bump' separated in time. The non-shifted wavelet correlation plot (Equation 2.81) is given by the continuous line in Figure 2.50b. $WCC_{g,h}(a,\tau)$ is plotted in Figure 2.50c, where we can see that maximum values are obtained across the plane with respect to time, for a delay of $T = 50$ seconds. These maximum values, indicated by the dashed line, are all equal to unity as the two signals are perfectly correlated (actually identical) across scales at this delay. The corresponding 3-D plot is presented in Figure 2.50d. The maximum is depicted on the plot by a dashed line. (It is also superimposed in Figure 2.50b as a dashed line.) In addition, the edge of this plot (at $\tau = 0$, marked by the arrow) is the same curve as that given by the continuous line in Figure 2.50b. We can see that only by employing the variable time delay, τ, can we see the high degree of correlation at $\tau = 50$ seconds for the higher frequencies that we would have missed by using only the non-shifted version of WCC. The example provided is very simple and the reader is urged to replicate the calculations with signals of increasing complexity and differences between them. Note that for the sinusoidal test signals of Figure 2.49a,b, $WCC_{g,h}(a,\tau)$ is equal to unity across the whole a–τ plane (not shown). For more information on the wavelet cross-correlation of Equation 2.82, the reader is referred to Rowley et al. (2007). See also Ge (2008), who used the squared form of Equation 2.81 and called it the 'wavelet linear coherence' and 'wavelet coherency'. (There is an additional note in the last section of this chapter on the diverse, and often confusing, nature of the nomenclature used in the literature.) We will come across many examples of the use of WCC throughout the book.

2.21.4 Phase Comparison Measures

The measure of the global mean phase difference between two signals with respect to the transform scale may be given by the *circular mean phase* (CMP):

$$CMP(a) = \tan^{-1}\left[\frac{\langle \sin(\Delta\phi_{g,h}(a,b)) \rangle}{\langle \cos(\Delta\phi_{g,h}(a,b)) \rangle}\right] \quad (2.83)$$

A corresponding *phase synchronization index* may be defined as

$$PSI(a) = \langle \sin(\Delta\phi_{g,h}(a,b)) \rangle^2 + \langle \cos(\Delta\phi_{g,h}(a,b)) \rangle^2 \quad (2.84)$$

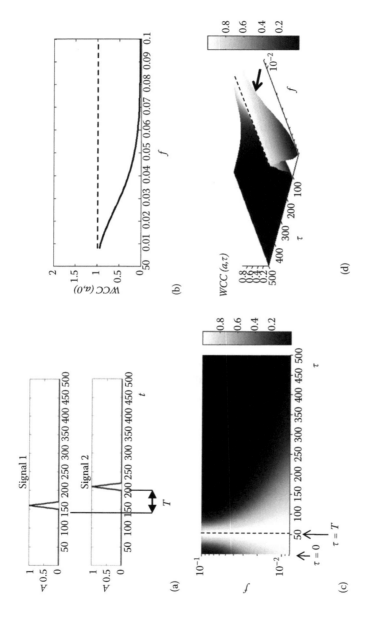

FIGURE 2.50 Two triangular bump signals and their wavelet cross-correlation: (a) The two signal with bumps separated by 50 seconds. (b) WCC(a,0) – this is the WCC of the non-shifted signals – WCC(a,T) is also plotted as a dashed line. (c) WCC(a,τ) – this is the wavelet cross-correlation the two signals shifted relative to each other by a time period τ. (d) 3-D plot of (c) illustrating that at $T = \tau$, the WCC is equal to unity across all scales – indicated by the dashed line (also plotted in [b]).

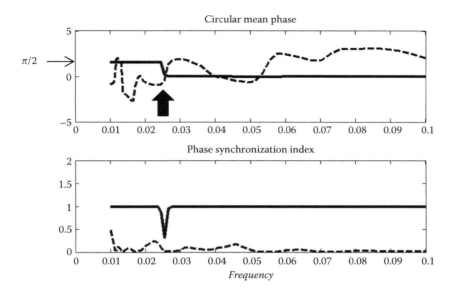

FIGURE 2.51 Phase measures: **(a)** Circular mean phase for the sinusoidal test signal (continuous line) and random signal (dashed line). **(b)** Phase synchronization indices corresponding to (a).

Here, $\langle\ \rangle$ indicates a temporal averaging over b only. The index lies in the range between 0 and 1. A uniform distribution of phase differences indicative of no coupling or random coupling, leads to a value of 0. A value of 1 indicates that the phase angle is invariant with time (i.e. that the phase difference is constant and equal to the CMP) and corresponds to a perfect level of coupling between the two transforms at that scale. Figure 2.51 shows an example of CMPs and phase synchronization indexes (PSIs) for the sinusoidal and random test signals of Figure 2.49. We can see that the circular mean phase is $\pi/2$ for the sinusoidal test signal in the region of the low-frequency component and that this switches to zero phase for the high-frequency component as this component dominates the transform domain. The PSI is equal to 1 for both components with a slight dip caused by the transition between the two components marked by the arrow in the plot. The random signal, however, exhibits an erratic CMP across scales and exhibits a low value of PSI as expected. For more information and examples of the use of these metrics, the reader is referred to Latka et al. (2005) and Rowley et al. (2007). Note that Bernjak et al. (2012) used the square root of Equation 2.84 and called it the 'phase coherence function'.

2.21.5 Wavelet Coherence

In order to probe the relationship between signal components more locally in the transform plane, we may define a squared *wavelet coherence* estimator as follows:

$$\text{WCH}^2_{g,h}(a,b) = \frac{\left|\left\langle T^*_g(a,b)T_h(a,b)\right\rangle\right|^2}{\left\langle\left|T_g(a,b)\right|^2\right\rangle\left\langle\left|T_h(a,b)\right|^2\right\rangle} \qquad (2.85)$$

where $\langle \rangle$ represents a localized smoothing operation in both time and wavelet scale performed on the constituent transform components. Torrence and Webster (1999), for example, used a smoothing function in time equal to the scale-dependent size of the Gaussian envelope of the Morlet wavelet that they employed, that is, $e^{-(1/2)(t^2/a^2)}$, and a smoothing in scale using a boxcar filter of $0.6a$ in extent. See also Camussi et al. (2008), Keissar et al. (2009a), Garg et al. (2013) and Tiwari et al. (2014b) for the practical applications of wavelet coherence in fluid turbulence, biosignals and financial time series.

2.22 BICOHERENCE AND CROSS-BICOHERENCE

The wavelet transform-based bispectrum, and related bicoherence, measure the quadratic phase coupling of signal components due to nonlinear interactions (van Milligen et al., 1995a,b). Considering first the interactions of components within a single signal, the *wavelet bispectrum* is defined as

$$\text{WBS}(a_1,a_2) = \int_\tau T(a_1,b)T(a_2,b)T^*(a_3,b)\,db \qquad (2.86)$$

where the summation of the transform components takes place over a time period τ and at three wavelet scales a_1, a_2 and a_3 which are related as follows:

$$\frac{1}{a_1}+\frac{1}{a_2}=\frac{1}{a_3} \qquad (2.87)$$

or in terms of wavelet frequencies: $f_1+f_2=f_3$. An example of the wavelet bispecturm of a phase-coupled signal is shown in Figure 2.52 where a dominant peak in the bispectrum may be observed at (0.03, 0.02).

The normalized, squared *wavelet bicoherence* (WBC) is defined as

$$\left[\text{WBC}(a_1,a_2)\right]^2 = \frac{\left|\int_\tau T(a_1,b)T(a_2,b)T^*(a_3,b)\,db\right|^2}{\int_\tau |T(a_1,b)T(a_2,b)|^2\,db \int_\tau |T(a_3,b)|^2\,db} \qquad (2.88)$$

The magnitude of wavelet bicoherence varies between 0 and 1 and indicates the degree of quadratic phase coupling between wavelet components at scales a_1, a_2 and a_3. As with the 'wavelet bispectrum' (WBS), the wavelet bicoherence may be mapped onto the (a_1,a_2)-plane or the (f_1,f_2)-plane, where f_1 and f_2 are the characteristic frequencies relating to the wavelet transform of the signal at scales a_1 and a_2.

Probing nonlinear interactions between two different signals g and h can also be performed using the *wavelet cross-bispectrum*:

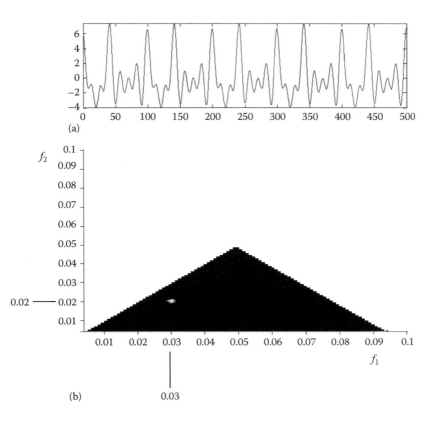

FIGURE 2.52 Wavelet bispectrum of test signal: (a) Part of the test signal. (Modified from the method of Keissar et al., 2009b.) (b) Magnitude of the wavelet bispectrum showing the dominant component as a localized white region. (Note the conversion from scale to characteristic frequency for the purpose of the plot.)

$$\text{WCrBS}(a_1, a_2) = \int_\tau T_g(a_1, b) T_g(a_2, b) T_h^*(a_3, b) db \tag{2.89}$$

and normalized, squared *wavelet cross-bicoherence*:

$$\left[\text{WCrBC}(a_1, a_2)\right]^2 = \frac{\left|\int_\tau T_g(a_1, b) T_g(a_2, b) T_h^*(a_3, b) db\right|^2}{\int_\tau |T_g(a_1, b) T_g(a_2, b)|^2 db \int_\tau |T_h(a_3, b)|^2 db} \tag{2.90}$$

which measures the amount of phase coupling between components at scales a_1 and a_2 in signal g and scale a_3 in signal h.

Wavelet bicoherence has been used in a number of areas including the analysis of plasma turbulence signals (van Milligen et al., 1995a,b); electron beams (Koronovskii and

Khramov, 2002); wind and wave signals (Gurley et al., 2003; Elsayed, 2006); pulmonary wheeze signals (Taplidou and Hadjileontiadis, 2007, 2010); cardiovascular signals (Keissar et al., 2009b); gas and oil pricing financial signals (Tonn et al., 2010); and condition monitoring signals (Li et al., 2014a). These references provide a good starting point for delving further into this topic and interesting variants of the measures are proposed by some of the authors. Some additional ideas concerning more flexible measures of bicoherence along non-stationary coupled ridge pairs are given by Addison and Watson (2015). Examples of bicoherence are also to be found later in this book: in Figures 4.40 (fluids), 5.19 (machines) and 6.46 (biomedical).

2.23 ENDNOTES

2.23.1 Chapter Key Words and Phrases

(You may find it helpful to jot down your understanding of each of them.)

wavelet	*wavelet transform*	*wavelet variance*
mother/analyzing wavelet	*dilation parameter*	*morlet wavelet*
mexican hat wavelet	*location parameter*	*wavelet transform ridges*
passband centre	*fourier transform*	*short-time Fourier transform (STFT)*
admissibility condition	*inverse wavelet transform*	*windowed Fourier atom*
admissibility constant	*wavelet transform modulus maxima*	*time–frequency atom*
energy spectrum	*scalogram*	*matching pursuit*
bandpass filter	*spectrogram*	*boundary effects*
transform phase difference	*modulus difference*	*modulus ratio*
secondary wavelet feature decoupling (SWFD)	*running wavelet archetyping (RWA)*	*wavelet rephasing*
reassignment	*synchrosqueezing*	*cross-wavelet transform (CrWT)*
wavelet cross-correlation (WCC)	*circular mean phase (CMP)*	*phase synchronization index (PSI)*
wavelet coherence (WCH)	*wavelet bispectrum (WBS)*	*wavelet bicoherence (WBC)*
wavelet cross-bispectrum (WCrBS)	*wavelet cross-bicoherence (WCrBC)*	

2.23.2 Additional Notes and Resources

There are many good sources of accessible information concerning the continuous wavelet transform. An introduction to the continuous wavelet transform including the Morlet wavelet, scale–frequency relationship, phase and so on has been given by Aguiar-Conraria and Soares (2011). Ashmead (2012) has provided a comprehensive treatment of the Morlet wavelet and Büssow (2007) has presented an algorithm for calculating the continuous transform with the Morlet wavelet. The use of tests to determine the statistical significance of the components appearing in the wavelet representation of a signal has been addressed by a number of researchers; see, for example, Torrence and Compo (1998), Grinsted et al. (2004), Maraun and Kurths (2004), Maraun et al. (2007) and Zhang and Moore (2012). Su et al.

(2011) investigated edge effects in the transform domain, and Zhang and Moore (2011) considered the influence of edge effects on significance testing. Heisenberg boxes are covered by many authors; see, for example, Mallat (2009) for more information. This reference also demonstrates succinctly how the wavelet transform may be rewritten as a convolution product and provides more detail on energy conservation and reconstruction formulae associated with transforms employing analytic wavelets. Further information on the application of ridges, modulus maxima, the complete Morlet wavelet and the complex Mexican hat wavelet can be found in the paper on low-oscillation complex wavelets by Addison et al. (2002a). These wavelets are useful in themselves when higher (or lower) temporal resolution is required from the wavelet function than provided by the standard Morlet. More material on modulus maxima and signal reconstruction can be found in the papers by Mallat and Hwang (1992) and Mallat and Zhong (1992), and a good account of the matching pursuit method is given in the original paper by Mallat and Zhang (1993). In this chapter, we considered the two-dimensional Mexican hat wavelet. Elsewhere, a two-dimensional Morlet wavelet has been presented by Kumar and Foufoula-Georgiou (1997). See also the discussion by Daubechies (1992, pp. 33–34) concerning the continuous wavelet in higher dimensions where rotations can be introduced into the definition of higher-dimensional wavelets. More comprehensive coverage of the two-dimensional continuous wavelet transform can be found in the book by Antoine et al. (2004) and more recently in the concise, lucid account by Watkins (2015).

It is worth reiterating that the nomenclature found in the literature for many of the terms discussed in this chapter can be confusing (cross-wavelet transform, wavelet coherence, bicoherence and so on) – both in terms of the appellations used and the mathematics. Often, the same word is employed in quite different contexts or quite different names used for the same parameter; for example, the cross-wavelet transform (CrWT in Section 2.21.2) has also been called the 'wavelet cross-spectrum', and the wavelet cross-correlation (WCC in Section 2.21.3) has also been called the 'wavelet linear coherence' and 'wavelet coherency'. So, beware when reading the literature, standardize the nomenclature for yourself. In addition, within this chapter, the mathematical form that most commonly appears in the literature has generally been used. Hence, for example, the difference in mathematical form between the squared wavelet coherence estimator (Equation 2.85) and the normalized squared cross-bicoherence (Equation 2.90).

2.23.3 Things to Try

The reader is encouraged to replicate the results in this chapter and extend them to other examples, both simple synthetic test signals and real signals. The coding of a wavelet transform 'engine' would be a first step in this endeavour whereby a range of scales should be prescribed over which to compute the transform. This could be undertaken using a discretized summation for the integral time domain expression of Equation 2.13 or, preferably, in the frequency domain using Equation 2.35. Note that if the wavelet's equation is known in the frequency domain (e.g. the Morlet wavelet of Equation 2.38), then this can be hard coded within the algorithm and computed over a pre-set range of scales. The reader is encouraged to try many of his or her own simple test signals as it will lead to a better understanding of how the transform works and, just as importantly, how it breaks.

It is suggested that the reader switch between scale a and characteristic frequency f on the transform axes – also, try logarithmic and linear scales for these axes and look to see how the morphology of the transform plot changes. Another simple task is to compute matching energies for the time domain signal and its wavelet transform. (It can be tricky to get an exact match for the CWT energy – watch out for edge effects!) Use the timescale version of the transform given by Equation 2.21, and then try it using the time–frequency version of the transform of Equation 2.26. Once you are comfortable with the transform computation, try performing the cross-wavelet transform of some simple test signals (regular and random) from which the circular mean phase and phase synchronization index can also be computed.

The reader may also want to attempt to derive the ridge amplitude and frequencies associated with simple test functions for himself or herself. Start with a sinusoid, compare its amplitude with its corresponding ridge in the transform domain – both for the original and rescaled transform – then change its amplitude and see how the amplitude of the scalogram component changes. Then try other, more complex, signals. Note that a number of equations for the Morlet wavelet, its Fourier transform and wavelet transform ridge amplitudes are provided in the literature (e.g. Carmona et al., 1997; Staszewski, 1997, 1998a; Wang et al., 2003; Harrop et al., 2002; Liebling et al., 2005; Tse and Lai, 2007; Abid et al., 2007; Dien, 2008; Aguiar-Conraria et al., 2008; Jänicke et al. 2009; Lilly and Olhede, 2010). These may differ from those presented in this chapter (Equation 2.65) by multiples of $\sqrt{2}, \sqrt{\pi}, \sqrt[4]{\pi}$ or other factors. This is often due to the definitions of the Fourier transform and/or the Morlet wavelet (e.g. normalized or not) used, whether the rescaled transform is employed and/or whether a real-only or complex sinusoid is considered in the analysis. It is worth noting that such minor mathematical differences between the written forms of equations are prevalent in the literature and therefore this applies to many of the other equations cited in this chapter.

The reader is encouraged to perform the Morlet wavelet transform on a complex sinusoid analytically, then try a Fourier transform of the wavelet function – what is the difference? If more adventurous, the reader may perform a wavelet transform of the individual sine and cosine components of the complex sinusoid. A symbolic mathematical calculator program is useful in this regard (less work!).

When performing wavelet analysis on signals, try combining methods; for example, try computing the CWT modulus for a signal using a Morlet wavelet, then perform a running wavelet archetype prior to extracting the ridge from the RWA in order to the determine instantaneous frequency. Compare this with the ridge extracted directly from the CWT modulus. Try different characteristic frequencies of the wavelet function and various RWA weights. (See an example of this kind of modular wavelet approach to signal analysis in Chapter 6, Figures 6.39 through 6.41 and in Addison, 2016, where the method is used to determine physiological parameters from noisy biosignals.) Finally, when applying these approaches to real-world signals, the reader is advised to try them out first on a few reasonably simple test signals – for example, a sinusoid with perhaps one or more of the following:

a varying frequency, varying amplitude, perhaps a bit of noise and maybe some periods of missing information. That way, you might understand better what the wavelet transform is really able to tell you about the signal you are interested in.

2.23.4 Final Note: The CWT as a 'Soft Tool' for Algorithm Development

The continuous wavelet transform is an integral part of numerous signal processing algorithms, finding use across multiple disciplines (as described in later chapters of this book). However, the author also suggests its 'upstream' use as a soft tool to aid the comprehension of complex signals; in effect, a manual pre-processing aid. There have been many examples in this chapter of the ability of the continuous wavelet transform to unfold information within a signal, making the underlying components more 'accessible' to visual interpretation. Hence, the CWT may provide a particularly useful technique for initiating and driving the development of final algorithms by providing rapid visualization of the underlying components of a signal. Regardless of whether or not it forms part of the final algorithmic solution, it can aid comprehension of signal characteristics, provide insight into the feasibility of the underlying task and facilitate the development of signal processing strategies for attacking the problem to be solved (Addison, 2014).

CHAPTER 3

The Discrete Wavelet Transform

3.1 INTRODUCTION

In this chapter, we consider the discrete wavelet transform, or DWT. We will see that, when certain criteria are met, it is possible to completely reconstruct the original signal using infinite summations of discrete wavelet coefficients rather than continuous integrals (as required for the continuous wavelet transform [CWT]). This leads to a fast wavelet transform for the rapid computation of the discrete wavelet transform and its inverse. We will then see how to perform a discrete wavelet transform on discrete input signals of finite length; the kind of signal we might be presented with in practice. We will also consider briefly biorthogonal wavelets, two-dimensional discrete wavelet transforms, wavelet packets and recent enhancements to the wavelet method to efficiently represent directional-dependent two-dimensional data.

3.2 FRAMES AND ORTHOGONAL WAVELET BASES

3.2.1 Frames

In Chapter 2, the wavelet function was defined at scale a and location b as

$$\psi_{a,b}(t) = \frac{1}{\sqrt{a}} \psi\left(\frac{t-b}{a}\right) \tag{3.1}$$

In this section, the wavelet transform of a continuous time signal, $x(t)$, is considered where discrete values of the dilation and translation parameters, a and b, are used.

A natural way to sample the parameters a and b is to use a logarithmic discretization of the a scale and link this, in turn, to the size of steps taken between b locations. To link b to a, we move in discrete steps to each location b which is proportional to the a scale. This kind of discretization of the wavelet has the form:

$$\psi_{m,n}(t) = \frac{1}{\sqrt{a_0^m}} \psi\left(\frac{t - nb_0 a_0^m}{a_0^m}\right) \qquad (3.2)$$

where:

The integer m controls the wavelet dilation
The integer n controls the wavelet translation
a_0 is a specified fixed dilation step parameter set at a value greater than 1
b_0 is the location parameter which must be greater than zero

The control parameters m and n are contained in the set of all integers, both positive and negative. It can be seen from Equation 3.2, that the size of the translation steps, $\Delta b = b_0 a_0^m$, is directly proportional to the wavelet scale, a_0^m.

The wavelet transform of a continuous signal, $x(t)$, using discrete wavelets of the form of Equation 3.2 is then

$$T_{m,n} = \int_{-\infty}^{\infty} x(t) \frac{1}{a_0^{m/2}} \psi\left(a_0^{-m} t - nb_0\right) dt \qquad (3.3a)$$

which can also be expressed as the inner product:

$$T_{m,n} = \langle x, \psi_{m,n} \rangle \qquad (3.3b)$$

where $T_{m,n}$ are the discrete wavelet transform values given on a scale–location grid of index m,n. For the discrete wavelet transform, the values $T_{m,n}$ are known as *wavelet coefficients* or *detail coefficients*. These two terms are used interchangeably in this chapter as they are in the general wavelet literature. To determine how 'good' the representation of the signal is in wavelet space using this decomposition, we can resort to the theory of *wavelet frames*, which provides a general framework for studying the properties of discrete wavelets. Wavelet frames are constructed by discretely sampling the time and scale parameters of a continuous wavelet transform as we have just done. The family of wavelet functions that constitute a frame are such that the energy of the resulting wavelet coefficients lie within a certain bounded range of the energy of the original signal, that is

$$AE \leq \sum_{m=-\infty}^{\infty} \sum_{n=-\infty}^{\infty} |T_{m,n}|^2 \leq BE \qquad (3.4)$$

where:

$T_{m,n}$ are the discrete wavelet coefficients
A and B are the frame bounds
E is the energy of the original signal given by Equation 2.19 in Chapter 2:

$$E = \int_{-\infty}^{\infty} |x(t)|^2 dt = \|x(t)\|^2$$

where our signal, $x(t)$, is defined to have finite energy. The values of the frame bounds A and B depend on both the parameters a_0 and b_0 chosen for the analysis and the wavelet function used. (For details of how to determine A and B, see Daubechies, 1992.) If $A = B$, the frame is known as 'tight'. Such *tight frames* have a simple reconstruction formula given by the infinite series:

$$x(t) = \frac{1}{A} \sum_{m=-\infty}^{\infty} \sum_{n=-\infty}^{\infty} T_{m,n} \psi_{m,n}(t) \tag{3.5}$$

A tight frame with $A(=B) > 1$ is redundant, with A being a measure of the redundancy. However, when $A = B = 1$, the wavelet family defined by the frame forms an *orthonormal basis*. If A is not equal to B, a reconstruction formula can still be written:

$$x'(t) = \frac{2}{A+B} \sum_{m=-\infty}^{\infty} \sum_{n=-\infty}^{\infty} T_{m,n} \psi_{m,n}(t) \tag{3.6}$$

where $x'(t)$ is the reconstruction which differs from the original signal $x(t)$ by an error which depends on the values of the frame bounds. The error becomes acceptably small for practical purposes when the ratio B/A is near unity. It has been shown, for the case of the Mexican hat wavelet, that if we use $a_0 = 2^{1/\nu}$ where $\nu \geq 2$ and $b_0 \leq 0.5$, the frame is nearly tight or 'snug' and for practical purposes it may be considered tight. (This fractional discretization, ν, of the power-of-two scale is known as a 'voice'.) For example, setting $a_0 = 2^{1/2}$ and $b_0 = 0.5$ for the Mexican hat leads to $A = 13.673$ and $B = 13.639$ and the ratio B/A equals 1.002. The closer this ratio is to unity, the tighter the frame. Thus, discretizing a Mexican hat wavelet transform using these scale and location parameters results in a highly redundant representation of the signal but with very little difference between $x'(t)$ and $x'(t)$. The nearly tight Mexican hat wavelet frame with these parameters ($a_0 = 2^{1/2}$ and $b_0 = 0.5$) is shown in Figure 3.1 for two consecutive scales m and $m+1$ and at three consecutive locations: $n = 0, 1$ and 2.

3.2.2 Dyadic Grid Scaling and Orthonormal Wavelet Transforms

A common choice for discrete wavelet parameters a_0 and b_0 are 2 and 1, respectively. This power-of-two logarithmic scaling of both the dilation and translation steps is known as the 'dyadic grid' arrangement. The *dyadic gid* is perhaps the simplest and most efficient discretization for practical purposes and lends itself to the construction of an orthonormal

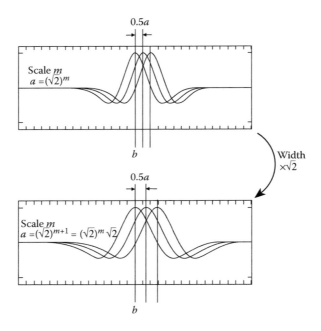

FIGURE 3.1 The nearly tight Mexican hat wavelet frame with $a_0 = 2^{1/2}$ and $b_0 = 0.5$. Three consecutive locations of the Mexican hat wavelet for scale indices m (top) and $m+1$ (lower) and location indices n, $n+1$, $n+2$. That is, $a = 2^m$ and $a = 2^{m+1}$, respectively, and three consecutive b locations separated by $a/2$.

wavelet basis. Substituting $a_0 = 2$ and $b_0 = 1$ into Equation 3.2, we see that the dyadic grid wavelet can be written as

$$\psi_{m,n}(t) = \frac{1}{\sqrt{2^m}} \psi\left(\frac{t - n2^m}{2^m}\right) \tag{3.7a}$$

or, more compactly, as

$$\psi_{m,n}(t) = 2^{-m/2} \psi\left(2^{-m}t - n\right) \tag{3.7b}$$

Note that this has the same notation as the general discrete wavelet given by Equation 3.2. From here on in this chapter, we will use $\psi_{m,n}(t)$ to denote only dyadic grid scaling with $a_0 = 2$ and $b_0 = 1$.

Discrete dyadic grid wavelets are commonly chosen to be orthonormal. These wavelets are both orthogonal to each other and normalized to have unit energy. This is expressed as

$$\int_{-\infty}^{\infty} \psi_{m,n}(t) \psi_{m',n'}(t) \, dt = \begin{cases} 1 & \text{if } m = m' \text{ and } n = n' \\ 0 & \text{otherwise} \end{cases} \tag{3.8}$$

That is to say, the product of each wavelet with all others in the same dyadic system (i.e. those which are translated and/or dilated versions of each other) are zero. This means that the information stored in a wavelet coefficient $T_{m,n}$ is not repeated elsewhere and allows for the complete regeneration of the original signal without redundancy. In addition to being orthogonal, orthonormal wavelets are normalized to have unit energy. This can be seen from Equation 3.8 as, when $m = m'$ and $n = n'$, the integral gives the energy of the wavelet function equal to unity. Orthonormal wavelets have frame bounds $A = B = 1$ and the corresponding wavelet family is an orthonormal basis. (A basis is a set of vectors, a combination of which can completely define the signal, $x(t)$. An orthonormal basis has component vectors which, in addition to being able to completely define the signal, are perpendicular to each other.) The discrete dyadic grid wavelet lends itself to a fast computer algorithm as we shall see.

Using the dyadic grid wavelet of Equation 3.7a, the *discrete wavelet transform* can be written as

$$T_{m,n} = \int_{-\infty}^{\infty} x(t) \psi_{m,n}(t) \, dt \tag{3.9}$$

By choosing an orthonormal wavelet basis, $\psi_{m,n}(t)$, we can reconstruct the original signal in terms of the wavelet coefficients, $T_{m,n}$, using the *inverse discrete wavelet transform* as follows:

$$x(t) = \sum_{m=-\infty}^{\infty} \sum_{n=-\infty}^{\infty} T_{m,n} \psi_{m,n}(t) \tag{3.10a}$$

requiring the summation over all integers m and n. This is actually Equation 3.5 with $A = 1$ due to the orthonormality of the chosen wavelet. Equation 3.10a is often seen written in terms of the inner product:

$$x(t) = \sum_{m=-\infty}^{\infty} \sum_{n=-\infty}^{\infty} \langle x, \psi_{m,n} \rangle \psi_{m,n}(t) \tag{3.10b}$$

where the combined decomposition and reconstruction processes are clearly seen: going from $x(t)$ to $T_{m,n}$ via the inner product $\langle x, \psi_{m,n} \rangle$ then back to $x(t)$ via the infinite summations. In addition, as $A = B$ and $A = 1$, we can see from Equation 3.4 that the energy of the signal may be expressed as

$$\int_{-\infty}^{\infty} |x(t)|^2 \, dt = \sum_{m=-\infty}^{\infty} \sum_{n=-\infty}^{\infty} |T_{m,n}|^2 \tag{3.11}$$

Before continuing, it is important to make clear the distinct difference between the DWT and the discretized approximations of the CWT covered in the last chapter. The discretizations of the continuous wavelet transform, required for its practical implementation, involve a discrete approximation of the transform integral (i.e. a summation) computed on a discrete grid of a scales and b locations. The inverse continuous wavelet transform is also computed as a discrete approximation. How close an approximation to the original signal is recovered depends mainly on the resolution of the discretization used and, with care, usually a very good approximation can be recovered. On the other hand, for the DWT, as defined in Equation 3.9, the transform integral remains continuous but is determined only on a discretized grid of a scales and b locations. We can then sum the DWT coefficients (Equation 3.10a) to get the original signal back exactly. We will see later in this chapter how, given an initial discrete input signal, which we treat as an initial approximation to the underlying continuous signal, we can compute the wavelet transform and inverse transform discretely, quickly and without loss of signal information.

3.2.3 Scaling Function and Multiresolution Representation

Orthonormal dyadic discrete wavelets are associated with *scaling functions* and their dilation equations. The scaling function is associated with the smoothing of the signal and has the same form as the wavelet, given by

$$\phi_{m,n}(t) = 2^{-m/2} \phi(2^{-m}t - n) \tag{3.12}$$

They have the property

$$\int_{-\infty}^{\infty} \phi_{0,0}(t)dt = 1 \tag{3.13}$$

where $\phi_{0,0}(t) = \phi(t)$ is sometimes referred to as the 'father scaling function' or 'father wavelet' (cf. mother wavelet). (Remember from Chapter 2 that the integral of a wavelet function is zero.) The scaling function is orthogonal to translations of itself, *but not to dilations of itself*. The scaling function can be convolved with the signal to produce *approximation coefficients* as follows:

$$S_{m,n} = \int_{-\infty}^{\infty} x(t)\phi_{m,n}(t)\,dt \tag{3.14}$$

From the last three equations, we can see that the approximation coefficients are simply weighted averages of the continuous signal factored by $2^{m/2}$. The approximation coefficients at a specific scale m are collectively known as the *discrete approximation* of the signal at that scale. A *continuous approximation* of the signal at scale m can be generated by summing a sequence of scaling functions at this scale factored by the approximation coefficients as follows:

$$x_m(t) = \sum_{n=-\infty}^{\infty} S_{m,n} \phi_{m,n}(t) \qquad (3.15)$$

where $x_m(t)$ is a smooth, scaling function–dependent version of the signal $x(t)$ at scale index m. This continuous approximation approaches $x(t)$ at small scales, that is, as $m \longrightarrow -\infty$. Figure 3.2a shows a simple scaling function, a block pulse, at scale index 0 and location index 0: $\phi_{0,0}(t) = \phi(t)$ – the father function – together with two of its corresponding dilations at that location. It is easy to see that the convolution of the block pulse with a signal (Equation 3.14) results in a local weighted averaging of the signal over the non-zero portion of the pulse. Figure 3.2b shows one period of a sine wave, $x(t)$, contained within a window. Figure 3.2c shows various approximations of the sine wave generated using Equations 3.14 and 3.15 with the scaling function set to a range of widths: 2^0–2^7. These widths are indicated by the vertical lines and arrows in each plot. Equation 3.14 computes the approximation coefficients $S_{m,n}$ which are, as mentioned for this simple block scaling function, the weighted average of the signal over the pulse width. The approximation coefficients are then used in Equation 3.15 to produce an approximation of the signal which is simply a sequence of scaling functions placed side by side, each factored by their corresponding approximation coefficient. This is obvious from the blocky nature of the signal approximations. The approximation at the scale of the window ($=2^7$) is simply the average over the whole sine wave which is zero. As the scale decreases, the approximation is seen to approach the original waveform. The simple block pulse scaling function used in this example is associated with the Haar wavelet, which we will come to shortly.

We can represent a signal $x(t)$ using a combined series expansion using both the approximation coefficients and the wavelet (detail) coefficients as follows:

$$x(t) = \sum_{n=-\infty}^{\infty} S_{m_0,n} \phi_{m_0,n}(t) + \sum_{m=-\infty}^{m_0} \sum_{n=-\infty}^{\infty} T_{m,n} \psi_{m,n}(t) \qquad (3.16)$$

We can see from this equation that the original continuous signal is expressed as a combination of an approximation of itself, at arbitrary scale index m_0, added to a succession of signal details from scales m_0 down to negative infinity. The *signal detail* at scale m is defined as

$$d_m(t) = \sum_{n=-\infty}^{\infty} T_{m,n} \psi_{m,n}(t) \qquad (3.17)$$

hence, we can write Equation 3.16 as

$$x(t) = x_{m_0}(t) + \sum_{m=-\infty}^{m_0} d_m(t) \qquad (3.18)$$

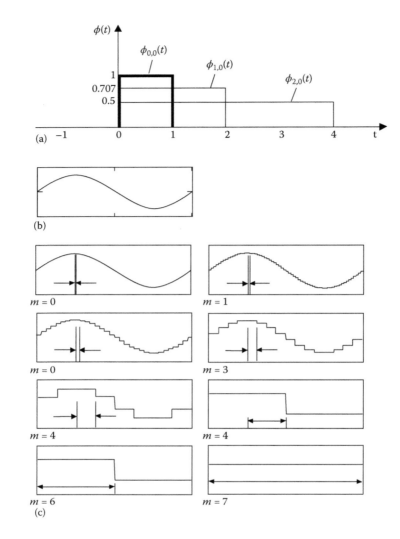

FIGURE 3.2 Smooth approximation of a sine wave using a block pulse scaling function: (a) Simple block scaling function shown at scale 1 (scale index $m=0$) and location $n=0$, that is, $\phi_{0,0}(t)$ (shown bold), together with its dilations at that location. (b) Sine wave of one period. (c) Selected smooth approximations, $x_m(t)$, of the sine wave at increasing scales. The width of one of the scaling functions $\phi_{m,n}(t)$ at scale index m is depicted in each figure by the arrows. Note that the example has been set up so that one period of the sinusoid corresponds to scale 2^7. Note also that at small scales, $x_m(t)$ approaches $x(t)$.

From this equation, it is easy to show that

$$x_{m-1}(t) = x_m(t) + d_m(t) \qquad (3.19)$$

which tells us that if we add the signal detail at an arbitrary scale (index m) to the approximation at that scale, we get the signal approximation at an increased resolution (i.e. at a smaller scale, index $m-1$). This is called a *multiresolution representation*.

3.2.4 Scaling Equation, Scaling Coefficients and Associated Wavelet Equation

The *scaling equation* (or *dilation equation*) describes the scaling function $\phi(t)$ in terms of contracted and shifted versions of itself as follows:

$$\phi(t) = \sum_k c_k \phi(2t - k) \qquad (3.20)$$

where $\phi(2t-k)$ is a contracted version of $\phi(t)$ shifted along the time axis by an integer step k and factored by an associated *scaling coefficient*, c_k. (Take note of the similar but different terminology – scaling *equation* and scaling *function*.) Equation 3.20 basically tells us that we can build a scaling function at one scale from a number of scaling equations at the previous scale. The solution to this two-scale difference equation gives the scaling function $\phi(t)$. For the sake of simplicity, in the rest of the chapter we concern ourselves only with wavelets of *compact support*. These have sequences of non-zero scaling coefficients which are of finite length. Integrating both sides of Equation 3.20, we can show that the scaling coefficients must satisfy the following constraint:

$$\sum_k c_k = 2 \qquad (3.21)$$

In addition, in order to create an orthogonal system, we the require that

$$\sum_k c_k c_{k+2k'} = \begin{cases} 2 & \text{if } k' = 0 \\ 0 & \text{otherwise} \end{cases} \qquad (3.22)$$

This also tells us that the sum of the squares of the scaling coefficients is equal to 2. The same coefficients are used in reverse with alternate signs to produce the differencing of the associated wavelet equation, that is

$$\psi(t) = \sum_k (-1)^k c_{1-k} \phi(2t - k) \qquad (3.23)$$

This construction ensures that the wavelets and their corresponding scaling functions are orthogonal. This wavelet equation is commonly seen in practice. In this chapter, however, we will consider only wavelets of compact support which have a finite number of scaling coefficients, N_k. For this case, we can define the wavelet function as

$$\psi(t) = \sum_k (-1)^k c_{N_k - 1 - k} \phi(2t - k) \qquad (3.24)$$

This ordering of scaling coefficients used in the wavelet equation allows for our wavelets and their corresponding scaling equations to have support over the same interval $(0, N_k - 1)$. (The ordering of Equation 3.23 leads to wavelet and scaling functions displaced from each other except for the Haar wavelet where $N_k = 2$.) Note that, if the number of scaling coefficients is not finite, we cannot use this reordering and must revert back to an ordering of the type given by Equation 3.23. We will stick to the ordering specified by Equation 3.24 in this text.

Often, the reconfigured coefficients used for the wavelet function are written more compactly as

$$b_k = (-1)^k c_{N_k - 1 - k} \quad (3.25)$$

where the sum of all the coefficients b_k is zero. Using this reordering of the coefficients, Equation 3.24 can be written as

$$\psi(t) = \sum_{k=0}^{N_k - 1} b_k \phi(2t - k) \quad (3.26)$$

From Equations 3.12 and 3.20 and examining the wavelet at scale index $m + 1$, we can see that for arbitrary integer values of m the following is true:

$$2^{-(m+1)/2} \phi\left(\frac{t}{2^{m+1}} - n\right) = 2^{-m/2} 2^{-1/2} \sum_k c_k \phi\left(\frac{2t}{2 \times 2^m} - 2n - k\right) \quad (3.27a)$$

which may be written more compactly as

$$\phi_{m+1,n}(t) = \frac{1}{\sqrt{2}} \sum_k c_k \phi_{m, 2n+k}(t) \quad (3.27b)$$

That is, the scaling function at an arbitrary scale is composed of a sequence of shifted scaling functions at the next smaller scale, each factored by their respective scaling coefficients. Similarly, for the wavelet function, we obtain

$$\psi_{m+1,n}(t) = \frac{1}{\sqrt{2}} \sum_k b_k \phi_{m, 2n+k}(t) \quad (3.28)$$

3.2.5 Haar Wavelet

The *Haar wavelet* is the simplest example of an orthonormal wavelet. Its scaling equation contains only two non-zero scaling coefficients and is given by

$$\phi(t) = \phi(2t) + \phi(2t-1) \tag{3.29}$$

that is, its scaling coefficients are $c_0 = c_1 = 1$. We get these coefficient values by solving Equations 3.21 and 3.22 simultaneously. (From Equation 3.21, we see that $c_0 + c_1 = 2$ and from Equation 3.22, $c_0 c_0 + c_1 c_1 = 2$.) The solution of the Haar scaling equation is the single block pulse shown in Figure 3.3a and defined as

$$\phi(t) = \begin{cases} 1 & 0 \le t < 1 \\ 0 & \text{elsewhere} \end{cases} \tag{3.30}$$

This is, in fact, the scaling function used in Figure 3.2 to generate the signal approximations of the sine wave. Reordering the coefficient sequence according to Equation 3.25, we can see that the corresponding Haar wavelet equation is

$$\psi(t) = \phi(2t) - \phi(2t-1) \tag{3.31}$$

The Haar wavelet is shown in Figure 3.3b and defined as

$$\psi(t) = \begin{cases} 1 & 0 \le t < \dfrac{1}{2} \\ -1 & \dfrac{1}{2} \le t < 1 \\ 0 & \text{elsewhere} \end{cases} \tag{3.32}$$

The mother wavelet for the Haar wavelet system, $\psi(t) = \psi_{0,0}(t)$, is formed from two dilated unit block pulses sitting next to each other on the time axis, with one of them inverted. From the mother wavelet, we can construct the Haar system of wavelets on a dyadic grid, $\psi_{m,n}(t)$. This is illustrated in Figure 3.3c for three consecutive scales. The orthogonal nature of the family of Haar wavelets in a dyadic grid system is obvious from Figure 3.3d, where it can be seen that the positive and negative parts of the Haar wavelet at any scale coincide with a constant (positive or negative) part of the Haar wavelet at the next larger scale (and all subsequent larger scales). In addition, Haar wavelets at the same scale index m on a dyadic grid do not overlap. Hence, it is obvious that the convolution of the Haar wavelet with any others in the same dyadic grid gives zero. Figure 3.3e shows three Haar wavelets which are not specified on a dyadic grid. The non-orthogonal nature of the Haar wavelets across scales, when specified in this way, is obvious from the plot. In addition, if these wavelets overlap each other along each scale, this also destroys orthogonality. (Although this is not

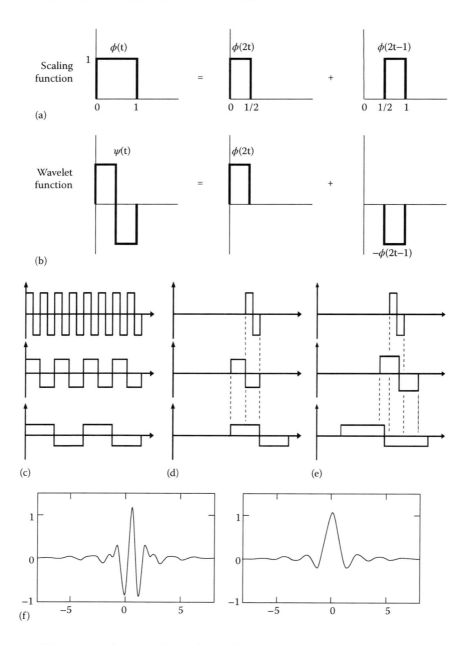

FIGURE 3.3 Discrete orthonormal wavelets: (a) The Haar scaling function in terms of shifted and dilated versions of itself. (b) The Haar wavelet in terms of shifted and dilated versions of its scaling function. (c) Three consecutive scales shown for the Haar wavelet family specified on a dyadic grid, for example, from top to bottom: $\psi_{m,n}(t)$, $\psi_{m+1,n}(t)$ and $\psi_{m+2,n}(t)$. (d) Three Haar wavelets at three consecutive scales on a dyadic grid. (e) Three Haar wavelets at different scales. This time, the Haar wavelets are not defined on a dyadic grid and are hence not orthogonal to each other. (f) A Meyer wavelet and associated scaling function (right).

the case for other orthonormal wavelets.) Finally, note that the Haar wavelet is of finite width on the time axis, that is, it has compact support. As was stated in the last section, wavelets which have compact support have a finite number of scaling coefficients and these are the type of wavelets we concentrate on in this chapter. Not all orthonormal wavelets have compact support. Figure 3.3f, for example, shows a Meyer wavelet which is orthonormal with infinite support, although, as with all wavelets, it is localized, decaying relatively quickly from its central peak.

3.2.6 Coefficients from Coefficients: Fast Wavelet Transform

From Equation 3.14, we can see that the approximation coefficients at scale index $m+1$ are given by

$$S_{m+1,n} = \int_{-\infty}^{\infty} x(t)\phi_{m+1,n}(t)dt \qquad (3.33)$$

Using Equation 3.27b, this can be written as

$$S_{m+1,n} = \int_{-\infty}^{\infty} x(t)\left[\frac{1}{\sqrt{2}}\sum_{k} c_k \phi_{m,2n+k}(t)\right]dt \qquad (3.34)$$

We can rewrite this as

$$S_{m+1,n} = \frac{1}{\sqrt{2}}\sum_{k} c_k \left[\int_{-\infty}^{\infty} x(t)\phi_{m,2n+k}(t)dt\right] \qquad (3.35)$$

The integral in brackets gives the approximation coefficients $S_{m,2n+k}$ for each k. We can therefore write this equation as

$$S_{m+1,n} = \frac{1}{\sqrt{2}}\sum_{k} c_k S_{m,2n+k} = \frac{1}{\sqrt{2}}\sum_{k} c_{k-2n} S_{m,k} \qquad (3.36)$$

Hence, using this equation, we can generate the approximation coefficients at scale index $m+1$ using the scaling coefficients at the previous scale.

Similarly, the wavelet coefficients can be found from the approximation coefficients at the previous scale using the reordered scaling coefficients b_k as follows:

$$T_{m+1,n} = \frac{1}{\sqrt{2}}\sum_{k} b_k S_{m,2n+k} = \frac{1}{\sqrt{2}}\sum_{k} b_{k-2n} S_{m,k} \qquad (3.37)$$

We can see now that, if we know the approximation coefficients $S_{m_0,n}$ at a specific scale m_0, then, through the repeated application of Equations 3.36 and 3.37, we can generate the approximation and detail wavelet coefficients at *all* scales larger than m_0. Notice that, to do this, we do not even need to know exactly what the underlying continuous signal $x(t)$ is, only $S_{m_0,n}$. Equations 3.36 and 3.37 represent the multiresolution *decomposition algorithm*. The decomposition algorithm is the first half of the *fast wavelet transform*-which allows us to compute the wavelet coefficients in this way, rather than computing them laboriously from the convolution of Equation 3.9. Iterating Equations 3.36 and 3.37 performs, respectively, a highpass and lowpass filtering of the input (i.e. the coefficients $S_{m,2n+k}$) to get the outputs ($S_{m+1,n}$ and $T_{m+1,n}$). The vectors containing the sequences $(1/\sqrt{2})c_k$ and $(1/\sqrt{2})b_k$ represent the filters: $(1/\sqrt{2})c_k$ is the *lowpass filter*, letting through low-signal frequencies and hence a smoothed version of the signal, and $(1/\sqrt{2})b_k$ is the *highpass filter*, letting through the high frequencies corresponding to the signal details. We will come back to the filtering process in more detail in later sections of this chapter.

We can go in the opposite direction and reconstruct $S_{m,n}$ from $S_{m+1,n}$ and $T_{m+1,n}$. We know already from Equation 3.17 that $x_{m-1}(t) = x_m(t) + d_m(t)$, we can expand this as

$$x_{m-1}(t) = \sum_n S_{m,n} \phi_{m,n}(t) + \sum_n T_{m,n} \psi_{m,n}(t) \tag{3.38}$$

Using Equations 3.27b and 3.28, we can expand this equation in terms of the scaling function at the previous scale as follows:

$$x_{m-1}(t) = \sum_n S_{m,n} \frac{1}{\sqrt{2}} \sum_k c_k \phi_{m-1,2n+k}(t) + \sum_n T_{m,n} \frac{1}{\sqrt{2}} \sum_k b_k \phi_{m-1,2n+k}(t) \tag{3.39}$$

Rearranging the summation indices, we get

$$x_{m-1}(t) = \sum_n S_{m,n} \frac{1}{\sqrt{2}} \sum_k c_{k-2n} \phi_{m-1,k}(t) + \sum_n T_{m,n} \frac{1}{\sqrt{2}} \sum_k b_{k-2n} \phi_{m-1,k}(t) \tag{3.40}$$

We also know that we can expand $x_{m-1}(t)$ in terms of the approximation coefficients at scale $m - 1$, that is

$$x_{m-1}(t) = \sum_n S_{m-1,n} \phi_{m-1,n}(t) \tag{3.41}$$

Equating the coefficients in Equation 3.41 with Equation 3.40, we note that the index k at scale index m relates to the location index n at scale index $m - 1$. In addition, location index n in Equation 3.40 is not equivalent to location index n in Equation 3.41 as the former corresponds to scale index m, with associated discrete location spacings 2^m, and the latter to scale index $m - 1$, with discrete location spacings 2^{m-1}. Hence, the n indices are twice as dense in the latter expression. The simplest way to proceed before equating the two expressions is to swap the indices k and n in Equation 3.40, which, after some algebra, produces the *reconstruction algorithm*:

$$S_{m-1,n} = \frac{1}{\sqrt{2}} \sum_k c_{n-2k} S_{m,k} + \frac{1}{\sqrt{2}} \sum_k b_{n-2k} T_{m,k} \tag{3.42}$$

where we have *reused* k as the location index of the transform coefficients at scale index m to differentiate it from n, the location index at scale $m - 1$. Hence, at the smaller scale, $m - 1$, the approximation coefficients can be found in terms of a combination of approximation and detail coefficients at the next scale, m. Note that if there are only a finite number of non-zero scaling coefficients ($=N_k$), then c_{n-2k} has non-zero values only when in the range 0 to $N_k - 1$. The reconstruction algorithm is the second half of the *fast wavelet transform* (FWT). Note that in the literature, the 'fast wavelet transform', 'discrete wavelet transform', 'decomposition/reconstruction algorithms', 'fast orthogonal wavelet transform', 'multiresolution algorithm', 'pyramid algorithm', 'tree algorithm' and so on, mean the same thing. It becomes even more confusing when other discretizations of the continuous wavelet transform are referred to as the 'discrete wavelet transform'. Take care!

3.3 DISCRETE INPUT SIGNALS OF FINITE LENGTH

3.3.1 Approximations and Details

So far, we have considered the discrete orthonormal wavelet transform of a continuous function $x(t)$, where it was shown how the continuous function could be represented as a series expansion of wavelet functions at all scales and locations (Equation 3.10a), or a combined series expansion involving the scaling and wavelet functions (Equation 3.16). In this section, and from here on, we will consider discrete input signals specified at integer spacings. To fit into a wavelet multiresolution framework, the discrete signal input into the multiresolution algorithm should be the signal approximation coefficients at scale index $m = 0$. Defined by

$$S_{0,n} = \int_{-\infty}^{\infty} x(t) \phi(t - n) \, dt \tag{3.43}$$

which, as we now know from Equations 3.36 and 3.37, will allow us to generate all subsequent approximation and detail coefficients, $S_{m,n}$ and $T_{m,n}$, at scale indices greater than

$m = 0$. In this section, we will assume that we have been given $S_{0,n}$. Section 3.4 considers further the question of discrete input data which may not be $S_{0,n}$.

In practice, our discrete input signal $S_{0,n}$ is of finite length N, which is an integer power of 2: $N = 2^M$. Thus, the range of scales that we can investigate is $0 < m < M$. Substituting both $m = 0$ and $m = M$ into Equation 3.16, and noting that we have a finite range of n which halves at each scale, we can see that the signal approximation scale $m = 0$ (the input signal) can be written as the smooth signal at scale M plus a combination of detailed signals as follows:

$$\sum_{n=0}^{2^{M-m}-1} S_{0,n}\phi_{0,n}(t) = S_{M,n}\phi_{M,n}(t) + \sum_{m=1}^{M}\sum_{n=0}^{2^{M-m}-1} T_{m,n}\psi_{m,n}(t) \qquad (3.44)$$

This is the form that we use to describe our finite length discrete signal in terms of its discrete wavelet expansion. The covering of a finite length time segment with wavelets is illustrated in Figure 3.4 for Daubechies D4 wavelets at two successive scales. The lower scale covers the time window using eight wavelets, and the larger scale uses four wavelets. One of the wavelets in each plot is shown bold for clarity. The wavelets shown which spill over the end of the window have been wrapped around back to the beginning. Known as *wraparound*, it is the simplest and one of the most common treatments of the boundary for a finite length signal and we will employ it throughout the rest of this chapter. However, note that by employing this method, we assume that the signal segment under investigation

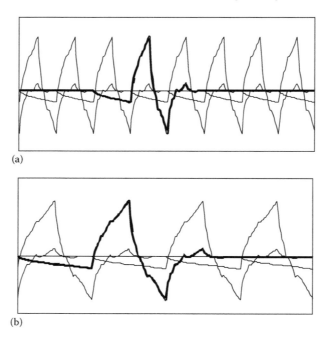

FIGURE 3.4 **Covering the time axis with dyadic grid wavelets:** (a) Eight Daubechies D4 wavelets covering the time axis at scale m. (b) Four Daubechies D4 wavelets covering the time axis at scale $m + 1$. These wavelets are twice the width of those in (a).

represents one period of a periodic signal and we are in effect pasting the end of the signal back onto the beginning. Obviously, if the signal is not periodic, and in practice it usually is not, then we create artificial singularities at the boundary which may result in large detail coefficients generated near to the boundary.

We can rewrite Equation 3.44 as

$$x_0(t) = x_M(t) + \sum_{m=1}^{M} d_m(t) \qquad (3.45)$$

where the mean signal approximation at scale M is

$$x_M(t) = S_{M,n} \phi_{M,n}(t) \qquad (3.46)$$

As the approximation coefficients are simply factored, weighted averages of the signal then, when wraparound is employed to deal with the boundaries, the single approximation component $S_{M,n}$ is related to the mean of the input signal through the relationship $\overline{S_{0,n}} = S_{M,n}/\sqrt{2^M}$ where the overbar denotes the mean of the sequence $S_{0,n}$. In addition, when wraparound has been used to deal with the boundaries, the mean signal approximation at the largest scale, $x_M(t)$, is a constant valued function equal to the input signal mean. (We will see why this is so in Section 3.5.1.)

The term on the far right of Equation 3.45 represents the series expansion of the fluctuating components of the finite length signal at various scales in terms of its detail coefficients. The detail signal approximation corresponding to scale index m is defined for a finite length signal as

$$d_m(t) = \sum_{n=0}^{2^{M-m}-1} T_{m,n} \psi_{m,n}(t) \qquad (3.47)$$

As we have seen (Equation 3.45), adding the approximation of the signal at scale index M to the sum of all detail signal components across scales $0 < m < M$ gives the approximation of the original signal at scale index 0. Figure 3.5a shows the details of a chirp signal with a short burst of noise added to the middle of it. A Daubechies D20 wavelet was used in the decomposition (see an example of this wavelet in Figure 3.15e). The original signal is shown at the top of the plot. Below the signal, the details for 10 wavelet scales, $d_1(t)$ to $d_{10}(t)$ are shown. The bottom trace is the remaining signal approximation $x_{10}(t)$. Adding together all these details plus the remaining approximation (which is the signal mean) returns the original signal. Two things are noticeable from the plot. First, there is a shift to the left of the large amplitude details with increasing scale, as we would expect, oscillation increases in frequency from left to right. The second thing to notice is that the high-frequency burst of noise is captured at the smallest scales, again as we would expect.

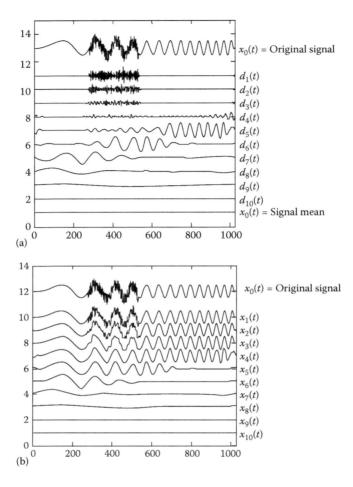

FIGURE 3.5 Multiresolution decomposition of a chirp signal containing a short burst of noise: (a) Signal details $d_m(t)$. (The signals have been displaced from each other on the vertical axis to aid clarity.) (b) Signal approximations $x_m(t)$.

We saw before in Equation 3.19 that the signal approximation at a specific scale was a combination of the approximation and detail at the next lower scale. If we rewrite this equation:

$$x_m(t) = x_{m-1}(t) - d_m(t) \tag{3.48}$$

and begin at scale $m - 1 = 0$, that of the input signal, we can see that at scale index $m = 1$ the signal approximation is given by

$$x_1(t) = x_0(t) - d_1(t) \tag{3.49a}$$

at the next scale ($m = 2$), the signal approximation is given by

$$x_2(t) = x_0(t) - d_1(t) - d_2(t) \quad (3.49b)$$

and at the next scale by

$$x_3(t) = x_0(t) - d_1(t) - d_2(t) - d_3(t) \quad (3.49c)$$

and so on, corresponding to the successive stripping of high-frequency information (contained within the $d_m(t)$'s) from the original signal. Figure 3.5b contains successive approximations $x_m(t)$ of the chirp signal. The top trace is the original signal $x_0(t)$. Subsequent smoothing of the signal takes place throughout the traces from the top to the bottom of the figure. As we saw in Equation 3.48, the difference between each of the approximation traces $x_{m-1}(t)$ and $x_m(t)$ is the detail component $d_m(t)$. We can view these differences in the detail component traces of Figure 3.5a.

We have glossed over much of the mathematical detail of multiresolution analysis here. Most mathematical accounts of the subject begin with a discussion of orthogonal nested subspaces and the signal approximations and details, $x_m(t)$ and $d_m(t)$, as projections onto these spaces. This tack has not been followed herein, see, for example, Mallat (2009), Blatter (1998), Sarkar et al. (1998) or Williams and Amaratunga (1994) for more detailed mathematics. In this chapter, we concentrate on the mechanics, rather than the mathematics, of multiresolution analysis.

3.3.2 Multiresolution Algorithm: An Example

Once we have our discrete input signal $S_{0,n}$, we can compute $S_{m,n}$ and $T_{m,n}$ using the decomposition algorithm given by Equations 3.36 and 3.37. This can be done for scale indices $m > 0$, up to a maximum scale determined by the length of the input signal. To do this, we use an iterative procedure as follows. First, we compute $S_{1,n}$ and $T_{1,n}$ from the input coefficients $S_{0,n}$, that is

$$S_{1,n} = \frac{1}{\sqrt{2}} \sum_k c_k \, S_{0,2n+k} \quad \text{and} \quad T_{1,n} = \frac{1}{\sqrt{2}} \sum_k b_k \, S_{0,2n+k} \quad (3.50a \text{ and b})$$

In the same way, we can then find $S_{2,n}$ and $T_{2,n}$ from the approximation coefficients $S_{1,n}$, that is

$$S_{2,n} = \frac{1}{\sqrt{2}} \sum_k c_k \, S_{1,2n+k} \quad \text{and} \quad T_{2,n} = \frac{1}{\sqrt{2}} \sum_k b_k \, S_{1,2n+k} \quad (3.51a \text{ and b})$$

Next, we can find $S_{3,n}$ and $T_{3,n}$ from the approximation coefficients $S_{2,n}$, and so on, up to those coefficients at scale index M, where only one approximation and one detail coefficient is computed: $S_{M,0}$ and $T_{M,0}$. At scale index M, we have performed a full decomposition of the finite-length, discrete input signal. We are left with an array of coefficients: a single approximation coefficient value, $S_{M,0}$, plus the detail coefficients, $T_{m,n}$, corresponding to discrete wavelets of scale $a=2^m$ and location $b=2^m n$. The finite time series is of length $N=2^M$. This gives the ranges of m and n for the detail coefficients as, respectively, $1<m<M$ and $0<n<2^{M-m}-1$. Notice that the range of n successively halves at each iteration as it is a function of scale index m for a finite length signal. At the smallest wavelet scale, index $m=1$, $2^M/2^1 = N/2$ coefficients are computed, at the next scale $2^M/2^2 = N/4$ are computed and so on, at larger and larger scales, until the largest scale ($m=M$) where only one ($=2^M/2^M$) coefficient is computed. The total number of detail coefficients for a discrete time series of length $N=2^M$ is then, $1+2+4+8+\ldots+2^{M-1}$, or

$$\sum_{m=0}^{M-1} 2^m = 2^M - 1 = N - 1.$$

In addition to the detail coefficients, the single approximation coefficient $S_{M,0}$ remains. This is related to the signal mean as we have seen, and is required in addition to the detail coefficients to fully represent the discrete signal. Thus, a discrete input signal of length N can be broken down into exactly N components without any loss of information using discrete orthonormal wavelets. In addition, no signal information is repeated in the coefficient representation. This is known as *zero redundancy*.

The decomposition of approximation coefficients into approximation and detail coefficients at subsequent levels can be illustrated schematically by the following:

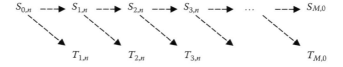

Figure 3.6 shows an alternative schematic of the same process, illustrating the decomposition and insertion of the approximation and detail coefficients at each iteration within the wavelet transform vector for an arbitrary input signal vector containing 32 components. The wavelet transform vector after the full decomposition has the form $\mathbf{W}^{(M)} = (\mathbf{S}_M, \mathbf{T}_M, \mathbf{T}_{M-1}, \ldots \mathbf{T}_m \ldots \mathbf{T}_2, \mathbf{T}_1)$ where $\mathbf{T}_m$ represents the sub-vector containing the coefficients $T_{m,n}$ at scale index m, where n is in the range 0 to $2^{M-m}-1$. We can halt the transformation process before the full decomposition. If we do this, say at an arbitrary level m_0, the transform vector has the form $(\mathbf{W}^{(m_0)} = \mathbf{S}_{m_0}, \mathbf{T}_{m_0}, \mathbf{T}_{m_0-1}, \ldots \mathbf{T}_2, \mathbf{T}_1)$ where m_0 can take the range $1 \leq m_0 \leq M-1$. In this case, the transform vector does not contain a single approximation component but rather the sequence of approximation components $S_{m_0,n}$. However, the transform vector always contains $N=2^M$ components. For example, we can see from Figure 3.6 that stopping the algorithm at $m=2$ results in $\mathbf{W}^{(2)} = (\mathbf{S}_2, \mathbf{T}_2, \mathbf{T}_1)$. Remember, the range of n is a function of scale index m; hence, this vector contains 8 $S_{2,n}$ components, 8 $T_{2,n}$ components and 16 $T_{1,n}$ components, matching the 32 components of the original input signal vector. The range of n is indicated below the full decomposition vector in

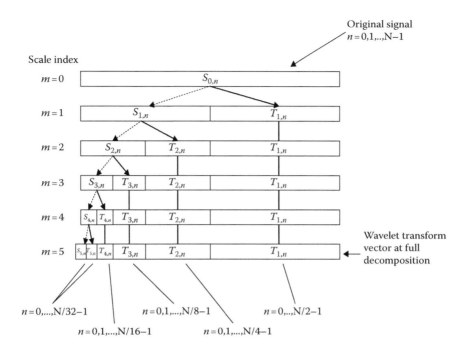

FIGURE 3.6 Schematic diagram of the Haar filtering algorithm: $\dashrightarrow S_{m,n}$ used to derive $S_{m+1,n}$; $\rightarrow S_{m,n}$ used to derive $T_{m+1,n}$; $T_{m,n}$ taken to next level as $T_{m,n}$, that is, no further manipulation.

Figure 3.6. Notice also that we can express the original input signal as the transform vector at scale index zero, that is, $\mathbf{W}^{(0)}$.

An example of a wavelet decomposition using a discrete wavelet is shown in Figure 3.7. The input signal is composed of a section of a sine wave, some noise and a flatline. The signal is decomposed using a Daubechies D6 wavelet. A member of this family is shown in Figure 3.7b. (We will look more closely at the Daubechies wavelet family in this chapter.) The discrete transform plot is shown in Figure 3.7c where the dyadic grid arrangement may be seen clearly. This plot is simply a discretized dyadic map of the detail coefficients, $T_{m,n}$, where the coefficients at larger scales have correspondingly longer boxes (as the wavelets cover larger segments of the input signal). In addition to the detail coefficients, $T_{m,n}$, the remaining approximation coefficient $S_{M,0}$ is added to the bottom of the plot. As we would expect, it covers the whole time axis. We can see from the transform plot that the dominant oscillation is picked up at scale index $m=6$ and the high-frequency noise is picked up within the middle segment of the transform plot at smaller scales. We can use the reconstruction algorithm (Equation 3.42) to get back the original input signal $S_{0,n}$ from the array of detail coefficients shown in Figure 3.7c. Alternatively, as with the continuous transform, we can reconstruct a modified version of the input signal by using only selected coefficients in the reconstruction. This is shown in Figure 3.7d,e where only the coefficients corresponding to scales $m=5$ to 8 are kept (the others are set to zero) and the signal is reconstructed. This has removed a significant amount of the noise from the signal although the

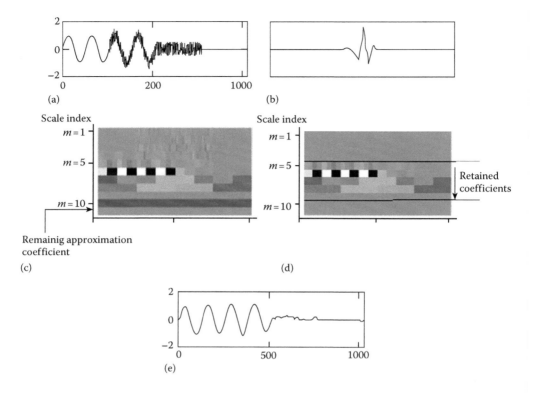

FIGURE 3.7 Discrete wavelet transform of a composite signal: (a) Original composite signal. (b) A member of the Daubechies D6 wavelet family. (c) Discrete transform plot. (Note dyadic structure – large positive coefficient values are white and large negative values are black.) (d) Coefficient removal. (e) Reconstructed signal using only retained coefficients in (d). The original composite signal (a) is composed of three segments: A sinusoid, uniformly distributed noise and a flatline. The signal is decomposed using Daubechies D6 wavelets (b) to give the dyadic array of transform coefficients plotted in (c). The coefficients corresponding to scales 5–9 are kept (d) and used to reconstruct the signal in (e). Note that a greyscale is used to depict the coefficient values, where the maximum value is white and the minimum value is black.

sinusoidal waveform is less smooth than the original in Figure 3.7a. We will come across more sophisticated ways to remove noise and other signal artefacts in Section 3.4.2.

3.3.3 Wavelet Energy

After a full decomposition, the energy contained within the coefficients at each scale is given by

$$E_m = \sum_{n=0}^{2^{M-m}-1} \left(T_{m,n}\right)^2 \tag{3.52}$$

A wavelet-based power spectrum of the signal may be produced using these scale-dependent energies. To do so, we require a frequency measure which is a reciprocal of the wavelet dilation, for example, the passband centre of the power spectrum of the wavelet. A wavelet power spectrum can then be produced for the signal which is directly comparable with both its Fourier and continuous wavelet counterparts (see Chapter 2, Section 2.9). Note that this topic is dealt with in much greater detail in Chapter 4, Section 4.2.1 in connection with the statistical measures used to analyze turbulent fluid signals.

The total energy of the discrete input signal:

$$E = \sum_{n=0}^{N-1} (S_{o,n})^2 \qquad (3.53)$$

is equal to the sum of the squared detail coefficients over all scales plus the square of the remaining approximation coefficient, $S_{M,0}$, as follows:

$$E = (S_{M,0})^2 + \sum_{m=1}^{M} \sum_{n=0}^{2^{M-m}-1} (T_{m,n})^2 \qquad (3.54)$$

In fact, the energy contained within the transform vector at all stages of the multiresolution decomposition remains constant. We can, therefore, write the conservation of energy more generally as

$$E = \sum_{i=0}^{N-1} \left(W_i^{(m)}\right)^2 \qquad (3.55)$$

where $W_i^{(m)}$ are the individual components of the transform vector $\mathbf{W}^{(m)}$, ordered as described in the previous section. When $m=0$, this equation corresponds to the summation of the component energies of the input signal (Equation 3.53) and when $m=M$ it corresponds to the summation of the energies within the coefficients at full decomposition (Equation 3.54).

3.3.4 Alternative Indexing of Dyadic Grid Coefficients

There are three main methods used in practice to index the coefficients resulting from the discrete wavelet transform. It is worth discussing these now. All three methods are popular in the scientific literature and appear often in many of the examples of the practical application of wavelet analysis in the subsequent chapters of this book. We will use the full decomposition of a 32-component input signal as an illustration.

Scale Indexing: The scale indexing system (m,n) corresponding to an input signal of length $N=2^M$ is shown schematically on a dyadic grid in Figure 3.8b for the discrete signal shown in Figure 3.8a. The lowest scale on the grid $m=1$ corresponds to a spacing of $2^1=2$ on the data set. The discrete input signal is at scale index $m=0$. We have already come across this grid structure in the plot of the transform coefficients in Figure 3.7c. Such plots

116 ■ The Illustrated Wavelet Transform Handbook

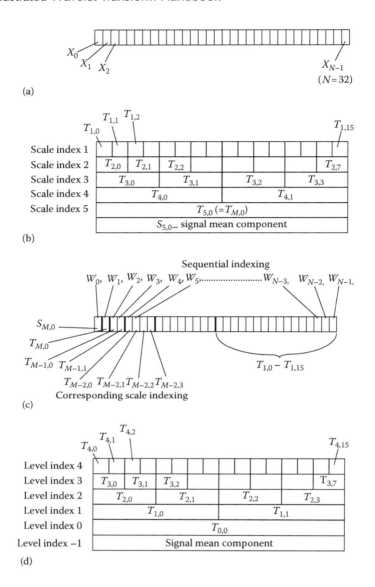

FIGURE 3.8 Schematic diagram of alternative indexing methods for dyadic grid wavelet transform coefficients: (a) Original signal containing 32 components (scale index $m = 0$). (b) Scale indexing, $T_{m,n}$. (c) Sequential indexing, $W_i^{(M)}$, where $i = 2^{M-n} + n$ (corresponding scale indexing is shown at the bottom of the plot). (d) Level indexing, $T_{l,n}$.

give a good visual indication of the covering of the timescale plane by the wavelets and their relative importance in making up the signal.

Sequential Indexing: Again, we have already come across this form of indexing (Figure 3.6). Figure 3.8c contains a schematic of the transform coefficients given in sequential format. We know that a signal N samples long produces N wavelet vector components. It makes sense, therefore, to find an ordering of these coefficients to fit into a vector of length N. The discrete time series, $S_{0,n}$, $n = 0, 1, 2, \ldots, N-1$, can be decomposed into $N-1$

detail coefficients plus one approximation coefficient where N is an integer power of 2: $N = 2^M$. After a full decomposition, these transform coefficients can be resequenced from the two-dimensional dyadic array $T_{m,n}$ to the transform vector $\mathbf{W}^{(M)}$ where the components $W_i^{(M)}$ have the same range as the original signal ($i = 0, 1, 2, \ldots, N-1$). The vector component index i is found from the dyadic grid indices m and n through the relationship $i = 2^{M-m} + n$. In addition, the superscripted M in parenthesis denotes a full decomposition of the signal over M scales. The transform vector components, $W_i^{(M)}$ are plotted in Figure 3.8c. The last half of the series in the figure represents the coefficients corresponding to the smallest wavelet scale (index $m = 1$). The next quarter, working backwards, represents the coefficients corresponding to scale index $m = 2$, the next eighth to the $m = 3$ coefficients and so on, back to $W_1^{(M)}$, which is the single coefficient for scale index $m = M$. The remaining component $W_0^{(M)}$ is the single approximation coefficient ($S_{M,0}$). As mentioned in Section 3.3.2, if we halt the decomposition algorithm at scale m_0 (before full decomposition) this results in an intermediate transform vector $\mathbf{W}^{(m_0)}$ which contains a number of approximation coefficients, $S_{m_0,n}$ at its beginning.

Level Indexing: Level indices, l, are often used instead of scale indices, m. The level index l is simply equal to $M-m$. In this case, the number of wavelets used to cover the signal at each level at a specific level is 2^l; for example, level $l = 0$ corresponds to a single wavelet and the scale of the whole time series, level $l = 1$ corresponds to two wavelets and the scale of one half of the time series and so on. The number of wavelets used at a level is simply 2^l. In addition, it is standard in this nomenclature to denote the remaining approximation coefficient at level $l = -1$. The wavelet array becomes $T_{l,n}$. Figure 3.8d shows the l,n indexing system. Level indexing is employed when the signal is specified over the unit interval and the analysis is set up in terms of resolution rather than scale (i.e. the lowest resolution $l = 0$ corresponds to the length of the whole signal, whereas the smallest scale $m = 0$ corresponds to the distance between each signal point).

3.3.5 A Simple Worked Example: The Haar Wavelet Transform

Now we will illustrate the methods just described using a Haar wavelet in the decomposition of a discrete input signal, $S_{0,n}$: $n = 0, 1, 2, \ldots, N-1$. To do so, we employ the decomposition algorithm given by Equations 3.36 and 3.37. The Haar wavelet has two scaling coefficients, $c_0 = 1$ and $c_1 = 1$, substituting these into Equation 3.36 we can obtain the approximation coefficients at the next scale through the relationship:

$$S_{m+1,n} = \frac{1}{\sqrt{2}} \left[S_{m,2n} + S_{m,2n+1} \right] \tag{3.56}$$

Similarly, through Equation 3.37 we can obtain the detail coefficients at subsequent scales using

$$T_{m+1,n} = \frac{1}{\sqrt{2}} \left[S_{m,2n} - S_{m,2n+1} \right] \tag{3.57}$$

Using these equations, we will perform the Haar wavelet decomposition of the (very) simple discrete signal, (1,2,3,4). As the signal contains only four data points, we can only perform two iterations of the Haar decomposition algorithm given by Equations 3.56 and 3.57. After two iterations, we expect to obtain four transform coefficients: three wavelet coefficients $T_{m,n}$, with two at scale index $m = 1$ ($T_{1,0}$, $T_{1,1}$), and one at scale index $m = 2$, ($T_{2,0}$), plus a signal mean coefficient at scale index $m = 2$, ($S_{2,0}$). This is illustrated through the iteration of the transform vector in Figure 3.9 and also through a schematic of the coefficients in Figure 3.10.

The first iteration of the decomposition algorithm gives

$$T_{1,0} = \frac{1}{\sqrt{2}}[1-2] = -\frac{1}{\sqrt{2}} \qquad T_{1,1} = \frac{1}{\sqrt{2}}[3-4] = -\frac{1}{\sqrt{2}}$$

$$S_{1,0} = \frac{1}{\sqrt{2}}[0+3] = \frac{3}{\sqrt{2}} \qquad S_{1,1} = \frac{1}{\sqrt{2}}[3+4] = \frac{7}{\sqrt{2}}$$

The transform coefficient vector, after this first iteration is then $\mathbf{W}^{(1)} = (S_{1,0}, S_{1,1}, T_{1,0}, T_{1,1}) = (3/\sqrt{2}, 7/\sqrt{2}, -1/\sqrt{2}, -1/\sqrt{2})$. The second iteration (only involving the remaining approximation coefficients $S_{1,0}$ and $S_{1,1}$) yields

$$T_{2,0} = \frac{1}{\sqrt{2}}\left[3/\sqrt{2} - 7/\sqrt{2}\right] = -2 \quad \text{and} \quad S_{2,0} = \frac{1}{\sqrt{2}}\left[3/\sqrt{2} + 7/\sqrt{2}\right] = 5$$

The transform coefficient vector after this second iteration is now $\mathbf{W}^{(2)} = (S_{2,0}, T_{2,0}, T_{1,0}, T_{1,1}) = (5, -2, -1/\sqrt{2}, -1/\sqrt{2})$. The signal mean is found from the remaining approximation

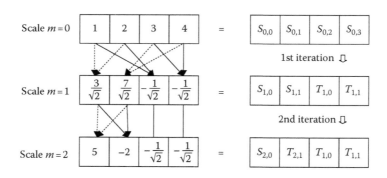

FIGURE 3.9 Iteration of the wavelet transform vector in the decomposition of a simple signal: ⇢ $S_{m,n}$ used to derive $S_{m+1,n}$; → $S_{m,n}$ used to derive $T_{m+1,n}$; $T_{m,n}$ taken to the next level as $T_{m,n}$, that is, no further manipulation.

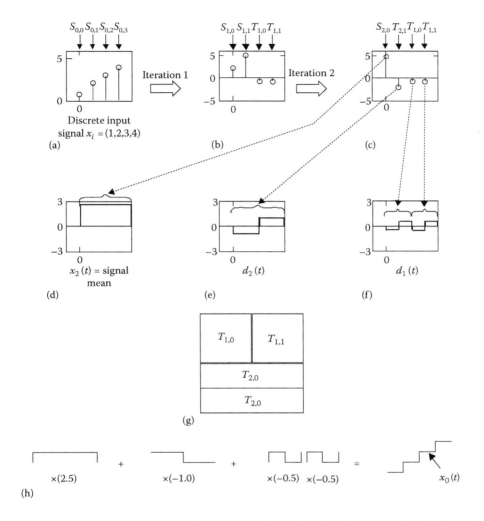

FIGURE 3.10 Haar decomposition and reconstruction of a simple ramp signal. The original discrete input signal is decomposed through an initial iteration to (b), the first decomposition, then further to (c), the second and final decomposition. From the coefficients at the full decomposition (c), the following can be constructed: The scale index 2 approximation (the signal mean) (d), the scale index 2 detail (e) and the scale index 1 detail (f). (g) The transform coefficient plot associated with the transform values given in (c). (h) Schematic diagram of the addition of the rescaled Haar wavelets to reconstruct the original signal. (Adding the reconstructions in (d), (e) and (f) returns the approximation of the signal at scale index $m = 0$.)

coefficient, that is, $S_{2,0}/(\sqrt{2})^2 = 2.5$. We can also see that the energy of the original discrete signal $(1^2+2^2+3^2+4^2=30)$ is equal to the energy of the transform coefficient vector after the first iteration $((3/\sqrt{2})^2 +(7/\sqrt{2})^2 +(-1/\sqrt{2})^2 +(-1/\sqrt{2})^2 = 30)$ and the second iteration, giving the full decomposition $(5^2 +(-2)^2 +(-1/\sqrt{2})^2 +(-1/\sqrt{2})^2 = 30)$; a result we expect from the conservation of energy condition expressed in Equations 3.52 through 3.55.

To return from the transform coefficients to the original discrete signal, we work in reverse. The inverse Haar transform can be written simply as

at even locations, $2n$:

$$S_{m,2n} = \frac{1}{\sqrt{2}}\left[S_{m+1,n} + T_{m+1,n}\right] \tag{3.58}$$

and at odd locations, $2n+1$:

$$S_{m,2n+1} = \frac{1}{\sqrt{2}}\left[S_{m+1,n} - T_{m+1,n}\right] \tag{3.59}$$

For the Haar wavelet transform, we can derive these reconstruction equations directly from Equations 3.56 and 3.57. Alternatively, we could derive them from the reconstruction algorithm of Equation 3.42.

We will perform the reconstruction on the transform vector that we have just computed:

$$\mathbf{W}^{(2)} = \left(S_{2,0}, T_{2,0}, T_{1,0}, T_{1,1}\right) = \left(5, -2, -\frac{1}{\sqrt{2}}, -\frac{1}{\sqrt{2}}\right)$$

The first iteration using the reconstruction pair (Equations 3.58 and 3.59) yields

$$S_{1,0} = \frac{1}{\sqrt{2}}\left(S_{2,0} + T_{2,0}\right) = \frac{3}{\sqrt{2}} \qquad S_{1,1} = \frac{1}{\sqrt{2}}\left(S_{2,0} - T_{2,0}\right) = \frac{7}{\sqrt{2}}$$

this results in $\left(S_{1,0}, S_{1,1}, T_{1,0}, T_{1,1}\right) = \left(\dfrac{3}{\sqrt{2}}, \dfrac{7}{\sqrt{2}}, -\dfrac{1}{\sqrt{2}}, -\dfrac{1}{\sqrt{2}}\right)$.

Iterating again gives

$$S_{0,0} = \frac{1}{\sqrt{2}}\left(S_{1,0} + T_{1,0}\right) = 1 \qquad S_{0,1} = \frac{1}{\sqrt{2}}\left(S_{1,0} - T_{1,0}\right) = 2$$

$$S_{0,2} = \frac{1}{\sqrt{2}}\left(S_{1,1} + T_{1,1}\right) = 3 \qquad S_{0,3} = \frac{1}{\sqrt{2}}\left(S_{1,1} - T_{1,1}\right) = 4$$

Hence, we get back the original signal $(S_{0,0}, S_{0,1}, S_{0,2}, S_{0,3}) = (1,2,3,4)$.

Figure 3.10 attempts to show visually the decomposition of the signal into the Haar wavelet components. Figure 3.10a contains the original signal, Figure 3.10b plots the coefficients after the first iteration of the decomposition algorithm and Figure 3.10c plots the coefficients after the second iteration of the algorithm. The coefficients contained in Figure 3.10c correspond in turn to a single scaling function at scale index $m=2$, a wavelet at scale index $m=2$ and two wavelets at scale index $m=1$. These are shown, respectively, in Figure 3.10d–f. The transform coefficient plot for this signal is equally simplistic, consisting of four coefficient values partitioned as shown in Figure 3.10g.

We can find the corresponding approximation and details of the signal by taking the inverse transform of the coefficients at each scale. First, the approximation coefficient is used to determine the signal approximation at the largest scale:

$$x_2(t) = S_{2,0}\phi_{2,0}(t) \tag{3.60}$$

where the scaling function $\phi_{2,0}(t)$, for the Haar wavelet, is simply a block pulse of length 4 and amplitude $1/(\sqrt{2})^2 = 1/2$, beginning at $t=0$. Thus, $x_2(t)$ is simply a block pulse of length 4 and magnitude $S_{2,0} \times 1/2 = 2.5$, which is, in fact, the signal mean. A plot of $x_2(t)$ is shown in Figure 3.10d. Similarly, we can obtain the signal detail component at scale index 2 as follows:

$$d_2(t) = T_{2,0}\psi_{2,0}(t) \tag{3.61}$$

This detail component (shown in Figure 3.10e) is simply a single Haar wavelet spanning the data set with coefficient value $T_{2,0}=-2$, and hence magnitude $-2/(\sqrt{2})^2 = -1$. Next, the detail signal component at scale index 1 is found from

$$d_1(t) = \sum_{n=0}^{1} T_{1,n}\psi_{1,n}(t) = T_{1,0}\psi_{1,0}(t) + T_{1,1}\psi_{1,1}(t) \tag{3.62}$$

which is simply two Haar wavelets at scale index 1, side by side, with amplitudes given by $T_{1,0}/(\sqrt{2}) = (-1/\sqrt{2})/(\sqrt{2}) = -0.5$ and $T_{1,1}/(\sqrt{2}) = (-1/\sqrt{2})/(\sqrt{2}) = -0.5$. The detail component $d_1(t)$ is plotted in Figure 3.10f. We already know from Equation 3.45 that the approximation of the signal at scale index 0 can be found by adding together all the detail components plus the signal approximation at scale index M, that is

$$x_0(t) = x_M(t) + \sum_{m=1}^{M} d_m(t) \tag{3.63a}$$

We have already seen an example of this partitioning of the signal into approximation and detail components in Figure 3.5 for a chirp signal. For the Haar case considered here, $M=2$, hence

$$x_0(t) = x_2(t) + d_2(t) + d_1(t) \tag{3.63b}$$

This is simply the addition of the approximation and detail components shown in Figure 3.10d–f. This is shown schematically in Figure 3.10h.

3.4 EVERYTHING DISCRETE

3.4.1 Discrete Experimental Input Signals

We will now consider the case where we have a discrete experimental signal collected, say, using some piece of experimental apparatus, and we want to perform a wavelet decomposition of it using the multiresolution algorithm. In addition, we would also like to represent the approximation and detail signal components discretely at the resolution of the input signal. This discrete signal, which we will denote x_i, is of finite length N, that is, $i = 0, 1,\ldots, N-1$. It has been acquired at discrete time intervals Δt (the sampling interval) to give the discrete time signal $x(t_i)$: $i = 0, 1, 2,\ldots, N-1$. The sampling of the signal provides a finite resolution to the acquired signal. This discretization of the continuous signal is then mapped onto a discrete signal x_i where the sampling interval has been normalized to 1. In doing so, we must remember Δt and add it when required, for example, in real applications when computing signal frequencies.

Common practice is to input the discretely sampled experimental signal, x_i, directly as the approximation coefficients at scale $m = 0$, and begin the multiresolution analysis from there. However, it is not correct to use the sampled time series x_i directly in this way. We should really use the approximation coefficients $S_{0,n}$ obtained from the original continuous signal at scale $m = 0$, as defined by Equation 3.43. In practice, we usually do not know what exactly $x(t)$ is. As $S_{0,n}$ is a weighted average of $x(t)$ in the vicinity of n, then it is usually reasonable to input x_i as $S_{0,n}$ if our signal is slowly varying between samples at this scale. That is, we simply set

$$S_{0,n} = x_n \tag{3.64}$$

where, at scale index $m = 0$, both the coefficient location index n and signal discretization index i have the same range (0 to $N-1$) and are equal to each other. In the rest of this chapter, we will assume that the sampled experimental signal has been input directly as $S_{0,n}$. The literature is full of studies of real data where this has been done and, if anything, it is the rule rather than the exception. Obviously, it does not make it any more correct! In fact, Strang and Nguyen (1996) call the use of the sampled signal directly within the transform a 'wavelet crime' and suggest various ways to preprocess the sampled signal prior to performing the analysis.

In practice, continuous approximations, $x_m(t)$, and details, $d_m(t)$, of the signal are not constructed from the multiresolution analysis, especially when inputting the signal x_i as the detail coefficients at scale index $m = 0$. Instead, either the approximation and detail coefficients, $S_{m,n}$ and $T_{m,n}$, are displayed at their respective scales or, alternatively, they are used

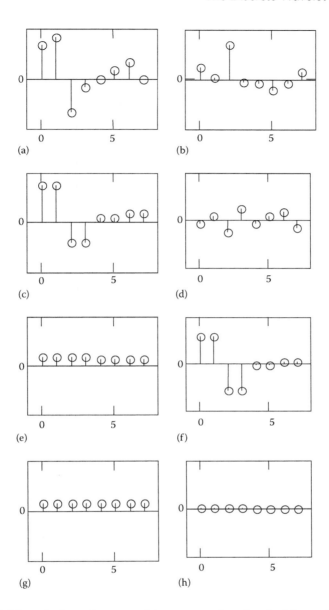

FIGURE 3.11 Multiresolution decomposition as scale thresholding: (a) $x_{0,i}$ $(=x_{1,i}+d_{1,i})$. (b) Wavelet coefficients $(S_{3,0}, T_{3,0}, T_{2,0}, T_{2,1}, T_{1,0}, T_{1,1}, T_{1,2}, T_{1,3})$. (c) $x_{1,i}$ $(=x_{2,i}+d_{2,i})$, $(=x_{0,i}-d_{1,i})$. (d) $d_{1,i}$. (e) $x_{2,i}$ $(=x_{3,i}+d_{3,i})$, $(=x_{1,i}-d_{2,i})$. (f) $d_{2,i}$. (g) $x_{3,i}$ (signal mean), $(=x_{2,i}-d_{3,i})$. (h) $d_{3,i}$.

to construct representations of the signal at the scale of the input signal ($m=0$). The latter is sometimes preferable as the scalings of the displays are visually comparable. As an example, let's consider the Haar wavelet decomposition of a simple discrete signal containing eight components: $x_i = (4, 5, -4, -1, 0, 1, 2, 0)$. The signal is shown in Figure 3.11a. We can see that the first half of the signal has a marked step from 5 to −4 whereas the second half of the signal appears much smoother. The transform vector after a full decomposition is

$$\mathbf{W}^{(3)} = (S_{3,0},\ T_{3,0},\ T_{2,0},\ T_{2,1},\ S_{1,0},\ T_{1,1},\ T_{1,2},\ T_{1,3}) = (2.475, 0.354, 7.000, -0.500, -0.707, -2.121,$$
$$-0.707, 1.414)$$
$$= \left(7/\left(\sqrt{2}\right)^3, 1/\left(\sqrt{2}\right)^3, 14/2, -1/2, -1/\sqrt{2}, -3/\sqrt{2}, -1/\sqrt{2}, 2/\sqrt{2}\right)$$

(You can find these values for yourself using the procedure described in Section 3.3.5.) This vector is shown schematically in Figure 3.11b. Performing the inverse transform on this vector leads us back to the original discrete input signal. Referring back to the schematic of the wavelet decomposition given in Section 3.3.2, we can similarly represent the reconstruction from the transform vector, $\mathbf{W}^{(3)}$, as

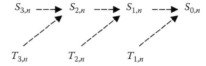

Remember that the original signal x_i was input into the multiresolution algorithm as $S_{0,n}$.

Let's now look at what happens to the reconstructed signal when we remove the smallest scale (i.e. highest frequency) detail coefficients in turn (Figure 3.11c–e). Removing the components, $T_{1,n}$, the modified transform vector becomes (2.475, 0.354, 7.000, −0.500, **0**, **0**, **0**, **0**) where the coefficients set to zero are shown bold. Performing the inverse transform on this vector with the coefficients at the smallest wavelet scale set to zero removes the high-frequency details at this scale and returns a smoother version of the input signal. We will denote this smooth version of the input signal, $x_{1,i}$. This operation is shown schematically as

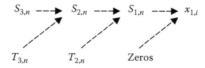

The reconstructed discrete signal becomes $x_{1,i}$ = (4.5, 4.5, −2.5, −2.5, 0.5, 0.5, 1, 1). That is, the signal is smoothed by the averaging of each pair of signal values as shown in Figure 3.11c. Note that we use the nomenclature $x_{1,i}$ for the approximation signal $x_1(t)$ expressed in terms of discrete coefficients at the scale of the original input signal (i.e. $m=0$). Remember from Section 3.2.3 that the approximation coefficients, $S_{m,n}$, provide discrete approximations of the signal at scale index m. By passing the coefficients through the reconstruction filter, we can express the contributions of these discrete approximations at the scale of the original signal (scale index 0). In the rest of this section, we will see how to express all the discrete approximations and details at the scale of the input signal with index $m=0$.

Now, we remove the detail coefficients associated with the next smallest scales from the transform vector to get (2.475, 0.354, **0**, **0**, **0**, **0**, **0**, **0**). Reconstructing the signal using this modified transform vector, we get $x_{2,i} = (1, 1, 1, 1, 0.75, 0.75, 0.75, 0.75)$ shown in Figure 3.11e. Again, we can show this schematically as

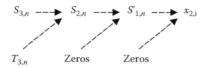

where we have put the dash on $S'_{1,n}$ to indicate that it has different valued components from $S_{1,n}$ in the full reconstruction as it is computed with the $T_{2,n}$ components set to zero. We can see that the nomenclature is beginning to get a bit awkward here as $x_{2,i}$ is, in effect, $S'_{0,n}$: a modified version of the input signal. In $x_{2,i}$, the subscript '2' relates to the largest scale of the original approximation coefficients used at the beginning of the reconstruction, and the index i reminds us that we are back at the input resolution of the original signal ($m = 0$) with N components.

Removing the last detail coefficient, the modified transform vector becomes (2.475, **0, 0, 0, 0, 0, 0, 0**) which, when used to reconstruct the signal, produces $x_{3,i} = (0.875, 0.875, 0.875, 0.875, 0.875, 0.875, 0.875, 0.875)$. This can be shown as

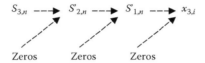

Actually, we have now removed all signal detail and all the components of vector $x_{3,i}$ are of constant value. The signal has been progressively smoothed until all that is left is a row of constant values equal to the signal mean (Figure 3.11g).

We can see by comparing Figure 3.11c,e,g that the high-frequency information has been successively stripped from the discrete input signal (Figure 3.11a) by successively removing the discrete details at scale indices $m_0 = 1$, 2 and 3 in turn. These discrete signal details are shown in Figure 3.11d,f,h. If we want to generate the details of the discrete signal at any scale using the multiresolution algorithm, we perform the inverse transform using only the detail coefficients at that scale (we could also subtract two successive discrete approximations). For example, to compute the detail at the lowest wavelet scale (index $m = 1$), which we will denote $d_{1,i}$, the following multiresolution operation is required:

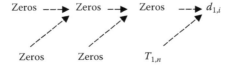

$d_{1,i}$ is in fact the contribution of the wavelets at scale index $m = 1$ (i.e. $\psi_{1,0}(t)$, $\psi_{1,1}(t)$, $\psi_{1,2}(t)$ and $\psi_{1,3}(t)$) to the original signal x_i expressed at discrete points at the scale of the input function $m = 0$ at locations $i = 0, 1, \ldots, N-1$. Similarly, we can calculate the contribution $d_{2,i}$ using the following operation:

```
Zeros  --→  Zeros  --→  S'₁,ₙ  --→  d₂,ᵢ
         ↗          ↗          ↗
      Zeros       T₂,ₙ       Zeros
```

where, obviously, the $S'_{2,n}$ vector has all of its components equal to zero as it comes from $S_{3,n}$ and $T_{3,n}$ with elements set to zero. $d_{2,i}$ is then the contribution of the wavelets at scale index $m = 2$ (i.e. which for this example come from the two wavelets $\psi_{2,0}(t)$ and $\psi_{2,1}(t)$) to the original signal x_i, given at discrete points at the scale of the input function $m = 0$, that is, at locations $i = 0$ to $N-1$. Similarly, beginning with $T_{3,n}$ we can calculate the contribution $d_{3,i}$.

In the previous example, we were effectively taking the information stored in the approximation and detail coefficients $S_{m,n}$ and $T_{m,n}$ and expressing it at scale index $m = 0$. The discrete approximation and detail contributions $x_{m,i}$ and $d_{m,i}$ are related to their continuous counterparts $x_m(t)$ and $d_m(t)$ through the scaling equation at scale $m = 0$ as follows:

$$x_m(t) = \sum_{i=0}^{N-1} x_{m,i}\, \phi_{0,i}(t) \left[= \sum_{n=0}^{2^{M-m}-1} S_{m,n}\, \phi_{m,n}(t) \right] \quad (3.65)$$

and

$$d_m(t) = \sum_{i=0}^{N-1} d_{m,i}\, \phi_{0,i}(t) \left[= \sum_{n=0}^{2^{M-m}-1} T_{m,n}\, \psi_{m,n}(t) \right] \quad (3.66)$$

Notice that the scaling function $\phi_{0,i}(t)$ is not only used to compute the continuous approximation from $x_{m,i}$ but also the detail components from $d_{m,i}$. This is, because the scaling coefficient sequence for the wavelet, b_k, has already been used in the initial stages of the reconstruction sequence to take $T_{m,n}$ to $S'_{m-1,n}$. Thereafter, the contributions can be expressed in terms of scaling functions, that is, from $S'_{m-1,n}$ to $S'_{m-2,n}$ and so on. You can verify Equations 3.65 and 3.66 using the reconstruction algorithm of Equation 3.42 (noting that one of its terms will be zero) and the scaling function relationship given by Equation 3.27b.

Figure 3.11c shows $x_{1,i}$, that is, where only the detail signals $d_{1,i}$ corresponding to the smallest scale wavelets ($m = 1$) have been removed (Figure 3.11d). Figure 3.11e shows $x_{2,i}$, where both detail signals $d_{1,i}$ and $d_{2,i}$ corresponding to the smallest and next smallest scale wavelets have been removed from the original discrete input signal; $d_{2,i}$ is shown in Figure 3.11f. Figure 3.11g shows $x_{3,i}$, where the detail signals $d_{m,i}$ corresponding to all of the wavelets scales ($m = 1, 2$ and 3) have been stripped from the original signal leaving only the signal mean component. This can be written as

$$x_{3,i} = x_{0,i} - \sum_{m=1}^{3} d_{m,i} \qquad (3.67a)$$

or, in general, over M scales as

$$x_{M,i} = x_{0,i} - \sum_{m=1}^{M} d_{m,i} \qquad (3.67b)$$

where $x_{0,i}$ is simply the input signal x_i. This equation stems directly from its continuous counterpart of Equation 3.45 using Equations 3.65 and 3.66. We have now seen how we can express everything discretely. We can use a suitable discrete input signal within the multiresolution algorithm and we can express the approximation and details of this signal at discrete points at the input resolution. And, as we will see in Section 3.5.1, even the wavelet and scaling functions are expressed discretely within the multiresolution algorithm, built up by the repeated iteration of the scaling coefficient vectors.

3.4.2 Smoothing, Thresholding and Denoising

Let us look again at the discrete input signal used in the previous section (Figure 3.12a). Once we have performed the full decomposition, we are free to alter any of the coefficients in the transform vector before performing the inverse. We can set groups of components to zero, as we did in the last section (Figure 3.11), or set selected individual components to zero. We can reduce the magnitudes of some components rather than set them to zero. In fact, we can manipulate the components in a variety of ways depending on what we want to achieve. Notice, however, that the transform vector contains a range of values: some large and some small. Let's see what happens if we throw away the smallest valued coefficients in turn and perform the inverse transforms. We start with the smallest valued coefficient 0.354. Setting it to zero, the modified transform vector becomes (2.475, **0**, 7.000, −0.500, −0.707, −2.121, −0.707, 1.414). Performing the inverse on this vector gives the reconstructed discrete signal (3.875, 4.875, −4.125, −1.125, 0.125, 1.125, 2.125, 0.125), as shown in Figure 3.12b. There is no discernible difference between the original discrete signal and the reconstruction. We now remove the next smallest valued coefficient (smallest in an absolute sense), the −0.5 coefficient. The transform vector becomes (2.475, **0**, 7.000, **0**, −0.707, −2.121, −0.707, 1.414) and the reconstructed signal is (3.875, 4.875, −4.125, −1.125, 0.375, 1.375, 1.875, −0.125). Again, we have to look carefully at Figure 3.12c to discern any difference between the original discrete signal and the reconstruction. Next, we remove the two −0.707 components to get (2.475, **0**, 7.000, **0**, **0**, −2.121, **0**, 1.414). Reconstructing, we get (4.375, 4.375, −4.125, −1.125, 0.875, 0.875, 1.875, −0.125). This reconstruction is shown in Figure 3.12d where we can now notice obvious smoothing between both the first and second signal point and the fifth and sixth. Next, setting the last transform vector component to zero, we reconstruct to get (4.375, 4.375, −4.125, −1.125, 0.875, 0.875, 0.875,

128 ■ The Illustrated Wavelet Transform Handbook

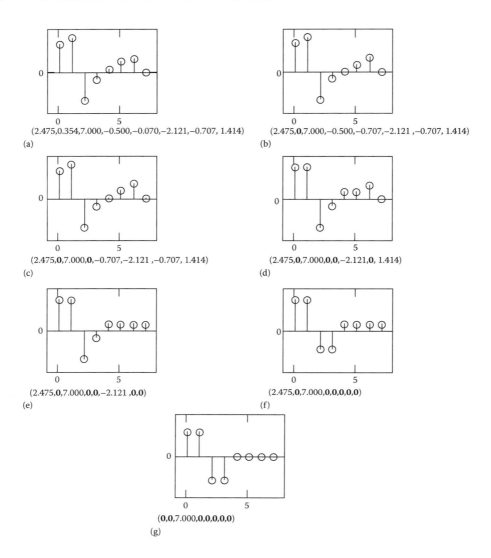

FIGURE 3.12 Signal reconstruction using thresholded wavelet coefficients. The coefficient vectors used in the reconstructions are given below each reconstructed signal. Note that (a) is the original signal as it uses all of the original eight wavelet coefficients.

0.875). This is shown in Figure 3.12e. Removing the −2.121 component leaves the coefficient vector as (2.475, **0**, 7.000, **0, 0, 0, 0, 0**) and the reconstruction becomes (4.375, 4.375, −2.625, −2.625, 0.875, 0.875, 0.875, 0.875) as shown in Figure 3.12f. Removing the 2.475 component (the signal mean component) leaves the coefficient vector with only a single component: (**0, 0**, 7.000, **0, 0, 0, 0, 0**). Reconstructing using this component leaves a single (wavelet-shaped) fluctuation contained within the first half of the signal (Figure 3.12g).

Can you see what has happened? The least significant components have been smoothed out first, leaving the more significant fluctuating parts of the signal intact. What we have actually done is to threshold the wavelet coefficients at increasing magnitudes. First, all the components whose magnitude was equal to or below 0.354 were removed; that is, 0.354

was the threshold. Then 0.5 was set as the threshold, then 0.707 and so on. By doing this, we removed the least significant influences on the signal first. Hence, the shape of the reconstructed signal resembles the original, even with a large number of coefficients set to zero. Compare this *magnitude thresholding* of the coefficients with the *scale-dependent smoothing* of Figure 3.11 which removed the components at each scale in turn, beginning with the smallest scales. Note that in practice we would normally deal with the signal mean coefficient separately, either retaining it regardless of magnitude or removing it at the beginning of the thresholding process.

We can define the scale-dependent smoothing of the wavelet coefficients as

$$T_{m,n}^{scale} = \begin{cases} 0 & m \geq m^* \\ T_{m,n} & m < m^* \end{cases} \quad (3.68)$$

where m^* is the index of the threshold scale. The transform vector in Figure 3.13a, shows a schematic diagram of scale-dependent smoothing for sequentially indexed coefficients W_i. (Assuming a full decomposition, we have dropped the M superscript from W_i^M.) In this case, the thresholding criterion is defined as

$$W_i^{scale} = \begin{cases} 0 & i \geq 2^{M-m^*} \\ W_i & i < 2^{M-m^*} \end{cases} \quad (3.69)$$

where the range of the sequential index i is from 0 to $N-1$ and N is the length of the original signal. Hence, $i = 2^{M-m^*}$ is the first location index within the transform vector where the coefficients are set to zero. (Remember that the smallest scale coefficients are placed at the right-hand end of the transform vector and hence have the highest index values.) Note that in practice the coefficients at the very largest scales are sometimes also set to zero to remove drift effects from the signal. The reconstructed signal using the scale thresholded coefficient vector with scale threshold m^* is simply the smooth approximation of the signal at scale m^* expressed at the scale of the input signal, $m=0$, that is, $x_{m^*,i}$. (See previous section.)

Magnitude thresholding is normally carried out to remove noise from a signal; to partition signals into two or more (and not necessarily noisy) components or simply to smooth the data. It involves the reduction or complete removal of selected wavelet coefficients in order to separate out the behaviour of interest from within the signal. There are many methods for selecting and modifying the coefficients. The two most popular are *hard* and *soft thresholding*. Unlike scale-dependent smoothing, which removes all small-scale coefficients below the scale index m^* regardless of amplitude, hard and soft thresholding remove, or reduce, the smallest amplitude coefficients regardless of scale. This is shown schematically in Figure 3.13b. To hard threshold the coefficients, a threshold, λ, is set which is related to some mean value of the wavelet coefficients at each scale, for example, standard

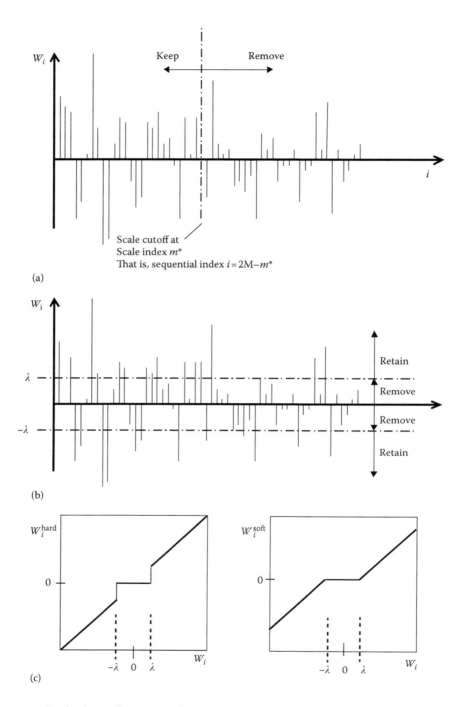

FIGURE 3.13 **Scale-dependent smoothing and coefficient thresholding:** (a) Scale-dependent smoothing. (b) Amplitude thresholding. (c) The relationship between the original coefficients and hard (left) and soft (right) thresholded coefficients.

deviation, mean absolute deviation and so on. Those coefficients above the threshold are deemed to correspond to the coherent part of the signal, and those below the threshold are deemed to correspond to the random or noisy part of the signal. Hard thresholding is of the form:

$$W_i^{\text{hard}} = \begin{cases} 0 & |W_i| < \lambda \\ W_i & |W_i| \geq \lambda \end{cases} \tag{3.70}$$

Hard thresholding makes a decision simply to keep or remove the coefficients. (Hard thresholding was performed in the example of Figure 3.12.) Soft thresholding recognizes that the coefficients contain both signal and noise and attempts to isolate the signal by removing the noisy part from all coefficients. Soft thresholding is of the form:

$$W_i^{\text{soft}} = \begin{cases} 0 & |W_i| < \lambda \\ \text{sign}(W_i)(|W_i| - \lambda) & |W_i| \geq \lambda \end{cases} \tag{3.71}$$

where all coefficients below the threshold, λ, are set to zero and all the coefficients whose magnitude is greater than λ are shrunk towards zero by an amount λ. Figure 3.13c shows a schematic diagram of the relationship between the original and thresholded coefficients for both hard and soft thresholding. The retained coefficients shown in the figure are kept as they are when hard thresholding is employed and their magnitude is reduced by λ when soft thresholding is employed.

Figure 3.14 shows examples of both hard and soft thresholding of a test signal composed of two sinusoids plus Gaussian white noise (Figure 3.14a,b). The wavelet coefficients obtained from a Daubechies D10 decomposition are shown in Figure 3.14c. These have been hard thresholded for thresholds set at $\lambda = 2$, 4, 6 and 8, respectively. The reconstructions corresponding to each of the thresholds are shown in Figure 3.14d–g. We can see that for low thresholds some of the high-frequency noise is retained, whereas for high thresholds the signal is excessively smoothed. From visual inspection, an optimum threshold would appear to lie somewhere in the region of $\lambda = 4$. One commonly used measure of the optimum reconstruction is the mean square error between the reconstructed signal and the original signal and, in fact, it is found to be minimum near to this value of the threshold. The corresponding soft thresholded reconstructions are shown in Figure 3.14h,i for $\lambda = 2$ and 4, respectively.

Often, we do not know the exact form of either the underlying signal or the corrupting noise. The choice of threshold is therefore non-trivial and we can apply a number of signal and/or noise-based criteria in producing a value pertinent to the problem under consideration. The threshold can, for example, be a constant value applied to the coefficients across all scales, some of the scales, or its value can vary according to scale. One of the most popular and simplest thresholds in use is the *universal threshold* defined as

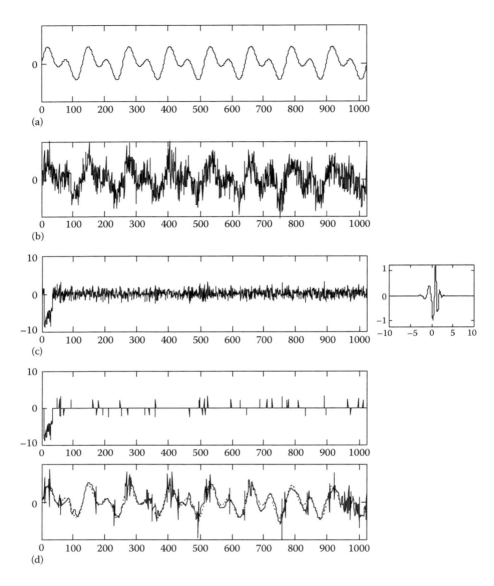

FIGURE 3.14 Hard and soft thresholding: (a) Original time series composed of two sinusoids, both with unit amplitude, one twice the periodicity of the other. (b) Time series in (a) with added Gaussian noise (zero mean and unit standard deviation). (c) Wavelet coefficients in sequential format derived using the Daubechies D10 shown. (d) Hard thresholded coefficients, $\lambda = 2$ (top), and corresponding reconstructed time series (bottom). (The original time series of figure (a) is shown dashed.)

$$\lambda_U = \left(2\ln N\right)^{1/2} \sigma \quad (3.72)$$

where $(2 \ln N)^{1/2}$ is the expected maximum value of a white noise sequence of length N and unit standard deviation and σ is the standard deviation of the noise in the signal under consideration. Thus, if the underlying coherent part of the signal is zero (i.e. the signal contains only noise), using λ_U as the threshold gives a high probability of setting all the

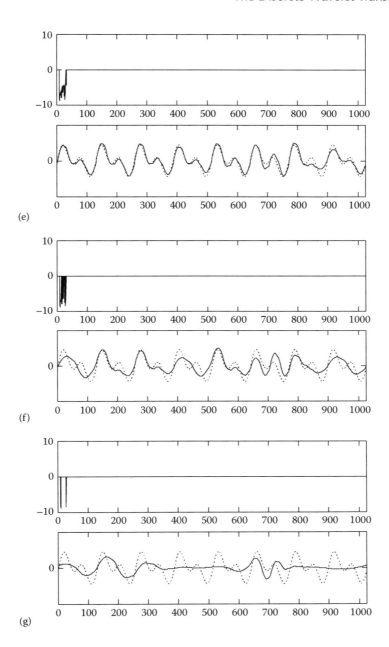

FIGURE 3.14 (CONTINUED) Hard thresholded coefficients: (e) $\lambda = 4$ (top), and corresponding reconstructed time series (bottom). (The original time series of figure (a) is shown dashed.) **(f)** Hard thresholded coefficients, $\lambda = 6$ (top), and corresponding reconstructed time series (bottom). (The original time series of figure (a) is shown dashed.) **(g)** Hard thresholded coefficients, $\lambda = 8$ (top), and corresponding reconstructed time series (bottom). (The original time series of figure (a) is shown dashed.)

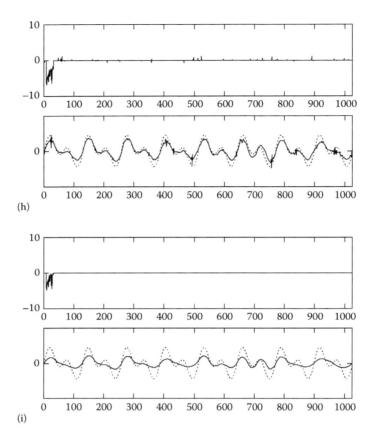

(h)

(i)

FIGURE 3.14 (CONTINUED) Hard thresholded coefficients: **(h)** Soft thresholded coefficients, $\lambda = 2$ (top), and corresponding reconstructed time series (bottom). (The original time series of figure (a) is shown dashed.) **(i)** Soft thresholded coefficients, $\lambda = 4$ (top), and corresponding reconstructed time series (bottom). (The original time series of figure (a) is shown dashed.)

coefficients to zero. For large samples, the universal threshold will remove, with high probability, all the noise in the reconstruction, but part of the underlying function might also be lost, hence the universal threshold tends to over-smooth in practice. (This method is often known as the *VisuShrink* method as the resulting smooth signal estimate is visually more appealing.) In addition, in practice it is usual to leave the coefficients at the largest scales untouched even if they do not pass the universal threshold. The example of the sinusoids plus noise shown in Figure 3.14 contained Gaussian noise with $\sigma = 1$ and $N = 1024$, the universal threshold for this data is then $\lambda_U = 3.723$. Within a wide range of practical data sizes (2^6–2^{19}), only in about one-tenth of the realizations will any pure noise variable exceed the threshold. In this sense, the universal thresholding method gives a 'noise-free' reconstruction.

Universal soft thresholding usually produces a signal reconstruction with less energy than that for hard thresholding using the same threshold value, as the retained coefficients are minimized towards zero. This can be seen by comparing Figure 3.14e with Figure 3.14j for $\lambda = 4$. Hence, it is often the case in practice that the universal threshold used for hard thresholding is divided by about 2 when employed as a soft threshold. Another problem encountered when implementing the universal threshold in practice is that we do not know the value of σ for our signal. In this case, a robust estimate $\hat{\sigma}$ is used, typically set to the median of absolute deviation (MAD) of the wavelet coefficients at the smallest scale divided by 0.6745 to calibrate with the standard deviation of a Gaussian distribution. Thus, the universal threshold becomes

$$\lambda_U = \frac{(2\ln N)^{1/2} \, MAD}{0.6745} = (2\ln N)^{1/2} \, \hat{\sigma} \qquad (3.73)$$

Many other thresholding methods have been proposed including the *minimax* method, 'Stein's unbiased risk estimate' (*SURE*) method, *hybrid* method, *cross-validation* methods, *Lorentz*' method and various *Bayesian approaches*. Some of these produce a global threshold of constant value across scales and others provide a scale-dependent threshold. We do not consider these in detail herein. More information is provided at the end of the chapter.

3.5 DAUBECHIES WAVELETS

As we saw with the Haar wavelet transform in Section 3.3.5, the coefficients are ordered in two distinct sequences: one acts as a smoothing filter for the data, the other extracts the signal detail at each scale. The Haar wavelet is extremely simple in that it has only two scaling coefficients and both are equal to unity. In this section, we will look at a family of discrete wavelets of which the Haar wavelet is the simplest member – *Daubechies wavelets*. The scaling functions associated with these wavelets satisfy the conditions given in Section 3.2.4, as all orthogonal wavelets do. In addition to satisfying these criteria, Daubechies required that her wavelets had compact support (i.e. a finite number, N_k, of scaling coefficients) and were smooth to some degree. The smoothness of the wavelet is associated with a moment condition, which can be expressed in terms of the scaling coefficients as

$$\sum_{k=0}^{N_k-1} (-1)^k c_k k^m = 0 \qquad (3.74)$$

for integers $m = 0, 1, 2, \ldots, N_k/2 - 1$. These wavelets have $N_k/2$ vanishing moments which means that they can suppress parts of the signal which are polynomial up to degree $N_k/2 - 1$. Or, to put it another way, Daubechies wavelets are very good at representing polynomial behaviour within the signal. Some examples of Daubechies wavelets, their scaling functions and associated energy spectra are given in Figure 3.15. The support lengths of Daubechies wavelets are $N_k - 1$, that is, the D2 (Haar) wavelet, as we already know, has a support length of 1, the D4 wavelet has a support length of 3, the D6 a support length of 5 and so on. We can see from Figure 3.15 that the scaling function lets through the lower frequencies and hence acts as a lowpass filter and the associated wavelet lets through the

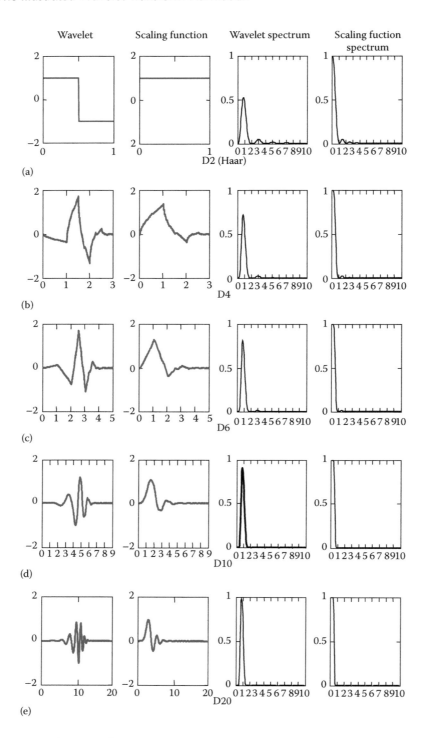

FIGURE 3.15 A selection of Daubechies wavelets and scaling functions with their energy spectra. All wavelets and scaling functions are shown only over their respective support – outside their support they have zero value. Note also that only the positive part of the spectrum is given; an identical mirror image is present at negative frequencies.

higher frequencies and hence acts as a highpass filter. In addition, we can see that the spectra are oscillatory in nature, with bumps decreasing in amplitude from the lower to the higher frequencies. The magnitude of the secondary bumps in the spectra reduce as the number of scaling coefficients, and hence the number of vanishing moments of the wavelet, increase.

In this chapter, we have already looked in detail at a Daubechies wavelet with only two scaling coefficients (the D2 or Haar wavelet). Let us now look at a Daubechies wavelet which has four scaling coefficients, the D4. The 'D' represents this particular family of Daubechies wavelets and the '4' represents the number of non-zero scaling coefficients, N_k. (Note that Daubechies wavelets are also often defined by the number of zero moments they have, which equals $Nk/2$, in which case the sequence runs D1, D2, D3, D4,..., and so on.)

From Equation 3.20, we know that the scaling equation for a four coefficient wavelet is

$$\phi(t) = c_0 \phi(2t) + c_1 \phi(2t-1) + c_2 \phi(2t-2) + c_3 \phi(2t-3) \tag{3.75}$$

and, from Equation 3.23, that the corresponding wavelet function is

$$\psi(t) = c_3 \phi(2t) - c_2 \phi(2t-1) + c_1 \phi(2t-2) - c_0 \phi(2t-3) \tag{3.76}$$

To find the values of the scaling coefficients for the D4 wavelet, we use Equations 3.21, 3.22 and 3.74. From Equation 3.21 we get

$$c_0 + c_1 + c_2 + c_3 = 2 \tag{3.77}$$

from Equation 3.22 we get

$$c_0^2 + c_1^2 + c_2^2 + c_3^2 = 2 \tag{3.78}$$

and from Equation 3.74 with $m=0$ we get

$$c_0 - c_1 + c_2 - c_3 = 0 \tag{3.79}$$

and again using Equation 3.74, this time setting $m=1$, we get

$$-1c_1 + 2c_2 - 3c_3 = 0 \tag{3.80}$$

Four scaling coefficients which satisfy Equations 3.77 through 3.80 are

$$c_0 = \frac{1+\sqrt{3}}{4} \quad c_1 = \frac{3+\sqrt{3}}{4} \quad c_2 = \frac{3-\sqrt{3}}{4} \quad c_3 = \frac{1-\sqrt{3}}{4}$$

and so do

$$c_0 = \frac{1-\sqrt{3}}{4} \quad c_1 = \frac{3-\sqrt{3}}{4} \quad c_2 = \frac{3+\sqrt{3}}{4} \quad c_3 = \frac{1+\sqrt{3}}{4}$$

One set leads to $\phi(t)$ and the other $\phi(-t)$. We will adopt the first set which are $c_0 = 0.6830127$; $c_1 = 1.1830127$; $c_2 = 0.3169873$; and $c_3 = -0.1830127$, respectively. The scaling coefficients for the Daubechies wavelet system for larger numbers of coefficients are found by numerical computation. The coefficients for Daubechies wavelets up to D20 are given in Table 3.1.

We can compute the scaling function from the D4 coefficients using Equation 3.75. To do this, we must rewrite it as

TABLE 3.1 Daubechies Wavelet Coefficients D2 to D20

D2	−0.34265671	0.01774979	0.04345268
1	−0.04560113	6.07514995e − 4	−0.09564726
1	0.10970265	−2.54790472e − 3	3.54892813e − 4
	−0.00882680	5.00226853e − 4	0.03162417
D4	−0.01779187		−6.67962023e − 3
0.6830127	4.71742793e − 3	**D16**	−6.05496058e − 3
1.1830127		0.07695562	2.61296728e − 3
0.3169873	**D12**	0.44246725	3.25814671e − 4
−0.1830127	0.15774243	0.95548615	−3.56329759e − 4
	0.69950381	0.82781653	−5.5645514e − 5
D6	1.06226376	−0.02238574	
0.47046721	0.44583132	−0.40165863	**D20**
1.14111692	−0.31998660	6.68194092e − 4	0.03771716
0.650365	−0.18351806	0.18207636	0.26612218
−0.19093442	0.13788809	−0.02456390	0.74557507
−0.12083221	0.03892321	−0.06235021	0.97362811
0.0498175	−0.04466375	0.01977216	0.39763774
D8	7.83251152e − 4	0.01236884	−0.35333620
0.32580343	6.75606236e − 3	−6.88771926e − 3	−0.27710988
1.01094572	−1.52353381e − 3	−5.54004549e − 4	0.18012745
0.89220014		9.55229711e − 4	0.13160299
−0.03957503	**D14**	−1.66137261e − 4	−0.10096657
−0.26450717	0.11009943		−0.04165925
0.0436163	0.56079128	**D18**	0.04696981
0.0465036	1.03114849	0.05385035	5.10043697e − 3
−0.01498699	0.66437248	0.34483430	−0.01517900
	−0.20351382	0.85534906	1.97332536e − 3
D10	−0.31683501	0.92954571	2.81768659e − 3
0.22641898	0.10084647	0.18836955	−9.69947840e − 4
0.85394354	0.11400345	−0.41475176	−1.64709006e − 4
1.02432694	−0.05378245	−0.13695355	1.32354367e − 4
0.19576696	−0.02343994	0.21006834	−1.87584156e − 5

$$\phi_j(t) = c_0\phi_{j-1}(2t) + c_1\phi_{j-1}(2t-1) + c_2\phi_{j-1}(2t-2) + c_3\phi_{j-1}(2t-3) \qquad (3.81)$$

where subscript j is the iteration number. Then, choosing an arbitrary initial shape for the scaling function $\phi_0(t)$, we find $\phi_1(t)$, then using $\phi_1(t)$, we find $\phi_2(t)$ and so on, iterating until $\phi_j(t) = \phi_{j-1}(t)$ (or at least until $\phi_j(t)$ is close enough to $\phi_{j-1}(t)$ for our purposes). Once we have an approximation for $\phi(t)$, we could define the wavelet directly using Equation 3.76. However, to perform the D4 wavelet transform of a discrete signal in practice we are not required to compute the wavelet or scaling functions directly, rather we simply employ the scaling coefficients within a multiresolution algorithm (Equations 3.36 and 3.37) in the same manner as we did for the Haar wavelet. In this case, the approximation coefficients are computed using

$$\begin{aligned} S_{m+1,n} &= \frac{1}{\sqrt{2}} \sum_{k=0}^{N_k-1} c_k\, S_{m,2n+k} \\ &= \frac{1}{\sqrt{2}} \left[c_0 S_{m,2n} + c_1 S_{m,2n+1} + c_2 S_{m,2n+2} + c_3 S_{m,2n+3} \right] \\ &= 0.483 S_{m,2n} + 0.837 S_{m,2n+1} + 0.224 S_{m,2n+2} - 0.129 S_{m,2n+3} \end{aligned} \qquad (3.82a)$$

We therefore take the product of a four-digit sequence of the signal by the scaling coefficient vector $(1/\sqrt{2})c_k$ (the lowpass filter) to generate the approximation component at the first scale. The mechanics of this is shown schematically in Figure 3.16. The four-digit sequence, $(1/\sqrt{2})c_k$, slides across the signal at scale index m in jumps of two, generating, at each jump, an approximation component at the next scale $(m+1)$. To generate the corresponding coefficients for the wavelet, we use the reconfigured scaling coefficient sequence $(1/\sqrt{2})b_k$ (the highpass filter), where $b_k = (-1)^k c_{N_k-1-k}$, as follows:

$$\begin{aligned} T_{m+1,n} &= \frac{1}{\sqrt{2}} \sum_{k=0}^{N_k-1} b_k\, S_{m,2n+k} \\ &= \frac{1}{\sqrt{2}} \left[c_3 S_{m,2n} - c_2 S_{m,2n+1} + c_1 S_{m,2n+2} - c_0 S_{m,2n+3} \right] \\ &= -0.129 S_{m,2n} - 0.224 S_{m,2n+1} + 0.837 S_{m,2n+2} - 0.483 S_{m,2n+3} \end{aligned} \qquad (3.82b)$$

As an example, we will now consider the D4 wavelet decomposition of the signal $x_i = (1, 0, 0, 0, 0, 0, 0, 0)$ which we input into the multiresolution algorithm as $S_{0,n}$. It has $N = 8$ components, so we iterate the transform decomposition algorithm three times. After the first iteration, the transform vector becomes (0.483, 0.000, 0.000, 0.224, −0.129, 0.000, 0.000, 0.837) containing four approximation components followed by four detailed coefficients,

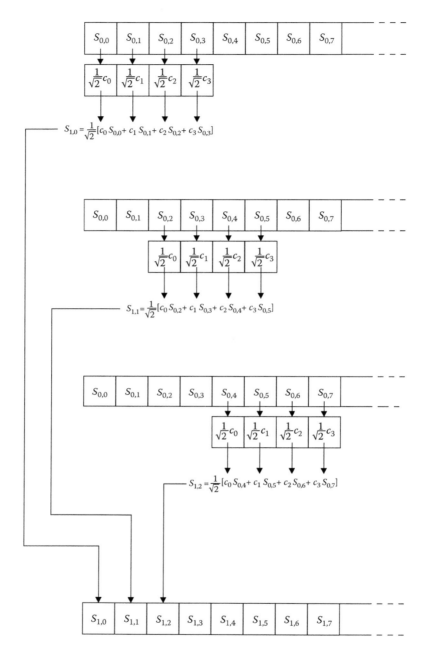

FIGURE 3.16 Filtering of the signal: Decomposition. The original signal at scale index $m = 0$ (i.e. $S_{0,n}$) is filtered to produce the approximation coefficients $S_{1,n}$. This is done by sliding the lowpass filter along the signal one step at a time. Subsampling removes every second value. The diagram shows only the retained values, that is, effectively the filter coefficient vector jumps two steps at each iteration. Next, the sequence $S_{1,n}$ is filtered in the same way to get $S_{2,n}$ and so on. The detail coefficients are found by using the same method but employing highpass wavelet filter coefficients.

that is, ($S_{1,0}$, $S_{1,1}$, $S_{1,2}$, $S_{1,3}$, $T_{1,0}$, $T_{1,1}$, $T_{1,2}$, $T_{1,3}$). Remember that we employ wraparound to deal with the edges of the signal. When either of the filter pairs is placed at $2n = 6$, the $2n + 2$ and $2n + 3$ locations are not contained within the signal vector ($i = 0, 1, \ldots, N-1$). Thus, the last two filter coefficients of 0.224 and −0.129 (approximation) or 0.837 and −0.483 (detail) are placed back at the start of the signal vector, the first of each coinciding with the unit value at the beginning of the signal, hence producing, respectively, the values 0.224 or 0.837. These values are placed at location $S_{1,3}$ and $T_{1,3}$, respectively. The next iteration is performed on the remaining four approximation components $S_{1,0}$, $S_{1,1}$, $S_{1,2}$, $S_{1,3}$ producing two approximation components followed by two detailed wavelet components. These are added to the four detailed coefficients obtained at the previous iteration to give the vector (0.204, 0.296, −0.171, 0.354, −0.129, 0.000, 0.000, 0.837). The third and final iteration produces (0.354, −0.065, −0.171, 0.354, −0.129, 0.000, 0.000, 0.837). This vector has the form ($S_{3,0}$, $T_{3,0}$, $T_{2,0}$, $T_{2,1}$, $T_{1,0}$, $T_{1,1}$, $T_{1,2}$, $T_{1,3}$), where the signal mean can be found from the first coefficient, that is, $S_{3,0}/2^{3/2} = 0.354/2^{3/2} = 0.125$.

We know that a Daubechies D4 wavelet has two vanishing moments; thus, it can suppress both constant signals and linear signals. This means that for an infinitely long constant or linear signal, all the coefficients are zero. However, end effects occur if the signal is of finite length. We will use the linear ramp signal (0, 1, 2,…, 31), shown in Figure 3.17a, as an example. The first iteration produces the transform vector:

$$\mathbf{W}^{(1)} = (S_{1,0}, S_{1,1}, S_{1,2}, \ldots S_{1,15}, T_{1,0}, T_{1,1}, T_{1,2}, \ldots T_{1,15})$$

$$= (0.897, 3.725, 6.553, 9.382, 12.21, 15.039, 17.867, 20.696, 23.524, 26.352, 29.181,$$

$$32.009, 34.838, 37.666, 40.495, 40.291, 0, 0, 0, 0, 0, 0, 0, 0, 0, 0, 0, 0, 0, 0, 0, -11.314)$$

where all the detail coefficients at the lowest scale, $T_{1,n}$, are zero except for the end coefficient $T_{1,15}$. This non-zero coefficient is caused by the signal wraparound where the wavelet filter vector $(1/\sqrt{2})b_k$ placed at the end of the signal encounters the sequence 30, 31, 0, 1. We can see that the computation for the corresponding detail component is

$$T_{1,15} = -0.129 \times 30 - 0.224 \times 31 + 0.837 \times 0 - 0.483 \times 1 = -11.314.$$

A histogram of the coefficients is shown in Figure 3.17b. Also notice the edge effect in the approximation coefficients where they all increase linearly in value except for the last one due to the wraparound with the scaling filter. The second iteration produces the transform vector:

$$\mathbf{W}^{(2)} = (3.804, 11.804, 19.804, 27.804, 35.804, 43.80, 52.196, 52.981,$$

$$0, 0, 0, 0, 0, 0, 1.464, -15.321, 0, 0, 0, 0, 0, 0, 0, 0, 0, 0, 0, 0, 0, 0, 0, 11.314)$$

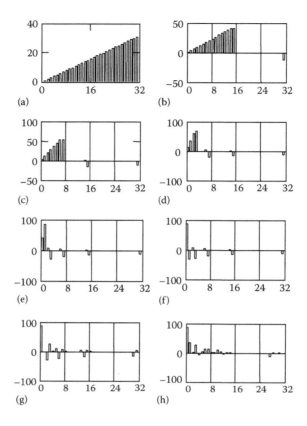

FIGURE 3.17 The multiresolution decomposition of a ramp signal using Daubechies wavelets: (a) Original signal. (b) Decomposition 1 (D4 wavelet). (c) Decomposition 2 (D4 wavelet). (d) Decomposition 3 (D4 wavelet). (e) Decomposition 4 (D4 wavelet). (f) Decomposition 5, full decomposition (D4 wavelet). (g) Full decomposition using D6 wavelets. (h) Full decomposition using D12 wavelets. (Note the change in scale of the vertical axes.)

where we see that the next set of detail coefficients ($T_{2,0}$ to $T_{2,7}$) are zero except those at the end (Figure 3.17c). There are two non-zero coefficients now due to the edge effect in the smooth coefficients at the previous scale. If we repeat the decomposition to the largest scale, we get the fully decomposed signal:

$$\mathbf{W}^{(5)} = (87.681, -31.460, 6.405, -29.530, 0, 0, 3.624, -21.149, 0, 0, 0, 0, 0, 0,$$
$$1.464, -15.321, 0, 0, 0, 0, 0, 0, 0, 0, 0, 0, 0, 0, 0, 0, 11.314)$$

As we get 'edge' coefficients at all scales, we get no zero coefficients at scales $m = 4$ and $m = 5$ where there are, respectively, only two ($T_{4,0}$, $T_{4,1}$) and one ($T_{5,0}$) coefficient.

The full D2 (Haar) decomposition of the ramp signal is

$$\mathbf{W}^{(5)} = (87.7, -45.3, -16.0, -16.0, -5.7, -5.7, -5.7, -5.7, -2.0, -2.0, -2.0, -2.0, -2.0,$$
$$-2.0, -2.0, -2.0, -0.7, -0.7, -0.7, -0.7, -0.7, -0.7, -0.7, -0.7, -0.7, -0.7,$$
$$-0.7, -0.7, -0.7, -0.7, -0.7, -0.7)$$

Notice that there are no zero coefficients as we would expect since the Haar wavelet has only one vanishing moment and therefore does not suppress linear signals.

On the other hand, all Daubechies wavelets with more than four scaling coefficients have more than two vanishing moments and all can therefore suppress linear signals. For example, a D6 wavelet can suppress mean, linear and quadratic signals. A D6 decomposition of the ramp signal, which employs the scaling coefficients $(c_0, c_1, c_2, c_3, c_4, c_5) = (0.470, 1.141, 0.650, -0.191, -0.121, 0.0498)$, is

$$\mathbf{W}^{(5)} = \big(87.681, -2.259, -29.201, 25.344, 0.149, 9.909, -22.224, 6.2484,$$
$$0, 0, 0, 0, 0, 4.22, -17.636, 4.766, 0, 0, 0, 0, 0, 0, 0, 0, 0, 0, 0, 0, 0, -15.175, 3.861\big)$$

Notice that this time the two end coefficients are non-zero. This is because the filter is longer. It now has six components and hence two end locations of the filter on the discrete signal overlap and wraparound to the beginning of the signal. A D8 decomposition would generate three end coefficients, a D10 would generate 4 and so on. The full decompositions of the ramp signal using both the D6 and D12 wavelet are shown in Figure 3.17f,g for comparison. The more scaling coefficients the wavelet has, the higher the number of its vanishing moments and hence the higher the degree of polynomial it can suppress. However, the more scaling coefficients that a wavelet has, the larger its support length and hence the less compact it becomes. This makes it less localized in the time domain and hence less able to isolate singularities in the signal (including edge effects). This is the trade-off which must be considered when selecting the best wavelet for the data analysis task.

3.5.1 Filtering

Let us revisit the filtering process described in the last section and shown in Figure 3.16. In signal processing, the approximation coefficients at resolution m, $S_{m,n}$, are convolved with the lowpass filter. This done by moving the filter along the signal one step at a time. The approximation coefficients are then *subsampled* (or *downsampled*) where every second value is chosen to give the approximation coefficient vector at scale $m+1$. The approximation coefficients at resolution m, $S_{m,n}$, are also convolved with the highpass filter and subsampled in the same way to give the detail signal coefficients at scale $m+1$. The detail components $T_{m+1,n}$ are kept and the approximation components $S_{m+1,n}$ are again passed through the lowpass and highpass filters to give components $T_{m+2,n}$ and $S_{m+2,n}$. This process is repeated over all scales to give the full decomposition of the signal. Each step in the decomposition filtering process is shown schematically in Figure 3.18a. The sequences $(1/\sqrt{2})c_k$ and $(1/\sqrt{2})b_k$ are contained, respectively, within the low and highpass filter

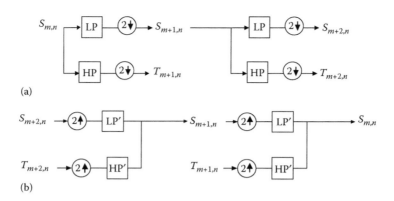

FIGURE 3.18 Schematic of the signal filtering: (a) Schematic diagram of the filtering of the approximation coefficients to produce the approximation and detail coefficients at successive scales. The subsample symbol ②↓ indicates to take every second value of the filtered signal. HP and LP are, respectively, the highpass and lowpass filters. In practice, the filter is moved along one location at a time on the signal, hence filtering plus subsampling corresponds to skipping every second value as shown in Figure 3.16. (b) Schematic diagram of the filtering of the approximation and detail coefficients to produce the approximation coefficients at successively smaller scales. The upsample symbol ②↑ indicates to add a zero between every second value of the input vector. The coefficients in the HP′ and LP′ filters are in reverse order to their counterparts used in the decomposition shown in (a). Figure 3.18a,b represents a sub-band coding scheme.

vectors used within the wavelet decomposition algorithm. The filter coefficients are sometimes referred to as 'taps'.

For signal reconstruction (Equation 3.42), the filtering process is simply reversed, whereby the components at the larger scales are fed back through the filters. This is shown in Figure 3.18b. The approximation and detail components at scale index $m+1$ are first *upsampled* (zeros are inserted between their values) and then passed through the low and highpass filters, respectively. This time, however, the filter coefficients are reversed in order and are shifted back along the signal. An example of the reconstruction filtering is shown in Figure 3.19 where the component $S_{m,5}$ is found from the sequences $S_{m+1,n}$ and $T_{m+1,n}$. Note that the leading (right-hand) filter coefficient defines the location of the $S_{m,n}$ coefficient. Hence, the computation of the $S_{m,0}$, $S_{m,1}$ and $S_{m,2}$ coefficients involve components of the filter vector lying off the left-hand end of the upsampled signal at scale $m+1$. These are simply wrapped back around to the right-hand end of the signal.

We can produce a discrete approximation of the scaling function at successive scales if we set all the values of the transform vector to zero except the first one and pass this vector repeatedly back up through the lowpass filter. This is illustrated in the sequence of plots contained in Figure 3.20. Figure 3.20a shows the initial transform vector, 64 components long, with only the first component set to unity, the rest set to zero, that is, $(1, 0, 0, 0,\ldots, 0)$. After the first iteration, four components can be seen. These are equal to $(c_0/\sqrt{2}, c_1/\sqrt{2}, c_2/\sqrt{2}, c_3/\sqrt{2}, 0, 0, 0, \ldots 0)$ (notice that this is the decomposition filter). The next iteration (Figure 3.20c) produces 10 non-zero values in the transform vector, the next 22 non-zero values and so on. The transform vector components take on the

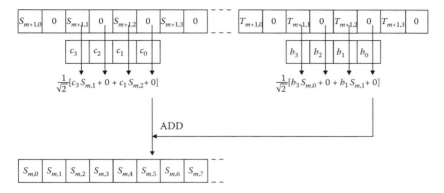

FIGURE 3.19 Filtering of the signal: Reconstruction. The smooth and detail coefficients at scale index m are passed back through the filters to reconstruct the smooth coefficients at the higher resolution $S_{m-1,n}$. This is done by sliding the lowpass and highpass filters along their respective coefficient sequences, with zeros inserted between the coefficients. (Refer back to Equation 3.42 where, remember, 'n' is the coefficient location index at the lower scale and 'k' is the index at the higher scale.)

appearance of the scaling function quite quickly. What we are actually doing is reconstructing $S_{0,n}$ based on the initial vector $(1, 0, 0, 0, \ldots, 0)$ equal to the following transform vectors: $(S_{1,0}, S_{1,1}, S_{1,2},\ldots, T_{1,0}, T_{1,1}, T_{1,2},\ldots)$ for Figure 3.20b; $(S_{2,0}, S_{2,1}, S_{2,2},\ldots, S_{1,0}, S_{1,1}, S_{1,2}, \ldots, T_{1,0}, T_{1,1}, T_{1,2},\ldots)$ for Figure 3.20c; $(S_{3,0}, S_{3,1}, S_{3,2},\ldots, S_{2,0}, S_{2,1}, S_{2,2},\ldots, S_{1,0}, S_{1,1}, S_{1,2},\ldots, T_{1,0}, T_{1,1}, T_{1,2},\ldots)$ for Figure 3.18d and so on. Notice that, as we employ wraparound to deal with the edge effects, the final signal shown in Figure 3.20g is a constant value equal to the signal mean. Figure 3.20h shows a high-resolution representation of a Daubechies scaling function generated by iterating nine times with an initial transform vector containing 4096 components.

You can verify that the constant signal of Figure 3.20g stems from the wraparound employed when reconstructing from the full decomposition wavelet vector with $S_{M,0}$ set to unity. In this case, for the first iteration, adding a zero to get $(S_{M,0},0)$ and running the reconstruction filter over both components leads to two approximation coefficients $(S_{M-1,0}, S_{M-1,1})$ equal to $\left((c_0+c_2)/\sqrt{2},(c_1+c_3)/\sqrt{2}\right)=(0.707,0.707)$. Adding zeros to this vector and iterating again produces four approximation coefficients $(0.5, 0.5, 0.5, 0.5)$ and so on until we obtain the full reconstruction after the sixth iteration which contains constant values of $S_{0,n}$ equal to $(1/\sqrt{2})^6 = 0.125$. We can see this is true for the general case as, from Equation 3.74 with $m=0$ together with Equation 3.21, it can be shown that

$$\sum_{k\text{ even}} c_k = \sum_{k\text{ odd}} c_k = 1 \tag{3.83}$$

Hence, the combined effect of wraparound plus the zeros inserted in the reconstruction process causes $S_{M,0}$ to be multiplied by both $(1/\sqrt{2})\Sigma_{k\text{ even}}c_k$ and $(1/\sqrt{2})\Sigma_{k\text{ odd}}c_k$ in turn, to

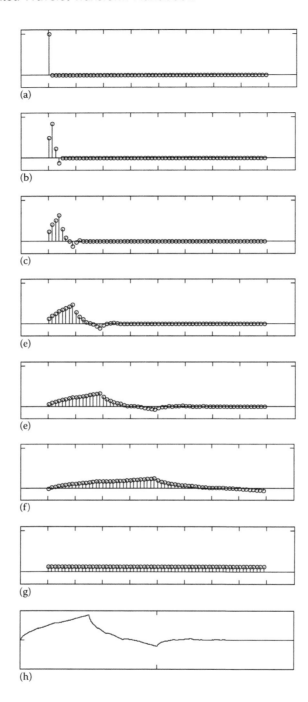

FIGURE 3.20 **The reconstruction of a Daubechies D4 scaling function:** (a) The transform vector $(1, 0, 0, 0, 0, 0, \ldots)$. (b) One iteration, that is, $S_{1,0} = 1$. (c) Two iterations, that is, $S_{2,0} = 1$. (d) Three iterations, that is, $S_{3,0} = 1$. (e) Four iterations, that is, $S_{4,0} = 1$. (f) Five iterations, that is, $S_{5,0} = 1$. (g) Six iterations of the transform vector with 64 components, that is, $S_{6,0} = 1$. This results in a constant signal equal to the mean due to wraparound. (h) After nine iterations of a transform vector $(1, 0, 0, 0, 0, \ldots)$ with 4096 components.

find $S_{M-1,0}$ and $S_{M-1,1}$ which both have the same value, simply equal to $S_{M,0}/\sqrt{2}$. This transform vector now has two constant values (the rest zero). The next iteration produces four constant non-zero values and so on. It is easy to prove to yourself that using Daubechies wavelets with different numbers of non-zero scaling coefficients, N_k, will always result in constant valued transform vector components.

If we repeat the reconstruction filtering, this time setting the first detail coefficient to unity, we can generate an approximation to a wavelet. Notice that, in doing so, only the first iteration of the reconstruction algorithm requires the reordered scaling coefficient sequence for the wavelet, b_k, subsequent reconstruction filtering then use the scaling coefficient sequence c_k. Figure 3.21 shows the wavelet reconstruction process. From these examples, where a single component was set to unity, we can see that the transform vector components, when passed back through the filters during the reconstruction process, increasingly spread their influence (information) over the transform vector in terms of either discrete scaling functions (if the original component was an approximation coefficient) or discrete wavelet function (if the original component was a detail coefficient). Figures 3.20 and 3.21 illustrate the discrete approximations of the wavelet and scaling

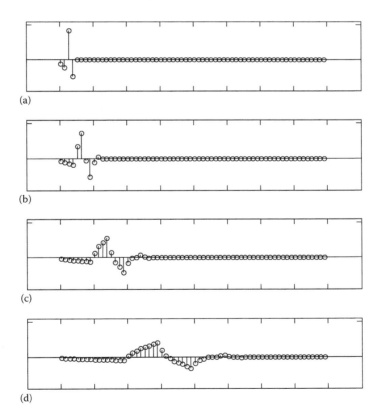

FIGURE 3.21 The reconstruction of a Daubechies D4 wavelet: **(a)** One iteration with only $T_{1,0}$ initially set equal to 1. **(b)** Two iterations with only $T_{2,0}$ initially set equal to 1. **(c)** Three iterations with only $T_{3,0}$ initially set equal to 1. **(d)** Four iterations with only $T_{4,0}$ initially set equal to 1.

functions at various scales implicit within the multiresolution algorithm. The discrete scaling and wavelet functions at scale 1 are, in fact, the filter coefficient vectors $(1/\sqrt{2})b_k$ and $(1/\sqrt{2})c_k$, respectively.

3.5.2 Symmlets and Coiflets

Looking back at Figure 3.15, we can see that Daubechies wavelets are quite asymmetric. To improve symmetry while retaining simplicity, Daubechies proposed *symmlets* as a modification to her original wavelets (also spelled symlets). Symmetric wavelet filters are desirable in some applications; for example, image coding, as it is argued that our visual system is more tolerant of symmetric errors. In addition, it makes it easier to deal with image boundaries. Daubechies came up with symmlets by 'juggling with their phase' during their construction. Figure 3.22 contains two examples of symmlets together with their scaling functions. They have $N_k/2-1$ vanishing moments, support length N_k-1 and filter length N_k. However, true symmetry (or antisymmetry) cannot be achieved for orthonormal wavelet bases with compact support with one exception: the Haar wavelet which is antisymmetric. *Coiflets* are another wavelet family found by Daubechies. They are also nearly symmetrical and have vanishing moments for both the scaling function and wavelet: the wavelet has $N_k/3$ moments equal to zero and the scaling function has $N_k/3-1$ moments equal to zero. They have support length N_k-1 and filter length N_k. Three examples of Coiflet wavelets are shown in Figure 3.23. The number of coefficients, N_k, used to define Coiflets increase in multiples of six.

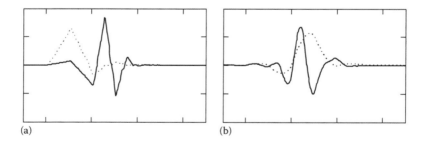

FIGURE 3.22 Symmlets and their associated scaling functions: (a) S6. (b) S10. (The scaling functions are shown dotted.)

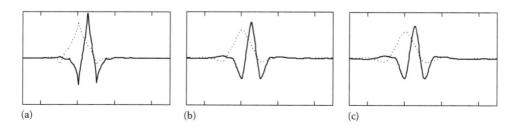

FIGURE 3.23 Coiflets and their associated scaling functions: (a) C6. (b) C12. (c) C18. (Scaling functions are shown dotted.)

3.6 TRANSLATION INVARIANCE

Translation invariance (or *shift invariance*) is an important property associated with some wavelet transforms but not others. It simply means that if you shift along the signal, all your transform coefficients simply move along by the same amount. However, for the dyadic grid structure of the discrete wavelet transform, this is clearly not the case: only if you shift along by the grid spacing at that scale do the coefficients become translation invariant at that scale and below. Even for the discretization of the continuous wavelet transform, the transform values are translation invariant only if shifted by any integer multiple of the discrete time steps. This is illustrated in Figure 3.24, where a simple sinusoidal signal is decomposed using the discrete orthonormal Haar wavelet and a continuous Mexican hat wavelet discretized at each time step. The original signal shown on the left-hand side of Figure 3.24a is composed of 10 cycles of a sinusoid made up of 1024 discrete data points. The middle plot in Figure 3.24a contains the Haar transform coefficients of the signal plotted in their dyadic grid formation. The right-hand plot contains the continuous wavelet transform using the Mexican hat wavelet. The coarse structure of the dyadic grid is evident when compared with the smooth, high-resolution Mexican hat transform plot.

We will assume that the signal continues indefinitely and shift the signal window along to see what happens to the transform plots. The variation in the coefficients of the Haar transform with the displacement of the time window along the signal can be seen by looking down the central plots in the figure. Shifting the window forward (or back) along the signal by a scale of 2^m produces the same coefficients in the transform plot at and below that scale shifted in time by 2^m. This can be observed in Figure 3.24b–d, where the size of the shift is indicated in the transform plot of Figure 3.24a. The coefficient level at and below which the coefficients remain the same as the original is indicated by the arrows at the left-hand side of each transform plot. Shifting the signal by an arbitray scale (not a power of two) leads to a completely different set of coefficients as can be seen in Figure 3.24e where the signal is shifted by an eighth of a sinusoidal cycle (=12.8 data points). The translation invariance of the continuous transform plots is obvious by scanning over the right-hand figures where, for any arbitrary shift in the signal, the wavelet transform values are simply translated with the shift.

We can make a translation invariant version of the discrete wavelet transform – known by a variety of names including the *redundant, stationary, translation invariant, maximal overlap* or *non-decimated* wavelet transform – we simply skip the subsampling part of the filtering process described in Section 3.5.1. This results in the same number of wavelet coefficients generated at each step, equal to the number of signal components N. The decomposition is the same as that shown in Figure 3.16 except that every value is retained as the filter moves one step at a time along the signal. In addition, the filters have to be stretched through the addition of zeros between coefficients; hence, this algorithm is called the *a trous* algorithm from the French *with holes*. An average basis inverse can be performed which gives the average of all possible discrete wavelet transform reconstructions over all possible choices of time origin in the signal. This is sometimes useful for statistical applications including denoising. In addition to being translation invariant, the redundant

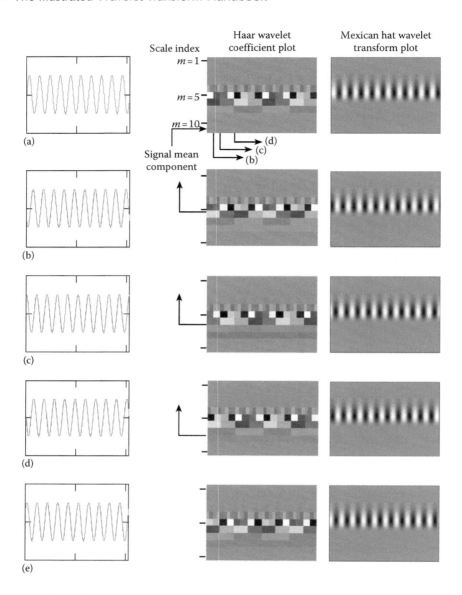

FIGURE 3.24 Translation invariance: (a) Original signal, 1024 data points. (b) Shift by 64 data points. (c) Shift by 128 data points. (d) Shift by 256 data points. (e) Shift by 1/8 cycle.

discrete wavelet transform is extendable to signals of arbitrary length. We do not consider the redundant discrete wavelet transform herein, rather the reader is referred elsewhere at the end of this chapter.

3.7 BIORTHOGONAL WAVELETS

For certain applications, real, symmetric wavelets are required. One way to obtain symmetric wavelets is to construct two sets of *biorthogonal wavelets*: $\psi_{m,n}$ and its *dual*, $\tilde{\psi}_{m,n}$. One set is used to decompose the signal and the other to reconstruct it. For example, we can decompose the signal using $\psi_{m,n}$ as follows:

$$T_{m,n} = \int_{-\infty}^{\infty} x(t)\psi_{m,n}(t)\, dt \qquad (3.84)$$

and perform the inverse transform using $\tilde{\psi}_{m,n}$:

$$x(t) = \sum_{m=-\infty}^{\infty} \sum_{n=-\infty}^{\infty} T_{m,n}\, \tilde{\psi}_{m,n}(t) \qquad (3.85)$$

Alternatively, we can decompose the signal using

$$\tilde{T}_{m,n} = \int_{-\infty}^{\infty} x(t)\tilde{\psi}_{m,n}(t)\, dt \qquad (3.86)$$

and reconstruct using

$$x(t) = \sum_{m=-\infty}^{\infty} \sum_{n=-\infty}^{\infty} \tilde{T}_{m,n}\, \psi(t) \qquad (3.87)$$

Biorthogonal wavelets satisfy the biorthogonality condition:

$$\int_{-\infty}^{\infty} \psi_{m,n}(t)\tilde{\psi}_{m',n'}(t)\, dt = \begin{cases} 1 & \text{if } m = m' \text{ and } n = n' \\ 0 & \text{otherwise} \end{cases} \qquad (3.88)$$

Using biorthogonal wavelets allows us to have perfectly symmetric and antisymmetric wavelets. Further, they allow certain desirable properties to be incorporated separately within the decomposition wavelet and the reconstruction wavelet. For example, $\psi_{m,n}$ and $\tilde{\psi}_{m,n}$ can have different numbers of vanishing moments. If $\psi_{m,n}$ has more vanishing moments than $\tilde{\psi}_{m,n}$, then decomposition using $\psi_{m,n}$ suppresses higher-order polynomials and aids data compression. Reconstruction with the wavelets, $\tilde{\psi}_{m,n}$, with less vanishing moments leads to a smoother reconstruction. This can sometimes be a useful property, for example, in image processing. Figure 3.25 shows three examples of compactly supported biorthogonal spline wavelets and their duals commonly used in practice, together with their associated scaling equations.

3.8 TWO-DIMENSIONAL WAVELET TRANSFORMS

In many applications, the data set is in the form of a two-dimensional array, for example, the heights of a machined surface or natural topography, or perhaps the intensities of an array of pixels making up an image. We may want to perform a wavelet decomposition of

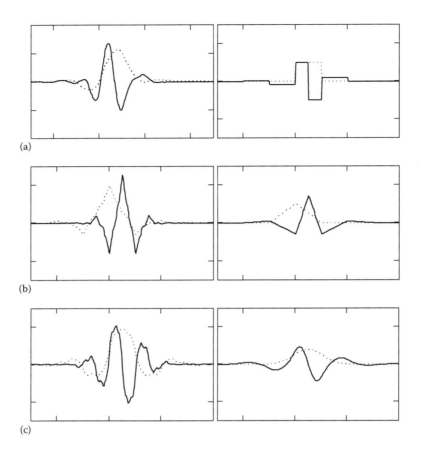

FIGURE 3.25 **Biorthogonal spline wavelets:** (a) Biorthogonal (1,5) spline wavelets: $\phi(t)$, $\psi(t)$ (left) and $\tilde{\phi}(t)$, $\tilde{\psi}(t)$ (right). (b) Biorthogonal (2,4) spline wavelets: $\phi(t)$, $\psi(t)$ (left) and $\tilde{\phi}(t)$, $\tilde{\psi}(t)$ (right). (c) Biorthogonal (3,7) spline wavelets: $\phi(t)$, $\psi(t)$ (left) and $\tilde{\phi}(t)$, $\tilde{\psi}(t)$ (right). (Scaling functions shown dotted.)

such arrays, either to compress the data or to carry out a wavelet-based parametric characterization of the data. To perform a discrete wavelet decomposition of a 2-D array we must use 2-D discrete wavelet transforms. We can simply generate these from tensor products of their 1-D orthonormal counterparts. The most common (and simplest) arrangement is to use the same scaling in the horizontal and vertical directions. These give square transforms. (Other forms are possible, for example, rectangular transforms where the horizontal and vertical scaling vary independently and also transforms which are not simply tensor products. These are outwith the scope of this text.) The two-dimensional Haar scaling and wavelet functions are

$$\text{2-D scaling function} \quad \phi(t_1, t_2) = \phi(t_1)\phi(t_2) \quad (3.89a)$$

$$\text{2-D horizontal wavelet} \quad \psi^h(t_1, t_2) = \phi(t_1)\psi(t_2) \quad (3.89b)$$

2-D vertical wavelet $\quad \psi^v(t_1,t_2) = \psi(t_1)\phi(t_2)$ (3.89c)

2-D diagonal wavelet $\quad \psi^d(t_1,t_2) = \psi(t_1)\psi(t_2)$ (3.89d)

Remember that in both the last chapter and this chapter, we have been using t as our independent variable, either temporal or (less common) spatial. In this section, t_1 and t_2 represent spatial coordinates.

The multiresolution decomposition of the 2-D coefficient matrices can be expressed as

$$S_{m+1,(n_1,n_2)} = \frac{1}{2}\sum_{k_1}\sum_{k_2} c_{k_1}c_{k_2} S_{m(2n_1+k_1, 2n_2+k_2)} \quad (3.90a)$$

$$T^h_{m+1,(n_1,n_2)} = \frac{1}{2}\sum_{k_1}\sum_{k_2} b_{k_1}c_{k_2} S_{m(2n_1+k_1, 2n_2+k_2)} \quad (3.90b)$$

$$T^v_{m+1,(n_1,n_2)} = \frac{1}{2}\sum_{k_1}\sum_{k_2} c_{k_1}b_{k_2} S_{m(2n_1+k_1, 2n_2+k_2)} \quad (3.90c)$$

$$T^d_{m+1,(n_1,n_2)} = \frac{1}{2}\sum_{k_1}\sum_{k_2} b_{k_1}b_{k_2} S_{m(2n_1+k_1, 2n_2+k_2)} \quad (3.90d)$$

where k_1, k_2 are the scaling coefficient indices and n_1, n_2 are the location indices at scale $m+1$. (Compare with the 1-D case of Equations 3.36 and 3.37.) We can simply use discrete versions of the 2-D wavelets at scale index 1 to perform the multiresolution analysis of the array. The discrete 2-D Haar scaling and wavelet functions in matrix form at scale index $m=1$ are

Scaling function $\quad$ Vertical wavelet

$$\frac{1}{2}\begin{bmatrix} 1 & 1 \\ 1 & 1 \end{bmatrix} \quad \frac{1}{2}\begin{bmatrix} 1 & -1 \\ 1 & -1 \end{bmatrix}$$

Horizontal wavelet $\quad$ Diagonal wavelet

$$\frac{1}{2}\begin{bmatrix} 1 & 1 \\ -1 & -1 \end{bmatrix} \quad \frac{1}{2}\begin{bmatrix} 1 & -1 \\ -1 & 1 \end{bmatrix}$$

Because these are at scale index 1, they are simply tensor products of the scaling and wavelet filter coefficients $c_k/\sqrt{2}$ and $b_k/\sqrt{2}$. These discrete functions are shown schematically in Figure 3.26. Notice the 1/2 factor before each matrix is simply the square of the $1/\sqrt{2}$ factor preceding the corresponding one-dimensional functions. For discrete scaling and wavelet functions at larger scales, this normalization factor becomes $1/2^m$. These four 2×2 matrices are required for the Haar wavelet decomposition of the 2-D array. The general idea of the 2-D wavelet decomposition is shown schematically in Figure 3.27. The original input array is represented by $\mathbf{X}_0$ defined at scale index $m = 0$. As with the one-dimensional case, its components are input as the approximation coefficients at scale index 0, that is, the matrix $\mathbf{S}_0$. After the first wavelet decomposition, a decomposition array is formed at scale index 1 which is split into four distinct submatrices: the vertical detailed components $\mathbf{T}_1^v$, the horizontal detailed components $\mathbf{T}_1^h$, the diagonal detailed components $\mathbf{T}_1^d$ and the remaining approximation components $\mathbf{S}_1$. The decomposition array is the same size as the original array. As with the discrete wavelet decomposition of 1-D signals, the detail coefficients are subsequently left untouched and the next iteration further decomposes only the approximation components in the submatrix $\mathbf{S}_1$. This results in detail coefficients contained within submatrices $\mathbf{T}_2^v$, $\mathbf{T}_2^h$ and $\mathbf{T}_2^d$ at scale index 2 and approximation coefficients within submatrix $\mathbf{S}_2$. This procedure may be repeated M times for a $2^M \times 2^M$ array to get a number of coefficient submatrices $\mathbf{T}_m^v$, $\mathbf{T}_m^h$ and $\mathbf{T}_m^d$ of size $2^{M-m} \times 2^{M-m}$, where $m = 1, 2, \ldots, M$.

Let us look at a very simple example, a matrix with a single non-zero component:

$$\mathbf{X}_0 = \begin{bmatrix} 0 & 0 & 0 & 0 \\ 0 & 1 & 0 & 0 \\ 0 & 0 & 0 & 0 \\ 0 & 0 & 0 & 0 \end{bmatrix} \quad (3.91)$$

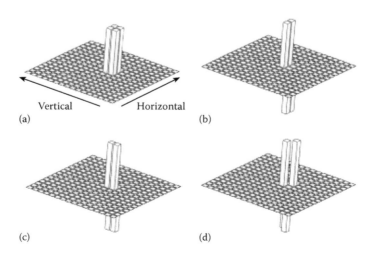

FIGURE 3.26 The two-dimensional discrete Haar wavelet at scale index 1: (a) Scaling function. (b) Horizontal wavelet. (c) Vertical wavelet. (d) Diagonal wavelet.

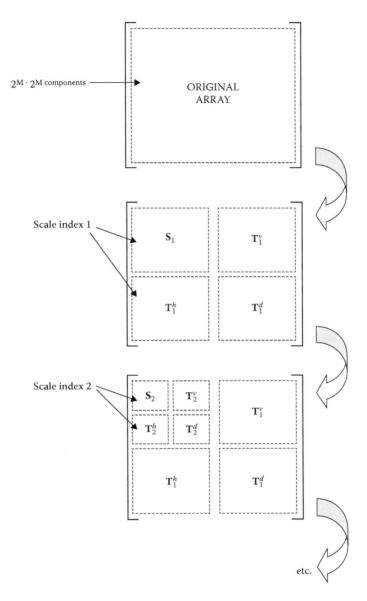

FIGURE 3.27 Schematic diagram of the matrix manipulation required to perform wavelet decomposition on a two-dimensional grid.

This array could represent a single pixel turned on in a display or perhaps a spike protruding from a flat surface. We can visualize it in Figure 3.28. The 2-D Haar multiresolution decomposition of this array is carried out by scanning over it with each of the four discrete wavelet matrices in turn. This is shown schematically in Figure 3.29. On the first iteration (from scale $m=0$ to 1), the scaling function 2×2 matrix is scanned over the input array. We require four of these matrices to cover this 4×4 array. The components of the original array are multiplied in turn by the scaling function array to give the resulting matrix product S_1 which is placed in the top right-hand quadrant of the first iteration

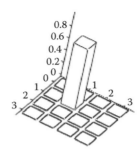

FIGURE 3.28 A two-dimensional signal containing a single spike.

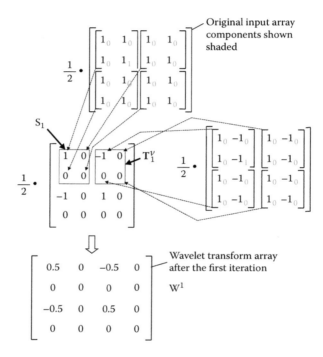

FIGURE 3.29 Schematic diagram of the matrix manipulation required to perform the Haar wavelet decomposition of the spike signal. Components of the discrete scaling and wavelet matrices are shown in bold, components of the original discrete signal are shown shaded.

matrix. Similarly, the vertical 2 × 2 wavelet matrix is scanned over the array and each value is placed in the top right-hand quadrant of the first iteration array. This is illustrated in the T_1^v matrix marked on the right hand of the figure. The procedure is repeated for the horizontal and diagonal wavelet components, placing the products in the bottom left and bottom right-hand quadrants, respectively. The normalization factor of 1/2 has been left out of the matrix manipulation until the end to aid clarity. Hence, after the first iteration, the resulting wavelet transform matrix is

$$\mathbf{W}^{(1)} = \begin{bmatrix} 0.5 & 0 & -0.5 & 0 \\ 0 & 0 & 0 & 0 \\ -0.5 & 0 & 0.5 & 0 \\ 0 & 0 & 0 & 0 \end{bmatrix} \quad (3.92)$$

The next iteration uses only the approximation components in the top right-hand quadrant (shown within the dotted box). The four component values of this array are then interrogated by each of the scaling and wavelet matrices in turn. This produces the second iteration matrix

$$\mathbf{W}^{(2)} = \begin{bmatrix} 0.25 & 0.25 & -0.5 & 0 \\ 0.25 & 0.25 & 0 & 0 \\ -0.5 & 0 & 0.5 & 0 \\ 0 & 0 & 0 & 0 \end{bmatrix} \quad (3.93)$$

This is the second and last iteration we can perform in the decomposition of the 4×4 (i.e. $M=2$) array. This matrix is made up of $\mathbf{T}_1^v$, $\mathbf{T}_1^h$, $\mathbf{T}_1^d$, $\mathbf{T}_2^v$, $\mathbf{T}_2^h$, $\mathbf{T}_2^d$ and, in the top left-hand corner, the approximation component $\mathbf{S}_2$ which is related to the mean of the original array. Notice that, at each iteration, the resulting matrix has the same energy as the original matrix where the energy is the sum of the squared components of the matrix, that is

$$E = \sum_{i=0}^{2^M-1} \sum_{j=0}^{2^M-1} (X_{o,i,j})^2 = \sum_{i=0}^{2^M-1} \sum_{j=0}^{2^M-1} (W_{i,j}^{(m)})^2 \quad (3.94)$$

where $X_{0,i,j}$ and $W_{i,j}^{(m)}$ are, respectively, the elements of the input and wavelet decomposition matrices located on row i and column j. E is equal to 1 for the array just considered. We can easily verify that E is equal to 1 for the intermediate decomposition matrix as well as the full decomposition matrix. You can confirm this conservation of energy property in all the examples which follow.

Let's at another simple example, the step shown in Figure 3.30a and given in matrix form as

$$\mathbf{X}_0 = \begin{bmatrix} 2 & 2 & 2 & 2 & 2 & 2 & 2 & 2 \\ 2 & 2 & 2 & 2 & 2 & 2 & 2 & 2 \\ 2 & 2 & 2 & 2 & 2 & 2 & 2 & 2 \\ 2 & 2 & 2 & 2 & 2 & 2 & 2 & 2 \\ 2 & 2 & 2 & 2 & 2 & 2 & 2 & 2 \\ 1 & 1 & 1 & 1 & 1 & 1 & 1 & 1 \\ 1 & 1 & 1 & 1 & 1 & 1 & 1 & 1 \\ 1 & 1 & 1 & 1 & 1 & 1 & 1 & 1 \end{bmatrix} \quad (3.95)$$

158 ■ The Illustrated Wavelet Transform Handbook

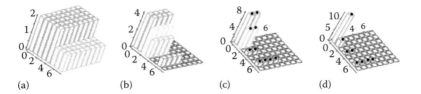

FIGURE 3.30 An 8 × 8 'step' array together with its associated Haar decompositions: (a) Original step. (b) Iteration 1. (c) Iteration 2. (d) Iteration 3. Note that black dots have been added to the non-zero coefficients in plots (c) and (d) to aid clarity.

This 8 × 8 array can be decomposed through three iterations using the Haar wavelet as follows:

$$\begin{bmatrix} 4 & 4 & 4 & 4 & 0 & 0 & 0 & 0 \\ 4 & 4 & 4 & 4 & 0 & 0 & 0 & 0 \\ 3 & 3 & 3 & 3 & 0 & 0 & 0 & 0 \\ 2 & 2 & 2 & 2 & 0 & 0 & 0 & 0 \\ 0 & 0 & 0 & 0 & 0 & 0 & 0 & 0 \\ 0 & 0 & 0 & 0 & 0 & 0 & 0 & 0 \\ 1 & 1 & 1 & 1 & 0 & 0 & 0 & 0 \\ 0 & 0 & 0 & 0 & 0 & 0 & 0 & 0 \end{bmatrix}$$

1st Decomposition
$W^{(1)}$ (3.96a)

$$\begin{bmatrix} 8 & 8 & 0 & 0 & 0 & 0 & 0 & 0 \\ 5 & 5 & 0 & 0 & 0 & 0 & 0 & 0 \\ 0 & 0 & 0 & 0 & 0 & 0 & 0 & 0 \\ 1 & 1 & 0 & 0 & 0 & 0 & 0 & 0 \\ 0 & 0 & 0 & 0 & 0 & 0 & 0 & 0 \\ 0 & 0 & 0 & 0 & 0 & 0 & 0 & 0 \\ 1 & 1 & 1 & 1 & 0 & 0 & 0 & 0 \\ 0 & 0 & 0 & 0 & 0 & 0 & 0 & 0 \end{bmatrix}$$

2nd Decomposition
$W^{(2)}$ (3.96b)

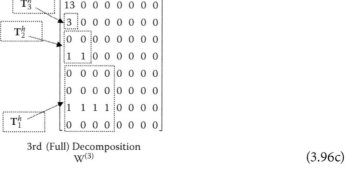

3rd (Full) Decomposition
$W^{(3)}$ (3.96c)

Notice that, after the three iterations, only the approximation component and the components in regions corresponding to the horizontal wavelet decomposition (labelled $\mathbf{T}_1^h$, $\mathbf{T}_2^h$ and $\mathbf{T}_3^h$) contain non-zero elements. This is because the original array only contains a single horizontal feature – the horizontal discontinuity or step. Another thing to notice is that the 64 non-zero elements of the original array have been transformed into only eight non-zero elements, that is, a compression of the data has taken place. The original array, together with the three levels of decomposition, are shown as histograms in Figure 3.30.

We can now use the detail coefficients to reconstruct the array at different scales. This is shown in Figure 3.31 and is performed using the detail coefficients at each scale of interest. In the same way as we did for the one-dimensional signals in Section 3.4, discrete detailed arrays $\mathbf{D}_m^h$, $\mathbf{D}_m^d$ and $\mathbf{D}_m^v$ can be reconstructed at the scale of the input array using each of the detail coefficients, $\mathbf{T}_m^h$, $\mathbf{T}_m^d$ and $\mathbf{T}_m^v$. A combined detail can be found at scale index m, given by

$$\mathbf{D}_m = \mathbf{D}_m^h + \mathbf{D}_m^d + \mathbf{D}_m^v \tag{3.97}$$

As with the one-dimensional case, the array can be represented as a sum of the discrete details plus an array mean.

$$\mathbf{X}_0 = \mathbf{X}_M + \sum_{m=1}^{M} \mathbf{D}_m \tag{3.98}$$

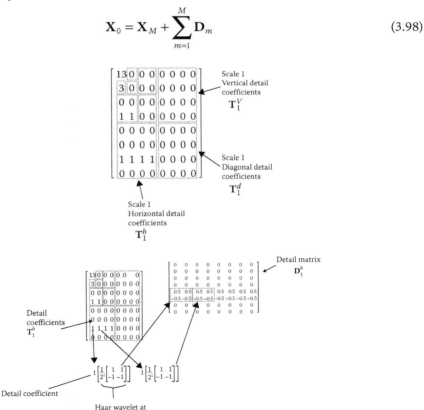

FIGURE 3.31 Schematic diagram of the matrix manipulation required to derive the detail matrices at scale index $m = 1$: (a) The three detail coefficient submatrices at scale index 1 for the step. (b) Computing the horizontal discrete detail from the detail coefficients at scale index 1.

where $\mathbf{X}_M$ is the smooth version of the input array at the largest scale index, M, expressed at the scale of the input array $m=0$. For example, to reconstruct the detail array components at the smallest scale ($m=1$), each corresponding component in each of the three coefficient submatrices, $\mathbf{T}_1^h$, $\mathbf{T}_1^d$ and $\mathbf{T}_1^v$, are used in conjunction with the corresponding Haar wavelets (horizontal, diagonal and vertical) to produce the array details at this level: $\mathbf{D}_1^h$, $\mathbf{D}_1^d$ and $\mathbf{D}_1^v$. The $m=1$ submatrices are shown in Figure 3.31a. These three 4×4 coefficient submatrices are each expanded into an 8×8 array corresponding to the horizontal, diagonal and vertical details of the original array using the corresponding discrete wavelets. This is shown schematically in Figure 3.31b for the horizontal detail at scale index 1, that is, $\mathbf{D}_1^h$. The expansion of the transform coefficients through the discrete Haar wavelet into the detailed components is shown explicitly for two of the $\mathbf{T}_1^h$ components in the figure. We can see that the detail matrices $\mathbf{D}_1^d$ and $\mathbf{D}_1^v$ will have all elements equal to zero as both $\mathbf{T}_1^d$ and $\mathbf{T}_1^v$ contain only zero elements. Hence, the combined detail component matrix is

$$\mathbf{D}_1 = \mathbf{D}_1^h + \mathbf{D}_1^d + \mathbf{D}_1^v = \begin{bmatrix} 0 & 0 & 0 & 0 & 0 & 0 & 0 & 0 \\ 0 & 0 & 0 & 0 & 0 & 0 & 0 & 0 \\ 0 & 0 & 0 & 0 & 0 & 0 & 0 & 0 \\ 0 & 0 & 0 & 0 & 0 & 0 & 0 & 0 \\ 0.5 & 0.5 & 0.5 & 0.5 & 0.5 & 0.5 & 0.5 & 0.5 \\ -0.5 & -0.5 & -0.5 & -0.5 & -0.5 & -0.5 & -0.5 & -0.5 \\ 0 & 0 & 0 & 0 & 0 & 0 & 0 & 0 \\ 0 & 0 & 0 & 0 & 0 & 0 & 0 & 0 \end{bmatrix}$$

(3.99)

Similarly, we can get the detail matrices at scale index 2, $\mathbf{D}_2^h$, $\mathbf{D}_2^d$ and $\mathbf{D}_2^v$, using the scale 2 coefficients, respectively, $\mathbf{T}_2^h$, $\mathbf{T}_2^d$ and $\mathbf{T}_2^v$ and the discrete wavelet at scale index 2. This is shown for $\mathbf{D}_2^h$ in Figure 3.32. Note that the normalization factor at scale 2 is $1/2^2$. The scale 3 coefficient submatrices, $\mathbf{T}_3^h$, $\mathbf{T}_3^d$ and $\mathbf{T}_3^v$, are simply [3], [0] and [0], respectively. It is easily seen that this gives the detail at the largest wavelet scale equal to

$$\mathbf{D}_3 = \begin{bmatrix} 0.375 & 0.375 & 0.375 & 0.375 & 0.375 & 0.375 & 0.375 & 0.375 \\ 0.375 & 0.375 & 0.375 & 0.375 & 0.375 & 0.375 & 0.375 & 0.375 \\ 0.375 & 0.375 & 0.375 & 0.375 & 0.375 & 0.375 & 0.375 & 0.375 \\ 0.375 & 0.375 & 0.375 & 0.375 & 0.375 & 0.375 & 0.375 & 0.375 \\ -0.375 & -0.375 & -0.375 & -0.375 & -0.375 & -0.375 & -0.375 & -0.375 \\ -0.375 & -0.375 & -0.375 & -0.375 & -0.375 & -0.375 & -0.375 & -0.375 \\ -0.375 & -0.375 & -0.375 & -0.375 & -0.375 & -0.375 & -0.375 & -0.375 \\ -0.375 & -0.375 & -0.375 & -0.375 & -0.375 & -0.375 & -0.375 & -0.375 \end{bmatrix}$$

(3.100)

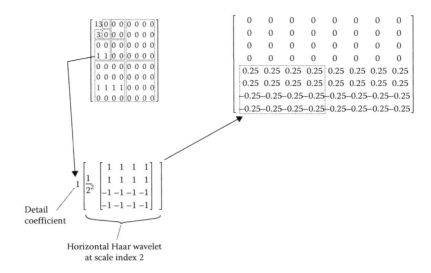

FIGURE 3.32 Schematic diagram of the matrix manipulation required to compute the horizontal detail at scale index $m = 2$.

where each component is either $3/2^3$ or $-3/2^3$. In fact, this matrix is simply a single large horizontal discrete Haar wavelet scaled by $3/2^3$. The approximation coefficient ($=13$) leads to a matrix of values all equal to the mean of the original array ($=13/2^3$), that is

$$\mathbf{X}_3 = \begin{bmatrix} 1.625 & 1.625 & 1.625 & 1.625 & 1.625 & 1.625 & 1.625 & 1.625 \\ 1.625 & 1.625 & 1.625 & 1.625 & 1.625 & 1.625 & 1.625 & 1.625 \\ 1.625 & 1.625 & 1.625 & 1.625 & 1.625 & 1.625 & 1.625 & 1.625 \\ 1.625 & 1.625 & 1.625 & 1.625 & 1.625 & 1.625 & 1.625 & 1.625 \\ 1.625 & 1.625 & 1.625 & 1.625 & 1.625 & 1.625 & 1.625 & 1.625 \\ 1.625 & 1.625 & 1.625 & 1.625 & 1.625 & 1.625 & 1.625 & 1.625 \\ 1.625 & 1.625 & 1.625 & 1.625 & 1.625 & 1.625 & 1.625 & 1.625 \\ 1.625 & 1.625 & 1.625 & 1.625 & 1.625 & 1.625 & 1.625 & 1.625 \end{bmatrix} \quad (3.101)$$

Figure 3.33 shows the reconstructions at each scale together with the mean component. From the figure, we can see that if we add them all together, $\mathbf{D}_1 + \mathbf{D}_2 + \mathbf{D}_3 + \mathbf{X}_3$, we get back the original input array $\mathbf{X}_0$.

Figures 3.34 and 3.35 show examples of much larger data sets, both using the Haar wavelet transform. Figure 3.34 shows an image ('Lena') together with its first two decomposition matrices, where the approximation and detail coefficient submatrices can be clearly seen. Figure 3.35 shows an example of the Haar decomposition of a more irregular array. The 128×128 array shown in Figure 3.35a contains the heights of a measured rough surface. The details of this array, at scale indices $m = 1$ to 7 are shown in Figure 3.35b–h. Figure 3.35i contains the summation of the first three details and Figure 3.35j shows the

162 ■ The Illustrated Wavelet Transform Handbook

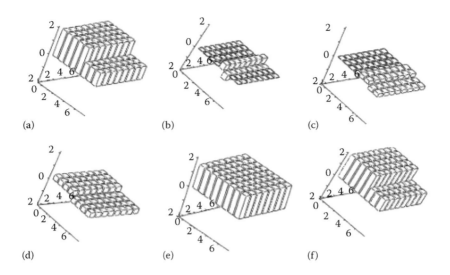

FIGURE 3.33 **The details of the step array:** (a) Original step array X_0. (b) D_1. (c) D_2. (d) D_3. (e) X_3. (f) Sum of (b), (c), (d) and (e) giving original signal (a).

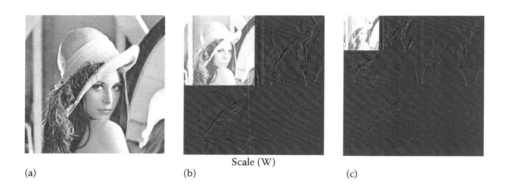

FIGURE 3.34 **An image and its first two decomposition matrices:** (a) Original array S_0. (b) Transform array $V^{(1)}$ containing submatrices S_1, T_1^v, T_1^h and T_1^d. (c) Transform array $V^{(2)}$ containing submatrices S_2, T_1^v, T_1^h, T_1^d, T_2^v, T_2^h and T_2^d.

resulting approximation at scale index $m = 3$ when these details are subtracted from the original array. The blocky nature of the Haar decomposition is noticeable from the plots. The 2-D Haar wavelet is very simple in its form and, as with their 1D counterparts, there are more complex forms of Daubechies wavelets and, of course, many other wavelet families to choose from. Some examples of these wavelets are given in Figure 3.36. Using these wavelets will result in an overlap of the wavelets at the array edge and therefore would require the use of wraparound or other methods to deal with the data edges.

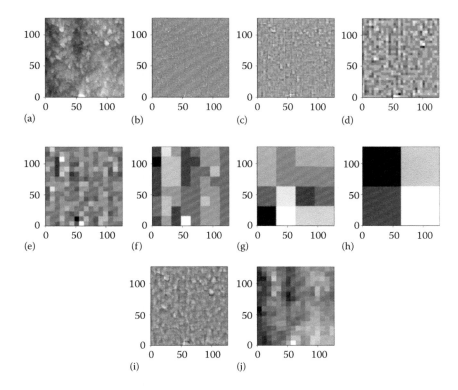

FIGURE 3.35 **Haar decomposition of a surface data set:** (a) Original data set 128×128 array of surface heights taken from a river bed. (b) Scale index $m=1$ discrete detail $\mathbf{D}_1$. (c) Scale index $m=2$ discrete detail $\mathbf{D}_2$. (d) Scale index $m=3$ discrete detail $\mathbf{D}_3$. (e) Scale index $m=4$ discrete detail $\mathbf{D}_4$. (f) Scale index $m=5$ discrete detail $\mathbf{D}_5$. (g) Scale index $m=6$ discrete detail $\mathbf{D}_6$. (h) Scale index $m=7$ discrete detail $\mathbf{D}_7$. (i) Sum of first three discrete details $\mathbf{D}_1+\mathbf{D}_2+\mathbf{D}_3$ (= [b] + [c] + [d]). (j) Scale index $m=3$ discrete approximation $\mathbf{X}_3 = \mathbf{X}_0 - (\mathbf{D}_1+\mathbf{D}_2+\mathbf{D}_3)$ (= [a]−[b]−[c]−[d]). (Greyscale used in all images: maximum = white, minimum = black.)

3.9 ADAPTIVE TRANSFORMS: WAVELET PACKETS

As we saw in Chapter 2 (Figure 2.29a), the resolution of the wavelet transform is not uniform in the time–frequency plane. The Heisenberg boxes expand in frequency and contract in time as fluctuations at smaller and smaller scales are explored. The short-term Fourier transform (STFT), on the other hand, covers the time–frequency plane with tiles of a constant aspect ratio (Figure 2.29c). We also looked briefly at matching pursuits which offer another way of extracting time–frequency information. In this section, we will consider another method which can adapt to the signal and hence allow for more flexibility in the partitioning of the time–frequency plane: the *wavelet packet transform*.

'Wavelet packet' (WP) transforms are a generalization of the discrete wavelet transform. Wavelet packets involve particular linear combinations of wavelets and the wavelet packet decomposition of a signal is performed in a manner similar to the multiresolution algorithm given for the discrete wavelet transform. The difference being that, in the WP signal decomposition, both the approximation and detailed coefficients are further

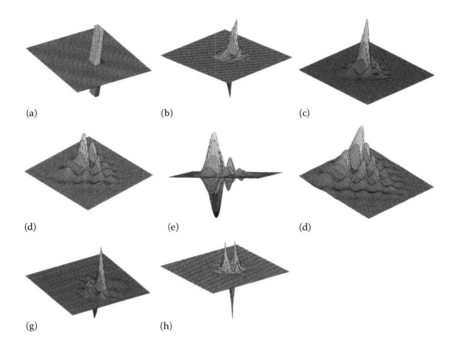

FIGURE 3.36 Examples of two-dimensional orthonormal wavelets: Daubechies, symmlets and Coiflets: **(a)** D2 Haar wavelet. **(b)** D4. **(c)** D8. **(d)** D12. **(e)** D12, another perspective. **(f)** D20. **(g)** Symmlet S8. **(h)** Coiflet C6. (Wavelets shown viewed at various angles and elevations.)

decomposed at each level. This leads to the decomposition tree structure depicted at the top of Figure 3.37. Compare this with the schematic of the wavelet decomposition given in Figure 3.6. At each stage in the wavelet algorithm, the detailed coefficients are simply transferred down, unchanged, to the next level. However, in the wavelet packet algorithm, all the coefficients at each stage, are further decomposed. In this way, we end up with an array of wavelet packet coefficients with M levels each with N coefficients. A total of N coefficients from this $M \times N$ array can then be selected to represent the signal. The standard wavelet transform decomposition coefficients are contained within the WP array, shown by the bold boxes in Figure 3.37. A new nomenclature is employed in the figure to indicate the operations that have been performed on each set of coefficients. S produces the approximation components of the previous set of coefficients by lowpass filtering, and T the detail components through highpass filtering. We simply add the letter S or T to the left-hand end of the coefficient name to indicate the most recent filtering procedure. $SSTS_n$, for example, corresponds to the original signal lowpass filtered, then highpass filtered and then passed twice through the lowpass filter. Notice also that the subscript contains only the location index n. The scaling index m is omitted as it is obviously equal to the number of letters S and T in the coefficient name. As with the original wavelet transform, the number of coefficients at each scale depends on the scale, with one coefficient in each coefficient group at the largest scale M and $N/2$ coefficients at the smallest scale $m = 1$. Hence, the coefficient index spans $n = 0, 1, \ldots, 2^{M-m}-1$.

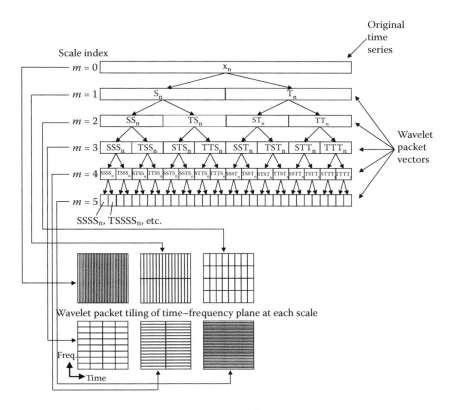

FIGURE 3.37 Schematic diagram of wavelet packet decomposition.

At each stage in the decomposition, the wavelet packet algorithm partitions the time–frequency plane into rectangles of constant aspect ratio. These become wider (in time) and narrower (in frequency) as the decomposition proceeds. This is shown schematically at the bottom of Figure 3.37 for each scale. A variety of tilings of the time–frequency plane is possible using the wavelet packet coefficients. For example, we could keep all the coefficients at a level and discard all the others. This would tile the plane in boxes of constant shape, just like one of those shown at the bottom of Figure 3.37. Other tilings are possible, and some examples of these are shown in the Figure 3.38. The standard wavelet transform is just one of all the possible tiling patterns. Figure 3.39 contains the wavelet packet coefficient selections corresponding to the tiling patterns of Figure 3.38.

The optimal or 'best' coefficient selection (hence tiling arrangement) is chosen to represent the signal based on some predefined criterion. This criterion is normally based on an information cost function which aims to retain as much information in as few coefficients as possible. The most common measure of information used is the *Shannon entropy measure*. This is defined for a discrete distribution p_i, $i = 0, 1, \ldots, N-1$, as

$$S(p) = -\sum_i p_i \log(p_i) \tag{3.102}$$

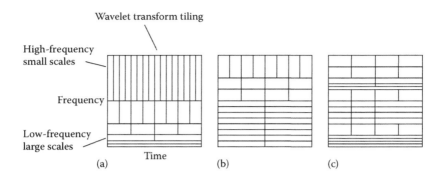

FIGURE 3.38 Schematic diagrams of allowable wavelet packet tiling of the time–frequency plane. The right-hand tiling is that used in the wavelet transform algorithm and corresponds to the components contained in the bold boxes shown in Figure 3.37.

where, for this case, p_i are the normalized energies (i.e. squared magnitudes) of the wavelet packet coefficients under consideration. Low entropies occur when the larger coefficient energies are concentrated at only a few discrete locations. The minimum possible entropy of zero occurs when $p_i = 1$ for only one value of i, the other probabilities being zero. In this case, all the information needed to represent the signal is condensed within a single coefficient. The maximum entropy occurs when there is an equal distribution of coefficient energies. In this case, $p_i = 1/N$ and the signal information is evenly spread throughout all the coefficients. We can see that p_i acts as a discrete probability distribution of the energies. (More information on the Shannon entropy measure together with an illustrative figure is given in Chapter 4, Section 4.2.4.)

The set of N wavelet packet coefficients which contain the least entropy are selected to represent the signal. That is, we want the signal information to be concentrated within as few coefficients as possible. To find these coefficients, the WP array, such as the one we saw in Figure 3.37, is inspected from the bottom upwards. At each scale, each pair of partitioned coefficient sets (the 'children') are compared with those from which they were derived (their 'parent'). If the combined children's coefficients have a smaller entropy than those of their parent then they are retained. If not, the parent's coefficients are retained. When the children are selected, their entropy value is assigned to their parent in order that subsequent entropy comparisons can be made further up the tree. This is shown schematically in Figure 3.40. Once the whole WP array has been inspected in this way, we get an optimal tiling of the time–frequency plane (with respect to the localization of coefficient energies). This tiling provides the best basis for the signal decomposition.

Figure 3.41 illustrates the wavelet packet method on a simple discrete signal composed of a sampled sinusoid plus a spike. The signal is 64 data points in length. A Haar wavelet is used to decompose the signal. Figure 3.41a shows the wavelet packet coefficients below the original signal for each stage in the WP algorithm. The coefficients are displayed as histograms. The bottom two traces contain the coefficients corresponding to the best wavelet packet basis and the 'traditional' discrete wavelet basis, respectively. The WP tiling of the coefficient energies in the time–frequency plane for each scale is given in Figure 3.41b.

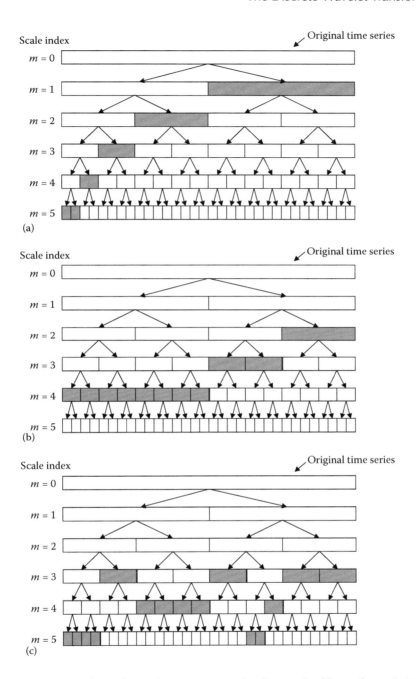

FIGURE 3.39 The choice of wavelet packet components leading to the filings shown in Figure 3.38.

The larger coefficients energies are shaded darker in the plot. In Figure 3.41c, the optimal WP tiling is compared with the wavelet transform tiling of the time–frequency plane. The plots in Figure 3.41d outline the 16 largest coefficients in both time–frequency planes of Figure 3.41c. These are used to reconstruct the signals shown in Figure 3.41e. The 16 largest wavelet packet coefficients contain 98.6% of the signal energy, whereas the 16 wavelet

168 ■ The Illustrated Wavelet Transform Handbook

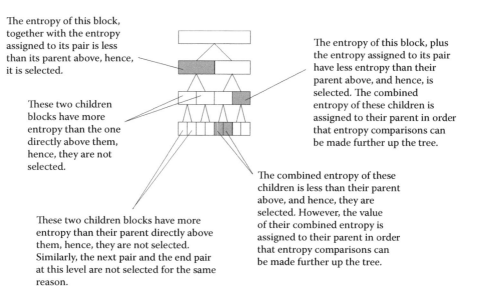

FIGURE 3.40 Wavelet packet coefficient selection.

transform coefficients contain 96.5% of the signal energy. The wavelet packet reconstruction using the selected coefficients is visibly smoother than the reconstruction using the traditional wavelet transform coefficients. Figure 3.42 contains the same signal as Figure 3.41; however, this time the WP decomposition is performed using a Daubechies D20 wavelet (refer to Figure 3.15). Using this wavelet results in a different tiling of the time–frequency plane for the WP method (compare the left-hand plots of Figure 3.42b and Figure 3.41c). Again, the 16 largest coefficients are used in the signal reconstruction. The oscillatory part of both reconstructions shown in Figure 3.42d are visibly smoother than their Haar counterparts in the previous figure (Figure 3.41e). We expect this as the D20 is more smoothly oscillatory than the Haar. Notice, however, comparing Figure 3.42c and Figure 3.41d, we see that the signal spike leads to a single high-frequency tile for both Haar decompositions but, respectively, five and four high-frequency tiles for the D20 wavelet. The more compact support of the Haar wavelet has allowed for a better localization of the signal spike, but it does make it less able than the D20 to represent the smooth oscillations in the signal. The energies of the reconstructed signals for the D20 decompositions using only the largest 16 coefficients are 99.8% (WP) and 99.7% (WT), indicating the data compression possibilities of the techniques.

3.10 'X-LETS': CONTOURLETS, RIDGELETS, CURVELETS, SHEARLETS AND SO ON

Over recent years, a number of enhanced wavelet-based methods have been formulated to more efficiently represent directionally dependent features in 2-D data. An example of this is the *contourlet transform*. The contourlet transform forms a discrete filter bank structure and seeks to capture the 'intrinsic geometric structures' associated with natural images that are key features in visual information where discontinuities along smooth

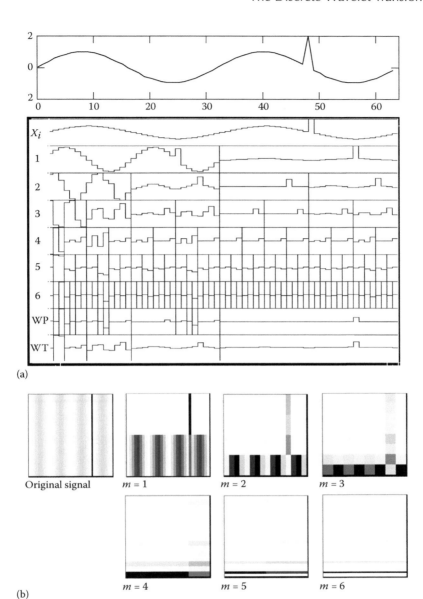

FIGURE 3.41 **Wavelet packet decomposition of a simple signal: (a)** Signal (top) with wavelet packet decomposition (below). The coefficient values are plotted as heights. The scale indices, $m=1$ to 6, are given down the left-hand side of the plot. Trace WP contains the best selection of wavelet packets and trace WT contains the wavelet transform decomposition for comparison. **(b)** The time–frequency tiling associated with each wavelet packet decomposition in (a). Larger coefficient values are plotted darker.

curves in the image are typical (Do and Vetterli, 2005). The successive refinements at a smooth contour in two dimensions are shown schematically in Figure 3.43 for the wavelet and contourlet scheme. The contourlet can, in fact, be seen as a discrete form of a particular *curvelet* transform – and the curvelet transform itself originated from the *ridgelet* transform proposed by Candès and Donoho (1999) as an optimal method for representing

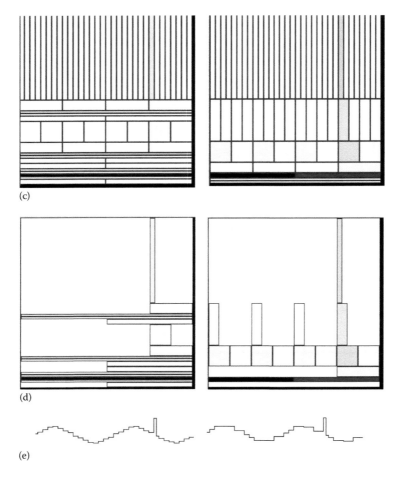

FIGURE 3.41 (CONTINUED) Wavelet packet decomposition of a simple signal: (c) The time–frequency tiling associated with the best wavelet packet decomposition (left) and wavelet decomposition (right). (d) The 16 largest coefficients from (c): Wavelet packet decomposition (left) and wavelet decomposition (right). (e) The reconstruction of the signal using only the 16 largest coefficients given in (d): Wavelet packet (left) and wavelet (right).

straight-line singularities in 2-D images. These were extended to the analysis of curved singularities through a block ridgelet transform (or first-generation curvelet transform) by the same authors in 2000. Curvelets were proposed to overcome the directional insensitivity of the 2-D discrete wavelet transform. Examples of some wavelets and curvelets at different scales are presented in Figure 3.44. *Shearlets* are yet another approach for efficiently representing directional features of two-dimensional data sets. These are distinctly different from the curvelets in that they employ shearing rather than rotation (Kutyniok and Sauer, 2007; Kutyniok et al., 2012). See also Donoho and Kutyniok (2009), who examined geometric separation of point-like and curve-like structures, such as those seen in astronomical images, using an orthonormal wavelet-shearlet dictionary and comparing it with a radial wavelet-curvelet dictionary.

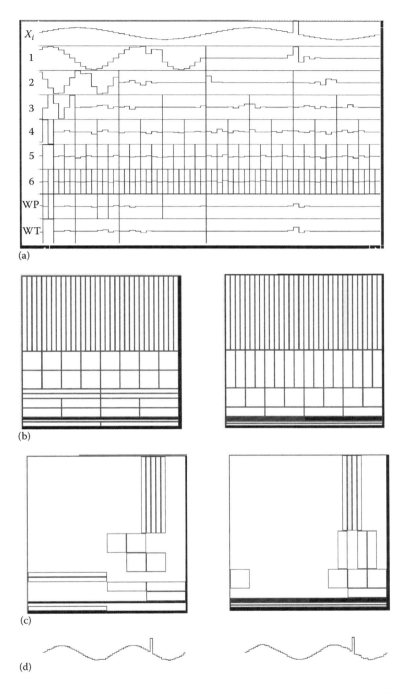

FIGURE 3.42 Wavelet packet decomposition of a simple signal using a Daubechies D20 wavelet: (a) The wavelet packet decomposition of the signal shown at the top of Figure 3.31a. (b) The time–frequency tiling associated with the best wavelet packet decomposition (left) and wavelet decomposition (right). (c) The 16 largest coefficients from Figure 3.31b: Wavelet packet decomposition (left) and wavelet decomposition (right). (d) The reconstruction of the signal using only the 16 largest coefficients given in Figure 3.31c: Wavelet packet (left) and wavelet (right).

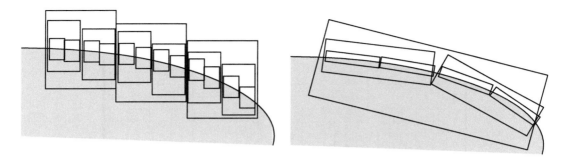

FIGURE 3.43 Schematic of wavelet and contourlet schemes – successive refinement at a smooth contour.

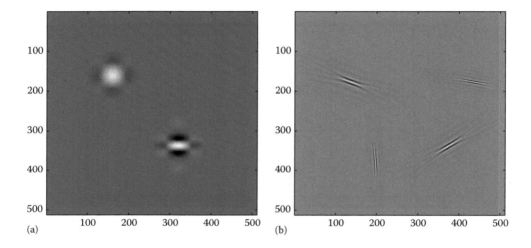

FIGURE 3.44 The elements of wavelets (a) and curvelets (b) on various scales, directions and translations in the spatial domain. (Reproduced with permission from figure 1 of Ma, J. et al., Multiscale geometric analysis of turbulence by curvelets. *Physics of Fluids*, 21, 075104-1–075104-19. Copyright 2009b, AIP Publishing LLC)

There are now a huge variety of functions and associated transforms in the literature that offer enhancements to the basic 2-D wavelet transform, especially to better represent directionally dependent data. A high-level two-page introduction to the curvelet is provided by Candes (2003), and more detailed introductions are given by Ma and Plonka (2010) and Fadili and Starck (2009). These authors also provide references to many of the other techniques that have been proposed to identify and restore geometrical properties; including contourlets, ridgelets, shearlets, bandlets, platelets, beamlets, shearable wavelets, wave atoms, surfacelets and so on. These methods are outside the general scope of this book and the reader is referred to the papers of these authors and the citations therein for further information. Another good place to begin is with the 'panorama on multiscale geometric representations' by Jacques et al. (2011), which intertwine 'spatial, directional and frequency selectivity'. This covers a wide variety of approaches and sets wavelet-based methods in the context of these other techniques.

3.11 ENDNOTES

3.11.1 Chapter Key Words and Phrases

(You may find it helpful to jot down your understanding of each of them.)

discrete wavelet transform
wavelet/discrete coefficients
wavelet frames
tight frame
orthonormal basis
dyadic grid
scaling function
approximation coefficients
inverse discrete wavelet transform
approximation coefficients
discrete approximation
continuous approximation
signal detail
multiresolution
scaling equation
scaling coefficients
compact support
fast wavelet transform
decomposition algorithm
reconstruction algorithm
wraparound
scale index

sequential indexing
level indexing
hard thresholding
soft thresholding
scale thresholding
Daubechies wavelets
subsampled
upsampled
symmlets
Coiflets
translation invariance
redundant/stationary/translation invariant/non-decimated/maximal overlap/discrete wavelet transform
biorthogonal wavelets
wavelet packet transforms
Shannon entropy measure
contourlet
ridgelet
curvelet
shearlet

3.11.2 Further Resources

Daubechies' (1992) seminal book provides a good grounding in all aspects of wavelet transforms, containing, among other things, informative text on wavelet frames and wavelet symmetry. It also provides more information on the construction of symmlets, Coiflets and biorthogonal wavelets. Other useful introductory papers on the discrete wavelet transform include those by Sarkar et al. (1998), Meneveau (1991a), Strang (1989, 1993) and the original paper on multiresolution analysis by Mallat (1989). The paper by Williams and Amaratunga (1994) contains a good explanation of the derivation of Daubechies D4 scaling coefficients and multiresolution analysis and clearly describes wavelet filtering using matrices. Strang and Nguyen (1996) cover the connection between wavelets and filter banks used in digital signal processing. Newland (1993) presents a more detailed account of the conditions that must be satisfied for discrete orthonormal wavelets.

This chapter has concentrated on two compact, discrete orthonormal wavelets – the Haar wavelet and Daubechies D4 wavelet. There are many other wavelets we have not considered, most notably the Shannon, Meyer and Battle–Lemarié wavelets. More information on these can be found in, for example, the book by Mallat (2009). Further examples of the multiresolution analysis of simple discrete signals can be found in the book by Walker (2008). Sweldens (1996, 1998) describes the lifting scheme method for constructing biorthogonal wavelets which are not necessarily translations or dilations of a single function. These, so-called second-generation wavelets, are easily implemented, work in the spatial domain for signals of arbitrary size and exhibit perfect reconstruction. A more recent treatise on the emerging engineering applications of wavelets is provided by Akansu et al. (2010), which also contains a description of the analog signal processing applications of the discrete wavelet transform. Also worth consulting are the texts by Van Fleet (2008) and Olkkonen (2011).

The early paper by Hess-Nielsen and Wickerhauser (1996) provides an in-depth account of wavelet packets. Some recent examples of the practical use of wavelet packets include the interrogation of electromyography (EMG) signals (Diab et al., 2012), the identification of coding regions in DNA (Liu and Luan, 2014), the analysis of multispectral imagery (Benedetto et al., 2013), feature extraction from acoustic emission signals for fault diagnosis in bearings (Han et al., 2015), the evaluation of the mass fraction of microcrystalline graphite/'polyvinyl alcohol' (PVA) composites (Wang et al., 2015), psycoacoustical sound quality recognition (Xing et al., 2016) and the work of Atto and Berthoumieu (2012), who addressed the analysis and interpretation of second-order random processes using the wavelet packet transform.

The chapter briefly covered the redundant wavelet transform, a variant of the discrete wavelet transform which produces N coefficients at each level and is translation invariant. This has been found useful in statistical applications; see, for example, the original work of Coifman and Donoho (1995), and Nason and Silverman (1995). The universal threshold method is described in detail in the paper by Donoho and Johnstone (1994) together with another global approach, the minimax thresholding method. Donoho and Johnstone (1995) also developed a scheme which uses the wavelet coefficients at each scale to choose a scale-dependent threshold. This technique is known as the 'SURE' or 'SureShrink' method after 'Stein's Unbiased Risk Estimate' on which it is based. Translation invariant denoising was considered by Coifman and Donoho (1995), who computed the estimated signal using all possible discrete shifts of the signal. See also Walker (2002), who has described treeadapted wavelet shrinkage: a technique which enables the selection of image-dominated wavelet coefficients at much lower thresholds than the VisuShrink threshold. The topic of thresholding is covered in detail by Jansen (2001) and we will come across the use of thresholding repeatedly in the subsequent application chapters of this book. In particular, thresholding is revisited in Chapter 4, Section 4.2.3 where another thresholding method, the Lorentz threshold, is explained. Note that Chapter 4 also contains the mathematical details of a number of statistical measures, including power spectra, skewness, flatness, intermittency and so on, derived from the discrete wavelet transform.

Note that all examples in this chapter have used wraparound to deal with the signal edges, that is, the part of the wavelet spilling over the end of the signal is placed back at the start. This is the simplest and one of the most common treatment of edge effects for a finite length signal, and it results in exactly the same number of decomposition coefficients as original signal components. However, it is not always the best method for the application. Other approaches have already been covered in Chapter 2 for the CWT (Figure 2.35). See also Peng et al. (2009), who has considered the energy leakage across frequency bands: a phenomenon inherent within the DWT method. Take particular care when using off-the-shelf wavelet software packages as they may employ other boundary techniques as the default setting, for example, zero padding, which results in slightly more than N coefficients resulting from the full decomposition. In addition, we have employed one version of the scaling coefficient reordering b_k (Equation 3.25). Again, this is very much software dependent and you will find alternative ordering employed in practice.

CHAPTER 4

Fluids

4.1 INTRODUCTION

A fluid is a non-rigid interconnected mass which may in general exhibit either laminar or turbulent flow. Laminar flows are characteristic of slow moving or highly viscous flows where the fluid particles move in an ordered fashion, sliding over themselves in sheets (or laminae, hence 'laminar'). Turbulent flows, on the other hand, are characteristic of fast-moving or low-viscosity flows, where small disturbances in the flow may quickly grow causing the fluid particles to move in an unpredictable fashion, mixing themselves up from one point in the flow to the next. Many of the flows of interest to scientists and engineers are turbulent: the flow of water in rivers and pipelines, the flow within hydraulic machinery such as turbines and pumps, atmospheric wind flows and ocean currents and the flow of air around buildings, moving vehicles and aircraft. Turbulence manifests itself as a multiscale cascading phenomenon, where fluctuations (eddies) over a large range of scales are superimposed on a mean flow. (Think, for example, of the buffeting experienced on a windy day.) The wavelet transform has emerged as a particularly powerful tool for the analysis of fluid signals (e.g. velocities, pressures, temperatures), both temporal and spatial, covering a variety of pertinent problems from the Kolgomorov scaling of high Reynolds number homogeneous turbulence to the nature of vortex-shedding downstream of bluff bodies. Many researchers have made use of the wavelet transform's ability to probe simultaneously both the spectral and temporal (or spatial) structure of turbulent fluid flows. This chapter begins with a basic outline of the wavelet-based statistical methods used extensively in the analysis of fluid flows. The chapter is then split into four remaining sections: the first details the wavelet analysis of jets, wakes and coherent structures in engineering flows, including synthetically generated flows and fluid–structure interactions; the second considers geophysical phenomena, including wind, waves, large-scale atmospheric oscillations and associated biological processes, rainfall and river flows; this is then followed by a short section on two-phase flows; and, finally, a number of other applications in fluids are summarized in the final section.

4.2 STATISTICAL MEASURES FOR FLUID TURBULENCE

Traditionally, turbulent statistical measures are often calculated in Fourier space; however, important temporal information is lost owing to the non-local nature of the Fourier modes. As a result, wavelets have been utilized to quantify the temporal and spectral distribution of the energy in new statistical terms, such as wavelet variance, skewness, flatness and so on. These statistical measures are generally computed for discrete orthonormal wavelet expansions which some authors believe are preferable because orthogonality reduces both the number of wavelet coefficients and suppresses undesired relationships between them. Wavelet-based statistics enable both scale and location-dependent behaviour to be quantified. In this section, we will look briefly at some basic wavelet-based statistics developed for the analysis of turbulent flow signals. We will consider mainly the manipulation of discrete transform coefficients $T_{m,n}$ generated from full decompositions using real valued, discrete orthonormal wavelet transforms of the type we covered in Chapter 3. In addition, we will assume that the mean has been removed from the signal and that it contains $N(=2^M)$ data points.

4.2.1 Moments, Energy and Power Spectra

The pth order statistical moment of the wavelet coefficients $T_{m,n}$ at scale index m is defined as

$$\left\langle T_{m,n}^{p}\right\rangle_{m} = \frac{\sum_{n=0}^{2^{M-m}-1}(T_{m,n})^{p}}{2^{M-m}} \tag{4.1}$$

where only the coefficients at scale m are used in the summation. The brackets $\langle ... \rangle_m$ denote the average taken over the number of coefficients at scale m, hence the 2^{M-m} term in the denominator. For example, the commonly used coefficient variance at scale index m is

$$\left\langle T_{m,n}^{2}\right\rangle_{m} = \frac{\sum_{n=0}^{2^{M-m}-1}(T_{m,n})^{2}}{2^{M-m}} \tag{4.2}$$

The wavelet coefficient variance is simply the average energy wrapped up *per coefficient* at each scale m. Do not confuse Equation 4.2 with P_m, the mean energy *per unit time* in the signal at scale m (given in Equation 4.7), also known as the 'scale-dependent power'.

A general dimensionless moment function can be defined as

$$F_{m}^{p} = \frac{\left\langle T_{m,n}^{p}\right\rangle_{m}}{\left(\left\langle T_{m,n}^{2}\right\rangle_{m}\right)^{p/2}} \tag{4.3}$$

where the *p*th order moment is normalized by dividing it by the rescaled variance. For example, the scale-dependent coefficient skewness factor is defined as the normalized third moment:

$$F_m^3 = \frac{\left\langle T_{m,n}^3 \right\rangle_m}{\left(\left\langle T_{m,n}^2 \right\rangle_m\right)^{3/2}} \quad (4.4)$$

and, similarly, the scale-dependent coefficient flatness factor is defined as

$$F_m^4 = \frac{\left\langle T_{m,n}^4 \right\rangle_m}{\left(\left\langle T_{m,n}^2 \right\rangle_m\right)^{2}} \quad (4.5)$$

The flatness factor gives a measure of the peakedness (or flatness) of the probability distribution of the coefficients at each level. It is well known that for Gaussian distributions the flatness factor is equal to 3. Values higher than 3 occur for distributions with more pronounced tails. The flatness factor increases as the flow signal becomes more intermittent (e.g. Meneveau, 1991a; Mouri et al., 1999; Nan, 2011). See also Farge et al. (2010) for a more generalized definition of ratios of moments.

We now consider the wavelet-based scale-dependent energy defined as

$$E_m = \sum_{n=0}^{2^{M-m}-1} (T_{m,n})^2 \Delta t \quad (4.6)$$

Notice that this equation is slightly different in form from that given in Chapter 3, Equation 3.52 (where an integer time step was assumed), as the sampling time, Δt, has now been added. The scale-dependent energy per unit time, or scale-dependent power, is $P_m = E_m/\tau$ where τ is the total time period of the signal. Hence, as $\tau = 2^M \Delta t$, it can be written as

$$P_m = \frac{\sum_{n=0}^{2^{M-m}-1} (T_{m,n})^2}{2^M} \quad (4.7)$$

Therefore, as long as a signal has zero mean, both the total energy and total power of the signal can be found by summing E_m and P_m, respectively, over all scale indices *m*.

We can construct a wavelet power spectrum for direct comparison with the Fourier spectrum as follows

$$P_W(f_m) = \frac{1}{\tau} \frac{2^m \Delta t}{\ln 2} \sum_{n=0}^{2^{M-m}-1} (T_{m,n})^2 \Delta t = \frac{1}{\tau} \frac{2^m \Delta t}{\ln 2} E_m \quad (4.8)$$

The term $(2^m\Delta t)/\ln 2$ stems from the dyadic spacing of the grid. The temporal scale of the wavelet at scale index m is equal to $2^m\Delta t$. The Haar wavelet is often used in fluid turbulence studies (Lovejoy and Shertzer, 2012; Nordbo and Katul, 2013) and this temporal scale is taken as its representative period, hence the associated frequency is $f_m = 1/(2^m\Delta t)$. (However, note that we can easily modify this expression to take into account a characteristic frequency of the mother wavelet, such as the spectral peak, f_p, or bandpass centre frequency, f_c. For example, by employing f_c, the scale-dependent frequency becomes $f_m = f_c/(2^m\Delta t)$ and hence, f_c would appear in the denominator of Equation 4.8.) A Taylor expansion of $f_m = 1/(2^m\Delta t)$ gives the discrete incremental change in frequency associated with the discrete change in scale index m:

$$\Delta f_m = f'_m \Delta m + f''_m \frac{(\Delta m)^2}{2!} + f'''_m \frac{(\Delta m)^3}{3!} + f''''_m \frac{(\Delta m)^4}{4!} + \ldots \quad (4.9a)$$

Truncating at the first term with $f_m = 1/(2^m\Delta t)$ and remembering also that the scale index is an integer (i.e. $\Delta m = 1$)), we get $\Delta f_m = \ln 2/(2^m\Delta t) = -\ln(2) \times f_m$, where the negative sign is ignored in practice. It is common in the literature for the Taylor expansion to be truncated at the first term in this way – even though it has a simple limit, that is

$$\Delta f_m = \frac{1}{2^m \Delta t}\left[-\ln(2) + \frac{\ln(2)^2}{2} - \frac{\ln(2)^3}{6} + \frac{\ln(2)^4}{24} + \ldots\right] = -\frac{1}{2}\frac{1}{2^m \Delta t} = -\frac{f_m}{2} \quad (4.9b)$$

Again, ignoring the negative sign, we see from the previous expression that the incremental discrete change in the frequency is equal to half the frequency itself. In fact, this is obvious without employing a Taylor expansion, that is

$$\Delta f_m = f_{m+1} - f_m = \frac{1}{2^{m+1}\Delta t} - \frac{1}{2^m \Delta t} = -\frac{1}{2}\frac{1}{2^m \Delta t} \quad (4.9c)$$

where, again, the incremental discrete change in the frequency is equal to half the frequency itself. This is, for the forward difference $f_{m+1} - f_m$ and for the backward difference, $f_m - f_{m-1}$, it is equal to f_m (try for yourself). So now we have three options for Δf_m ($f_m/2$, $\ln(2) f_m$ and f_m) which are all equally valid. In the rest of this text, we will use $\Delta f_m = \ln(2) f_m$ (Katul et al., 1994; Katul and Parlange, 1995; Kulkarni et al., 1999). The addition of the $(2^m\Delta t)/\ln(2)$ term in Equation 4.8 is therefore required for the discrete summation over the frequency range to equal the total power in the signal. The total length of signal τ is present in the denominator of Equation 4.8 as the area under the power spectrum represents the average energy per unit time. As $\tau = 2^M \Delta t$, we can rewrite the power spectrum as

$$P_W(f_m) = \frac{\Delta t}{\ln(2) 2^{M-m}} \sum_{n=0}^{2^{M-m}-1} (T_{m,n})^2 \quad (4.10a)$$

or simply in terms of the wavelet coefficient variance as

$$P_W(f_m) = \frac{\Delta t}{\ln(2)} \langle T^2_{m,n} \rangle_m \qquad (4.10b)$$

or in terms of the scale-dependent signal power as

$$P_W(f_m) = \frac{1}{\ln(2) f_m} P_m \qquad (4.10c)$$

Compare the wavelet power spectrum for the dyadic grid orthonormal transform given by Equations 4.8 and 4.10a–c with the definition of the power spectrum for the continuous wavelet, given in Chapter 2, Section 2.9 as

$$P_W(f) = \frac{1}{\tau f_c C_g} \int_0^\tau |T(f,b)|^2 db \qquad (4.11)$$

where C_g is the admissibility constant for the particular wavelet used, f_c is a characteristic frequency of the mother wavelet defined at scale $a = 1$, the frequency f is equal to f_c/a and the derivative $df = -da/a^2$.

Figure 4.1 contains a vortex-shedding signal taken downstream of a cylinder in an open channel flow together with associated wavelet and Fourier-based power spectra. Both continuous Mexican hat and discrete Daubechies D4 wavelets have been used to construct the wavelet spectra. Notice the distinct peak in the Fourier spectrum at the vortex-shedding frequency of 0.133 Hz. The continuous Mexican hat spectrum also peaks at this value, although smearing of the spectrum around the maximum is evident. The Mexican hat spectrum was produced using a fine resolution, translation-invariant discretization of the temporal location parameter, b, and a fractional power-of-two scale for the wavelet scale, a. Close inspection of the plot corresponding to the discrete orthonormal Daubechies wavelet transform, however, reveals the coarser resolution due to the Dyadic grid structure, that is, integer power-of-two translations and dilations. Other examples of wavelet power spectra occur later in this chapter: see for example, Figures 4.14 and 4.27.

One commonly used statistical measure of the energy distribution across scales is the normalized variance of the wavelet energy. This is called the 'fluctuation intensity' and is defined as

$$FI_m = \frac{\left[\langle T^4_{m,n} \rangle_m - \left(\langle T^2_{m,n} \rangle_m \right)^2 \right]^{1/2}}{\langle T^2_{m,n} \rangle_m} \qquad (4.12)$$

which measures the standard deviation of the variance in coefficient energies at scale index m. It is also sometimes referred to as the 'coefficient of variation', or 'CV' (e.g. Katul et al., 1994 – also see Figure 4.28 – and Kulkarni et al., 1999). It follows from Equation 4.12 that

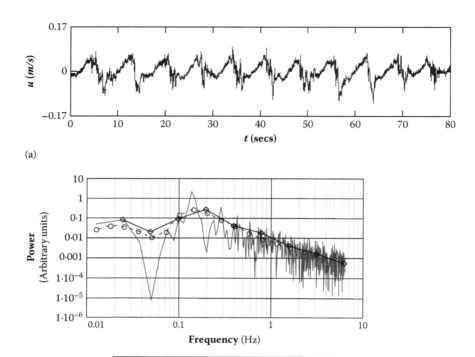

FIGURE 4.1 Power spectra of a vortex-shedding velocity signal: (a) Velocity signal taken downstream of a cylinder in open channel flow. (b) Fourier and wavelet (Daubechies D4 and Mexican hat) power spectra of the signal in (a). (From Addison, P. S. et al., *ASCE Journal of Engineering Mechanics*, 127(1), 58–70, 2001. With permission from ASCE.)

skewness and flatness measures may also be found for the scale-dependent energies (e.g. Yee et al., 1996). The fluctuation intensity provides a measure of the variability of the signal energy at scale index m. Another way to present this variability is by constructing the dual spectrum, which combines the power spectrum, $PW(f_m)$, and the fluctuation intensity converted to suitable units:

$$D_W(f_m) = \frac{2^m \Delta t \, E_m}{\ln(2) \, \tau}\left[1 + FI_m\right] = P_W(f_m)\left[1 + FI_m\right] \quad (4.13)$$

The dual spectrum is a plot of both $PW(f_m)$ and $DW(f_m)$ which provides information concerning both the contribution to the energy at various scales and its associated spatial variability (i.e. its variance – see Meneveau, 1991a,b) An example of a dual spectrum is given in Figure 4.2. It is derived from the streamwise velocity signal taken downstream of a cylinder in a wind tunnel experiment using Lemarie–Meyer–Battle (LMB) wavelets. The slight increase in the relative variance of the local energies at the larger wave numbers is due to the increasingly intermittent nature of the kinetic energy distribution here. This is confirmed by the flatness factor plot for the same signal given in Figure 4.3. Note that the

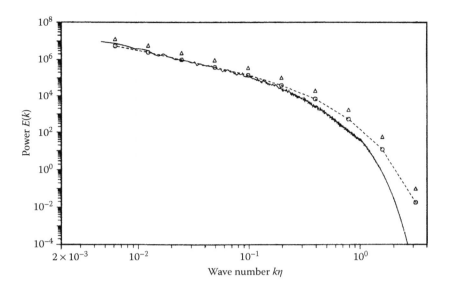

FIGURE 4.2 Dual spectrum of the turbulent flow behind a cylinder (streamwise one-dimensional spectrum). Solid line – usual Fourier spectrum. Circles – wavelet spectrum. Triangles – wavelet mean energy at every scale to which one standard deviation (computed from the spatial fluctuations) has been added. (After Meneveau, C., *Journal of Fluid Mechanics*, 232, 469–520, 1991a. With kind permission from Cambridge University Press.)

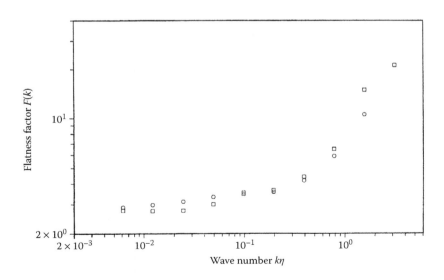

FIGURE 4.3 Flatness factors of the wavelet coefficient computed from laboratory data. The circles are for the boundary layer flow and the squares are for the turbulent wake (cylinder). (After Meneveau, C., *Journal of Fluid Mechanics*, 232, 469–520, 1991a. Reproduced with kind permission from Cambridge University Press.)

wave number is used in these plots, where the wave number is the spatial frequency, k, of the flow structures, which has a similar form to the temporal angular frequency ω. That is, the reciprocal of the length scale, y, multiplied by 2π:

$$k_m = \frac{2\pi}{y} \qquad (4.14)$$

The denominator, y, represents the physical distance between structures. We can estimate y from velocity signals for flows which have relatively low turbulence intensities compared with their mean advective velocities using Taylor's frozen flow hypothesis. This states that, if the turbulent field is changing slowly enough with respect to the mean velocity, then measuring data at a point as the turbulent field advects past is equivalent to taking a linear section through the field. Thus, characteristic time periods, p, in the velocity signal can be converted to spatial separations $y = up$, where u is the mean advective velocity. If, as before, we set the temporal scale $2^m \Delta t = p \; (=1/f_m)$ then $y = u 2^m \Delta t = 2^m \Delta y$, where Δy is the spatial increment set by the mean velocity and the sampling time. In addition, the scale index frequency is related to the wave number through the mean velocity and the 2π factor as $k_m = (2\pi/up) = (2\pi/u)f_m$. The power spectrum in terms of wave number k_m is then

$$P_W(k_m) = \left[\frac{u\,\Delta t}{2\pi \ln(2)\, 2^{M-m}}\right]\left[\sum_{n=0}^{2^{M-m}-1}(T_{m,n})^2\right] \qquad (4.15a)$$

As $\Delta y = u\Delta t$, this can be rewritten as

$$P_W(k_m) = \frac{\Delta y}{2\pi}\frac{1}{\ln(2)2^{-m}} P_m \qquad (4.15b)$$

where $P_m = E_m/\tau$ is the scale-dependent energy per unit time, or scale-dependent power, given in Equation 4.7. It can also be written (most commonly) in terms of the coefficient variation as

$$P_W(k_m) = \frac{\Delta y}{2\pi \ln(2)}\left\langle T_{m,n}^2\right\rangle_m \qquad (4.16)$$

Remember that $\left\langle T_{m,n}^2\right\rangle_m$ is the sum of the coefficient energies at scale m normalized by 2^{M-m}, that is, the energy per coefficient, hence the disappearance of the 2^{-m} factor in the denominator going from Equations 4.15b to 4.16, as P_m is the sum of the coefficient energies at scale m normalized (in a global sense) by 2^M.

The wavelet co-spectrum or cross-spectrum between two variables u and v is defined as

$$P_W(k_m) = \left[\frac{u\,\Delta t}{2\pi \ln(2)}\right]\left[\frac{\sum_{n=0}^{2^{M-m}-1}(T_{m,n})_u (T_{m,n})_v}{2^{M-m}}\right] \qquad (4.17)$$

where $(T_{m,n})_u$ denotes the wavelet coefficients for variable u. (If $u = v$, then Equation 4.17 reduces to the standard power spectral density function defined by Equation 4.15a.)

4.2.2 Intermittency and Correlation

The intermittency at each scale can be viewed directly using the intermittency index proposed by Farge (1992). This allows the investigator to visualize the uneven distribution of energy through time at a given wavelet scale. The intermittency index, $I_{m,n}$, is defined as

$$I_{m,n} = \frac{(T_{m,n})^2}{\langle T_{m,n}^2 \rangle_m} \tag{4.18}$$

$I_{m,n}$ is the ratio of local energy to the mean energy at temporal scale $2^m \Delta t$. For example, a constant value of $I_{m,n} = 1$ for all m and n means that there is no flow intermittency at all, whereas a value of 10 at a specific set of indices m and n means that at that location in the signal there is 10 times more energy contained within the coefficient at that location than for the temporal mean at that scale. Figure 4.4 contains a plot of the intermittency indices for the vortex-shedding signal shown in Figure 4.1a. High-magnitude values of $I_{m,n}$ can be observed intermittently at the lower scales.

The correlation between the scales can be measured using the pth moment scale correlation R_m^p, defined as

$$R_m^p = 2^{M-m} \sum_{n=0}^{2^{M-(m-1)}-1} B_{m,\left[\frac{n}{2}\right]}^p B_{m-1,n}^p \tag{4.19}$$

where $B_{m,n}^p$ is the pth order moment function (defined in Equation 4.20) and $[n/2]$ requires that the integer part only be used. In order to pair all the coefficients at the smaller scale

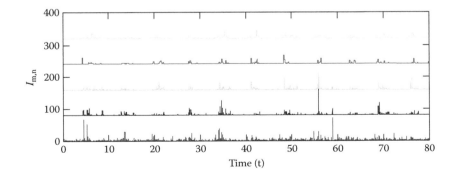

FIGURE 4.4 Intermittency indices according to wavelet scale for the flow downstream of a cylinder in an open channel. The traces from bottom to top correspond to scale indices $m = 1, 2, 3, 4$ and 5. The index plots have been displaced by 100 units to aid viewing. (From Addison, P. S. et al., *ASCE Journal of Engineering Mechanics*, 127(1), 58–70, 2001. With permission from ASCE.)

(index $m-1$) with those at the larger scale (index m), the sum is taken over the number of coefficients at the smaller scale, for example, from $n = 0$ to $2^{M-(m-1)} - 1$. $B_{m,n}^p$ is the pth order moment function defined as

$$B_{m,n}^p = \frac{(T_{m,n})^p}{\sum_{n=0}^{2^{M-m}-1} (T_{m,n})^p} \quad (4.20)$$

Note that this has a similar form to the intermittency index when $p = 2$, except that $B_{m,n}^p$ has a normalized sum at each scale, that is

$$\sum_n B_{m,n}^p = 1 \quad (4.21)$$

whereas the sum of the intermittency indices at scale m is equal to the number of coefficients at that scale, that is

$$\sum_n I_{m,n} = 2^{M-m} \quad (4.22)$$

Figure 4.5c shows the scale correlation that exists within the turbulent velocity signal shown in Figure 4.5a taken from within the atmospheric boundary layer by Yamada and Ohkitani (1991). At large scales, there is no obvious correlation; however, at small scales there is a general increase in all the pth moment scale correlations. This is in contrast with the pth moment correlations shown in Figure 4.5d for the phase-randomized signal of Figure 4.5b. Phase randomization of a signal is performed by randomizing the phases of the Fourier components of the signal, then taking the inverse Fourier transform. It destroys the correlations between levels in the signal and provides a useful benchmark signal for detecting such correlations. Moriyama et al. (1998) also examined scale correlation in a study of the density fluctuations in granular flows through pipes at various flow rates. They found that low-density flows exhibited a Gaussian distribution for the wavelet coefficients at all scales. However, for higher density flows the distribution became noticeably non-Gaussian. Figure 4.6 shows examples of the two probability density function (PDFs). The authors found no correlation between scales for the low-density flows (Figure 4.6a and Figure 4.7a) and concluded that the time series signal from these flows are equivalent to a random signal. Extended tails were, however, found for the probability distributions corresponding to the high-density flows (Figure 4.6b) which contain significant correlation across scales (Figure 4.7b). (Note that more recent work on two-phase flows is reviewed in Section 4.5.)

4.2.3 Wavelet Thresholding

Thresholding techniques are used extensively in the analysis of fluid flows to partition the signal into a coherent and 'more random' turbulent part, sometimes referred to as

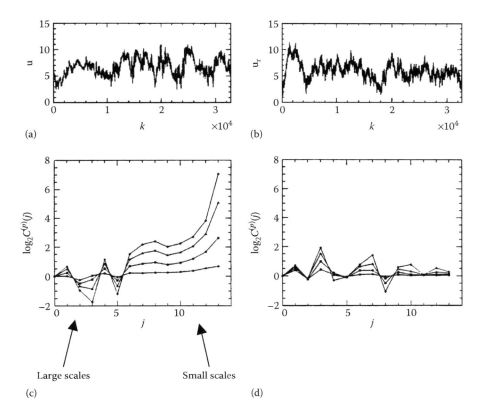

FIGURE 4.5 The pth moment correlations between scales m and $m-1$: (a) Original signal. (b) Phase-randomized signal. (c) pth moment correlations of original signal (a). $p = 1$ (squares); $p = 2$ (circles); $p = 3$ (triangles); $p = 4$ (diamonds). (d) pth moment correlations of phase-randomized signal (b). Note that authors use levels j instead of scale m where $j = M - m$ and the nomenclature $C^{(p)}(j)$ instead of R_m^p – that is, large levels j correspond to small scales. (From Yamada, M., and Ohkitani, K., *Progress of Theoretical Physics*, 86(4), 799–815, 1991.)

the 'strong' and 'weak' signal components, respectively. We saw a variety of thresholding methods in Chapter 3 (refer to Chapter 3, Section 3.4.2) and many of these have been employed in the analysis of turbulent fluid signals. Examples of wavelet thresholding used in the analysis of both spatial and temporal fluid signals can be found in Hagelberg and Gamage (1994), Turner and Leclerc (1994), Higuchi et al. (1994), Farge et al. (1996), Katul and Vidakovic (1996, 1998), Briggs and Levine (1997), Hagelberg et al. (1998), Katul et al. (1998), Szilagyi et al. (1999) and Kailas and Narasimha (1999). More recent applications of thresholding include those by Grizzi and Camussi (2012), Camussi et al. (2010), Nejadmalayeri et al. (2014), Cavalieri et al. (2010), Stephenson et al. (2014), Koenig et al. (2010) and Okamoto et al. (2011), all of which are described in more detail in later sections of this chapter.

Figure 4.8 shows an example of the partitioning of the vortex-shedding signal of Figure 4.1. The partitioning is performed using both hard thresholding and scale-dependent thresholding. The original coefficients from a Daubechies D4 decomposition are shown in

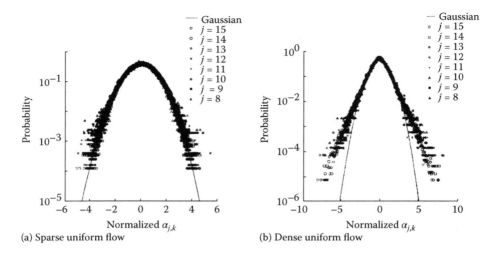

FIGURE 4.6 **PDFs of wavelet coefficients for sparse-uniform flows.** The abscissa stands for the wavelet coefficient which is normalized to have unit variance. The ordinate for the probability of finding the coefficient. (j = level index.) (From figures 6 and 9 of Moriyama, O. et al., *Journal of the Physical Society of Japan*, 67(5), 1603–1615, 1998. Reproduced with permission from the Physical Society of Japan.)

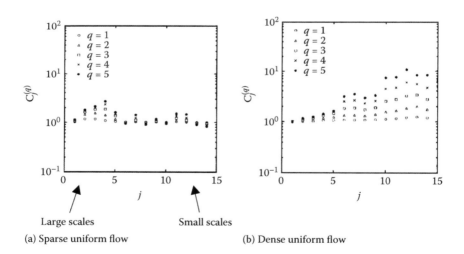

FIGURE 4.7 **Correlation between adjacent scales of the sparse and dense uniform flows.** (Note that level indexing is used: j = level index. q = moment. $C_j^{(q)}$ is the qth scale moment correlation. $C_j^{(q)}$ is equivalent to R_m^p given in the text for pth moment scale indexing.) **(a)** Sparse uniform flow. **(b)** Dense uniform flow. (From Moriyama, O. et al., *Journal of the Physical Society of Japan*, 67(5), 1603–1615, 1998. Reproduced with permission from the Physical Society of Japan.)

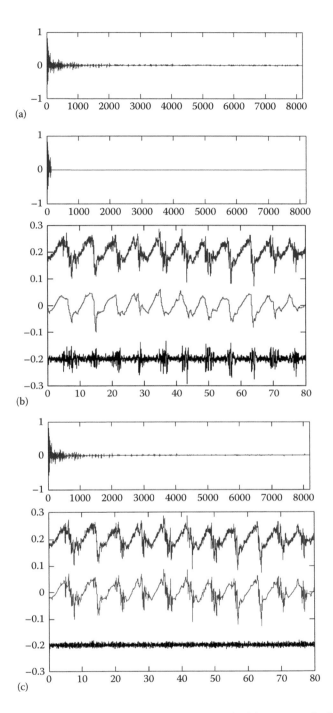

FIGURE 4.8 Scale and hard thresholding of a vortex-shedding signal: (a) Original coefficients. **(b)** Scale thresholded coefficients (top) and associated partitioning of signal (bottom). (Top trace = original signal; middle trace = strong signal component; bottom trace = weak signal component.) **(c)** Hard-thresholded coefficients (top) and associated partitioning of signal (bottom). (From Addison, P. S. et al., *ASCE Journal of Engineering Mechanics*, 127(1), 58–70, 2001. With permission from ASCE.)

sequential format in Figure 4.8a. The signal was 8192 data points in length. Figure 4.8b (top plot) shows the small-scale coefficients set to zero where the threshold scale index was set to $m^* = 6$. Subsequent wavelet reconstruction using the remaining coefficients gives the strong signal. The weak signal is reconstructed from the coefficients below m^*. The bottom plot of Figure 4.8b contains the original signal plotted together with the weak and strong parts of the signal. Figure 4.8c plots both the hard thresholded coefficients and corresponding signal partitions obtained using Donoho and Johnstone's universal threshold with $\hat{\sigma}$ derived from the median absolute deviation of the wavelet coefficients at the smallest scale. Comparing Figure 4.8b and Figure 4.8c, we can see that scale thresholding smoothes the signal in the strong part and leaves remnants from the vortex-shedding process in the weak part. On the other hand, hard thresholding retains much of the high-frequency components of large amplitude in the strong part of the signal and removes the vortex-shedding fluctuations from the weak part, leaving a more evenly distributed noisy weak signal.

Lorentz thresholding has been suggested as another method for setting a global threshold for the analysis of coherent structures within turbulent fluid signals (Katul and Vidakovic, 1996; Katul et al., 1998). In contrast to other methods, the Lorentz threshold does not assume a probabilistic structure for the wavelet coefficients. It uses the fact that the energy in the wavelet domain is not evenly distributed over the coefficients. If we plot the proportion of energy loss against the removal of each of the smallest energy coefficients in turn, we obtain a Lorentz curve. A schematic of a Lorentz curve is shown in Figure 4.9. As the energy is not evenly distributed throughout the coefficients for turbulent signals, this curve is convex. If the energy were distributed evenly, we would get the diagonal line shown in the figure. The tangent to the Lorentz curve with the same slope as the diagonal locates the point on the curve where the energy lost by removing a single coefficient is equal to the average energy in the coefficients. At this point, the gain (in parsimony) by thresholding an additional wavelet coefficient is smaller than the loss in energy. This tangent corresponds to P_0 and L_0. The example in Figure 4.9 shows that thresholding at this point removes 75% ($P_0 = 0.75$) of the coefficients but removes only 20% ($L_0 = 0.2$) of the energy.

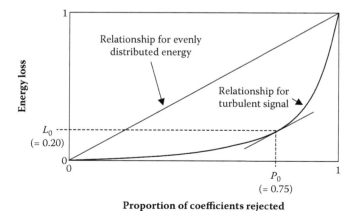

FIGURE 4.9 A schematic of the Lorentz curve used in wavelet thresholding. The optimal proportion P_0 is determined from the tangent parallel to the diagonal.

An example of Lorentz thresholding is given in Figure 4.10. Figure 4.10a,b both contain a section of a turbulent time series shown together with its strong and weak components found through hard thresholding using the Lorentz threshold. The data was acquired by a laser Doppler anemometer within a turbulent channel flow downstream of a bluff block obstacle. (For more information, see Addison et al., 2001). A Haar and symmlet (12) wavelet were used respectively in Figure 4.10a,b. For the Haar decomposition, only 244 coefficients out of 4096 making up the original signal were used in the strong reconstruction. This represents only 6% of the coefficients obtained from the decomposition. However, 95.1% of the signal energy is contained within this signal. The weak signal is composed of the remaining coefficients and represents only 4.9% of the signal energy. The smoother symmlet wavelet uses even fewer coefficients in the reconstruction and contains slightly more energy in the strong signal reconstruction. The thresholded symmlet coefficients are shown in Figure 4.10d where it can be seen that most of the large amplitude coefficients are to be found at lower scales. Figure 4.10e contains a blow-up of the first quarter of the signal showing the acquired data points together with the wavelet estimates of the strong signal. The blocky nature of the Haar reconstruction is obvious from the plot.

4.2.4 Wavelet Selection Using Entropy Measures

If we have no preset requirements for the wavelet used in the analysis of a signal, such as vanishing moments or smoothness, then an entropy measure can be employed for the selection of the most suitable wavelet. The Shannon entropy is defined for a discrete probability distribution p_i: $i = 1, 2, \ldots, N$.

$$S(p) = -\sum_i p_i \log(p_i) \tag{4.23}$$

where $\sum_i p_i = 1$. (Refer to Chapter 3, Section 3.9, where we looked briefly at the role of entropy measures in selecting the 'best' set of wavelet packet coefficients.) The maximum entropy possible from a distribution occurs when the data set has an equal probability distribution p_i at every i: that is, when the information is evenly spread across the signal. Any other distribution results in an $S(p)$ less than the maximum. The more clustered the distribution, the lower the entropy. The minimum entropy occurs when all the information is contained in a single location, that is, at only one value of i, where $p_i = 1$ (see Figure 4.11). This entropy measure is extended to the wavelet coefficient energies where we usually want to contain as much information from the signal in as few wavelet coefficients as possible. Hence, we look for the wavelet which gives us the maximum entropy for the squared coefficients. To utilize the Shannon entropy measure in the selection of the optimal wavelet, the normalized wavelet coefficient energies $\overline{T}^2_{m,n}$ are used, where

$$\overline{T}^2_{m,n} = \frac{T^2_{m,n}}{\sum_m \sum_n T^2_{m,n}} \tag{4.24}$$

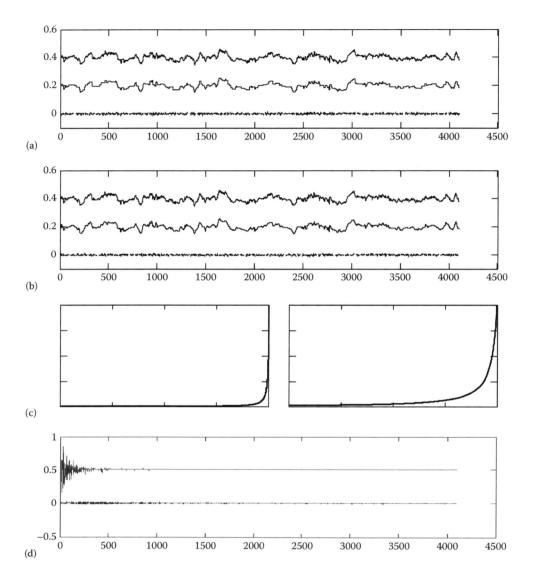

FIGURE 4.10 Lorentz thresholding of a turbulent velocity signal: (a) Original signal (top) with strong signal (middle) and weak signal (bottom). Lorentz thresholding used with a Haar wavelet. The strong signal contains 95.1% of the energy of the original signal, using only 6% (244 out of 4096) of the coefficients. (The mean has been removed from the original signal.) (b) As (a) but using a symmlet (12) wavelet. The strong signal contains 96.3% of the energy of the original signal, using only 4.7% (193 out of 4096) of the coefficients. (c) The Lorentz curve corresponding to the symmlet decomposition. The whole curve is shown on the left and the last eighth of the curve on the right. (d) The strong and weak coefficients for the symmlet decomposition.

(e)

FIGURE 4.10 (CONTINUED) Lorentz thresholding of a turbulent velocity signal: (e) Close-up of the strong signal reconstructions using the Haar (top) and symmlet (bottom) wavelets. (The reconstructed signal is shown as a bold line and the acquired data points are shown as light grey circles.) The velocity signal was acquired within a uniform channel (water) flow using a laser Doppler anemometer. Lorentz thresholding carried out following the method of Katul and Vidakovic. (From Katul, G., and Vidakovic, B., *Boundary-Layer Meteorology*, 77(2), 153–172, 1996.)

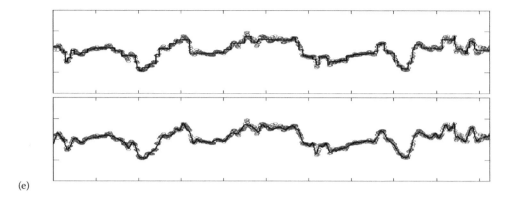

where P_i probability associated with box i
N = Number of boxes, here equal to 4
Base 10 logarithms used
Also define $P_i \log(P_i) = 0$ when $P_i = 0$

FIGURE 4.11 Schematic illustration of Shannon entropy measure. Three signals each containing four data points – (0.25, 0.25, 0.25, 0.25), (0.80, 0.10, 0.10, 0.00), (1.00, 0.00, 0.00, 0.00) – are shown with each data point contained within a box. The signals have been normalized so that the sum of their components equals 1, analagous to a discrete probability distribution.

The denominator is the total energy in the signal defined in terms of the wavelet coefficients, hence $\sum_{m,n} \overline{T}^2_{m,n} = 1$. The normalized wavelet coefficient energy $\overline{T}^2_{m,n}$ is the relative proportion of the total energy contained within the coefficient $T_{m,n}$. We then define the Shannon entropy measure in terms of the normalized wavelet coefficient energies as

$$S(\overline{T}^2) = -\sum_m \sum_n \overline{T}^2_{m,n} \log(\overline{T}^2_{m,n}) \qquad (4.25)$$

The Shannon entropy measure may be used to select the best wavelet in the analysis of turbulent signals (Briggs and Levine, 1997; Katul and Vidakovic, 1996).

4.3 ENGINEERING FLOWS

4.3.1 Experimental Flows: Jets, Wakes, Turbulence and Coherent Structures

Coherent structures are large-scale organized motions that exist in turbulent fluid flows and which influence several fluid-related processes including mixing, noise, vibrations, heat transfer and drag. Much of the research carried out in this area centres on the twin problems of separating them from background turbulence and characterizing their properties. Wavelet analysis has become an increasingly important tool in the toolbox of methods used in their interrogation.

Figure 4.12 shows three velocity signals taken within a low Reynolds number pulsed flow downstream of a pipe orifice plate (Addison, 1999). The axisymmetric, periodic vortices shed from the orifice plate produce the regular oscillatory velocity–time series shown in Figure 4.12a. This highly organized motion breaks down to a more complex flow regime as it advects downstream (Figure 4.12b,c). The Mexican hat wavelet transform plots relating to the velocity signals are shown as both contour plots and surfaces in Figure 4.13. The regular oscillatory nature of the initial vortex shedding is clearly seen in the smoothly undulating transform plot of Figure 4.13a. In addition, high-frequency background turbulent activity can also be seen at the smaller wavelet scales towards the bottom of the plot. Slightly further downstream, the vortices begin to interact with each other in a process of merging and disintegration. The subsequent development of larger-scale structures within the flow field can be seen in the transform plots of Figure 4.13b,c. The Fourier and wavelet power spectra associated with the transform plots of Figure 4.13 are given in Figure 4.14. The appearance of the wavelet spectrum as a smoothed version of the Fourier spectrum is evident in the plots. This is most obvious in Figure 4.14a, where there is a marked smearing of the strong spectral peak (corresponding to vortex shedding) in the wavelet spectrum.

Near-field pressure fluctuations in a subsonic jet have been interrogated by Grizzi and Camussi (2012). They employed an iterative algorithm to determine the optimum threshold for partitioning the signal into its acoustic and hydrodynamic contributions. The

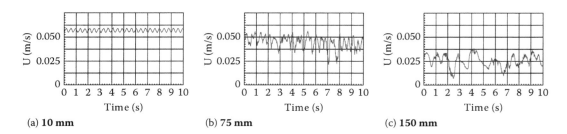

FIGURE 4.12 Velocity time series taken at various centreline locations downstream of an orifice plate in a pipe. (From Addison, P. S., *Journal of Mechanical Engineering Science*, 213(3), 217–229, 1999. Reproduced with permission of Sage Publications Ltd.)

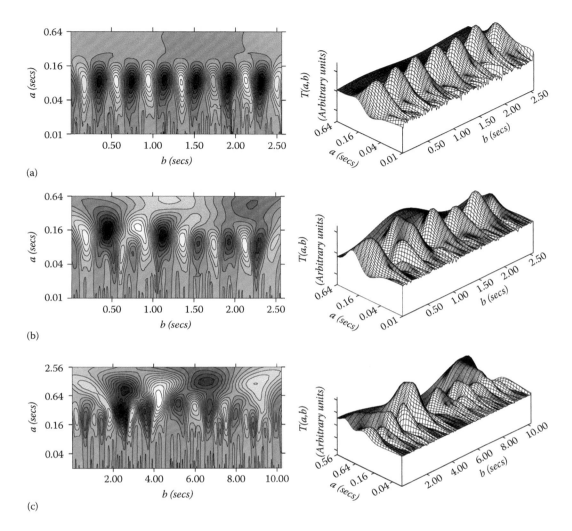

FIGURE 4.13 Wavelet transform plots of the velocity time series taken downstream of an orifice plate: (a) 10 mm downstream ($0 \leq b \leq 2.56$ s: $0.01 \leq a \leq 0.64$ s). (b) 75 mm downstream ($0 \leq b \leq 2.56$ s: $0.01 \leq a \leq 0.64$ s). (c) 150 mm downstream ($0 \leq b \leq 10.24$ s: $0.02 \leq a \leq 2.56$ s – note change in the scale ranges of the axes from plots a and b). (From Addison, P. S., *Journal of Mechanical Engineering Science*, 213(3), 217–229, 1999. Reproduced with permission of Sage Publications Ltd.)

method requires the simultaneous acquisition of two pressure signals using a microphone pair. These are partitioned into acoustic and hydrodynamic components using an initial guess for the threshold and an inverse transform of the partitioned coefficients. A cross-correlation procedure is then applied which correlates the acoustic and hydrodynamic components using separate convection and propagation velocity criteria. The peaks of the cross-correlations are compared with an signal-to-noise ratio (SNR) and, if this criterion is not satisfied, the threshold is incremented and the procedure repeated. Figure 4.15b contains an example of the cross-correlation between the acoustic and hydrodynamic pairs derived by the authors using their method where it can be seen that the acoustic

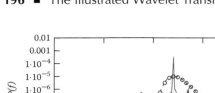
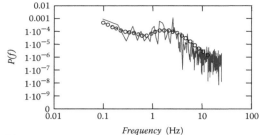
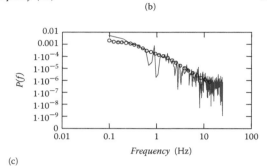

FIGURE 4.14 Wavelet and Fourier spectra of the velocity time series taken downstream of an orifice plate. (The ordinate is the turbulent energy in arbitrary units and the abscissa is the frequency in hertz. Circles indicate the wavelet power spectrum curve.) **(a)** 10 mm downstream. **(b)** 75 mm downstream. **(c)** 150 mm downstream. (From Addison, P. S., *Journal of Mechanical Engineering Science*, 213(3), 217–229, 1999. Reproduced with permission of Sage Publications Ltd.)

correlations happen over a much shorter timescale, as expected. A number of other jet experiments have also benefited from wavelet analysis: for example, Bernardini et al. (2014) studied the vortex shedding at an airfoil subject to a pulsed jet flow using a series of phase-averaged wavelet transform plots which demonstrated the convection downstream of the broadband frequency content in the flow; Koenig et al. (2010) investigated the sound field radiated by a high Mach number subsonic jet using two threshold intermittency metrics to filter the pressure signals based on scalogram topology; Grassucci et al. (2010) separated near-field pressures signatures of a free jet into intermittent and non-intermittent components using a wavelet filter based on a local intermittency measure criterion; Jaunet et al. (2012) used a wavelet-based technique to reconstruct a laser Doppler velocimetry (LDV) signal from a screeching rectangular jet flow; Bayraktar and Yilmaz (2011) performed a continuous wavelet transform (CWT)-based analysis to reveal the dominant frequencies in a transverse jet flow experiment; and Low et al. (2011) examined the simultaneous near-field and far-field measurements of a jet subject to both open- and closed-loop forcing using Mexican hat wavelets. The wavelet maps generated in this latter study provided an insight into how matched and residue events in the near field correspond to sound pressure in the far field. More recently, Wang et al. (2016) have considered the flow structures generated by two water jets impinging on a stationary pool. They observed almost identical flow structures in the wavelet-transformed data (both at small and large scales) appearing at distinctly different time periods. In a related area, the pressure fluctuations within the

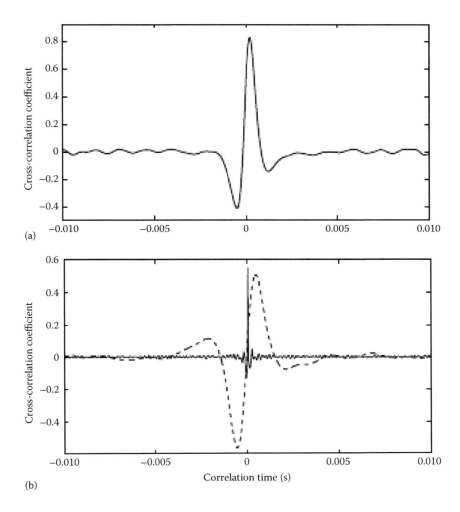

FIGURE 4.15 Cross-correlation coefficient computed between two microphones separated by a distance of 14 mm positioned at $x/D = 10$, $y/D = 5$ and for $M = 0.5$: (a) Cross-correlation coefficient between the original signals. (b) Cross-correlation coefficients between the acoustic (solid curve) and hydrodynamic (dashed curve) parts obtained through the wavelet method introduced in the present work. The resulting convection speeds determined from the peak of the cross-correlations are 345 and 55 ms^{-1} for the acoustic and hydrodynamic components, respectively. (From Grizzi, S., and Camussi, R., *Journal of Fluid Mechanics*, 698, 93–124, 2012. Reproduced with kind permission from Cambridge University Press.)

nozzle of a jet have been investigated using Morlet-based wavelet methods by Baars et al. (2011) and Baars and Tinney (2013).

In addition to jets, several studies have considered turbulent recirculation flows downstream of backward-facing steps and cavities. A wavelet coherence analysis of both a shallow cavity flow and a flow within a tunnel with a modified roof was undertaken by Camussi et al. (2008). They developed a post-processing method to select pressure signal events with strong local in-time coherence and demonstrated the ability of the wavelet transform to 'highlight' the convection of selected flow structures. Lee and Sung (2001) performed an 'unsteady wavelet analysis' on separated backward-facing step flows and, later, Liu et al. (2005) used wavelet-based multiresolution autocorrelation to diagnose the changing flow structure in a reattaching flow downstream of a backward-facing step with local forcing of the flow taking place. The recirculating flows generated by a street canyon experimental model have been studied using both the Mexican hat and Morlet wavelets by Kellnerova et al. (2011). Schram et al. (2004) have employed 2-D Mexican hat wavelets in their interrogation of the eddy structure within a backward-facing step flow using particle image velocimetry (PIV), and a Morlet wavelet approach was used by Vikramaditya and Kurian (2013) to demonstrate that the trailing wall coefficients were significantly higher in magnitude than those for the front wall in supersonic cavity flow.

Numerous research groups have attempted to use wavelets to understand the flow physics of wake flows: the flows around moving or stationary solid bodies. Alam and Sakamoto (2005) elucidated the multistable flow patterns and lock-in phenomenon from fluctuating pressures measured simultaneously on the surfaces of two staggered cylinders in a wind tunnel experiment by comparing the scalograms of the two signals. They concluded that wavelet analysis is a 'very useful tool' for analyzing multistable flow. In later work by the same group, Alam and Zhou (2008) considered the wake flow generated by two cylinders in tandem aligned with the flow. A number of upstream cylinder diameters were considered in the study. This time, they employed a Morlet-based cross-wavelet transform to investigate the lock-in phenomenon (the occurrence of the same frequency downstream of both cylinders) which occurred intermittently for some flow configurations. The cross-wavelet transform was used to compare the flow signals acquired from a location between the two cylinders and another behind the two cylinders and they found intermittent locking behaviour where the vortices shed from the downstream cylinder locked in with the vortices shed from its upstream counterpart. Endres and Möller (2009) demonstrated the 'almost steady-state behaviour' of an upstream disturbance within the turbulence signal acquired from within a square array of tube banks. In later work, Indrusiak and Möller (2011) investigated the Strouhal number changes in a transient wake behind a cylinder caused by accelerating and decelerating flows. They found that wavelet analysis is a valuable tool in interrogating the signals from these flows. Figure 4.16a shows the continuous wavelet scalograms of wake velocity data during the accelerating, constant and decelerating flow regimes. The dominant frequency of vortex shedding is obvious in this Morlet wavelet-based representation of the signal. Figure 4.16b shows a detail of the accelerating part. The authors also decomposed the signal using discrete wavelets to isolate the scale-dependent contributions to the signal. Later, De Paula and Möller (2013) utilized continuous and discrete wavelet transforms in an experimental study of bistable flow on two parallel circular cylinders. They classified the data according to a representative probability density function in a mixture model approach and adopted an energy model to describe the bistable behaviour of the velocity components.

A comprehensive examination of the 3-D nature of the wake downstream of a circular cylinder at various angles of inclination from 0° (perpendicular to the flow) to 45° was undertaken by Razali et al. (2010). They applied orthogonal wavelet decomposition to the

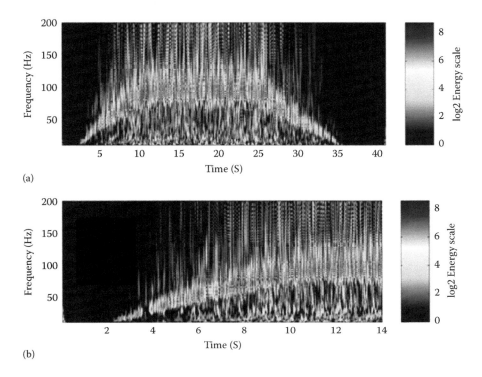

FIGURE 4.16 (a) Wavelet spectrum of the wake velocity data and (b) detail showing the accelerating part. (Original in colour.) (From Indrusiak, M. L. S., and Möller, S. V., *Experimental Thermal and Fluid Science*, 35(2), 319–327, 2011.)

velocity and vorticity signals and found that the most significant contributions to the measured velocity variances were from the organized large-scale structures at the vortex-shedding frequency band, followed by intermediate scale structures. Analysis of the vorticity variances indicated that the increase in streamwise vortices was at the expense of spanwise vortices. Wake flows generated by non-cylindrical elements have also been the subject of experimental study. Buchner and Soria (2013) performed multidimensional wavelet decomposition of dynamic stall events exhibited by a flat plate. They plotted maps of scale-dependent decomposed particle image velocity (PIV) data across the flow field. These maps incorporated both streamwise and spanwise information which elucidated features of the three-dimensionality of the dynamic stall vortex system. See also Iungo and Lombardi (2011), who employed Morlet wavelets to highlight the vortex structures in a wake flow generated by a triangular prism.

In other work, the pressure fluctuations of the tip leakage flow of an aerofoil was studied by Camussi et al. (2010) using both PIV and hot wire anemometry (HWA). They utilized thresholding of wavelet-based local intermittency measures (LIMs) to filter out pertinent pressure signals corresponding to high-energy pressure events (Figure 4.17). These were ensemble averaged to construct pressure time signatures (Figure 4.18). They employed both 'auto-conditioning', where the pressure signal itself was thresholded to trigger a determination of pressure time events, and 'cross-conditioning' where the trigger signal

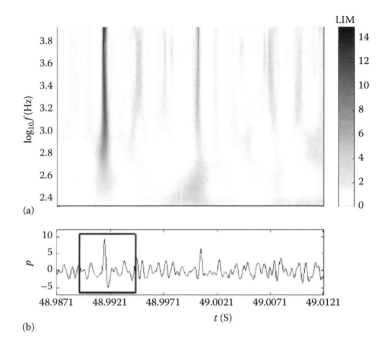

FIGURE 4.17 Example of selection procedure: (a) The large value of LIM indicates that an event occurs at the probe location in $t_0 \sim 48.992$ s. Once the event is detected, a portion of the pressure signal (b) centred on t_0 is extracted to perform the conditional average. The signal considered is taken from a pressure probe in the tip region. (From Camussi, R. et al., *Journal of Fluid Mechanics*, 660, 87–113, 2010. Reproduced with kind permission from Cambridge University Press.)

was always the pressure signal but the conditioned signal could be pressure at the wall or velocity (and where the velocity could be a HWA velocity time series or a two-dimensional PIV velocity field). The approach allowed the detection of causally linked intermittent perturbations in various flow regions and the assessment of their contributions to the far field.

4.3.2 Computational Fluid Dynamics: Simulation and Analysis

Significant steps have been made in the development of wavelet approaches in computational fluid dynamics (CFD). The articles cited in this section, and a few elsewhere in this chapter, deal with numerical simulations of turbulent flows, often comparing them with experiments. A detailed review of wavelet-based numerical algorithms for CFD is presented by Schneider and Vasilyev (2009). They recognized that dynamically active scales in turbulent flows are not distributed homogeneously in space or time and that this flow intermittency is suited to the wavelet's ability to isolate localized turbulent structures such as shocks, flame fronts and vortices. The authors reported on a number of methods for solving the Navier–Stokes equations, including pure wavelet methods, adaptive multiresolution methods, Lagrangian wavelet methods, space-time wavelet methods and wavelet-optimized adaptive methods.

There are now a considerable number of groups using wavelet techniques for the simulation of turbulent flows. Figure 4.19 contains some visualizations of intense vorticity regions

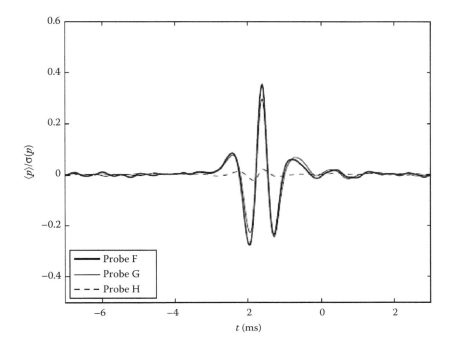

FIGURE 4.18 Averaged pressure time signatures obtained in the trailing-edge region ($x/c = 0.975$) for the reference gap configuration: Effect of the distance from the gap (z): probe F($z/c = 0.005$), probe G ($z/c = 0.015$) and probe H ($z/c = 0.03$). The amplitudes are normalized with respect to the standard deviation of the original signals (σ_p). (From Camussi, R. et al., *Journal of Fluid Mechanics*, 660, 87–113, 2010. Reproduced with kind permission from Cambridge University Press.)

generated by Okamoto et al. (2011) using the wavelet-based coherence vorticity simulation (CVS) method of 3-D homogeneous turbulence. CVS decomposes the flow into coherent and incoherent contributions by wavelet filtering the vorticity field. Nejadmalayeri et al. (2014) have recently described a new framework for spatiotemporally adaptive turbulence simulations using variable-fidelity representations. This was achieved by combining hierarchical wavelet-based computational modelling with a wavelet thresholding filter (WTF) comprising spatiotemporally varying thresholding. The authors described the work as the first successful attempt to develop a hybrid wavelet-based adaptive approach to turbulence modelling. This combines three different wavelet-based approaches – the CVS method, stochastic adaptive large eddy simulation (SCALES) and wavelet-based direct numerical simulation (WDNS) – allowing two-way transition between the approaches within the hybrid method to resolve all significant flow processes. In other work, Tabor and Baba-Ahmadi (2010) have compared a discrete wavelet method, among several techniques, for generating inlet flows for large eddy simulation (LES) models. Inlet conditions are the turbulent flows generated prior to (i.e. upstream of) the flow problem to be considered in the simulation. Accurate inlet flows are essential, as they have a significant impact on the flow dynamics generated within the problem domain. The authors point out that the wavelet scheme allows the synthesis of velocity fluctuations with local correlation, rather than the

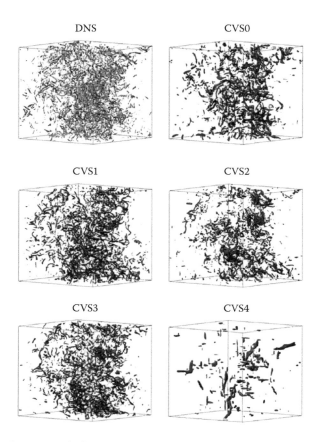

FIGURE 4.19 Visualization of the intense vorticity (ω) regions for DNS and (CVS0–4) at $t/\tau = 3.78$. Isosurfaces of vorticity are shown for $|\omega| = M + 4\sigma$, where M and σ denote the mean value and standard deviation of the modulus of each vorticity field. (Original in colour.) (From Okamoto, N. et al., *Multiscale Modeling and Simulation*, 9(3), 1144–1161. Copyright 2011, Society for Industrial and Applied Mathematics. Reprinted with permission. All rights reserved.)

infinite correlation generated by harmonic fluctuations. See also Keylock et al. (2011), who used gradual wavelet reconstruction (Keylock, 2010) in their work on inlet flow conditions.

A number of authors have utilized wavelet techniques to probe the nature of the synthetic turbulence generated by CFD schemes (both wavelet based and non-wavelet based). Courbebaisse et al. (2011) computed the modulus maxima skeleton plots of pressure field signals from a lid-driven cavity flow. The flow was generated using both LES and a direct numerical solution (DNS) method and the two flow fields were compared. Figure 4.20 contains the modulus maxima for both laminar and turbulent LES flow regions. The longest maxima lines were collected and plotted (Figure 4.21) where it can be seen that the laminar modulus maxima are of relatively constant amplitude whereas the turbulent modulus maxima exhibit a distinct slope. This slope, of −2/3, corresponds to very large signal features associated with turbulent bursts which occur when a pair of counterrotating vortices is produced. The corresponding plot (not shown) derived from the DNS modulus maxima contains a slope of −5/3 linked to turbulent scales and to the interaction between small and

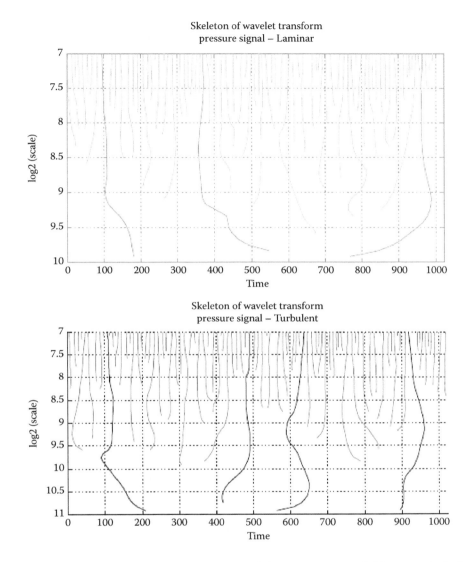

FIGURE 4.20 Skeleton of the wavelet transform of the LES pressure field successively for both the laminar (top) and turbulent (bottom) regimes. (From Courbebaisse, G. et al., *Computers and Fluids*, 43(1), 38–45, 2011.)

large eddies. In other work in this field, Seena and Sung (2011) employed both 1-D Morlet and 2-D Mexican hat wavelet analysis in educing dynamic structure in turbulent open cavity flows generated by an LES model. They plotted the vorticity pressure and streamlines within the cavity at various wavelet scales. Wang et al. (2013a) performed LES of inertial particle dispersion in turbulent flow over a backward-facing step (BFS) and calculated wavelet-based skewness and flatness indices to characterize the strong intermittencies within the flow field. Salpeter and Hassan (2012) studied the jet–flow interaction with a staggered rod bundle and Kerhervé et al. (2012) employed a fourth-order Paul wavelet as part of their investigation of the mechanism associated with downstream radiation in

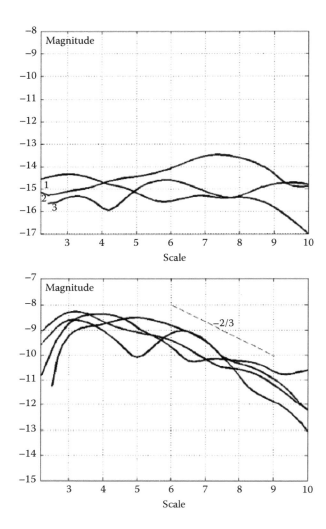

FIGURE 4.21 Amplitude of the wavelet transform along the longest filaments of the skeleton of the LES pressure field successively for both the laminar (top) and turbulent (bottom) regimes. (From Courbebaisse, G. et al., *Computers and Fluids*, 43(1), 38–45, 2011.)

subsonic jets. (The Paul wavelet was used because of its suitability for the analysis of jet noise [Koenig, 2011].)

Wavelet-based anisotropy measures have been employed by Jacobitz et al. (2010) to characterize the anisotropic properties of homogeneous DNS turbulence exhibiting mean shear and system rotation. These measures included directional energy components and flatness factors for the velocity components. They found that wavelet-based directional energy measures agreed with conventional measures. They also considered wavelet-based geometrical measures of scale-dependent helicities. They concluded that cases of growing turbulent kinetic energy (TKE) were characterized by a tendency towards local two-dimensionalization while decaying cases exhibited a preference for swirling motion. In later work, Jacobitz et al. (2012) investigated whether their findings held for a wider class

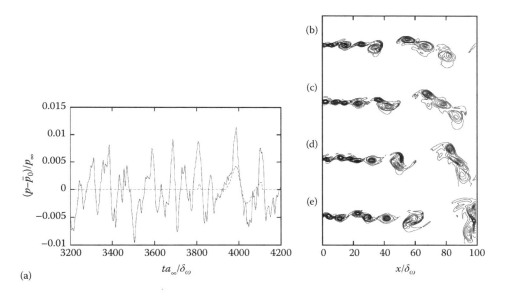

FIGURE 4.22 (a) Wavelet filtering, (——) original and (- - -) filtered pressures. (b)–(e) Vorticity contours at $ta_\infty/\delta_\omega == 3899.3$, $ta_\infty/\delta_\omega == 3916.1$, $ta_\infty/\delta_\omega == 3932.9$ and $ta_\infty/\delta_\omega = 3949.7$. Contours range from $0.014\Delta U/\delta_\omega$ to $0.7\Delta U/\delta_\omega$. (Reproduced with permission from Cavalieri, A. V. et al., Intermittent sound generation and its control in a free-shear flow. *Physics of Fluids*, 22, 115113-1–115113-14, 2010. Copyright 2009, American Institute of Physics.)

of homogeneous turbulent flows. Cavalieri et al. (2010) investigated numerically generated free-shear flows using Paul wavelets to determine the physical mechanism responsible for the generation of intermittent sound. They produced filtered signals by thresholding the CWT representation of the original pressure signals, which led them to conclude that intermittent triple vortex merging events had a considerable contribution to the overall sound produced. The results of their filtering operation are presented in Figure 4.22a. The filtered pressure signal (dashed line) is zero except for a localized segment corresponding to a triple vortex merging as shown in Figure 4.22b–e. Other reported studies of interest in this area include those by Gandhi et al. (2011), who generated a DWT wave number spectrum for use in their study of natural convection which employed both CFD simulation and particle image velocimetry (PIV) measurements; Wilczek et al. (2011), who performed a wavelet analysis of the conditional vorticity budget in fully developed homogeneous isotropic turbulence in DNS flows; Khujadze et al. (2011), who considered the extraction of the coherent contribution to the DNS boundary layer vorticity field using a new adaptive 3-D wavelet transform in order to identify the wavelet coefficients that retain the coherent flow structures; Abdilghanie and Diamessis (2013), who applied 1-D and 2-D wavelets in their analysis of internal gravity waves generated by a towed sphere in an internally stratified fluid; Wu et al. (2013), who used discrete and continuous wavelets in an investigation of the turbulent structure of DNS flows containing drag-resistant polymer solutions, comparing them with those without polymers; and Zhao et al. (2010), who considered the addition of turbulence to fluid flows for use in animations for movies and computer games.

4.3.3 Fluid–Structure Interaction

The interaction of a fluid with a structure is a common phenomenon which manifests itself, in an engineering context, as a number of problems including wind loading on buildings, flow-induced vibration of bridge decks and aircraft aerofoils, water loading of bridge piers, buffeting of bluff bodies and so on. Hajj and Tieleman (1996) suggested using the wavelet transform to characterize the intermittent nature of wind events to model pressure variations on low-rise structures, detailing the advantages of using a wavelet-based approach over a conventional Fourier approach to the problem. They illustrated their ideas briefly using Daubechies D4 wavelet to decompose a sample wind velocity time series. In later work using the Morlet wavelet transform, Hajj et al. (1998) found correlations between energetic events in the atmospheric wind and low-pressure peaks that occur at pressure taps placed over a large area of a low-rise building. Another early paper by Gurley and Kareem (1999) discussed the applications of both the discrete and continuous wavelet transforms to earthquake, wind and ocean engineering; including fluids engineering problems such as the transient response of buildings to wind storms, the analysis of bridge responses to vortex shedding and the correlation between pressure measured at a building rooftop and upstream. Later papers by Kareem and Kijewski (2002) and Gurley et al. (2003) included more sophisticated wavelet methods, such as wavelet bicoherence, to characterize the random nature of wind fluctuations in space and time and their interactions with structures. More recently, Kijewski-Correa and Bentz (2011) have outlined a wavelet analysis framework for diagnosing impulse-like responses embedded in narrowband wind-induced accelerations. This involves the use of a customized wavelet which identifies impulses as concentrated energy bands in the scalogram. Applying their technique to the investigation of acceleration data from a tall building subject to transient wind effects, they observed companion pairs of sway and torsional modes that exchanged energy in the course of the events. They discussed their findings in light of design philosophies for tall buildings. See also Yi et al. (2013), who employed wavelet transforms representations together with empirical mode decomposition in their examination of the dynamic characteristics of the wind-induced responses of a 420 m high 'super-tall' building in Hong Kong during the passage of typhoons.

Kareem and Wu (2013) have reviewed several wavelet techniques for the analysis of wind-induced effects on bluff bodies in turbulent flows, including scalograms and co-scalograms, and the wavelet-based PDFs of turbulent flows. Work by the same research group (Wang et al., 2013b) employed stationary wavelet transforms and Hilbert transforms in the simulation of full-scale thunderstorm downburst winds on structures. In later work, the same group (Spence et al., 2014) described their use of wavelet ridges to detect frequency coalescence of building vibrations over time, where separate frequency responses can be seen to merge and then separate out. Figure 4.23 shows the wavelet scalogram ridges over 200 seconds for one of their case studies, illustrating the frequency coalescence phenomenon where the two ridges merge then split apart again.

A number of authors have considered simpler structural elements in their investigations of fluid–structure interactions, such as cylinders, prisms and plates. Hamdan et al.

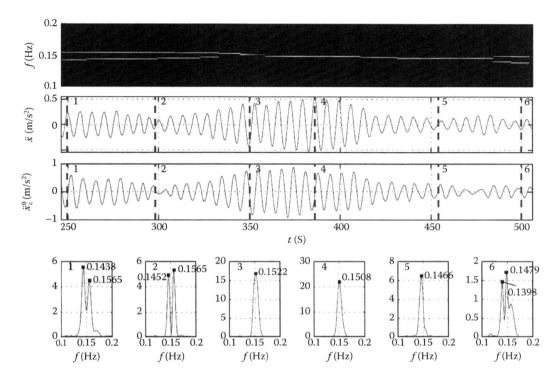

FIGURE 4.23 Case study, top floor corner acceleration response: Ridges of the wavelet scalograms, time history of the translational contribution $\ddot{x}$, time history of the rotational contribution $\ddot{x}_c^\theta$ and instantaneous spectra showing frequency coalescence. (From Spence, S. M. et al., *Probabilistic Engineering Mechanics*, 35, 108–117, 2014.)

(1996) compared the analytical properties of four wavelets – the Morlet, Mexican hat, eighth derivative of Gaussian and Daubechies tight frame wavelets – in the elucidation of vibration signals from a cylinder in a cross-flow. Jubran et al. (1998a,b) extended this work to cover the chaotic nature of vortex shedding from a cylinder in a cross-flow. Alam et al. (2003) used Morlet wavelet scalograms to examine the fluctuating lifts of a downstream cylinder with tripping rods to disrupt the oncoming flow and reduce drag, and Alam et al. (2006) also performed further two cylinder experiments with a T section to disrupt the flow. In the latter work, wavelet phase was also plotted. Figure 4.24 contains a series of wavelet transform modulus plots for the cross-flow hydrodynamic time series from a cylinder subjected to a flow of increasing velocity where the flow field was numerically modelled by solving Reynolds-averaged Navier–Stokes and k-w turbulence equations. The Morlet wavelet–based analysis, which generated this figure, was employed by Zhao et al. (2012) to illustrate the changing composition of the frequency response caused by the vortex-shedding vibration. The dominant components can be seen to be multiples of the fundamental frequency. The authors found that only a single frequency component occurred for low-velocity flows (e.g. Figure 4.24a) whereas for higher flow velocities the vibration modes altered very frequently (e.g. Figure 4.24g). The same group have also

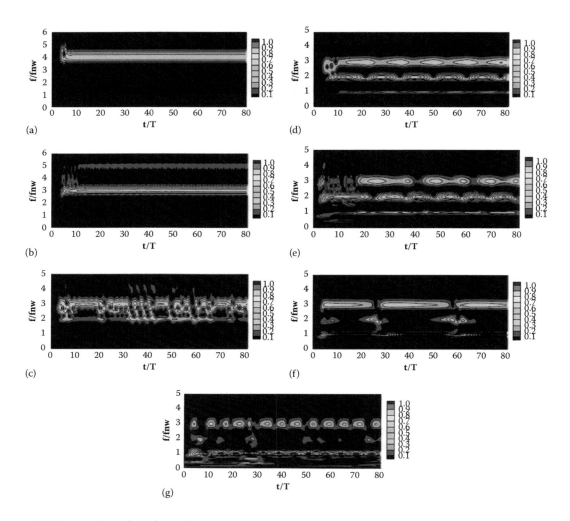

FIGURE 4.24 Results of amplitude contour by wavelet analysis in the time–frequency domain for KC = 20: (a) $V_r = 5$, (b) $V_r = 8$, (c) $V_r = 10$, (d) $V_r = 14$, (e) $V_r = 16$, (f) $V_r = 20$ and (g) $V_r = 35$. V_r is the reduced velocity. (Original in colour.) (From Zhao, M. et al., *Ocean Engineering*, 41, 39–52, 2012.)

utilized wavelet transform plots (real part) to explore the numerically simulated vortex-induced vibration of two cylinders with different diameters (Zhao and Yan, 2013) and in rigidly coupled cylinders of the same diameter in low Reynolds number flows (Zhao, 2013).

The cross-wavelet transform was employed by Nemes et al. (2012) to determine the phase between body position and lift force in an experimental investigation of the flow-induced vibration of a square cylinder at varying angles of attack. Their results revealed the influence of the shear layer separation on the fluid driving force experienced by the body which differs from a circular cylinder. They superimposed contours of equal wavelet energy in their plots of shedding frequencies at various flow velocities and angles of attack. Examples of these plots are shown in Figure 4.25. The symbols on the plots represent frequency power spectrum peaks and their size indicates the relative energy contained by the dominant frequency of oscillation. The authors proposed this kind of plot as a useful visualization aid

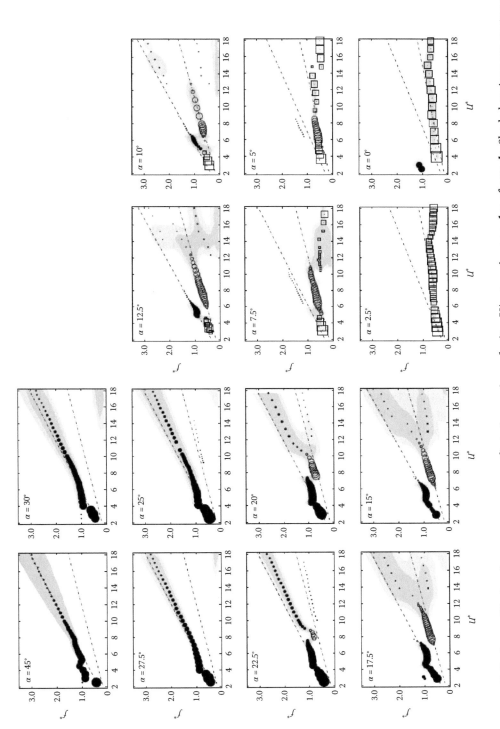

FIGURE 4.25 Cylinder oscillation frequency response, f^*, to increasing velocity, U^*, at various angles of attack. Shaded regions are contours of relative normalized intensity in wavelet energy. Filled circles represent shedding frequencies, open circles represent frequencies associated with a higher branch and open squares represent lower frequencies including those associated with galloping. The size of the symbols relate the power magnitude of the peaks. Dashed lines represent the shedding frequency of the stationary body at the specified angle of attack (α) and its first subharmonic. (From Nemes, A. et al., *Journal of Fluid Mechanics*, 710, 102–130, 2012. Reproduced with kind permission from Cambridge University Press.)

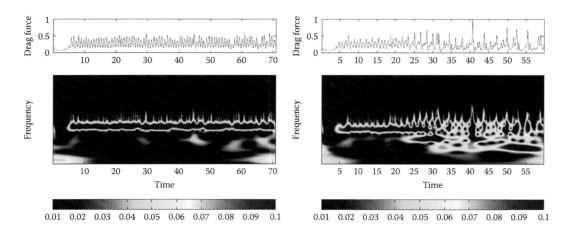

FIGURE 4.26 Wavelet analyses of the drag force for case $u^* = 20$ (left) and $u^* = 22$ (right). The modulus of the complex-valued Morlet wavelet coefficients are plotted as a function of position and scale. The original signal is plotted on the top. (Original in colour.) (From Engels, T. et al., *Computers and Structures*, 122, 101–112, 2013.)

of the rate of transfer of spectral energy between competing frequency peaks as a function of reduced velocity. Thus, the shrinking symbol size and increasing contoured energy area as velocity increases in the figure for $\alpha = 45°$ indicates the energy being distributed over an increasingly wider band. At high angles of attack, frequencies associated with vortex-induced vibration behaviour were observed, whereas at low angles of attack (0°–10°), only low-frequency oscillations typical of galloping behaviour were noted. Synchronized multiple modes can be discerned for lower angles of attack ($\alpha = 15°$–20°). The authors also plotted out scalograms exhibiting vortex-shedding periods interspersed with aperiodic behaviour at various flow velocities. They found that the aperiodicity became more dominant as the velocity increased.

Le and Nguyen (2008) and Le et al. (2010) applied the cross-wavelet transform and wavelet coherence in a study of the relationship between coherent structures in turbulent velocity signals and induced pressures on a bluff body in a wind tunnel. They found that the coherence was higher for the pressure signal than the turbulence signal. In other work, the time evolution of the drag force on a slender structure in a free stream exhibiting flutter instability has been investigated by Engels et al. (2013). Plots of wavelet scalograms of the fluctuating drag force clearly demonstrated both periodic and aperiodic (chaotic) behaviour. Examples of periodic and chaotic behaviour are shown in Figure 4.26a,b, respectively, for two different reduced free-stream velocities (u^*). In other work, Huang et al. (2014) have employed Morlet wavelet transforms to extract instantaneous frequencies and amplitudes of the aerodynamic forces on a flat plate from a numerically generated time series.

4.4 GEOPHYSICAL FLOWS

The meteorological community has been particularly active in the application of wavelet-based methods to the analysis of fluid flows. An early paper by Meyers et al. (1993) provided an introduction to the use of wavelets (specifically the Morlet wavelet) in oceanography and meteorology. They illustrated their discussion with examples of the wavelet decomposition of simple signals into modulus and phase before applying wavelets to the analysis of Yanai waves. They commented both on the 'non-trivial task' of interpreting the phase of complicated signals and on the edge effects of the data on the transform plots. Since then, numerous papers have appeared covering a wide area of geophysical flow processes including many novel advanced wavelet techniques.

4.4.1 Atmospheric Processes: Wind, Boundary Layers and Turbulence

Several workers have considered the power spectra of time series measurements acquired in the atmospheric boundary layer. Some researchers have investigated the power spectra of the measurements and the deviation from the expected $-5/3$ Kolmogrov scaling in the inertial subrange. Others have concentrated more on the detection and interrogation of coherent structures in the flow. Figure 4.27 shows a plot of the power spectra (Haar, Wavelet and Fourier) of three signals (u,w velocities and temperature) acquired within the atmospheric boundary layer by Katul et al. (1994). There is good agreement between the Fourier and wavelet spectra. The $-5/3$ signature of the inertial subrange is evident in all spectra, especially for that of the u velocity where it extends over a large portion of the curve. Figure 4.28 contains a plot of the fluctuation intensity (called the 'coefficient of variation', 'CV', by the authors) for the three signals (refer to Equation 4.12). An increase in the fluctuation intensity with increasing wave number indicates increasing turbulent energy activity at smaller scales. Also noticeable in the plot is that the fluctuation intensity for the temperature signal is much larger than those of the velocities, possibly indicating that the temperature is not simply advected by the flow field, even at small scales. Szilagyi et al. (1996) examined the effect of turbulent intermittency on the shape of the power spectrum of an atmospheric boundary layer turbulent time series. Using Daubechies D4 wavelets, they showed that the local wavelet spectrum in the inertial subrange was sensitive to intermittency in the flow. They defined the strength of the intermittency in terms of the wavelet variance. For regions of weak intermittency, the local wavelet spectrum was found to have a slope flatter than that of Kolmogorov's $-5/3$ law, whereas regions of strong intermittency exhibited slopes greater than $-5/3$. The average slope tended to the $-5/3$ law.

The large-scale intermittent structures involved in the exchange of heat and mass within and above natural surfaces have been investigated through the wavelet decomposition of a variety of signals by a number of research groups. This includes the early work of Chen et al. (1997), who decomposed temperature signals using the Mexican hat wavelet in an attempt to detect ramp structures in turbulent flow signals acquired above a variety of surfaces; Gao and Li (1993), who considered wavelet variances from a Mexican hat wavelet decomposition of thermal and velocity fields at an atmosphere–forest interface to identify coherent structures; Qiu et al. (1995), who employed representative sections of flow signals as the basic shape for pseudo-wavelets in a study of turbulence patterns above three different vegetation layers – an orchard canopy, forest canopy and maize canopy; Collineau and Bruinet (1993a,b), who used four continuous wavelets (Haar, Mexican Hat, a wave shape and a ramp shape) to detect coherent motions in a forest canopy; and Lu and Fitzjarrald

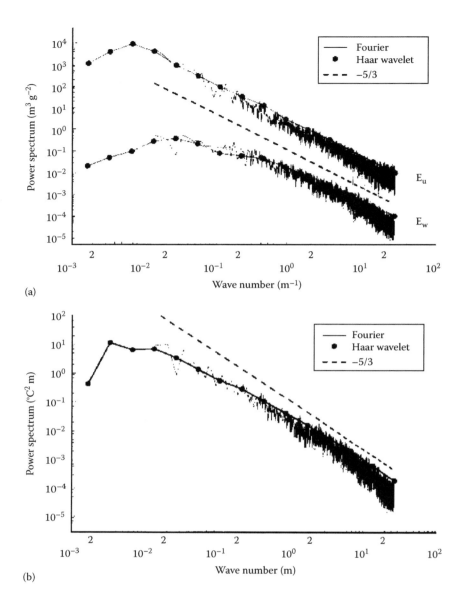

FIGURE 4.27 Fourier and wavelet power spectra: (a) Comparison between Fourier (solid line) and Haar wavelet (closed circle) power spectra for the longitudinal (U) and vertical velocity (W). The U spectrum is shifted by two decades to permit comparison with the W spectrum at small wave numbers. Taylor's hypothesis is used to convert the time domain to the wave number domain. The −5/3 power law (dotted line) predicted by K41 is also shown. (b) Same as for (a) but for temperature. (Reproduced with permission from Katul, G. G. et al., Intermittency, local isotropy, and non-Gaussian statistics in atmospheric surface layer turbulence. *Physics of Fluids*, 6(7), 2480–2492, 1994. Copyright 2009, American Institute of Physics.)

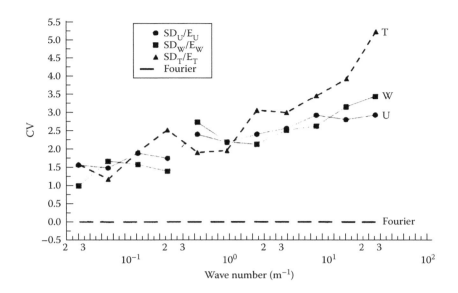

FIGURE 4.28 Fourier and wavelet power spectra. The coefficient of variation (CV) as a function of wave number for longitudinal and vertical velocity as well as temperature. The dotted line is the CV assumed by Fourier analysis. (Reproduced with permission from Katul, G. G. et al., Intermittency, local isotropy, and non-Gaussian statistics in atmospheric surface layer turbulence. *Physics of Fluids*, 6(7), 2480–2492, 1994. Copyright 2009, American Institute of Physics.)

(1994), who used the peaks in the scale-dependent wavelet variance of the continuous Haar wavelet to identify coherent structures above a mid-latitude deciduous forest. In later work, the cross-wavelet power (a normalized cross-wavelet transform) and a global wavelet spectrum measure were utilized by Bolzan and Vieira (2006) to probe vertical wind and temperature time series measured in the Amazonian forest. They found that ramp-like structures promoted an increase in interaction among the variables. More recently, Seto et al. (2013) employed Morlet wavelets to generate wind and velocity spectra in their study of the turbulence generated due to the passage of a fire front over various terrain types. They stated that wavelet analysis provides a better, smoothed, global spectral estimate than Fourier power spectra, which requires binning and smoothing routines. Zhu et al. (2010) employed normalized wavelet transform spectra to investigate the role of turbulent eddies in vertical transport in the unsteady, inhomogeneous surface layer of hurricanes during landfalls. They found that the Morlet-based method illustrated the role of eddies in generating fluxes and turbulent kinetic energy (TKE). Also, using a wavelet co-spectral decomposition of fluxes into their positive and negative contributions, Zhang et al. (2014a) demonstrated that scalar exchanges take place above a forest canopy mostly through the action of large-scale eddies.

A wavelet technique to remove high-frequency fluctuations of background turbulence, within and above a tall spruce forest, without altering the low-frequency coherent motion has been developed by Thomas and Foken (2005). Their scheme first applied a wavelet filter based on a biorthogonal wavelet, then a Morlet wavelet to determine the characteristic scales of the structures and finally a Mexican hat wavelet to detect individual coherent structures. They found that the maxima in wavelet variance plots were useful in determining the characteristic timescales (event durations) for coherent structures in both synthetic (sinusoidal and ramp functions) and real field data, comprising a variety of turbulence signals (see Figure 4.29). Using Thomas and Foken's methodology, Kang et al. (2014) have re-examined the assumption that turbulent coherent structures have correlated phases. They found that many structures, although clearly organized in space and time, do not necessarily exhibit phase correlation.

García-Lorenzo and Fuensalida (2006) interrogated generalized scintellation detection and ranging technique (G-SCIDAR) measurements of atmospheric turbulent layers using a fully automated Morlet wavelet analysis to determine their position, direction and altitude. They validated their results against balloon measurements. A number of different structures occurring in stable atmospheric boundary layers – including a solitary wave, gravity wave, density current and a low-level jet – were investigated by Ferreres et al. (2013) using a Morlet wavelet-based approach. Their analysis of tower-based wind data highlighted the

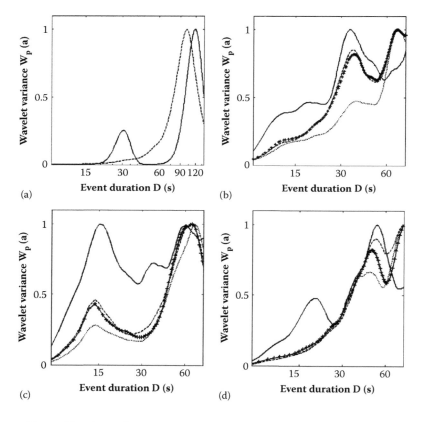

FIGURE 4.29 **Normalized wavelet variance versus event duration D of (a)** the generated signals: Sine functions (solid) and ramp signal (dashed) and real turbulent, **(b)** Data A, **(c)** Data B, **(d)** Data D: vertical wind (solid), potential sonic temperature (dashed), carbon dioxide (crosses) and specific humidity (dotted). (From Thomas, C., and Foken, T., *Theoretical and Applied Climatology*, 80, 91–104, 2005.)

various roles of turbulence and coherent structures in the transfer of heat, moisture and CO_2 in the nocturnal boundary layer. In another study, a wavelet-based technique that corrects for the high-frequency losses that occur in scalar concentration measurements has been reported by Nordbo and Katul (2013). Their method sought to correct the scalar concentration, turbulent fluctuation time series directly prior to the computation of a turbulent flux or a velocity–scalar co-spectrum using the time–frequency localities of orthonormal wavelets. In doing so, it also adjusts other flow statistics such as variances and integral timescales. The authors state that the approach does not correct for attenuation but is superior to transfer function methods and may be used in conjunction with other non-wavelet-based correction schemes. Other work in this area includes the short-term wind forecasting technique developed by Catalao et al. (2011), which combined wavelet decomposition, particle swarm and an adaptive-network-based fuzzy interference system; the comparison of three different wavelet neural networks for wind speed forecasting by Yao et al. (2013); the generation of normalized Morlet-based co-spectra of isoprene and temperature with vertical wind in a study of airborne flux measurements of biogenic volatile organic compounds above Californian oak forests by Karl et al. (2013); the incorporation of wavelet coherence analysis by Wiebe et al. (2011) in their examination of vertical wind velocities and pressures acquired by a floating pontoon over a coral reef; and Risien et al.'s (2004) Morlet wavelet–based investigation of the variability in wind stress data over an ocean upwelling system – which leads us nicely into the next section regarding ocean processes, beginning with ocean waves.

4.4.2 Ocean Processes: Waves, Large-Scale Oscillations, Ocean–Atmosphere Interactions and Biological Processes

A clear and concise study of the grouping characteristics of wind waves using the Morlet wavelet transform has been presented by Liu (2000a). He used the transform plot to identify local wavegroups in a time series of surface elevation corresponding to wind-generated waves measured in near-shore areas of the Great Lakes and found a linear relationship between group energy and duration. He also showed that mean maximum group wave height and significant wave height differ by around 17% and stressed the implications this has for engineering design. Finally, he used his results to illustrate the non-stationarity of the data, linking his findings to an earlier study concerning the characteristics of waves on the Atlantic Ocean (Liu, 1994). Liu's 2000a paper is also interesting in that it contains a table linking traditional, Fourier and wavelet statistics. In a related paper, Liu (2000b) made a strong argument for the superiority of the wavelet method over Fourier methods for the analysis of wind waves as it allows for the non-stationary aspects to be captured. He plotted an equivalent constant energy (over time) wavelet energy plot and suggested a non-stationarity index to capture the time variability of the signal energy. The nature of ocean waves has, in fact, been the subject of study by many authors using a variety of sophisticated wavelet-based techniques. The nonlinear wind–wave interaction during a Mistral event has been tackled by Elsayed (2006) using wavelet bicoherence (Chapter 2, Section 2.22). The work demonstrated that phase coupling occurs between wind speed and wave height signals over a range of frequencies which change over time due to the non-stationary nature of the time series. In more recent work, Chuang et al. (2013) used a Morlet wavelet transform of ocean surface acceleration signals from a buoy to accurately synthesize sea surface elevations. (See also, Ge, 2007, who applied wavelet power significance tests to wave elevation time series and the corresponding comment by Zhang and Moore, 2012.) A new technique for separating a 2-D wave field into incident and reflected waves using a Morlet wavelet transform has been described by Ma et al. (2010). And, in later work, Ma et al. (2011) extended their method to separate obliquely incident and reflected irregular waves, where a minimum Shannon wavelet entropy criterion was employed to determine the optimum central frequency of the Morlet wavelet.

The impact of surface waves on vessel hulls (wave slam) has been considered by Amin et al. (2013) using a tunable Morlet wavelet analysis (with a slightly different formulation than that of Chapter 2, Equations 2.36 and 2.37). They employed both a low- and high-oscillation wavelet to identify slamming events in terms of instance of occurrence and the frequency and damping of the subsequent whipping response. Figure 4.30 shows the transform of a hull strain gauge signal containing two consecutive slams using a Morlet wavelet with high temporal resolution. The plot indicates that a single slamming event may contain more than one impact as indicated by the ridges separated from the first impact. Figure 4.31 contains the corresponding transform using a Morlet function with a high-frequency resolution which is more useful for finding modal parameters associated with the slamming events. Other work concerning ocean waves includes the detection of breaking wave events in wave time series by Liu and Babanin (2004), the examination of

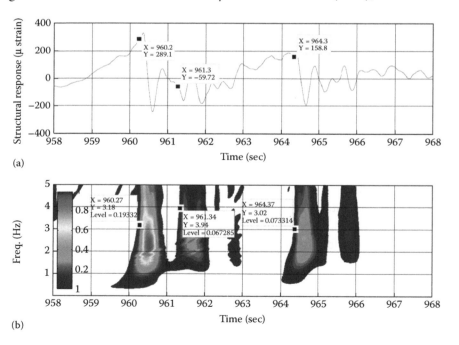

FIGURE 4.30 (a) Strain gauge response and (b) its wavelet transform by high time-resolution Morlet $f_b = 0.3$, $f_c = 1$. (Original figure in colour.) (From Amin, W. et al., *Ocean Engineering*, 58, 154–166, 2013.)

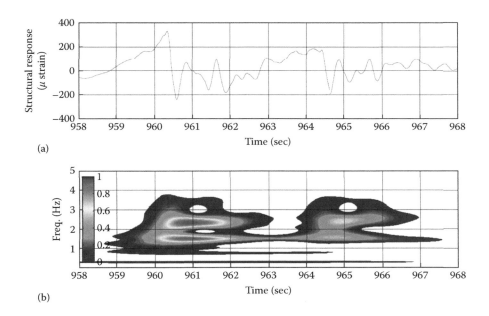

FIGURE 4.31 (a) Strain gauge response and (b) its wavelet transform by high-frequency resolution Morlet $f_b = 4$, $f_c = 1$. (Original figure in colour.) (From Amin, W. et al., *Ocean Engineering*, 58, 154–166, 2013.)

short-term temporal variability of energy in ocean wave surface elevation time records by Nolan et al. (2007), the Morlet-based analysis of freak waves by Lee et al. (2011) and the projection of extreme typhoon waves by Chun et al. (2013).

Larger scale and longer period waves have been the topic of research by several other groups. The interaction between short ocean swell and transient long waves was studied by Kaihatu et al. (2012). They found that unlike Fourier-based analysis, their wavelet analysis yielded behaviour that is quite different between cases with and without the long wave, where the dissipation characteristics of the combined short–long wave signals deviate considerably from that of swell alone. They went on to apply wavelet bicoherence (Chapter 2, Section 2.22) to the signals in order to determine the relative strength of nonlinear energy exchange between spectral components. And while the results were 'interesting', they suggested that further study is required before definitive conclusions can be drawn. Other investigations of longer period waves include the investigation of abnormal, tsunami-like waves by Yoo et al. (2010), the observations of tsunamis of meteorological origin ('meteotsunamis') in the Western Mediterranean by André et al. (2013) and along the Southwest Australian coast by Pattiaratchi and Wijeratne (2014). Very long period Yanai waves in the upper water column of the central equatorial Indian Ocean have been the subject of an investigation by David et al. (2011), and the relationship between Rossby wave passage in the Southern Atlantic and chlorophyll concentration anomalies has been analyzed using wavelet phase relationships between the two signals by Gutknecht et al. (2010). In the latter study, the spatial variation of the phase relationship between the two signals for wavelengths between 400 and 1100 km was examined. This was performed by extracting the cross-wavelet transform

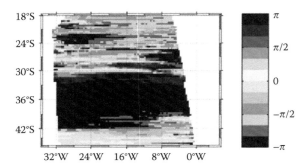

FIGURE 4.32 Phase relationships between sea level anomalies (SLA) and chlorophyll a concentration anomalies (CHLA): (phase [CHLA] − phase [SLA]). Spatial phase relationships are computed using cross-wavelet analysis. Phases for data having a maximum coherency >0.75 are extracted from each local wavelet power spectrum in the cone of influence for wavelengths between 400 and 1100 km. If there is more than one maximum coherency in the spectral domain, the point in the physical space is excluded to retain only unambiguous phases. (Original figure in colour.) (From Gutknecht, E. et al., *Journal of Geophysical Research: Oceans*, 115, C05004, 1–16, 2010.)

phase at various locations and mapping it out in longitude and latitude. A spatial plot of the phase relationship results are shown in Figure 4.32. The authors noted that the results were in good agreement with earlier work utilizing a Fourier-based cross-spectral analysis. See also, Timmermans et al. (2010), who employed wavelet analysis within their investigation of bottom-intensified motions in the deep ocean to reveal vertical excursions of periods around 50 days. These excursions were consistent with a bottom-trapped topographic Rossby wave.

El Niño is a large-scale climatic disturbance occurring every 3–7 years in the tropical Pacific Ocean linked to a warming in sea surface temperatures (SSTs). It has an 'opposite phase' of cooling called 'la Niña' and both together are known as the 'El Niño Southern Oscillation', or 'ENSO'. The application of wavelet techniques to the analysis of ENSO has now been carried out by several research groups. Jänicke et al. (2009) have described various techniques to visualize the wavelet analysis of the temporal evolution of El Niño from an entire multivariate two-dimensional climate data set consisting of several variables and extending over 250 years. Figure 4.33 contains an example of their display of wavelet-based characteristics for sea surface temperature (SST) and mean sea level pressure (MSLP) superimposed on a map of the world. Figure 4.34 contains the cross-wavelet coherence maps of water levels and El Niño and North Atlantic Oscillation (NAO) indices from the work of Karamperidou et al. (2013) which focused on the implications for coastal resources of sea level and climatic variability. Note the arrows superimposed on the coherence plots indicating the relative phase between the two signals.

A number of advanced wavelet analysis methods have recently emerged for interrogating large-scale climatic phenomena. Partial wavelet coherence (PWC) and multiple wavelet coherence (MWC) were employed by Ng and Chan (2012) to analyze the possible ENSO-related impact of the large-scale atmospheric factors on tropical cyclone activity over the Western Pacific. Wavelet-based geometrical and topological significance testing was demonstrated by Schulte et al. (2014) on North Atlantic Oscillation and El Niño time series and

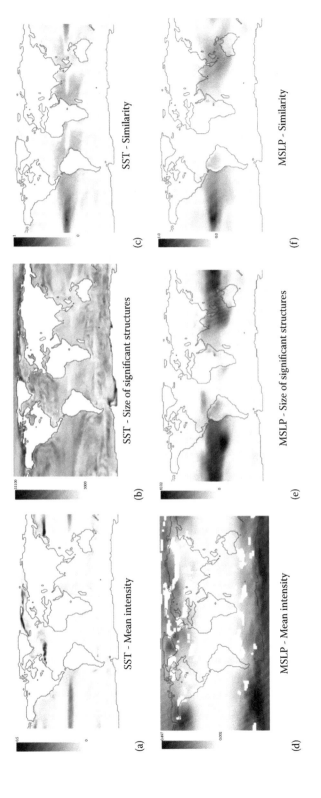

FIGURE 4.33 **Display of wavelet characteristics:** (**a, d**) Mean intensity of the significant structures in each power spectrum. White areas indicate positions whose power spectra feature no significant structures. (**b, e**) Size of the significant structures given by the number of included pixels. (**c, f**) Similarity fields for a position in the Nino3 region. (From Janicke, H. et al., *IEEE Transactions on Visualization and Computer Graphics*, 15, 1375–1382, 2009.)

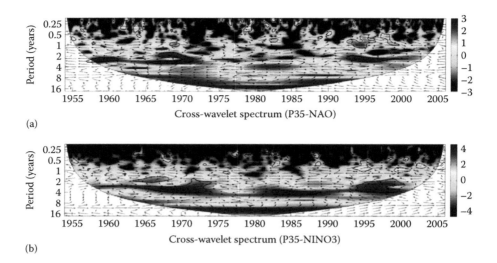

FIGURE 4.34 Cross-wavelet power spectrum between P35 water levels and (**a**) the NAO index, and (**b**) the NINO3 index. Inset arrows in the cross-wavelet plot show the relative phasing of the two time series, with arrows pointing to the (left) right denoting (anti) correlation. Note the similarity in cross-wavelet power patterns (reversal of correlation sign, high power at the reversal of Atlantic Multidecadal Oscillation [AMO] phase, etc.) with the relationship between Key West levels and the NAO and NINO3 indices. (Original figure in colour.) (From Karamperidou, C. et al., *Regional Environmental Change*, 13(S1), S91–S100, 2013.)

the Arctic Oscillation has been studied by Zhang et al. (2014b) using a Haar-based analysis to extract climatic time series information from background red noise.

A few authors have investigated the connection between El Niño and biological processes. Examples of this include the work of Hashizume et al. (2013), who performed a cross-wavelet coherence analysis of global and local climatic variables and cholera dynamics; Chaves et al. (2014), who found patterns of association between a tropical vector-borne disease, cutaneous Leishmaniasis, the sand fly vector abundance and the El Niño Southern Oscillation using a cross-wavelet coherence technique; Lee et al. (2010), who performed cross-wavelet analysis of the Southern Oscillation index and water temperature in their study of the long-term relationship between the marine environment, krill and salps in the Southern Ocean; Morris et al. (2014), who performed a wavelet coherence analysis between Buruli ulcer microbacterium infection and ENSO signals; and, Harrison and Chiodi (2013), who decomposed Darwin sea level pressures as part of their examination of the effect of ENSO on tropical fisheries.

In work of a different vein concerning large-scale ocean flow processes, Lilly and Olhede (2009) developed a method for the analysis of multivariate oscillations by inspecting novel bivariate transform ridges in order to examine the time-varying properties of ocean surface drifter trajectories. Their approach allows the decoupling of vortex motions from the background 'residual' signal. The technique is illustrated in Figures 4.35 and 4.36. A number of ocean drifter trajectories from the eastern subtropical Atlantic are shown in Figure 4.35a. The times series for the eastern and northern velocities corresponding to one

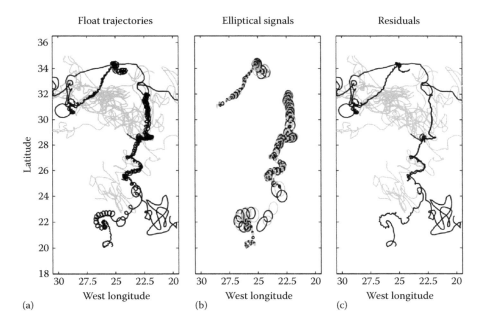

FIGURE 4.35 **Multivariate wavelet ridge analysis applied to oceanographic data.** Panel **(a)** shows a set of position records from 27 freely drifting subsurface oceanographic floats in the eastern subtropical Atlantic. The original data (a) is decomposed into a set of modulated bivariate oscillations, represented in **(b)** as snapshots of ellipses at successive times, together with the residuals shown in **(c)**. The ellipses alternate between black and grey, and are shown at twice their actual size for clarity. Time series for which a modulated oscillation is detected are plotted in (a) and (c) as black lines, while others are plotted as grey lines. (From Lilly, J. M., and Olhede, S. C., *Conference Record of the 43rd Asilomar Conference on Signals, Systems and Computers, November 1–9*, 452–456. Pacific Grove, CA. Copyright 2009, IEEE.)

of these trajectories are shown in Figure 4.36a. These signals can be observed to contain modulated components of increasing frequency. A wavelet transform of each signal is computed and the modulus of the corresponding bivariate transform is shown in Figure 4.36b. The bivariate transform modulus is derived from the sum of the squares of the transform moduli of the individual signals. The ridge of the bivariate transform may then be used to reconstruct a time-varying ellipse representative of the local drifter oscillation. These are shown in Figure 4.35b. A residual between the estimated modulated oscillations and the observed time series may also be constructed as shown in Figure 4.35c. Further work in this area is described in Lilly and Olhede (2010, 2012) and Lilly et al. (2011); the latter paper concerns the analysis of Lagrangian trajectories generated by a numerical model of an unstable baroclinic jet.

4.4.3 Rainfall and River Flows

The detection of forward, symmetric and inverse cascades from rainfall time series was carried out by Molini et al. (2010), who computed correlation coefficients across wavelet scales to probe causality across rainfall scales. They found that extreme events appeared

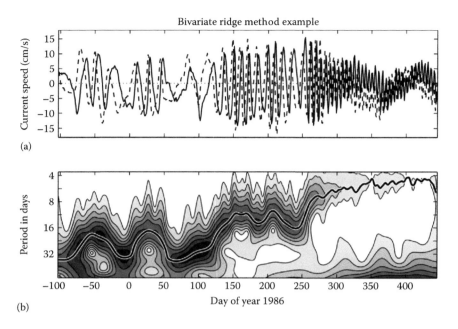

FIGURE 4.36 Example of multivariate wavelet ridge analysis. A bivariate position signal, differentiated in time for presentational clarity, is plotted in (a). The solid curve represents eastward velocity and the dashed curve northward velocity. The norm of the bivariate wavelet transform $w_\psi(t,s)$ of the position signal, shown in (b), has units of kilometers. The y-axis is logarithmic and shows the scale of the wavelet transform expressed in periods, $2\pi s/\omega_\psi$, with units of days. The contours range from 0 to 65 km with a spacing of 5 km. The heavy curve is a single unbroken ridge resulting from the amplitude ridge algorithm. (From Lilly, J. M., and Olhede, S. C., *Conference Record of the 43rd Asilomar Conference on Signals, Systems and Computers, November 1–9*, 452–456. Pacific Grove, CA. Copyright 2009, IEEE.)

to be driven by micro-scale features whereby fine scale features coalesce to form intense events at later times and scales. In their method, they employed a fourth-order Gaussian derivative wavelet and integrated locally using a bump function to produce local wavelet variance measures. Kolmogrov scaling was observed within the inertial subrange and produced autocorrelation plots from these measures along scales and cross-correlation maps between scales. They found evidence of mainly forward causal cascades in rainfall time series. These causal relationships tended to vanish when the rainfall was aggregated at larger timescales (i.e. hours and longer). Other recent work on rainfall includes that of Okonkwo et al. (2014), who characterized the association of West African jet streams to ENSO events and rainfall and Braga et al. (2014), who probed the role of Equatorial Pacific and Tropical Atlantic sea surface temperatures in modulating rainfall in Brazil. Also of interest in this area is the work of Quiroz et al. (2011), who formulated an orthonormal wavelet reconstruction technique to generate an approximation to daily rainfall data from satellite-based vegetation greening data. The absence of rain has also been considered: a hybrid wavelet linear genetic programming (or gene-wavelet) model was developed by Mehr et al. (2014) for drought forecasting. The model used wavelet-processed historical El

Niño and drought index time series as inputs and was compared with neuro-wavelet and fuzzy-wavelet drought forecasting models.

Particular emphasis has been placed on the analysis and prediction of streamflows by several groups. A good example of this is the work concerning wavelet coherence colour maps employed by Carey et al. (2013) to assess the daily to seasonal coupling of rainfall on streamflow variability in northern catchments. They found that rainfall and streamflow were decoupled during the winter months for catchments with cold winters, with strong coupling resuming during and immediately after the spring snowmelt. This is seen as bands of high- and low-value regions in their squared coherence plots (Figure 4.37). Brackets have been added to the plot (originally in colour) to show where the high- and low-coherence regions occur at the top of the plot. These regions alternate markedly on a once per year basis, clearly demonstrating the coupling during summer months and decoupling during winter months.

It was hypothesized by Lane (2007) that wavelet-based analysis may be utilized in the evaluation of rainfall run-off models. He conducted an analysis of measured and model hydrographs by considering differences in wavelet power and wavelet phase. Lane presented a strong argument against a priori objections to the use of a wavelet transform approach for this task. He developed wavelet power and phase error metrics and argued that because of the localization property and the low sensitivity of wavelet power and phase

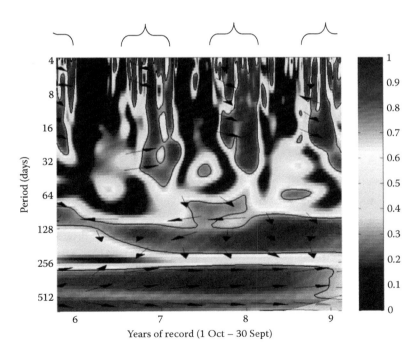

FIGURE 4.37 (See colour insert.) **Close-up of squared wavelet coherence between precipitation and discharge for Krycklan for ~3 years of record.** (Brackets have been added to the top of the plot to indicate where the high-coherence regions occur as the original figure is in colour.) (From Carey, S. K. et al., *Water Resources Research*, 49(10), 6194–6207, 2013.)

to wavelet function used, his results were of sufficient merit to warrant further exploration of the topic. A Morlet wavelet–based method for the synthetic generation of streamflow data for the Pearl River basin in China has been proposed by Niu and Sivakumar (2013), who found that it provided reliable information in terms of the preservation of spectral properties of the original time series. A flow prediction technique using a wavelet neural network was constructed by Krishna and Satyaji (2011) for the daily flows of the Malaprabha River basin in India. Other work in this area includes that by Adamowski et al. (2013), who used a Morlet-based wavelet and cross-wavelet analysis in their investigation of annual streamflows in Canadian provinces; Briciu (2014), who demonstrated the semidiurnal influence of the moon on non-tidal inland rivers; Mengistu et al. (2013), who employed wavelet transform and coherence analysis in their study of stationary components within water yield signals and also the strong correlation between water yield and global climate oscillation indices; Pasquini et al. (2013), who analyzed mean monthly lake water levels using the real part of the Morlet wavelet transform incorporating a critical limit criterion; and Markovic and Koch (2014), who used global wavelet spectra to identify low-frequency oscillations present in the monthly discharge at gauging stations in the Elbe and Saale Rivers within the Elbe River basin, finding statistically significant oscillations at around 7 and 14 years.

The effect of the microtopography of a river bed on the generation of turbulent flow structures was investigated by Hardy et al. (2009), who employed a Morlet-based method to determine the temporal length scales of the dominant coherent flow structure generated by the gravel bedform. They found that the large-scale coherent flow structures in the outer layer were the result of flow–topography interactions in the near bed region. See also the work by Nyander et al. (2003) which developed a novel form size distribution for characterizing the global topography of river bed surfaces. Note, however, that much of the research concerning river turbulence deals with two-phase flows which are considered in the next section.

4.5 TWO-PHASE FLOWS

Two-phase flows contain any of the two phases of matter: solid, liquid and gas. They occur in many situations in science and nature. (We have come across one obvious two-phase flow already – rain!) Sediment transport aspects of free surface flows (rivers and oceans) is a major topic in this area and many groups have tackled this flow phenomenon using a variety of wavelet methods. For example, Shugar et al. (2010) utilized wavelet coherence measures to illustrate the important characteristics of the transport of suspended sediment over alluvial sand dunes. Using stacked wavelet coherence plots corresponding to backscatter records at various heights above the dune, they could determine the evolution of low-frequency flow structures initiating close to the bed, growing with height and then breaking up as they were advected downstream. Figure 4.38 contains an example of one of their plots illustrating the interaction and evolution of a coherent structure as it rises and is advected downstream. Singh et al. (2012) conducted a series of flume experiments of sediment transport using an experimental channel and three flow rates, where Mexican hat–based wavelet cross-correlation was performed between the temporal bed

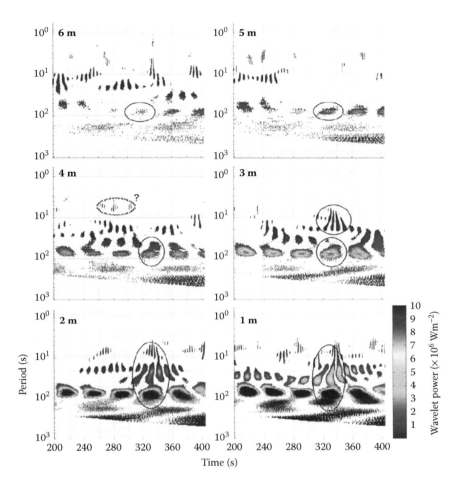

FIGURE 4.38 Detail of lower 6 m of a vertical stack of local wavelet power spectra of calibrated suspended sediment concentration. Wavelet plots are labelled by height above the bed. Circles highlight the evolution of a wavelet packet through the water column. (Original figure in colour.) (From Shugar, D. H. et al., *Sedimentology*, 57, 252–272, 2010.)

elevation and instantaneous Reynolds stress to probe the two-phase flow dynamics. They found that the cross-correlation increased with scale, with a saturation level occurring at around 40 minutes (at least for the lower and higher discharges – see Figure 4.39). This led them to suggest that integrated quantities at timescales larger than this could be used for predictive modelling processes, such as bedload transport, indicating that larger scales are more correlated than smaller scales. In this vein, see also Rajaee et al. (2010), who employed a wavelet neuro-fuzzy model to predict daily suspended sediment load in the Pecos River, New Mexico, and found that it produced a reasonable prediction of extreme values.

Brandner et al. (2010, 2015a) have used wavelet transforms to identify coherence and quantify frequency content in cavitating flows (around a sphere and about a jet in cross-flow) using a time series of pixel intensity from high-speed photography. The earlier work employed a second-order derivative of Gaussian wavelet, whereas the later investigation

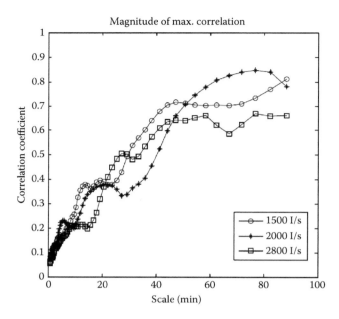

FIGURE 4.39 Plot of maximum cross-correlation coefficient obtained between temporal bed elevation and instantaneous Reynolds stress series as a function of scale using wavelet analysis. (Original figure in colour.) (From Singh, A. et al., *Journal of Geophysical Research: Earth Surface*, 117, F04016, 1–20, 2012.)

employed a Morlet wavelet to determine shedding frequencies and a derivative of Gaussian wavelet to identify individual shedding events. In other work by the same group, Brandner et al. (2015b) have identified the product bubble diameters from a single breakup of millimetre-sized rising bubbles encountering a turbulent shear layer from wavelet analysis of measured acoustic signals. In their study of two-phase bubbly flows, Sathe et al. (2010) constructed a series of decomposed particle image velocimetry (PIV) velocity field images which were generated using a 2-D-DWT. The images contained the superimposed air bubbles within the decomposed velocity fields. The authors commended the 'excellent filtering capabilities' of the wavelet method which allowed them to obtain the bubble slip velocity and the wave number spectrum of the liquid velocity field. An approach for enhancing tomographic images used to visualize a solid gas (pulverized coal/air) two-phase flow has been developed by Chen et al. (2012a). The quality of the image was greatly enhanced using the wavelet-based method. Electrical capacitance tomography was also employed by Kreitzer et al. (2012) in their examination of transient two-phase refrigerant flows where they decomposed the associated pressure signal data using a discrete wavelet transform. More recently, Lu et al. (2014) applied wavelet flatness factors in an investigation to show that small-scale vortex structures and intermittency in a simulated (LES) forced homogeneous isotropic turbulence are inhibited due to the presence of polymers. Other work in the area includes that of Li et al. (2014b), who used wavelet time–frequency decomposition within their investigation of nitrogen–water two-phase flows in a vertical channel; Nguyen et al. (2010), who developed a CWT approach for two-phase flow pattern identification from void fraction signals generated by a vertical air–water flow; Tsakiroglou et al. (2010), who analyzed two-phase flow through porous media, in order to correlate flow characteristics with heterogeneity (where both an experimental soil sample and numerical simulations were considered); and Wang et al. (2013a), mentioned earlier in the Section 4.3.2, who performed numerical (large eddy) simulations of inertial particle dispersion in turbulent flow over a backward facing step. See also the review concerning the analysis of pressure fluctuations of gas–solid fluidized beds by van Ommen et al. (2011), where much of the key wavelet work in this area is cited. This includes their use in denoising, fractal analysis, analysis of bubbly flows and in decomposing the three main flow components – individual particle motion, particle clusters and voids.

4.6 OTHER APPLICATIONS IN FLUIDS

Van Milligen et al. (1995a,b) introduced the concept of wavelet bicoherence as a 'new turbulence analysis tool' in their paper concerning the analysis of a tokomak plasma. They made the point that wavelet bicoherence detects phase coupling between short-lived wavelets rather than modes (as with Fourier bicoherence) and as such is better adapted to the examination of turbulence data. Example bicoherence plots from the work of van Milligen are shown in Figure 4.40. The van Milligen papers represent early examples of the wavelet analysis of plasmas and, in particular, of the practical application of wavelet bicoherence. There are now many papers considering plasmas and, more generally, magnetohydrodynamics from a wavelet perspective. Some of the more recent examples include those by Giri et al. (2014), Noskov et al. (2012), Yoshimatsu et al. (2011, 2013) and Sarma et al. (2013) – the latter is also mentioned in the next paragraph in relation to the analysis of chaotic flow phenomena. A comprehensive review of the use of wavelets in the study of magnetohydrodynamics and plasma turbulence has recently been provided by Farge and Schneider (2015).

The fractal and chaotic nature of fluid flows is a topic of interest across a range of flow phenomena. For example, Singh et al. (2009) employed wavelet-based multifractal analysis in their study of sediment transport data; Mosdorf et al. (2011) considered the multifractal properties of bubble paths; and Sona et al. (2014) generated multifractal spectra associated with the turbulent flow of a molten salt. More information on multifractals, including their application to other areas, is provided in Chapter 7, Section 7.2.3. Tardu (2011) considered the chaotic synchronization of wall turbulence through forcing induced by nearby coherent vortices and Sarma et al. (2013) employed continuous wavelet transforms to extract a chaotic oscillation signal from a turbulent DC glow discharge plasma which was then probed to determine its correlation dimension and largest Lyapunov exponent (both are tests for chaos). More information on chaos is provided in Chapter 5, Section 5.2.2.

Finally, it is worth mentioning that a number of other areas concerning fluid flow are now benefiting from the application of wavelet analysis. These include the following examples:

- Alcantara et al. (2010) studied the relationship between water surface temperature and heat flux in a tropical hydroelectric reservoir using the cross-wavelet transform, wavelet coherence and phase maps.

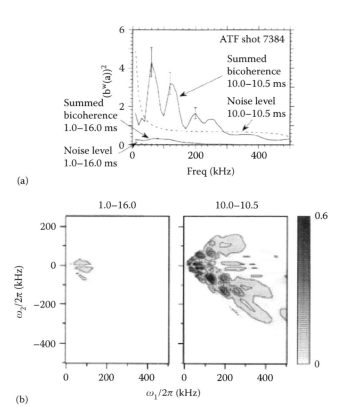

FIGURE 4.40 (a) Summed bicoherence for two time windows, 1.0–16.0 and 10.0–10.5 ms. The noise level for both cases is indicated by the dashed lines. The two peaks at 60 and 120 kHz in the short-time section indicate the detection of structure in the turbulence. (b) Bicoherence graphs for the time windows 1.0–16.0 and 10.0–10.5 ms. A structure is clearly visible in the latter. (Original figure in colour.) (Reprinted figure 3 with permission from van Milligen, B. Ph. et al., *Physical Review Letters*, 74, 395–398. Copyright 1995 by the American Physical Society.)

- Nourani et al. (2014) reviewed wavelet-artificial intelligence models in hydrology.

- Hariharan and Kannan (2014) reviewed the application of wavelet methods to the solution of reaction–diffusion equations.

- Lewalle (2010) performed a single-scale formulation of the Navier–Stokes equations using a Mexican hat family of Hermitian wavelets.

- Wang et al. (2013c) investigated both open- and closed-loop pressure control on a wind turbine blade using both the Morlet and Mexican hat wavelets.

- Stephenson et al. (2014) and Stephenson and Tinney (2014) analyzed the sound emanating from a manoeuvring helicopter. In the latter paper, they detailed a technique for isolating the blade vortex pressure signal by performing an inverse transform of filtered coefficients obtained by thresholding according to both frequency and amplitude criteria.

- Tsang et al. (2008) investigated the dynamic stall behaviour of an airfoil, and in particular considered the lift and drag time series of pre-stall, light-stall and full-stall regions.
- Johnson and Lind (2010) characterized aircraft wing rock phenomena with variations in the size and configurations of the vertical tail where they employed wavelet transform of roll rate and aileron deflection time series.

Note also that there are a few references to fluid flows contained in other chapters of this book. Many of these are in Chapter 6, concerning blood flow, and in particular Section 6.6 which covers blood flow and blood pressure.

CHAPTER 5

Engineering Testing, Monitoring and Characterization

5.1 INTRODUCTION

Wavelet analysis has been usefully applied to a variety of pertinent problems in engineering. In this chapter, we review a selection of these, including the examination of vibration modes, nonlinear oscillations and transient oscillations; the interrogation of non-destructive testing signals and the condition monitoring of rotating machinery; the assessment of machining processes; and the characterization of surfaces and fibrous materials. As with the fluid problems described in the previous chapter, the choice of the most appropriate wavelet for use in the analysis of a particular engineering problem depends very much on the nature of the data itself. Both discrete and continuous (usually complex) wavelets have been used to monitor rotating machinery such as gears, shafts and bearings. Continuous wavelets are favoured when high temporal resolution is required at all scales and, in their complex form, are well suited to the free vibrations of plates and beams. The temporal records of such vibrations may quickly exhibit a high degree of complexity due to the superposition of multiple wavegroups. Complex continuous wavelets are able to unfold these signals in time and frequency, allowing for the decoupling of vibration modes. Discrete wavelets may be favoured when, for example, a small number of data are required as input to a classifier such as a neural network. Most surface characterization work has employed discrete wavelets, whose coefficients are used to determine scale-dependent surface characteristics. Finally, as we might expect, discrete wavelet transform coefficients are particularly useful for signal compression problems in engineering.

5.2 DYNAMICS

5.2.1 Fundamental Behaviour

A number of studies have been carried out over recent years concerning the application of wavelet-based analytical techniques to the investigation and modelling of dynamical systems. Applications include the evaluation of dynamic properties and system characteristics, the modelling and control of dynamical behaviour, and the partitioning or decoupling of multiple responses. Early work worth consulting in this area is that by Staszewski (1997, 1998a), who employed a Morlet wavelet for the detection of system nonlinearities through the identification of damping and stiffness parameters for multi-degree-of-freedom (MDOF) dynamic systems during transient testing. This wavelet is very effective for this application as it has good support in both frequency and time, which allows the decoupling of the system's various modes of vibration. Figure 5.1a shows one of the signals analyzed in the study resulting from the impulse response of a two-degrees-of-freedom model system. Figure 5.1b shows the ridges of the modulus plot and Figure 5.1c shows the wavelet

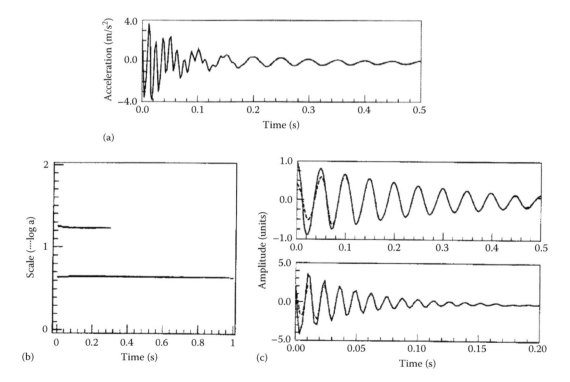

FIGURE 5.1 (a) Impulse response function for well separated modes. (b) Ridges of the wavelet transform scalogram. (c) Comparison of the real parts of the wavelet transform skeletons (dashed lines) obtained from the ridges given in the preceding figure and the theoretical impulse response function (solid line). Top trace: first mode (20 Hz); bottom trace: second mode (78 Hz). (From Staszewski, W. J., *Journal of Sound and Vibration*, 203(2), 283–305, 1997.) Reproduced with kind permission of Academic Press Ltd.

transform skeletons obtained from these ridges. The decoupling of the modes is evident in the latter figure. The reconstruction of the modes and subsequent damping parameter estimation is found to be better for the skeleton reconstructions than a simple reconstruction based on the wavelet coefficients within a certain frequency range. The ability of the method to separate modes that are closer in frequency is also tackled in the paper. Around the same time, Ruzzene et al. (1997) also employed a Morlet wavelet–based approach to identify natural frequencies and damping ratios of MDOF systems. They analyzed a synthetic 4-degrees-of-freedom model before applying the technique to the acceleration response of a bridge excited by road traffic. A skeleton ridge approach was also used by Wang et al. (2003) to analyze the dynamic characteristics of a number of model systems, including a torsion damper and a forced bilinear spring, and, in later work, Tjahjowidodo et al. (2007) compared wavelet skeleton and Hilbert transform–based approaches for the identification of backlash in mechanical systems. In this latter investigation, the researchers achieved an improvement in nonlinear modal parameter estimation using a Morlet-based wavelet technique. In subsequent work, Tjahjowidodo (2012) focused on deriving the equivalent damping and stiffness parameters in mechanical systems with hysteretic frictional elements using the skeleton method.

More recently, Staszewski and Wallace (2014) developed a wavelet-based frequency response function for the modal analysis of time-variant systems, which are often resistant to full interrogation by classical schemes. They applied their Morlet-based technique to a number of simulated single-degree-of-freedom systems before considering an experimental system where a sliding steel cylindrical mass was run across a beam subjected to Gaussian white noise excitation. They performed classical frequency response function (FRF) analysis in the Fourier domain prior to utilizing their wavelet-based approach. Figure 5.2 contains the experimental time series together with the classical FRF, where it is difficult to estimate the relevant natural frequencies according to the authors. Figure 5.3

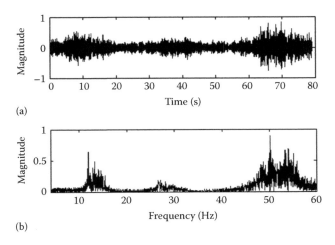

FIGURE 5.2 Classical vibration analysis for the time-variant system: **(a)** Normalized time response and **(b)** normalized FRF. (From Staszewski, W. J., and Wallace, D. M., *Mechanical Systems and Signal Processing*, 47, 35–49, 2014.)

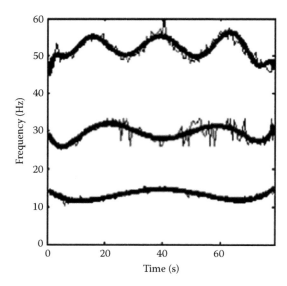

FIGURE 5.3 Varying natural frequencies for the time-variant system. Solid thick line: analytical values; solid thin line: values extracted from the wavelet-based FRF. (From Staszewski, W. J., and Wallace, D. M., *Mechanical Systems and Signal Processing*, 47, 35–49, 2014.)

contains the ridges of the wavelet-based FRF computed from the same experimental signal. The time-variant nature of the second, third and fourth vibration modes is clearly seen in the figure. The wavelet result closely matches analytical values computed for the system (also plotted in the figure). However, this scheme of computing the wavelet-based FRF required time–frequency averaging, an issue which was addressed in subsequent papers by the same group (Dziedziech et al., 2015a,b).

Modal parameter analysis using the reassigned wavelet scalogram has been suggested by Sun et al. (2006) and, more recently, Peng et al. (2011), who examined synthetic signals from a model two-degrees-of-freedom linear mechanical system. The scalogram and reassigned scalograms generated in this latter study are shown in Figure 5.4, clearly showing the two modes. They found no difference in the original and reassigned scalograms for determining the modal parameters. The authors then extended their work to signals with added white noise and found that the reassigned scalograms performed better than their conventional counterparts in estimating modal parameters for modes with weak amplitudes. They attributed this improvement to the concentration of energy in the transform domain associated with the reassignment technique, which provides an increased signal-to-noise ratio (SNR). The work also considered the effect of distortions at the edges of the scalograms and reassigned scalograms and suggested a number of approaches for mitigating against these effects. The characteristic fork-shaped distortion exhibited at the edges of the reassigned scalogram ridge is obvious in Figure 5.4b. The authors stated that this is typical for reassigned scalograms. Boltezar and Slavic (2004) also provided details on modifications to the wavelet transform to reduce edge effects in their work on determining the damping characteristics of the free response of an experimental steel beam. They suggested

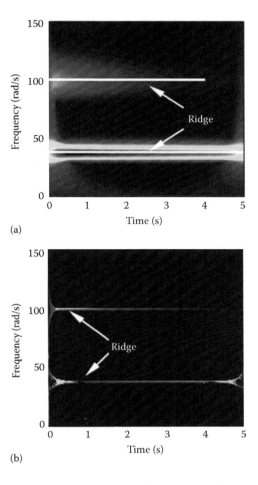

FIGURE 5.4 (a) The scalogram and the ridge and (b) the reassigned scalogram and the ridge. (Reprinted from Peng, Z. K. et al., *Shock and Vibration*, 18, 299–316, 2011. Copyright 2011, with permission from IOS Press and the authors.)

three methods for this – the reflected window method, the equal-area window method and the adaptive wavelet function method – and found that these techniques produced more reliable results than the original transform for the same length of signal considered.

Perez-Ramirez et al. (2016) recently developed a synchrosqueezed methodology for identifying modal parameters in structures from ambient vibrations and applied it to a numerical model, an experimental steel frame model and signals from a reinforced-concrete highway bridge structure. They found it effective for identifying the natural frequencies and damping ratios for signals which were non-stationary and embedded within a high degree of noise. A number of other investigations have been conducted in this area of fundamental dynamical behaviour, including that by Remick et al. (2014), who considered the high-frequency dynamic instability of a nonlinear system of forced coupled oscillators, superimposing the wavelet spectra on the associated Hamiltonian frequency–energy plots (FEPs); Makris and Kampas (2013), who investigated the effective period in bilinear systems subject to earthquake loading; Noel et al. (2014), who interrogated the dynamics of a strongly nonlinear spacecraft structure; Banfi and Ferrini (2012), who performed cross-correlation and phase analysis of a forced damped cantilever signal; Awrejcewicz et al. (2011), who examined the transition to multifrequency regimes of single-layered Euler–Bernoulli beams; Mahdavi and Razak (2015), who performed a comparative study of structural dynamics using Haar and Chebyshev wavelets; Tarinejad and Damadipour (2014), who suggested a combined frequency domain decomposition and wavelet approach for the determination of natural frequencies, mode shapes and damping parameters in structural systems; and Srivastav et al. (2009), who employed wavelet phase space partitioning in their work on an information-theoretic measure for anomaly detection in complex mechanical systems.

5.2.2 Chaos

Nonlinear oscillator systems are capable of the most fascinating behaviour, known as 'chaotic motion' or simply 'chaos', whereby even simple nonlinear systems can behave in a seemingly unpredictable manner under certain operating conditions (Addison, 1997). The realization that real systems can exhibit this type of non-periodic response has prompted much research in the area over recent years, with a number of researchers in the field concerning themselves with the ability of wavelet-based techniques to aid in the analysis of chaotic oscillations.

Both Daubechies and Morlet wavelets were employed in early work on the topic by Staszewski and Worden (1999), who used them to analyze time series data sets containing a variety of features, including coherent structures (fluid turbulence), fractal structures (the devil's staircase and the Mandelbrot–Weierstrass function), chaos (Duffing, Henon, Lorenz and Rössler systems) and noise (Gaussian white). They considered the simple nonlinear damped-spring system of the Duffing oscillator and examined its chaotic response, contrasting its modulus and phase plots with non-chaotic cases. Their paper is well worth consulting for the many clearly presented diagrams used to illustrate the discussion. More recently, Clemson and Stefanovska (2012) have considered the time series from the chaotic, stochastic and non-autonomous Duffing system and have suggested that, by including time in its representation of the signal, the wavelet transform may be well suited to distinguishing non-autonomous from stochastic systems.

The chaotic behaviour of structural elements, such as plates and beams, has been reported by Awrejcewicz and coworkers in a series of papers. For example, the chaotic vibration of infinite-length flexible panels was studied by Awrejcewicz et al. (2009), where the advantages and disadvantages of a number of different wavelets were considered and the Morlet wavelet was recommended for this type of analysis. A numerical model of a system comprising a plate and two beams has also been described by Awrejcewicz et al. (2013). This system, shown in Figure 5.5a, was subject to harmonic excitation and exhibited chaotic behaviour for the chosen system parameters. In Figure 5.5b, the frequency spectra together with the 2-D and 3-D Morlet-based wavelet plots are shown next to the time series and phase portraits for each element (plate signal at the top). Interestingly, the vibration characteristics of the two beams are different.

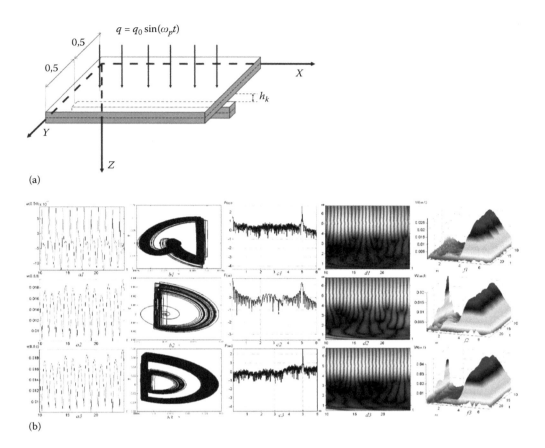

FIGURE 5.5 (a) Two-layered system consisting of a plate and a beam and (b) contact interaction of the plate and two symmetrically located beams for $q_0 = 0.2$. (Original in colour.) (From Awrejcewicz, J. et al., *Latin American Journal of Solids and Structures*, 10, 163–174, 2013.)

A number of papers have appeared concerning the nature of chaotic electrical circuits. Hanbay et al. (2008) demonstrated a wavelet-based preprocessing method for feature extraction prior to neural network modelling of a chaotic Chua's electrical circuit, and Murguia and Campos-Canton (2006) performed wavelet analysis on three experimental chaotic circuits (Chua's circuit, an electrical Rőssler circuit and a chaotic generator). More recently, Setoudeh et al. (2014) investigated the chaotic behaviour of a memristor-based electrical circuit using a new energy distribution analysis based on discrete wavelet transforms.

Other models exhibiting chaos have also been assessed using wavelet analysis; for example, Hramov and Koronovskii (2005) considered the synchronization behaviour of both coupled Rőssler and Lorenz systems using a Morlet-based approach. They advocated the wavelet method as it allows the universal consideration of different types of behaviour of coupled oscillator systems, including 'complete', 'lag', 'phase' and 'generalized' synchronizations as well as 'non-synchronized oscillations'. Postnikov (2007) showed that timescale synchronization of chaotic systems with ill-defined phase may be achieved using wavelet transforms. In the study, Morlet wavelets of varying central frequencies were employed, and it was shown that for central frequencies of greater than around 2.5π the phase difference remains bounded in a coupled Rőssler system. (See also Chapter 2, Section 2.11 and Addison et al. 2002a, for more details of these wavelets.) More recently, Postnikov and Lebedeva (2010) developed a modified continuous wavelet transform technique for the decomposition of strong nonlinear oscillations, which has advantages over traditional wavelet-based matching pursuit schemes. The method was demonstrated on a chaotic Rőssler system. See also Feng et al. (2012), who considered the phase synchronization of fractional-order Rőssler systems, and Rong-Yi and Xaio-Jing (2011), who illustrated phase space reconstruction of chaotic attractors (both in the Rőssler system and an epileptic EEG signal) using wavelet decomposition.

In other work, Benitez et al. (2010) have developed a scale index parameter to measure a signal's non-periodicity. They tested this on a Bonhoeffer–Van der Pol oscillator system, comparing it with the maximal Lyapunov exponent – a parameter traditionally used for detecting chaotic motion. Gao et al. (2010) and Tung et al. (2011) compared a nonlinear adaptive denoising approach with a number of wavelet-based thresholding techniques and chaos-based projective filtering for reducing noise in a chaotic Lorenz system. And, in an experimental study, Behera et al. (2014) interrogated the chaotic behaviour of a bouncing ball using both discrete and continuous wavelet analysis. They found that the wavelet method was useful in capturing both the transients and non-stationary periodic behaviour, including the phase synchronization of different modes.

5.3 NON-DESTRUCTIVE TESTING OF STRUCTURAL ELEMENTS

Non-destructive testing (NDT) covers a wide range of techniques concerned with the interrogation of underlying structural integrity and which do not impair in any way the intended performance of the structure during and after examination. Also known as non-destructive evaluation (NDE) or non-destructive inspection (NDI), the methods include visual inspection, ultrasonic testing, radiographic testing, pressure testing, vibration analysis, acoustic emission testing and so on. In this section, we consider the non-destructive testing of fixed structural elements. (The evaluation of rotating machinery is considered in the next section of this chapter.)

Sonic echo testing is a common technique employed in the non-destructive testing of structural elements. It involves striking the test specimen (e.g. a structural element or material specimen) with an instrumented hammer which records both the input pulse (strike) and subsequent response of the specimen. This response is interpreted as an indirect measurement of the specimen's integrity. A typical velocity trace from such a test on a foundation pile is shown in Figure 5.6 (Watson et al., 1999). A schematic of the sonic pulse transmission through the pile and the scalogram corresponding to the velocity trace is shown in the figure. For such a heavily damped system, there are rarely multiple longitudinal reflections, and the frequency dependence of the group velocities is negligible. Therefore, the temporal isolation of the signal features are more important than their frequency decoupling, and hence a 'Mexican hat' was used in this early work as it is more temporally compact than the standard Morlet ($5 \leq \omega_0 \leq 6$) often applied in the study

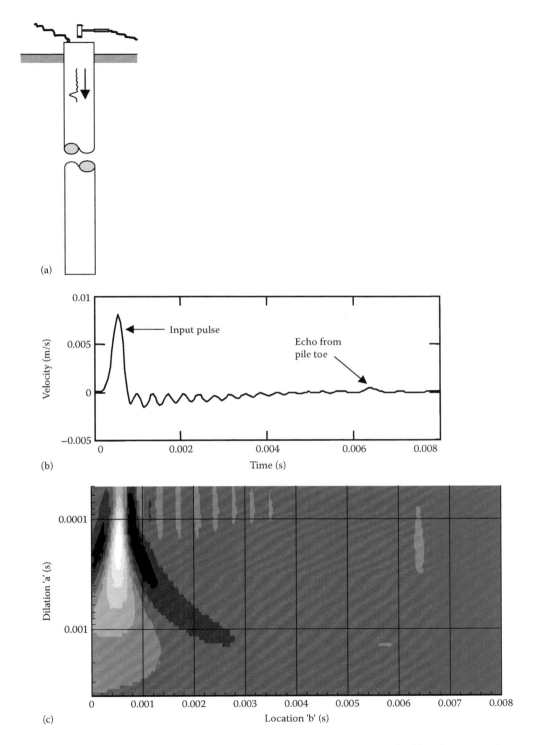

FIGURE 5.6 Wavelet transform decomposition of a sonic echo signal: (a) Schematic of sonic echo testing of a foundation pile, **(b)** a Fourier-filtered velocity trace and **(c)** corresponding wavelet transform plot of an 11 m pile in stiff/very stiff clay.

of free beam and plate vibrations. The pile was 11 m long and the velocity of the stress wave through the pile was 3800 ms^{-1}. Thus, we would expect to see the reflection of the end of the pile occur $2 \times 11/3800 = 0.0058$ s after the initial impulse. The pile toe reflection shows up particularly well in the top right-hand quadrant of the scalogram, as it has a distinctively different shape and appears lower down the scalogram from the initial oscillations, which occur at a dilation around $a = 10^{-4}$. These oscillations, seen to occur in the top left-hand quadrant of the scalogram just after the input initial pulse, are known as 'ringdown' and are in fact the surface oscillations of the pile head due to the hammer impact. Figure 5.7 contains the reconstructed traces of both wavelet- and Fourier-filtered signals. Simple scale-dependent wavelet filtering was employed where the transform components at a scales less than 0.0001 were set to zero and an inverse wavelet transform performed. The wavelet-filtered trace is shown in the top left quadrant of Figure 5.7. The Fourier low-pass filter cutoff frequency was set to 2.25 kHz, as is the case in practice. The Fourier-filtered trace is shown in the top-right quadrant of Figure 5.7. The two lower plots in the figure show a zoomed-in section of the upper traces in the region of the pile toe reflection. Comparing the filtered traces of Figure 5.7, it can be seen that the wavelet filtering separates the ringdown oscillations from the pile toe feature much more effectively than Fourier filtering.

Figures 5.8 and 5.9 illustrate a more sophisticated filtering approach for removing erroneous ringdown artefacts from the pile signal based on the modulus maxima of the scalogram (Watson and Addison, 2002). Figure 5.8b contains the finite element (FE)-generated velocity trace of an 11 m pile in stiff clay. A schematic of the pile is given in Figure 5.8a. A simulated defect in the form of a reduction in section ('necking') is present approximately one-third of the way down the pile. The locations of both the input pulse and echo from the pile discontinuity are highlighted in the figure. The transform plot associated with the wavelet decomposition of the trace is shown in Figure 5.8c. Modulus maxima were found from the original wavelet transform using a simple algorithm which scans across the transform plot scale by scale and identifies local maxima and minima. The modulus maxima obtained for the scalogram of Figure 5.8c is shown in Figure 5.9a. We can see from the modulus maxima plot that the large-scale input pulse feature in the signal contains a maxima line extending from high to low frequencies. The other maxima lines do not extend as far down into the low-frequency range. A close approximation to the signal can be reconstructed using only the maxima lines where the energy contained in the whole scalogram is reapportioned to the maxima lines in the reconstruction. The reconstruction using all the maxima lines in Figure 5.9a is shown in Figure 5.9b. Figure 5.10 illustrates the filtering of the initial signal using the modulus maxima. This is done in an anticlockwise manner from Figure 5.10a–d. The maxima lines are thresholded at a frequency of 310 Hz, shown in Figure 5.10b. All maxima lines which do not extend down from the higher bandpass frequencies to this threshold level are removed (Figure 5.10b). Those which do extend down to and beyond the threshold are retained. Only the retained lines (Figure 5.10c) are used to reconstruct the signal (Figure 5.10d). The threshold is chosen to be lower than the ringdown artefact in the signal, hence the maxima lines from the ringdown artefact will fall below the threshold and be removed. In addition, noise, which also manifests itself as

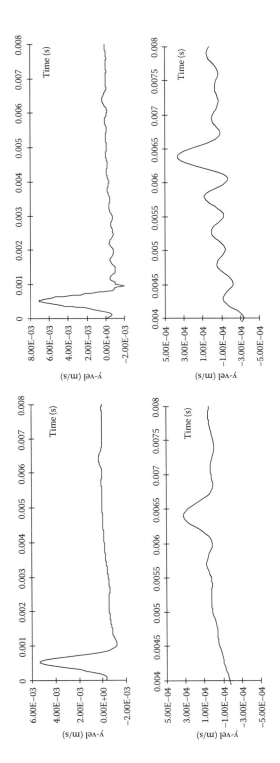

FIGURE 5.7 **Wavelet and Fourier filtering of the sonic echo pile signal:** Wavelet-filtered (left) and Fourier-filtered (right) traces for finite element–generated pile test data shown in the previous figure. (Reprinted from Watson, J. N., et al., *Shock and Vibration*, 6, 267–272, 1999. Copyright (1999), with permission from IOS Press.)

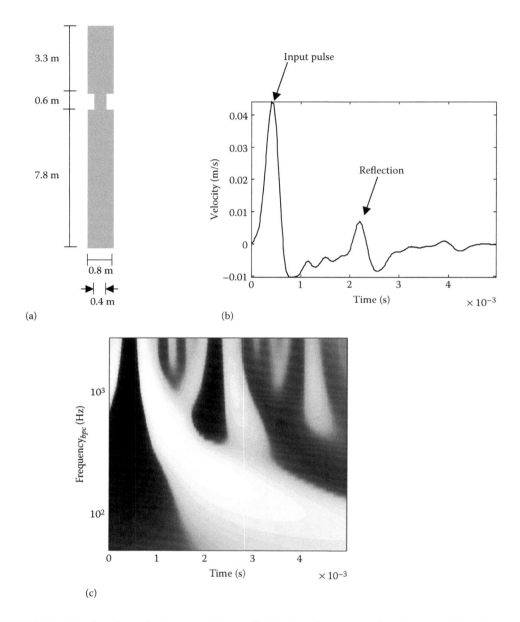

FIGURE 5.8 Sonic echo velocity trace from pile head and associated scalogram: (a) Schematic of a foundation pile (a necking fault has been modelled approximately one-third of the way down the pile), (b) sonic echo velocity trace take from the pile head and (c) wavelet transform plot of the signal in (b) (large positive components in black, large negative components in white). (From Watson, J. N., and Addison, P. S., *Mechanics Research Communications*, 29, 99–106, 2002.)

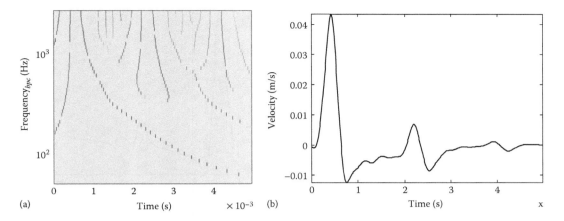

FIGURE 5.9 The reconstruction from only the scalogram modulus maxima lines: (a) Modulus maxima plot-derived lines from the scalogram in previous figure and (b) reconstructed trace using only the modulus maxima lines in (a). (From Watson, J. N., and Addison, P. S., *Mechanics Research Communications*, 29, 99–106, 2002.)

modulus maxima restricted to high-frequency regions, is also removed from the signal. The resultant reconstructed trace is shown in Figure 5.10d, which illustrates how all ringdown has been eliminated while retaining the pertinent signal features. Notice that the retained features still contain their high-frequency components and are not excessively smoothed, as would be the case if bandpass filtering using Fourier techniques had been employed.

Figure 5.11 illustrates the use of the Morlet wavelet with low central frequencies in the analysis of a highly oscillatory sonic echo signal where the pile toe is not obvious in the time domain (Addison et al., 2002a). When using central frequencies ω_0 less than 5 ($f_0 < 0.8$), the complete Morlet wavelet given by Equation 2.36 in Chapter 2 must be used. Morlet wavelets with low central frequencies result in analyses that are more 'temporal' than 'spectral' in that they are better at locating short-duration temporal features than those with higher values of ω_0 (refer to Chapter 2, Section 2.12). This can be seen in Figure 5.11, where the pile toe can be located in the wavelet transform scalogram plots only at lower values of ω_0. The location of the pile toe is indicated both in the time signal and the lowest scalogram plot, which corresponds to the complete Morlet wavelet with $\omega_0 = 1.5$ (i.e. $f_0 = 0.238$). It is interesting to note that, although they do appear to be particularly useful for certain tasks, the literature still contains surprisingly very little on the use of complex wavelets with few oscillations such as the complete Morlet wavelet of low central frequency or the complex Mexican hat (Addison et al., 2002a).

The use of wavelet transform reassignment for the identification of highly localized pertinent features in sonic echo time series was investigated by Addison et al. (2006). In the study, reassigned scalograms (Chapter 2, Section 2.20) were compared with their original counterparts. In addition, low-oscillation complex wavelets (Chapter 2, Section 2.11) were utilized in the work. Figure 5.12 contains a sonic echo signal with its scalogram and reassigned scalogram generated using a low-oscillation complex Morlet-based transform with $\omega_0 = 1.0$ (i.e. enhancing temporal resolution). The primary and secondary reflection

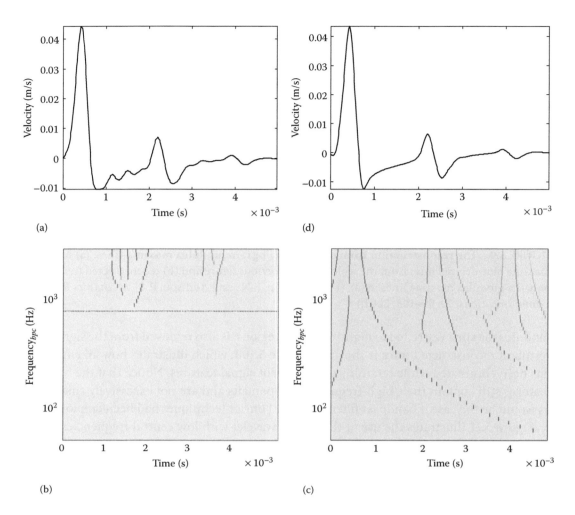

FIGURE 5.10 The partitioning of the modulus maxima lines for signal filtering: (a) Original data, (b) discarded modulus maxima lines, (c) retained modulus maxima lines and (d) filtered data. (From Watson, J. N., and Addison, P. S., *Mechanics Research Communications*, 29, 99–106, 2002.)

components corresponding to the pile toe are indicated in the plot. The investigation found that low-oscillation wavelets appeared to be better for detecting short-duration reflection features in the reassigned scalogram.

Other work on impact integrity testing involving wavelet analysis includes further research into the integrity testing of piles (installed bridge piles in Taiwan) by Ni et al. (2012a), the assessment of stiffness degradation in composite plates by filtered noisy impact testing (Lee et al., 2007), the predictive modelling of fatigue crack detection in thick steel bridge structures (Gresil et al., 2013), the optimization of damage detection in steel plates during noisy impact tests (Rus et al., 2006), the diagnosis of defects in steel beams using a wavelet-based damage index vector (Rizzo et al., 2009) and a combined wavelet and neural network–based fuzzy interference system (Escamilla-Ambrosio et al., 2011).

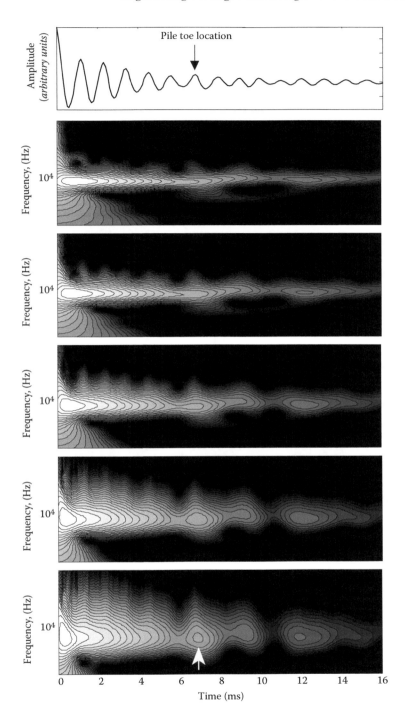

FIGURE 5.11 A complete Morlet wavelet analysis of a sonic echo signal. The figure contains a sonic echo signal taken from a pile together with a sequence of scalograms generated using a complete Morlet wavelet decomposition of the signal with the central frequency set to (from top to bottom) $\omega_0 = 5.5, 4.5, 3.5, 2.5$ and 1.5. (From Addison, P. S. et al., *Journal of Sound and Vibration*, 254(4), 733–762, 2002a.)

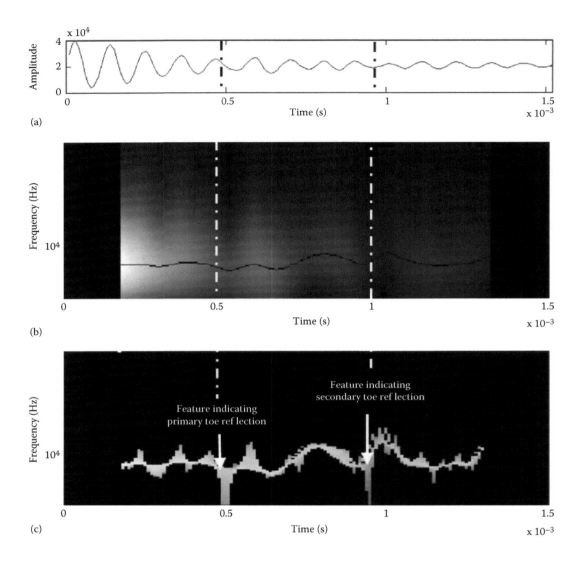

FIGURE 5.12 (a) Wavelet reassignment of a pile test signal at $\omega_0 = 1.0$ showing the first 1.5 ms of sonic echo signal, (b) the wavelet scalogram of the signal and (c) the reassigned scalogram derived from the scalogram in (b) showing features attributed to primary and secondary toe reflections. (From Addison, P. S. et al., *Mechanical Systems and Signal Processing*, 20(6), 1429–1443, 2006.)

Damage detection through the vibrational analysis of individual structural elements and whole structures has been carried out by a number of research groups. The method assumes that damage-induced changes in the physical properties of the element (mass, damping, stiffness, etc.) will cause underlying changes in modal properties (natural frequencies, modal damping, modal shapes, etc.). For example, a 2-D CWT approach was developed by Fan and Qiao (2009), where the wavelet decompositions of the fundamental mode shape highlighted regions of damage on a plate clamped on one side. They found the technique superior to two other damage detection algorithms based on strain energy and gapped smoothing methods. In work also concerning damaged plates, B-spline wavelets of fractional order were employed by Katunin and Przystalka (2014) to detect damaged regions in both numerically modelled and experimental plates. The damaged regions of the three plates considered are shown schematically in Figure 5.13a, with the corresponding numerical and experimental results given in Figure 5.13b,c. The authors also investigated different optimization algorithms for finding the optimal wavelet parameters for the structural diagnosis of the plates. They found that the results were similar, with an evolutionary algorithm performing best. In other vibrational analysis work, damage detection in cantilevered beams with 'notch' defects of varying depth has been studied by Rucka (2011) and Lepik (2012), a damage index based on wavelet energy has been developed by

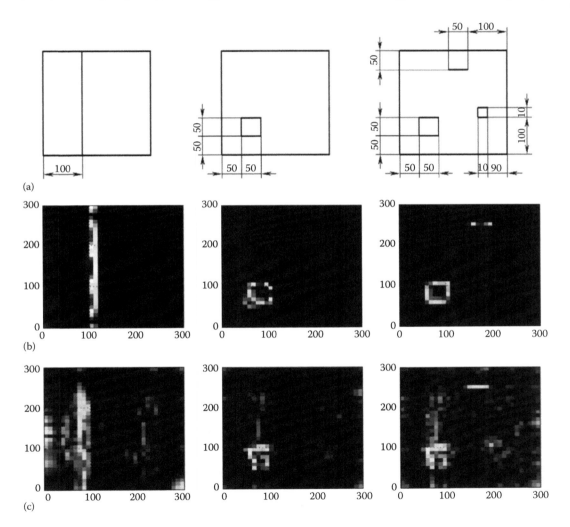

FIGURE 5.13 (a) The cases of damages considered in the analysis, (b) results of damage identification based on numerical data and (c) results of damage identification based on experimental data. (Original in colour.) (From Katunin, A., and Przystałka, P., *Engineering Applications of Artificial Intelligence*, 30, 73–85, 2014.)

Rajeev and Wijesundara (2014) for use in concentrically braced frame structures subjected to inelastic cyclical loading, and the dynamic testing of concrete slabs on a shaker table using a Morlet-based analysis to determine the characteristic scale bands associated with the different damage mechanisms has been carried out by Zitto et al. (2015).

A number of researchers have used the signals derived from in-service loading of structural elements to monitor their structural health. Hester and Gonzalez (2012) investigated the higher scales in Mexican hat–based CWT plots of the acceleration response to vehicles on a bridge. They found that the coefficients at these scales were of significantly higher value at the location of the damaged section than at other locations. This is shown in Figure 5.14, where the transform surface is sliced at five sections. Zooming into the higher scales, as shown in Figure 5.14c, shows that the $x(t)/L=0.3$ normalized location has the highest coefficient values. This was the nearest location to the damaged section of the bridge (which was at $x(t)/L=0.33$). Around the same time, Ni et al. (2012b) employed wavelet multiresolution

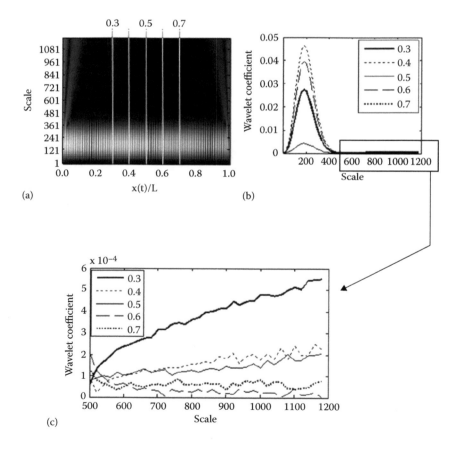

FIGURE 5.14 Vertical sections through wavelet transform surface: (a) Wavelet transform surface when $\delta=0.2$ at the 1/3 point, (b) absolute value of wavelet coefficients for vertical sections at different positions of the moving load and (c) absolute value of wavelet coefficients at different positions of the moving load for scale range between 500 and 1200. (From Hester, D., and González, A., *Mechanical Systems and Signal Processing*, 28, 145–166, 2012.)

analysis to separate out the low-frequency temperature effect signal from strain measurement data acquired from a suspension bridge. Later, Gokdag (2013) found that DWT denoising was superior to CWT denoising when used in a crack identification scheme for bridge structures under vehicle loading, and Zhong et al. (2014) proposed a damage prognosis system based on wavelet neural networks, testing it both on a steel-beam model and then on data from a suspension bridge. In other research studies, Pakrashi et al. (2010) found that wavelet-transformed phase spaces for damaged and undamaged model bridge–vehicle interaction tests were distinctly different at higher scales, and Li et al. (2009) evaluated earthquake-induced damage to structures during earthquake excitation using Morlet wavelets and data from a scale model of a three-storey structure with reinforced concrete walls.

Other work on the non-destructive evaluation of structural elements includes laser-induced ultrasonic wave–based monitoring of railroad track using DWTs (Rizzo et al., 2010), the processing of reflection records of transient steep pressure waves to monitor pipeline integrity using Fourier and wavelet approaches (the latter based on the Daubechies D10 wavelet; Hachem and Schleiss, 2012) and a nonlinear ultrasonic modulation method for reference-free crack detection in cracked aluminium plates and an aircraft fitting lug (Lim et al., 2014). Pulsed eddy current analysis has also been a topic of interest, and work in this area includes the CWT analysis of pulsed eddy current signals by Gombarska and Smetna (2011) and the DWT image fusion–based analysis of eddy current images, at two different frequencies, of a steel cladding tube by Balakrishnan et al. (2012). Finally, a comprehensive structural health-monitoring system for spacecraft damage due to space debris or meteorites has been developed by Yun et al. (2011). The system uses a damage metric computed from the DWT decomposition of Lamb waves generated in the structure using a piezoelectric transducer. An acoustic sensor detects impact events. This, in turn, switches on an impedance sensor to detect if damage has occurred and then the Lamb wave detector to determine the severity and location of the damage.

5.4 CONDITION MONITORING OF ROTATING MACHINERY

The condition monitoring of rotating machinery attempts to detect and diagnose faults from vibration signals usually picked up from the machine casing. Left undiagnosed, developing machine faults may ultimately lead to catastrophic failure. In this section, we begin with some of the more recent applications of wavelet techniques to gear diagnostics, where the early detection of gear failure is a prime concern. Then, we take a look at wavelet-based detection and diagnosis of signals from other rotating machinery components, such as shafts, bearings and blades.

5.4.1 Gears

The manifestation of gear faults is often more obvious in the wavelet domain and thus is the basis for much work in the area of fault diagnosis. Figure 5.15 shows the wavelet transform of a signal from an internal combustion engine gear-testing rig from the investigation conducted by Vernekar et al. (2014). The band at around 34 Hz is at twice the crank shaft rotation frequency. The higher frequency component evident intermittently across the Morlet wavelet transform domain at 117 Hz – the gear-meshing frequency – is indicative of a gear fault.

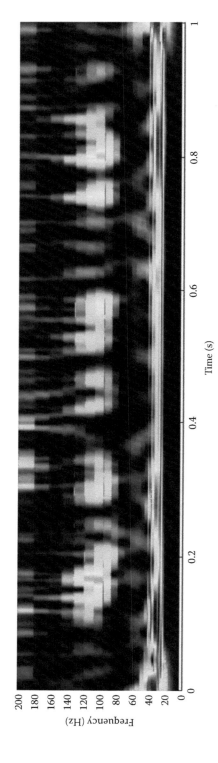

FIGURE 5.15 Time–frequency plot of the CWT of the fault-induced gear vibration signal. (From Vernekar, K. et al., *Procedia Materials Science*, 5, 1846–1852, 2014.)

Adaptive Morlet wavelets have been employed by Elbarghathi et al. (2013), where the wavelet properties may be optimized using an entropy optimization technique. When applied to the analysis of signals from a two-stage helical gearbox with varying tooth faults, the method compared favourably with another where the Morlet bandwidth and frequency parameters were optimized using a kurtosis maximization approach. The use of Hermitian wavelets (derivative of Gaussian functions) for fault detection in gearboxes has been advocated by Li et al. (2011a). These low-oscillation wavelets were used to diagnose localized gear crack faults. Their ability to extract transients from strong noise signals was demonstrated on a pair of spur gears with a transverse crack defect in the root of the driving gear. The authors stated that the technique excels in extracting transients, which are often the indicators of an incipient defect in a gear system.

The fusion of wavelet-based classifier inputs from vibration and acoustic signals has been demonstrated by Khazaee et al. (2014) as a precise method for fault detection in planetary gearboxes. Signals from gears exhibiting four conditions were monitored: healthy, worn tooth face, broken ring gear and cracked ring gear. A number of features were extracted from the wavelet coefficients, including mean value, standard deviation, skewness and kurtosis. Combining the wavelet-based input data from the two vibration signals increased the individual classification accuracies of the artificial neural network classifier used in the study from 86%–88% to a combined accuracy of 98%. Jedlinski and Jonak (2015) have compared support vector machines and multilayer perceptron networks with wavelet-based inputs. They found that the latter produced a marginally better classification (92% compared with 90%). Li et al. (2011b) combined wavelet denoising with autoregressive modelling and principal component analysis in a scheme for multifault diagnosis of defective gears. Multifault classification has also been tackled by Bordoloi and Tiwari (2014) using support vector machine optimization and wavelet-based vibration data.

Wavelet-based demodulation of gear vibration signals has been developed by Dien (2008) using Morlet wavelets. The scheme employs logarithmic projections of wavelet envelopes extracted from wavelet surface plots. Figures 5.16 and 5.17a,b contain, respectively,

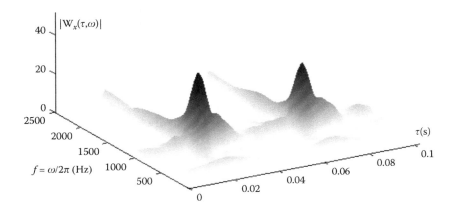

FIGURE 5.16 Wavelet plot of the signal. (From Dien, N. P., *Technische Mechanik*, 28(3–4), 324–333, 2008.)

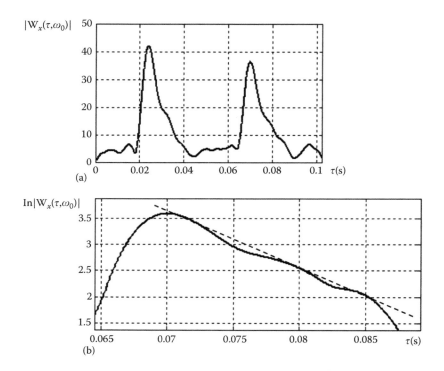

FIGURE 5.17 (a) The wavelet envelope extracted from the wavelet plot in the previous figure and (b) the wavelet envelope (with logarithmic scale) for evaluating the damping ratio. (From Dien, N. P., *Technische Mechanik*, 28(3–4), 324–333, 2008.)

a wavelet surface, the corresponding wavelet envelope and the logarithmic wavelet envelope from an experimental signal acquired by a laser Doppler vibrometer. The experiment comprised a spur gear pair system with the driving gear cracked at the tooth root. The damping ratio was determined from the downslope in the logarithmic plot (as shown in Figure 5.17b), which the authors could then use to model gear pair vibrational systems.

An enhanced synchrosqueezing technique (Chapter 2, Section 2.20) was developed by Li and Liang (2012) to analyze signals from faulty gearboxes. Their generalized synchrosqueezing transform (GST) performs an initial mapping of the vibration signal onto an analytic signal before performing the CWT and synchrosqueezing steps. They compared their approach with a number of other time–frequency tools: the Morlet CWT, the reassigned Morlet CWT, reassigned pseudo-Wigner–Ville distribution and the Hilbert–Huang transform (HHT). Figure 5.18 contains examples of the method applied to a signal from a gearbox with a chipped pinion. The Morlet-based CWT, HHT and original synchrosqueezing transform are also shown for comparison. The authors concluded that the GST yields better representation of the time–frequency signal analysis without excessive computational requirements.

Combet et al. (2012) have developed an instantaneous wavelet bicoherence (WB) measure for local tooth damage detection. The measure is based on the integration of the modulus of the WB of the signal within a specific (square) frequency region. Figure 5.19

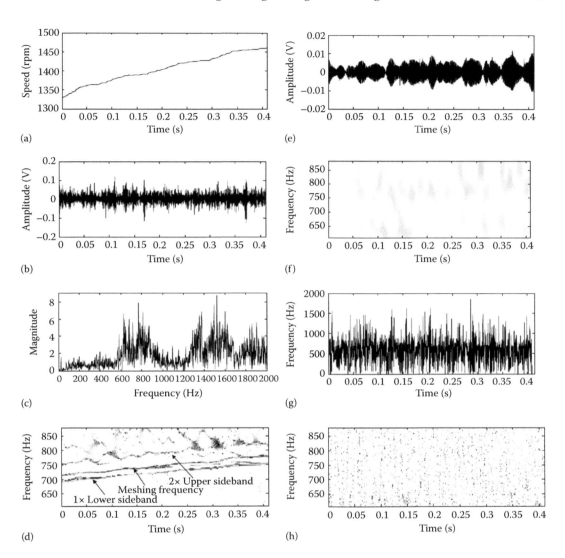

FIGURE 5.18 **Fault diagnosis for the gear box with chipped pinion:** (a) Rotation speed signal, (b) time domain waveform of the vibration signal, (c) spectrum of the vibration signal, (d) time-frequency representation (TFR) obtained using the GST, (e) reconstructed meshing frequency and side-band components, (f) TFR generated by Morlet CWT, (g) TFR generated by HHT (the first IMF is employed) and (h) TFR obtained by the synchrosqueezing approach. (From Li, C., and Liang, M., *Mechanical Systems and Signal Processing*, 26, 205–217, 2012.)

contains the wavelet bicoherence maps for three locations on the gear, and Figure 5.20 contains the integrated wavelet bicoherence feature for an unpitted and pitted gear. In the method, the locally averaged WB feature is thresholded to enable fault detection. The threshold was determined experimentally to avoid triggering false alarms. The researchers found that their technique was robust with superior detection capabilities.

254 ■ The Illustrated Wavelet Transform Handbook

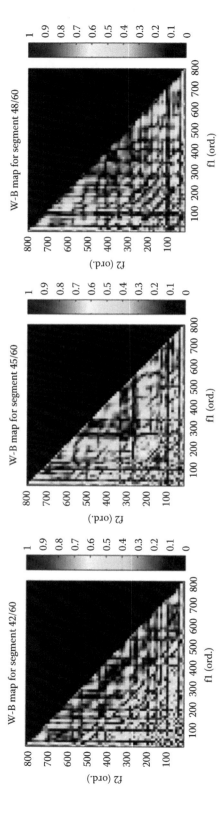

FIGURE 5.19 **The wavelet bicoherence map estimated at three different locations on the gear: (a)** On the meshing tooth previous to damage, **(b)** at the damaged tooth position and **(c)** on the meshing tooth next to the damage. (From Combet, F. et al., *Mechanical Systems and Signal Processing*, 26, 218–228, 2012.)

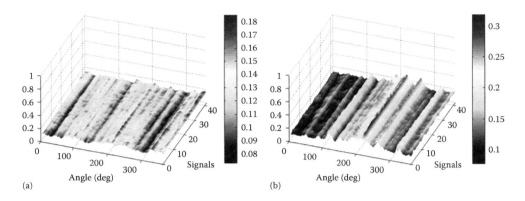

FIGURE 5.20 The locally averaged WB integrated feature for the (a) Unpitted and (b) pitted cases (integration bandwidth is 200–400 orders and window length parameter is $\beta = 8$). (From Combet, F. et al., *Mechanical Systems and Signal Processing*, 26, 218–228, 2012.)

Rafiee et al. (2010) studied over 300 mother wavelet candidates for bearing and gear fault diagnosis and concluded that the Daubechies 44 wavelet had the most similar shape to the bearing and gear signal. The same group (Rafiee and Tse, 2009) also developed an improved CWT technique for gearbox vibration signals where the autocorrelation of the continuous wavelet coefficients was employed. Other work in this area includes the monitoring of gear vibration signals using Limpel–Ziv complexity and approximate entropy methods from nonlinear dynamics as diagnostic tools by Kedadouche et al. (2015); an adaptive spectral kurtosis filtering algorithm based on the Morlet-based transform developed by Liu et al. (2014); the analysis of burst signals from defective gears using the analytic wavelet transform of the undecimated wavelet transform signal representation by Jena et al. (2013); the study of acoustic emission data in low-speed, heavy-duty gears using second-generation wavelets by Gao et al. (2011); and the investigation of DWT-based root mean square and kurtosis indicators for crack propagation assessment in spur gears by Tian et al. (2012). Additional coverage of the subject has been provided by Yan et al. (2014) in their high-level overview of the role of wavelet analysis in the fault diagnosis of rotating machinery.

5.4.2 Shafts, Bearings and Blades

A more standardized approach to visualizing the current status of rotor dynamics has been proposed by Lim and Loeng (2013) which takes into account the time shift, wavelet edge distortion and system noise suppression. Examples of their Coiflet-based analysis are shown in Figure 5.21 for minor rubbing and unbalance faults. For minor unbalance, the wavelet features became distorted at larger scales (when compared with the baseline fault-free signal) and distinct 'rubbing signals' were found in the transform plot for the minor blade rub. The researchers also found that major blade rubbing and severe unbalance were easily differentiated in the wavelet domain.

An autocorrelation function indicator was employed by Xie et al. (2012) to determine the optimal Morlet CWT coefficients, which were indicative of the fault characterization frequencies of a cooling fan bearing. They found the method superior to both DWT- and

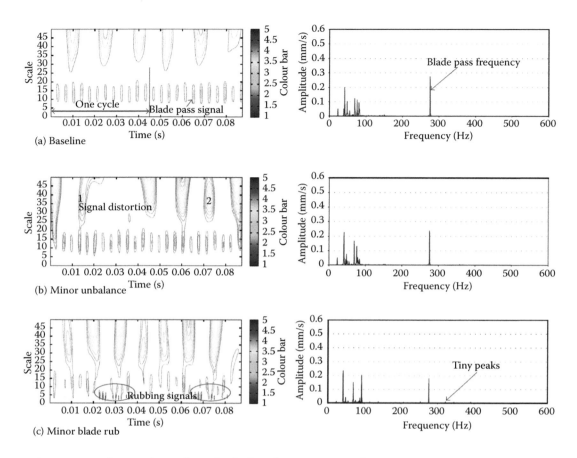

FIGURE 5.21 Comparison of results for baseline and minor fault conditions. (From Lim, M. H., and Leong, M. S., *Advances in Mechanical Engineering*, 625863, 1–8, 2013.)

FFT-based approaches. Jena et al. (2012) performed time-marginal integration (TMI) of the Morlet wavelet transform of vibration and acoustic signals from defective bearings. Figure 5.22 shows examples of burst signals associated with the defect together with the associated wavelet transform and TMI signal (derived by integrating across the frequency in the transform plane). The authors went on to estimate the radial bearing inner race defect width using the technique. An 'exact' wavelet transform has been developed by Tse et al. (2004) where, at each time frame, the algorithm generates an adaptive daughter wavelet to match the signal as exactly as possible. They successfully applied it to motor pump drive systems with bearing faults.

The use of reassigned scalograms to monitor rub impact faults in rotor systems transitioning into chaotic motion was studied by Ma et al. (2009a). Figure 5.23 shows an example of the reassigned wavelet scalogram associated with a chaotic region (left-hand plot) and a post-chaotic, periodic region (right-hand plot) from their investigation of a system with a crack coupled with a rub impact fault. They found that the oil film instability played a dominant role in the fault. Earlier work on the use of reassigned scalograms by Peng et al.

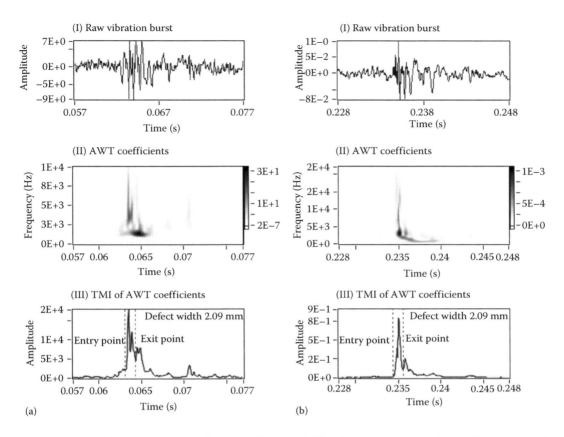

FIGURE 5.22 (a) Single acoustic burst analysis and (b) vibration burst analysis: (I) Raw signal burst, (II) AWT spectrum and (III) TMI of AWT spectrum. (Original in colour.) (From Jena, D. et al., *Measurement Science Review*, 12(4), 141–148, 2012.)

(2002) provides much of the background for such methods and includes an analysis of experimental data from rub impact, oil whirl and coupling misalignment.

A number of papers have emerged in recent years specifically targeting the use of wavelet schemes for the condition monitoring of wind turbines. For example, a CWT energy-tracking scheme for detecting both electrical and mechanical faults in wind turbines has been proposed by Yang et al. (2010). The approach reduced the calculation required to extract features from lengthy online data and is applicable to turbines operating at fixed or variable speeds. The same group (Watson et al., 2010) also performed wavelet-based analysis of data from operational wind turbines comprising doubly fed induction generators. They found sufficient change in the CWT power amplitude over the monitoring period consistent with independent vibration monitoring to conclude that rotor eccentricity due to misalignment had been detected. Also in this area is the brief discussion of a DWT-based neural network method based on LabVIEW by An et al. (2013) and the comprehensive review of the condition monitoring of wind turbines by Marquez et al. (2012), where wavelet analysis is considered among a plethora of other signal processing techniques.

258 ■ The Illustrated Wavelet Transform Handbook

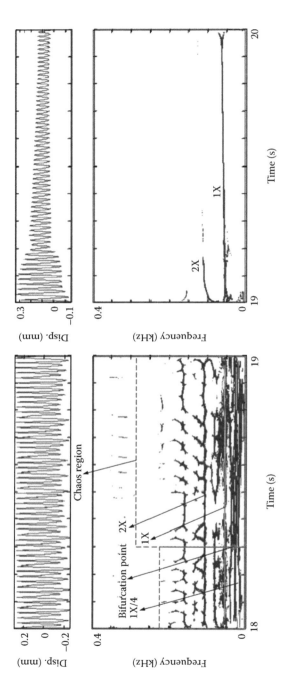

FIGURE 5.23 Reassigned wavelet scalograms with crack and rub impact faults in different rotational speed ranges. (Original in colour.) (From Ma, H. et al., *Journal of Sound and Vibration*, 321, 1109–1128, 2009.)

Other work in this area includes the development of a back-propagation neural network with inputs from wavelet binary images applied to unbalanced and misaligned shafts and defective bearings by Kaewkongka et al. (2001), the rolling element fault diagnosis study comparing two wavelet types (Meyer and Morlet) and three machine-learning methods (support vector machines, artificial neural networks and self-organized maps) by Kankar et al. (2011), the wear diagnosis of marine engines by Lim et al. (2012) and the wavelet analysis of rundown voltages to detect induction motor rotor bar faults by Supangat et al. (2007). Finally, it is worth consulting the comprehensive review by Chen et al. (2016) which covers the application of wavelet transforms to all aspects of fault diagnosis in rotating machinery.

5.5 MACHINING PROCESSES

The objective of any machining process is the efficient production of a part of specific shape with acceptable dimensional accuracy and surface quality. The monitoring of machining processes is therefore an important problem in manufacturing engineering. There is a considerable economic incentive to develop reliable monitoring techniques; hence, considerable research effort has centred on this problem over recent years. In an early paper concerning the use of wavelet transforms for the examination of both chatter in turning processes and tool wear in drilling processes, Wu and Du (1996) stated that previous success in the area had been limited by the inherent problems associated with monitoring signals, which are typically affected by process defects (e.g. chatter and wear), working conditions, process noise and sampling noise. Over the past couple of decades, numerous researchers have attempted to tackle the problem using ever-more sophisticated wavelet-based methodologies.

Wavelet-based investigation of motor currents has been carried out by a number of researchers. For example, stator motor current was analyzed by Han et al. (2006) in order to diagnose faults in induction motors. Their technique employed discrete wavelets, neural networks and genetic algorithms and showed considerable promise when applied to a range of signals from motors with a variety of faults, including a broken rotor bar, a faulty bearing, an unbalanced rotor, a bowed rotor and misalignment. Spindle motor currents were assessed for use in monitoring CNC milling machines by Mota-Valtierra (2011), where the fifth level of a Daubechies wavelet–based multiresolution analysis was used for data reduction prior to classification using an artificial neural network. The researchers found a correlation between cutting force and tool wear using the method. Both motor spindle and feed current was analyzed by Li et al. (2000) using wavelet and fuzzy techniques to detect tool breakage and wear conditions. Wavelet bicoherence was employed by Li et al. (2014a) to analyze servomotor torque signals, derived from three current signals, from a high-precision vertical machining centre. They found that a maximum eigenvalue of the biphase randomization wavelet bicoherence (MEBRWB) exhibited robust experimental performance when compared with more traditional features.

Cutting force signals from a dynamometer during the milling of Inconel material were studied using wavelet and Hilbert–Huang transforms by Litak et al. (2013). This material exhibits rapid work hardening and is difficult to machine. Figure 5.24 shows the wavelet

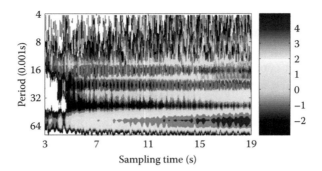

FIGURE 5.24 Wavelet power spectrum P_w for the force F_x in the process of milling. In the right panel, a colour logarithmic scale of P_w is shown. (Original in colour.) (From Litak, G. et al., *Latin American Journal of Solids and Structures*, 10, 133–140, 2013.)

transform of the cutting force time series where a Morlet wavelet was used ($\omega_0 = 6$). Four harmonics are visible in the plot, emerging over time, where all four are synchronized with the quarter-period oscillation. The authors state that this could be the effect of non-linearities and reflect a regenerative effect. In other work on force signals, both multifractal and Morlet-based wavelet approaches to the analysis of cutting thrust forces in stainless steel turning has been undertaken by Litak and Rusinek (2011).

A number of workers have considered multiple signals in their attempts to develop enhanced tool condition monitoring techniques. For example, Beruvides et al. (2013) investigated microdrilling operations using both force and vibration signals. Although they also considered time domain data and the Fourier and Hilbert–Huang transforms, they found that a combination of wavelet transforms and neural networks yielded the most suitable means for the prediction of tool usage. Jemielniak et al. (2012) utilized acoustic emission, cutting force and vibration signals in their work concerning tool wear monitoring, where 97 signal features were computed for each signal (including 84 based on wavelet packet decomposition). They considered each sensor signal separately and in combination, discussing the relative merits of each approach. Sensor fusion pattern recognition has been employed by Segreto et al. (2013), combining force, acoustic emission and vibration signals to assess residual stress in Inconel machining. In their described method, sensor fusion feature vectors incorporating wavelet-based features were input into a neural network for data processing. Force, vibration and acoustic emission signals were also examined using a segmented hidden Markov model by Geramifard et al. (2012). They applied Daubechies wavelet decomposition to the force signals and a discrete Meyer wavelet to the vibration and AE signals from a vertical milling machine.

Pal et al. (2011) fed wavelet-based cutting strain and motor current features into both radial basis and back-propagation neural networks in their study of tool wear monitoring. They tried Daubechies, symlet and Coiflet wavelets and noted that any of these wavelet functions could be used successfully in the analysis. Similarly, Jemielniak and Kossakowska (2010) reported that all four wavelet types they tried (Daubechies, symlet, Coiflet and biorthogonal) in a tool wear study using acoustic emission signals had at least one variant which came close to the best performing one. Other work of interest includes the combination of wavelet analysis and independent component analysis by Shao et al. (2011) for the blind source separation of various signal components in power signals associated with milling processes, the use of wavelet and power spectral density by Chen et al. (2013) in the identification of spindle unbalance in machine tools, the use of support vector regression and wavelet packets by Benkedjouh et al. (2015) for the health assessment and life prediction of cutting tools, and the wavelet-based chatter detection tool for plunge grinding by Ahrens et al. (2013). The reader is also directed to the comprehensive review of the many techniques for advanced monitoring of machining operations by Teti et al. (2010). This sets wavelet methods within a wider context of advanced signal processing and also touches on sensor systems, decision support systems and industrial experience. A review more specific to wavelet techniques for the analysis of sensor signals used in tool condition monitoring has been carried out by Kunpeng et al. (2009). Finally, a wide-ranging review of the application of image-processing schemes to tool condition monitoring has been conducted by Dutta et al. (2013), which includes a number of wavelet-based studies in this area.

5.6 CHARACTERIZATION OF SURFACES AND FIBROUS MATERIALS

The characterization of engineering surfaces is pertinent to a number of engineering fields providing quantitative information on the formation process of the surface – for example, the manufacturing process used to form a machine component or the fracture process causing a rugged crack surface. Surface topography is one of the most important factors affecting the performance of manufactured components. It can be related to a number of pertinent engineering aspects, such as wear, lubrication, friction, corrosion, fatigue, coating, paintability and so on.

Using wavelet-based image denoising of photolithographed surfaces, Facco et al. (2009) have developed a method for monitoring the roughness of the side walls of semiconductors by scanning electron microscope images. They found that the Daubechies 8 wavelet responded well to their requirements as it introduced very limited phase distortion, maintained a faithful localization in the spatial domain and decorrelated the signal in a sensitive manner for both the image smooth features and discontinuities. In another study by Zawada-Tomkiewicz and Sciegienka (2011), the Meyer-based discrete wavelet analysis of belt-ground machine surface images was found to reveal changes in surface topography on a range of roughness and microroughness scales. An automated marble plate classification system (AMPCS) has been developed by Topalova (2012), who compared three different training sets as input with a neural network classifier. Their investigation demonstrated that the DWT-based training set outperformed the discrete cosine transform and extracted texture histograms. Zelelew et al. (2014) characterized the macrotexture properties of asphalt pavements and found that their wavelet approach, employing a normalized energy parameter, was better suited to this task than the traditional mean profile depth approach. Alhasan et al. (2016) analyzed 30 road profiles, comprising a wide range of surface roughness and two pavement types, using a quarter-car simulation and Morlet-based CWT decomposition. They reported that the technique was advantageous in that relatively short segments of pavement data could be analyzed, which supports real-time assessment. Another example of the diverse use of wavelet transforms for surface characterization is the analysis of the fractal properties of cracked concrete surface profiles by Dougan et al. (2000), who compared wavelet methods with both Fourier-based power spectral approaches and other traditional schemes for determining fractal parameters.

A technology for the unique identification of individual mechanical components from the characteristics of an applied grinding imprint has been developed by Dragon et al. (2011). They created grinding imprints on the components and extracted profiles from depth images taken using a confocal white-light microscope. These profiles were wavelet transformed using a Daubechies wavelet. The method is shown schematically in Figure 5.25. Features in the wavelet transform were then identified from local maxima of the absolute values and used for comparison with other specimens. They achieved a profile false-detection probability of 10^{-20}, which is as strong as human fingerprinting for identification. Figure 5.26 shows two transforms of a profile subject to corrosion. The larger-scale structures have changed but there are still identifiable small-scale structures that can be used to identify the specimen over time; thus, the authors concluded that the technique is also robust to corrosion-generated perturbations of the surface.

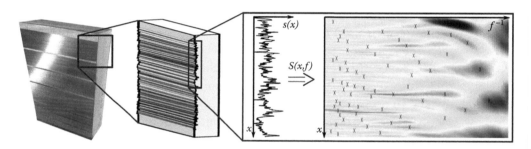

FIGURE 5.25 **Approach:** Profile $s(x)$ of a grinding imprint image is obtained and characteristic features are extracted in the space–frequency domain $S(x,f)$. (Original in colour.) (From Dragon, R. et al., *Pattern Recognition*, Springer, Berlin Heidelberg, 2011.)

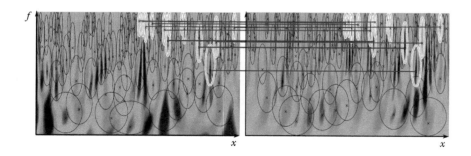

FIGURE 5.26 **Detected (black ellipses) and matching (white ellipses) features in the continuous wavelet space $S(x,f)$ between two profiles with partially similar local (right half) but different global structures (lower half).** (Original in colour.) (From Dragon, R. et al., *Pattern Recognition*, Springer, Berlin Heidelberg, 2011.)

Other work in this area includes the use of cross-correlation for carriage error identification by Mu et al. (2012) using Daubechies wavelets, the evaluation of the surface finish of carbon fibre–reinforced plastic by Palmer and Hall (2012) using wavelet texture analysis, the fast algorithm for automated surface quality control in real time by Sheybani et al. (2012), the study of 12 different mother wavelets to determine the optimal one for characterizing the surface texture of roundness profiles of machine parts by Adamczak and Makieła (2011), the surface characterization of turned metal pieces using both the Morlet and Mexican hat wavelet transforms by Brol and Grzesik (2009), the use of dual-tree complex wavelets in the investigation of 3-D nanoscalar surfaces by Ren et al. (2014), and the use of anisotropic Morlet and Mexican hat wavelets (Neupauer et al., 2005) in the analysis of the dominant parameters of synthetic rough surfaces generated from the profilometry of a roughed turbine blade by Wu and Ren (2014). It is also recommended that the reader consult the comprehensive review of the technological shifts that have taken place over recent years in surface metrology by Jiang and Whitehouse (2012), who set wavelet methods in the context of a range of other techniques for the filtering of surfaces in order to extract features of interest.

Finally, it is worth commenting on a number of research groups who have concentrated on the analysis of fibrous materials. Facco et al. (2010) found that a Daubechies D8 wavelet transform–based approach to the automatic characterization of nanofiber membrane assemblies (including pore size distribution, fibre diameter distribution and permeability) outperformed a grey-level co-occurrence matrix (GLCM) scheme. See also Tomba et al. (2010) for further comment on this work. A scheme for classifying cashmere and superfine merino wool fibres has been developed by Zhang et al. (2010) using a 2-D dual-tree complex wavelet transform to extract features that represent cuticular scale height, shape and interval. This study is further covered in the comprehensive summary of the application of wavelet methods to textile feature extraction by Wang et al. (2012b), which provides an excellent starting point from which to delve deeper into this area.

5.7 OTHER APPLICATIONS IN ENGINEERING

5.7.1 Compression

Early work by Staszewski (1998b) in this area considered the use of wavelet-based algorithms for the compression of vibration signals, comparing Fourier-based compression with wavelet-based compression (using Daubechies D4 and D20 wavelets) for a variety of signals. They found that for periodic signals, less Fourier coefficients than wavelet coefficients were required for the same mean-squared error of the reconstructed signal. However, the situation reversed for transient signals with less wavelet coefficients required. More recent work in this area includes that by Park et al. (2010), who developed an aggressive data reduction method in order to resolve data rate issues pertinent to damage detection in structural health monitoring. Their technique employed wavelet-based feature detection and obtained a high degree of data reduction while retaining the subset of features that contribute most to the goal of damage detection. In their investigation of vibration data from plane engines, Oltean et al. (2013) demonstrated that the wavelet basis provided better performance in compression in dealing with highly non-stationary signals than the discrete cosine and lapped orthogonal transforms. See also the comparison of wavelet and 'principled' independent component analysis data compression schemes for structural seismic responses undertaken by Yang and Nagarajaiah (2014).

In their study of fatigue life assessment, Putra et al. (2014) found marked reductions in signal length while retaining most of the pertinent statistical information to characterize amplitude changes in strain signals using a Morlet-based approach. Compressive sampling in structural health monitoring has been addressed by Bao et al. (2010) using the Haar wavelet basis. They employed their method for the sampling of a bridge accelerometer signal installed for structural health monitoring on a river highway bridge. They concluded that the compression ratios achieved were not high because the vibration data used in civil engineering structures are not naturally sparse in the wavelet basis. The reader is also referred to the study of the compressed sensing of motor-bearing vibration signals by Chen et al. (2014a), who compared three traditional wavelet schemes for extracting impulse components corresponding to defects – VisuShrink, SureShrink and translation-invariant hard thresholding – with a new scheme – sparse extraction of impulse by adaptive dictionary.

5.7.2 Control

Numerous authors have incorporated wavelet transforms within control algorithms. Merry et al. (2006), for example, utilized iterative learning control to significantly enhance the performance of systems that carry out repetitive tasks. The wavelet-based filtering they developed removed the non-repetitive parts of the tracking error. The wavelet components of the repetitive parts were then used to reconstruct a filter error signal which was input to the learning scheme. They found the method was able to remove the non-repetitive disturbances almost entirely. Jeong et al. (2006) developed wavelet-based statistical process control procedures for complex functional data used in manufacturing processes to characterize quality or reliability performance. Du et al. (2013) developed a hybrid approach for the recognition of control chart patterns by integrating the Haar-based wavelet transform and improved particle swarm optimization. They focused particularly on non-random control chart patterns including upward and downward shifts, increasing and decreasing trends, cyclical patterns and systematic patterns. They found that the approach's main merit was in improving the performance of recognizing concurrent chart patterns.

A continuous-time decentralized wavelet and neural identification and control scheme for robot manipulators has been proposed by Vazquez et al. (2015) based on the structure of a recurrent high-order neural networks model with a Morlet wavelet activation function. They validated their recurrent wavelet first-order neural network (RWFONN) technique in real time using experimental results. Song et al. (2013) developed a wavelet transform and neural network–based energy management control strategy for a hybrid wind power subsystem and found the method capable of compensating for power variation differences as well as maintaining a stable DC voltage. A Morlet-based cross-wavelet transform was employed by Sivalingam and Hovd (2011) to determine the source of oscillations in control loops. They found that they could differentiate between valve stiction, model–plant mismatch and oscillatory disturbances using the approach. Other work in this area includes the use of wavelet and neural network–based voice control systems for wheelchairs (Al-Rousan and Assaleh, 2011), a neural controller and robust compensator for a chaotic system which contained a functional-linked wavelet neural network by Hsu (2013), and the control of the synchronous motor drive in electric scooters using a hybrid wavelet neural network employing the first derivative of Gaussian wavelets by Lin (2014). See also the review of hybrid control systems and control strategies for smart structures by Fisco and Adeli (2011), where a number of wavelet-based control algorithms are discussed in detail.

5.7.3 Electrical Systems and Circuits

We have already covered the application of wavelet transforms to chaotic electrical circuits in Section 5.2.2 and also made brief mention of detecting mechanical and electrical faults in wind turbines in Section 5.4.2. A few further examples of their application to electrical systems and circuits are presented in this section. For example, Zhai (2015) has performed an analysis of load forecasting using wavelet-based estimation of the Hurst (fractal) exponent; amplitudes and frequencies of non-integer harmonics in electrical power signals have been identified to a very high accuracy by Tse and Lai (2007) using Morlet wavelet transforms; and the cross-wavelet transform has been employed by Dey et al. (2010) for feature extraction from noisy partial discharge pulses occurring in an acrylic resin insulator. Other applications to electrical circuits and systems include fault classification using magnetic signature recognition (Sartori and Sevegnani, 2010), real-time estimation of electrical variables using an improved recursive wavelet transform method (Rahmati et al., 2015), the implementation of wavelets in continuous-time analogue circuits (Karel et al., 2012), and a Haar wavelet–based time–frequency representation of circuit currents to find worst-case supply/ground voltage fluctuations (Ferzli et al., 2010).

5.7.4 Miscellaneous

Wu et al. (2013b) have utilized a wavelet transform trend-based key extraction scheme and a fuzzy vault-based key extraction protocol for encrypting body area network signals used in wireless health monitoring. A dynamic wavelet fingerprinting technique was employed by Bertoncini et al. (2010) in the development of an ultrasonic periodontal probe for use in the detection of gum disease. The approach produces a binary (black/white) image of the wavelet coefficients, which are then interrogated using image analysis tools, and where the extracted features are fed into a number of different classification algorithms. They found that this new probe closely matched the results of manual probing. A measure of driver fatigue from steering wheel movement signals (the raw steering angle time series) has been developed by Krajewski et al. (2010). They found that their wavelet-based approach provided additional information when compared with a Fourier method. Vehicle axle spacing and velocities have been successfully extracted from strain signals acquired from a bridge soffit by Chatterjee et al. (2006), and the transient nonlinear dynamic behaviour of disc brake squeal has been analyzed by Sinou (2010). The wavelet-based engineering testing of soil samples is discussed by Arroyo (2007) and the vibrations generated by open pit and deep cavern blasting are compared by Lu et al. (2011). (Note that geophysical vibrations are considered in more detail in Chapter 7, Section 7.4.1.)

CHAPTER 6

Medicine

6.1 INTRODUCTION

Nowhere has the wavelet transform made more of an impact in recent years than in the analysis of biosignals. These often comprise multiple information streams, are highly nonstationary and exhibit multiple forms of noise: perfect for wavelet analysis! For this reason, it is now a fundamental signal processing tool used regularly in the investigation of many different biosignals (Addison et al., 2009). This chapter reviews the numerous areas where the wavelet transform has made an impact. We begin with the electrocardiogram (ECG) where attempts have been made using wavelet methods to determine its characteristic points, detect abnormalities, characterize heart rate variability (HRV) and probe a variety of arrhythmias. We also discuss the issues that need to be considered for real-time analysis and hardware implementation, including compression techniques. The next section considers the wavelet as a diagnostic tool for neuroelectric waveforms including evoked potentials (EPs), event-related potentials (ERPs), seizures and the role of the electroencephalogram (EEG) in sleep studies. The pulse oximeter photoplethysmogram (PPG or 'pleth') is considered next, where methods to extract additional useful physiological information from this ubiquitous signal are described. Medical sounds are then examined, including cardiovascular sounds, lung sounds, swallowing sounds, snoring, speech and the acoustic response of the ear. There then follows a section on the analysis of blood flow and blood pressure signals. The subsequent section on medical images considers the application of wavelet transforms to ultrasonic, radiographic and optical images. The final section contains a selection of other medical applications of the wavelet transform, with particular emphasis on EMG, posture, gait and activity signals, as well as the analysis of multiple biosignals.

6.2 ELECTROCARDIOGRAM

Muscular contraction is associated with electrical changes known as 'depolarization'. The electrocardiogram (ECG) is a measure of this electrical activity associated with the heart. The ECG is measured at the body surface and results from electrical changes associated with the activation first of the two small heart chambers, the atria, and then of the two larger heart chambers, the ventricles. The contraction of the atria manifests itself as the 'P' wave in the ECG and contraction of the ventricles produces the feature known as the 'QRS' (Q wave, R wave, S wave) complex. The subsequent return of the ventricular mass to a rest state – repolarization – produces the 'T' wave. Repolarization of the atria is, however, hidden within the dominant QRS complex. Figure 6.1 shows a schematic of the ECG waveform for normal sinus rhythm. The wavelet transform of the ECG varies according to the timing and morphology of the beat. This is shown clearly in the example in Figure 6.2 which contains a segment of ECG from a pacemaker patient switching between intrinsic conduction and ventricular pacing (Stiles et al., 2004). Analysis of the local morphology of the ECG and its time-varying properties has led to the development of a variety of wavelet-based clinical diagnostic tools (Addison, 2005). In the rest of this section, the application of the wavelet transform to the interrogation of the ECG is reviewed in detail.

6.2.1 ECG Beat Detection and Timings

The detection of the P wave, QRS complex and T wave in an ECG is a problem still demanding the attention of those working in the field due to the time-varying morphology of the signal subject to physiological conditions and the presence of noise. Over the years, a number of wavelet-based techniques have been proposed to detect these features. Early work by Senhadji et al. (1995) compared the ability of wavelet transforms (based on three different wavelets: Daubechies, spline and Morlet) to recognize and describe isolated cardiac beats. Sahambi et al. (1997a,b) used a first-order derivative of the Gaussian function as a wavelet for the characterization of ECG waveforms. They applied modulus maxima–based wavelet analysis to detect and measure various parts of the signal, specifically the location of the onset and offset of the QRS complex and P and T waves. The group then undertook the same analysis for signals containing added baseline drift and high-frequency (HF) noise. They also computed intra-beat timing intervals to provide the relative positions of the components in the ECG which are important in delineating the electrical activity of the heart.

A 3-D plot of the modulus maxima of an ECG is shown in Figure 6.3 from Romero Legarreta et al. (2005a) which uses a continuous wavelet transform (CWT) with high-resolution scales (i.e. much finer resolution than the powers of 2 used in the dyadic grid of the discreet wavelet transform [DWT] or stationary wavelet transform [SWT]). The use of highly resolved scales allows the geometry of the maxima lines to be explored in detail, either through visual interpretation of the figure or using powerful processing methods to extract useful information from the maxima line set (such as the R-wave detection algorithm described by the authors). Burke and Nasor (2001) have provided plots of

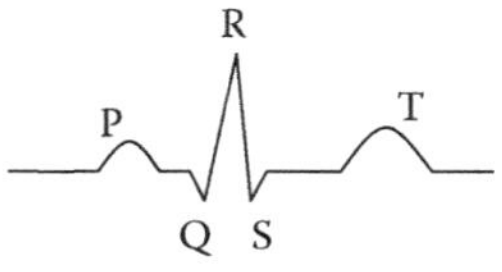

FIGURE 6.1 A schematic of the ECG exhibiting normal sinus rhythm. (Note that the shape of each feature can vary depending on the configuration of the ECG leads.)

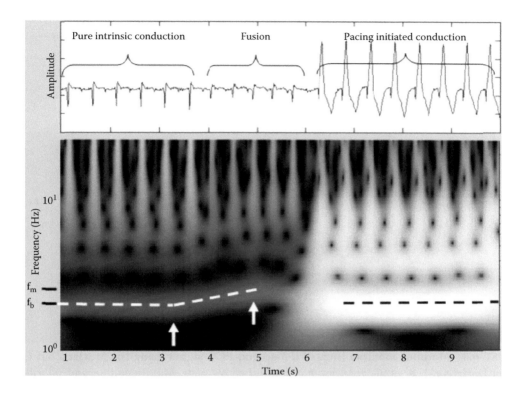

FIGURE 6.2 Scalogram from the patient whose pacemaker was set to DDD pacing mode and the PR interval began to exceed the programmed paced AV interval as the atrial pacing rate was increased. The ECG in the upper part of the figure shows intrinsically conducted beats, followed by 'fusion' beats where intrinsic ventricular depolarization is competing with ventricular paced beats and finally ventricular paced beats alone without competing intrinsically conducted beats. The ridge of the wavelet band associated with the beat frequency is shown dashed on the plot. The ridge increases in frequency over the fusion region between the two arrows. This band reverts back to the pre-fusion intrinsic conduction frequency when conduction is no longer competing. (From Stiles, M. K. et al., *Annals of Noninvasive Electrocardiology*, 9(4), 316–322, 2004.)

characteristic timings of ECG beat components versus cardiac cycle times derived using a Mexican hat modulus maxima technique. They went on to fit second-order equations to the relationships and compared these with traditional Bazett formulae which characterize the duration of intra-beat components in terms of cycle times. A number of more recent papers contain good examples of wavelet-based beat detection methods including R wave and QRS identifiers (Faezipour et al., 2009; Sumathi and Sanavullah, 2009; Sasikala and Wahidabanu, 2010; Talbi et al., 2011, 2012; Behbahani and Dabanloo, 2011; Madeiro et al., 2012). A supervised learning algorithm for R-wave spike detection employing Mexican hat wavelets has been described by de Lannoy et al. (2008). Dumont et al. (2010) improved the delineation of beats using quadratic spline wavelets in an evolutionary optimization process. Using multilead ECG components, Almeida et al. (2009) developed an ECG delineation strategy using an optimally derived lead for the specific purpose of delineating QRS and T boundaries. A Q–T interval measurement technique has been described by

270 ■ The Illustrated Wavelet Transform Handbook

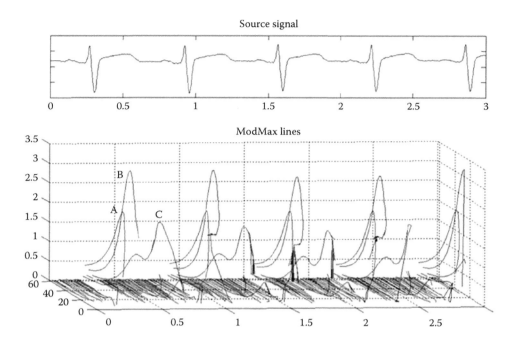

FIGURE 6.3 Modulus maxima lines of a 5-second ECG signal using the Mexican hat CWT. (From Legarreta, I. R. et al., *International Journal of Wavelets, Multiresolution and Information Processing*, 3(1), 19–42, 2005a. Copyright 2005, World Scientific.)

Ghasemi et al. (2010) employing a continuous (Haar) wavelet transform. More recent work by Noriega et al. (2012) has developed approaches for tackling the effect of respiration on delineating the end of the T wave using quadratic spline wavelets and Yochum et al. (2016) have reported on the delineation of QRS, T and P waves from a 12-lead ECG using a Daubechies wavelet–based approach.

The fetal ECG (fECG) is an important signal for providing useful information on the health of the fetal heart, including the monitoring of fetal distress during pregnancy and delivery. However, extraction of the fECG involves decoupling it from the more dominant maternal signal and other undesired disturbances. A number of wavelet methods have been proposed for this purpose. (An early paper by Mochimaru et al., 2002 contains some good explanatory scalograms of the maternal and fetal ECG.) Figure 6.4 contains a segment of ECG containing both the maternal and fetal heartbeat signals. The maternal and fetal R waves detected using the low-complexity algorithm of Rooijakkers et al. (2012) are denoted in the top plot of the figure. The technique employs an adaptive threshold setting based on the characteristics of previous beats and the expected maximum and minimum R–R intervals. The lower plot shows the amplitude of the CWT modulus at a single scale upon which the beat-dependent thresholds used to identify the maternal R waves are superimposed. The relatively small amplitude of these tiny fetal signal features compared with the maternal R and P waves is obvious in the plot. Castillo et al. (2013) tackled the noise suppression problem in the fECG using wavelet processing, while Zheng et al. (2010) employed a method using spline wavelets and comb filters. More recently, Wu et al. (2013a)

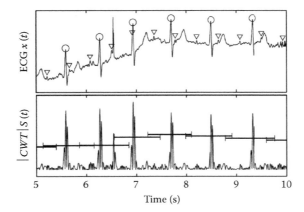

FIGURE 6.4 (Top) Example of an ECG segment, where O and ∇ indicate the maternal and fetal R peaks, respectively. (**Bottom**) The horizontal bars in the preprocessing output indicate the threshold level and length of each segment. (From Rooijakkers, M. J. et al., *Physiological Measurement*, 33, 1135–1150, 2012.)

have described an approach for adaptive filtering using both the abdominal ECG and a reference thoracic ECG. A 3-D phase space analysis to detect the fECG has been developed by Karvounis et al. (2009) which employs wavelet-based multivariate denoising. Jafari and Chambers (2005) have provided details of a scheme using blind source separation in the wavelet domain for fECG extraction and, in later work, Immanuel et al. (2012) and Zhou and Yang (2012) have described fECG detection techniques that combine wavelets and independent component analysis (ICA).

All methods for beat detection and timing analysis require some form of denoising of the raw ECG and many authors have suggested the utilization of wavelet-based approaches specifically for this task including Abi-Abdallah et al. (2006), Singh and Tiwari (2006), Sørensen et al. (2010), Kayhan and Ercelebi (2011) and Üstündağ et al. (2012).

6.2.2 Detection of Abnormalities

Premature ventricular contractions (PVCs) are commonly observed ectopic beats which exhibit wide complexes and occur sooner than expected, given the normal R–R interval. Their presence, especially if occurring often, may indicate increased morbidity and/or mortality in certain patient groups. A plot of a segment of ECG containing normal (N) and PVC (P) beats, together with its wavelet decomposition, is given in Figure 6.5. This is from the work of Shyu et al. (2004), who used a combination of fuzzy neural networks and quadratic spline wavelet decomposition to classify PVC beats with high accuracy. They also found that results could be improved by incorporating a method of normalization of the R wave to a standard amplitude based on the average from the training set. More recently, a PVC detection algorithm incorporating a wavelet-based R-wave detector was described by Chang et al. (2014), who also provided details of its field-programmable gate array (FPGA) hardware implementation. Uchaipichat et al. (2013) have detailed a technique which uses metrics derived from the wavelet power spectrum of individual beats to

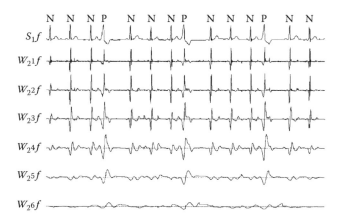

FIGURE 6.5 The WTs of ECG signals: After the WT there are distinct differences in the area under the QRS complex and QRS duration between VPC and normal heartbeats at every scale. (Note that here 'VPC' is used by the authors instead of the more common 'PVC'.) (From Shyu, L. Y. et al., *IEEE Transactions on Biomedical Engineering*, 51(7), 1269–1273. Copyright 2004, IEEE.)

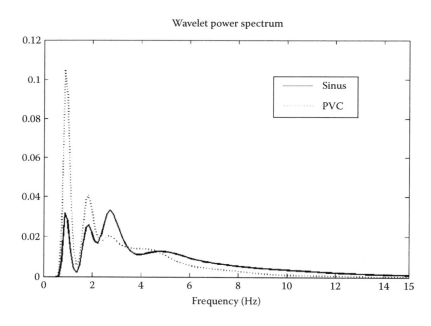

FIGURE 6.6 Wavelet power spectrum density. (From Uchaipichat, N. et al., *Proceedings of the World Congress on Engineering* (Vol. 2), WCE, London, 1316–1319, 2013. This is a reproduction of a figure in the WCE Congress. Copyright the International Association of Engineers.)

distinguish between sinus beats and PVC beats. Figure 6.6 contains power spectra, derived as described in Chapter 2 using a Morlet wavelet, corresponding to normal sinus and PVC beats. The beats were classified according to four metrics derived from individual wavelet power spectra of the waveforms: median frequency, variance, skewness and kurtosis. All four metrics resulted in statistically significant differences in their distributions and correspondingly high values of classification sensitivities and specificities. It was found that a combination of all four features provided the best performance.

Tuteur (1989) was one of the first proponents of the wavelet transform as an analysis tool for medical signal processing, using a complex Morlet wavelet to detect abnormalities in ECGs. In particular, Tuteur was interested in abnormalities known as 'ventricular late potentials' (VLPs). These represent low-amplitude electrical activity due to delayed electrical conduction by the ventricle muscles. VLPs occur in the ECG after the QRS complex and are often masked by noise. They have been used as a marker to identify patients at risk from certain types of life-threatening arrhythmias. Tuteur added a synthetic VLP-like segment of signal to ECG data from a child with a cardiac defect and demonstrated that wavelet decomposition could highlight this very subtle feature in the signal. Legarreta et al. (2005b) have attempted to detect VLPs prior to the onset of ventricular tachyarrhythmias, Gadaleta and Giorgio (2012) employed Coiflet-wavelet denoising of the ECG in a method to detect VLPs and Tsutsumi et al. (2011) have developed an enhanced predictor of the onset of lethal ventricular arrhythmias using a wavelet-based scheme to extract high-frequency information from within the QRS complex in addition to VLPs.

T wave alternans (TWAs) are beat-to-beat variations in the shape of the T wave and are considered markers of abnormal ventricular function, which may be associated with ventricular tachycardia (VT) and ventricular fibrillation (VF). A TWA detector algorithm utilizing the CWT has been proposed by Romero et al. (2008). In this method, a boxed region in the wavelet domain, corresponding to the T wave, was defined based on the location of the R wave (also detected in wavelet space). Four characteristic parameters were computed in the selected region: the magnitude, frequency and timing of the maximum coefficient within the box, and the overall energy of the T wave component within the box. The four parameters were separated into two groups: one containing the parameter values for the odd-numbered beats and the other corresponding to the even-numbered beats. A statistical analysis was then performed to determine whether significant alternate morphologies occurred within the signal indicative of TWAs. Figure 6.7 shows an example signal (top) and the corresponding scalogram (middle) from which the T wave regions were selected. (An example of a selected region is indicated by the box in the plot.) The extracted regions are shown enlarged and concatenated in the bottom plot. These regions were used to calculate the four characteristic parameters in the method described previously. The algorithm was tested on synthetic data achieving a high sensitivity and specificity. Wan et al. (2013) incorporated the algorithm of Romero et al. (2008) in their work, which also considered a time domain approach. A technique which employs transform ridges to detect TWAs has been reported by Bakhshi et al. (2012), and Janusek et al. (2011) employed wavelet denoising within an algorithm to detect T waves of myocardial ischemia patients and projected the resulting TWA magnitudes onto body surface maps.

In addition to the work centring on specific types of anomalous beats, some research groups have attempted to classify a range of anomalous beat types simultaneously. Ince et al. (2009) proposed a method of using both global and patient-specific training data to develop an ECG classifier which can adapt to significant intra-patient variation in ECG patterns. Morphological ECG information was extracted using a translation-invariant

274 ■ The Illustrated Wavelet Transform Handbook

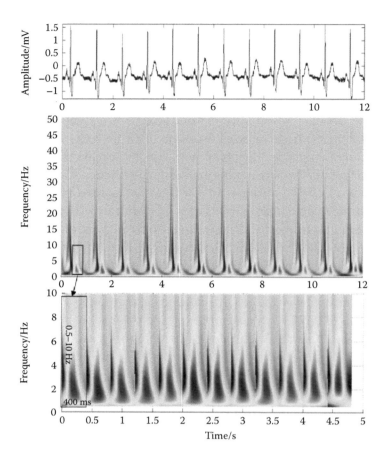

FIGURE 6.7 Time and frequency intervals extracted from the CWT of the whole ECG to delimit T waves. **Top plot:** ECG. **Middle plot:** Corresponding scalogram. **Lower plot:** Segments of wavelet transform localized to the T wave information. (From Romero, I. et al., *IEEE Transactions on Biomedical Engineering*, 55(11), 2658–2665. Copyright 2008, IEEE.)

dyadic wavelet transform employing quadratic spline wavelets. The morphological features extracted in this way were then preprocessed using principal component analysis to reduce the dimensionality of the feature vectors before use within the classification technique. The classifier was trained to recognize ventricular and supra-ventricular ectopic beats from normal beats. In a similar vein, Faezipour et al. (2010) have described an approach for adapting an ECG beat classifier using an additional local classifier to profile the normal cardiac behaviour of individual patients. Fiducial ECG points, derived from a symlet-based undecimated wavelet analysis of the signal, were used as input to the classifier. Daamouche et al. (2012) proposed a wavelet optimization approach to ECG classification where they applied the method to classify a range of beats types: 'normal sinus', 'atrial premature', 'ventricular premature', 'right bundle branch block', 'left bundle branch block' and 'paced'. They found that their optimized wavelet performed better than two standard wavelets but the processing time to reach convergence was high. Saxena et al. (2002) have described the use of two different wavelets to detect QRS (quadratic spline) and P and T wave (Daubechies D6) features from an ECG and suggested its use for arrhythmia detection or heart rate variability determination. Banerjee and Mitra (2012) briefly detailed a method of using Morlet-based, cross-wavelet analysis to compare a 'standard' beat waveform template with test beats, including normal and abnormal beats. They found useful distinguishing characteristics in both the QRS and T wave regions. Thomas et al. (2015) utilized dual-tree complex wavelet–based features to classify ECG beat types, gaining high sensitivities in the results. Ant colony optimization was applied by Korurek and Nizam (2010) to cluster arrhythmic beat types where the features used were derived from both the time and wavelet domains. In addition, principle component analysis (PCA) was employed to reduce the dimension of the wavelet coefficient data prior to input into a classifier. Zhu et al. (2013) have detailed an algorithm for the classification of arrhythmic beat types using a modified maximum margin clustering technique which uses wavelet-based ECG features sets and Javadi et al. (2011) have illustrated the use of an ensemble of neural network modules to classify ECG beat types.

6.2.3 Heart Rate Variability

Rather than consider the detailed morphology of the ECG waveform, many researchers have focused on the temporal variability of the heartbeat. To do this, they monitor the timing interval between beats, taken between each R point on the QRS complex, and plot this R–R interval against time to give the heart rate (HR). The minute fluctuations present in the R–R intervals have been used for assessing the influence of the autonomic nervous system on the heart rate. In addition, long-range correlations and power law scaling have been found through the analysis of heartbeat dynamics. Much of the current work concerning heart rate variability (HRV) focuses on its use as a marker for the prediction and diagnosis of heart disease and the assessment of heart function as a method for characterizing the cardiorespiratory dynamics associated with the autonomic nervous system. Early work by Wiklund et al. (1997) applied adaptive wavelet transforms (wavelet packets and cosine packets) to the study of the regulation of HRV by the autonomic nervous system. Thurner et al. (1998) employed both Daubechies D10 and Haar wavelets in the analysis of human heartbeat intervals, reporting 100% accuracy in differentiating normal patients from those with heart failure in a standard 27 patient data set. Figure 6.8 contains a wavelet transform scalogram plot of a heart rate (HR) signal from a healthy subject taken from the review of HRV measurements and the prediction of ventricular arrhythmias by Reed et al. (2005). The traditional high-frequency (HF), low-frequency (LF) and very-low-frequency (VLF) bands are plotted across the transform surface. In the review, the authors state that the CWT has become the most favoured tool as it does not contain the cross-terms associated with other time–frequency methods and its frequency-dependent windowing allows for arbitrarily high resolution of the high-frequency signal components (unlike the short-time Fourier transform [STFT]). In fact, a number of researchers have found a wavelet approach to be an optimal choice for deriving HRV information, including Belova et al. (2007) and Li et al. (2011d). A DWT-based hard thresholding technique was developed by Keenan (2008) for the detection and correction of ectopic beats in order to more accurately determine HRV. Keenan compared

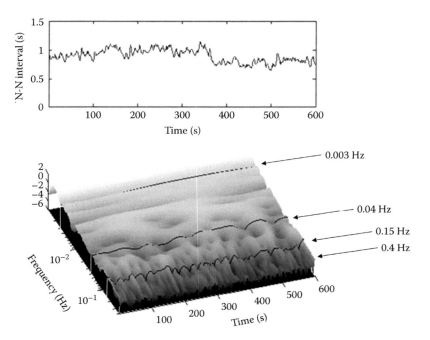

FIGURE 6.8 The HR signal from a healthy subject **(top)** together with its associated wavelet transform **(below)**. The boundaries of the HF, LF and VLF regions are plotted across the transform surface. (From Reed, M. J. et al., *Quarterly Journal of Medicine*, 98, 87–95, 2005.)

the method with linear interpolation based on simulated data and data from heart failure patients. More recently, Neto et al. (2016) developed a Morlet-based technique to identify sympathetic and parasympathetic alterations to the autonomic nervous system caused by the administration of drugs.

Numerous other research groups have chosen the wavelet transform for their analysis of HRV due to its ability to cope with relatively non-stationary signals. An investigation of the fetal autonomic nervous system using CWT methods in the analysis of the fetal ECG HRV has been reported by David et al. (2006). The use of wavelet transform modulus maxima techniques to probe the multifractal (see also Chapter 7, Section 7.2.3) nature of the HRV signal and associating it with laser Doppler flow signals acquired at the peripheries has been described by Humeau et al. (2010). Kheder et al. (2009) performed a brief study of HRV signals from subjects undertaking two different meditation exercises. They compared the results with a non-meditation control, finding a distinct difference in the LF and HF components between the control and the two meditation signals. Other studies include those by Pichot et al. (2002), who developed a wavelet-based method of analysis of HRV to quantify fatigue in the workplace; Magini et al. (2012), who employed a wavelet decomposition of HRV in a study of anxiety disorders; Bilgin et al. (2009), who developed a multilayer perceptron neural network to determine the sympathovagal balance in ventricular tachyarrhythmia patients; and Burri et al. (2006), who investigated HRV prior to ventricular arrhythmias using DWT methods.

A special case of HRV, or more accurately a subcomponent of HRV, is the respiratory modulation of the heartbeat cycle. The changes in thoracic pressure during the respiratory cycle induce changes in blood flows and heart rate. Thus, over a breath, the heart rate varies: first increasing during inhalation then decreasing during exhalation. This natural variation, known as 'respiratory sinus arrhythmia' (RSA), is sometimes so prominent that it can be seen by eye in the ECG. A number of researchers have used the wavelet transform to extract respiratory information from the ECG. Ponomarenko et al. (2005) investigated the synchronization of relative phases of HRV and blood pressure signals using Morlet wavelet transforms. Kenwright et al. (2008) also investigated the synchronization between ECG and respiratory signals (acquired using a piezo band), concluding that exercise acts as a perturbation to the system which reduces the cardiorespiratory interaction between components. Keissar et al. (2009a) performed wavelet transform coherence computations on simultaneously collected ECG and respiratory signals. (Refer to Chapter 2, Section 2.21.5). Figure 6.9 contains a wavelet coherence plot that they constructed from the two signals using a Morlet wavelet with $\omega_0 = 20$, together with a frequency-averaged coherence plot (averaged over the high-frequency band between 0.15 and 0.40 Hz). They found that most coherence was exhibited within this band. However, this also contained some areas of weak or no coherence (below the threshold level shown in Figure 6.9b) due to sudden changes in breathing or heart rate (at 600 s and 900 s in the figure). Changes in posture also led to loss of coherence in the HF band but elevated coherence in a lower-frequency band (0.04–0.07 Hz) in seven out of the eight patients studied. They proposed the application of the method for the analysis of dynamic linear coupling between physiological signals, and as a marker for the autonomic nervous system (ANS) function.

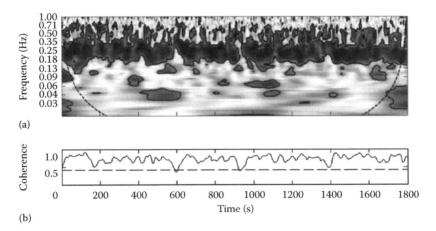

FIGURE 6.9 C1 subject: Example of typical HR–Resp wavelet transform coherence: (a) The coherence map. The dark contoured areas are above the significance level, hence regarded as areas of coherence. The left and right curves define the COI (cone of influence – the region outside of which the edge effect is significant and the values may be considered dubious). (b) The frequency average time-dependent coherence of the HF band. The dashed lines are the significance level. $\omega_0 = 20$. (From Keissar, K. et al., *Philosophical Transactions of the Royal Society of London A: Mathematical, Physical and Engineering Sciences*, 367, 1393–1406, 2009. With permission from the Royal Society.)

Numerous groups have employed wavelet analysis of the ECG in the characterization of sleep disorders – Roche et al. (2003) used discrete wavelets to predict sleep apnea and, later, the same group (Sforza et al. 2007) utilized time, spectral and wavelet analysis to generate markers of sleep fragmentation. Khandoker et al. (2009) extracted wavelet-based features from both HRV signals, determined from the R–R interval, and ECG-derived respiration (EDR) signals, determined from the QRS amplitude. These were fed into a number of candidate classifiers and their ability to recognize patients with obstructive sleep apnea was tested. They found that a support vector machine (SVM) outperformed other classifiers in identifying patients with obstructive sleep apnea syndrome (OSAS). Mendez et al. (2010) compared wavelet analysis with empirical mode decomposition for the automated screening of obstructive sleep apnea. They found that the two methods gave comparable results for accuracy, with the wavelet approach performing marginally better (89% vs. 85% accuracy).

6.2.4 Cardiac Arrhythmias

A number of wavelet-based techniques have been proposed for the detection, classification and analysis of the arrhythmic ECG. Note that Section 6.2.2 has already touched on this subject, dealing with the identification of individual abnormal/arrhythmic beats whereas here a more holistic approach is considered. Using a symlet wavelet–based neural network classification method, Wiggins et al. (2003) separated the sinus rhythms of healthy patients from the sinus rhythms of patients who did exhibit an arrhythmia at some other time. Arumugam et al. (2009) classified three life-threatening ventricular arrhythmias (ventricular fibrillation [VF], ventricular tachycardia [VT] and ventricular flutter [VFL]) using a wavelet neural network. Legarreta et al. (2004a,b) investigated HRV prior to and around ventricular tachycardia events and, as cited in Section 6.2.2, Legarreta et al. (2005b) attempted to detect VLPs prior to the onset of ventricular tachyarrhythmias. In other work, Ozbay (2009) has proposed a method to recognize 10 different arrhythmia types. The technique extracts signal features using a complex wavelet transform, which are input into a complex-valued artificial neural network (ANN) in order to preserve valuable phase information, which is missing from traditional discrete wavelet transform-based approaches.

Figure 6.10 contains three beats of a normal sinus rhythm of a pig heart together with its (Morlet) wavelet energy scalogram, shown as both a contour plot (Figure 6.10b) and a three-dimensional surface relief (Figure 6.10c). Note that the logarithm of the energy is plotted in the figures as it allows for features with large differences in their energy to be made visible in the same plot. The QRS complex of the waveform manifests itself as the conical structures in Figure 6.10b. These converge to the high-frequency components of the RS spike. The P and T waves are also labelled in the plot. In addition, a continuous band is evident in the plot at a frequency of around 1.7 Hz which corresponds to the beat frequency of the sinus rhythm. The 3-D morphology of the signal in wavelet space is shown in Figure 6.10c. Figure 6.11 shows a portion of a porcine ECG exhibiting ventricular fibrillation. The 3-D morphology of the energy scalogram reveals the presence of organized structure contained within the VF signal as seen by the three distinct undulations within

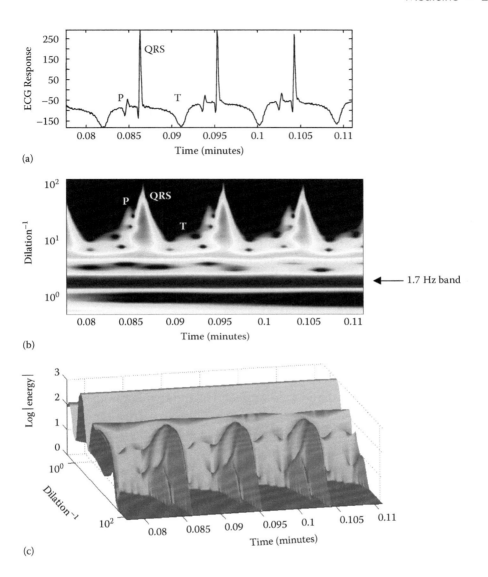

FIGURE 6.10 Wavelet transform of ECG exhibiting sinus rhythm: (a) Single channel porcine ECG showing sinus rhythm. **(b)** The corresponding energy scalogram of the temporal location against the bandpass frequency of the wavelet. **(c)** The 3-D landscape plot of (b). (Original in colour.) (From Addison, P. et al., *IEEE Engineering in Medicine and Biology Magazine*, 19, 104–109. Copyright 2000, IEEE.)

the 10 Hz band of the scalogram. This transient modulation of the 10 Hz band is located by the arrow in the plot. Figure 6.12 shows another portion of porcine VF together with the 2-D contour plot of the energy scalogram. Distinct high-frequency spiking of a periodic nature can be clearly seen within the scalogram. This regular structure is not at all evident from the ECG trace, nor is it evident using short-time Fourier transform analysis due to its fixed window width (see Figure 6.15). The spiking becomes more prominent with increasing downtime (duration of VF). (It is worth noting that this periodic spiking has also been observed in segments of human VF – see Figure 6.13.) A global view of the porcine VF

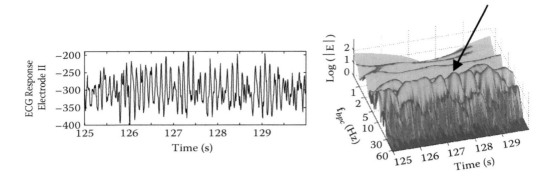

FIGURE 6.11 **Wavelet transform of ECG exhibiting ventricular fibrillation:** (a) Single channel ECG showing VF. (b) The corresponding 3-D energy scalogram of the temporal location against the bandpass frequency of the wavelet. The arrow points to a region of coherent modulation of the 10 Hz band. (From Addison, P. et al., *IEEE Engineering in Medicine and Biology Magazine*, 19, 104–109. Copyright 2000, IEEE.)

signal in wavelet space is given in Figure 6.14 which contains an energy scalogram for a 5 minute period of VF followed by a 2.5 minute period of cardiopulmonary resuscitation (CPR). The onset of CPR is distinguished by the large amplitude horizontal band appearing at a low frequency at 5 minutes. Distinct banding can be seen in the scalogram over the first 5 minutes: a high-frequency band at around 10 Hz and two lower energy bands at lesser frequencies labelled A, B and C. After the onset of CPR, an increase in the frequency of all three bands can be observed in the scalogram. The results depicted in Figures 6.11 through 6.14 demonstrate that VF, previously thought to represent disorganized and unstructured electrical activity of the heart, in fact contains a rich underlying structure hidden to traditional Fourier techniques (Addison et al., 2000; Watson et al., 2000). Figure 6.15 illustrates the shortcomings of traditional STFT analysis in detecting signal features of short duration. The figure contains a scalogram and a spectrogram corresponding to the rhythmic ECG shown in Figure 6.15a. The spectrogram is generated from an STFT with a 3.4 second Hanning window – typical for this type of analysis. The smearing and, hence, loss of local information across the spectrogram over these timescales is evident in the plot.

In a pilot study concerning the analysis of pressure traces and ECG corresponding to pig hearts exhibiting VF (Addison et al., 2002b), some evidence was found to suggest that wavelet phase information may be used to interrogate the ECG for underlying low-level mechanical activity in the atria. Figure 6.16a,b show the pressure in the aorta and an ECG corresponding to an episode of VF in a pig heart. The ECG has a typical random or unstructured appearance. The aorta pressure trace, however, reveals regular low-amplitude spikes. On opening the chest of this animal and observing the heart directly, it became apparent that the ventricles were fibrillating, but the atria were contracting independently in a coordinated manner. The irregular activity of the much larger ventricular muscle mass completely obscured this atrial activity in the standard ECG recording shown in Figure 6.16b. The wavelet energy scalogram for this signal is plotted below the ECG. (A Morlet wavelet was used in the study.) The high-amplitude band at around 8–10 Hz

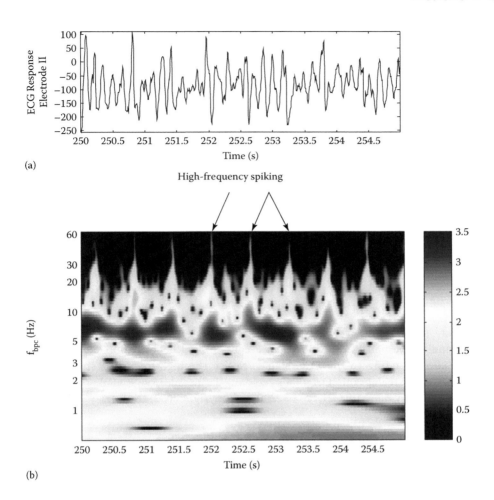

FIGURE 6.12 (See colour insert.) **Wavelet transform of ECG exhibiting ventricular fibrillation:** (a) Single channel ECG showing a region of VF. (b) The corresponding scalogram of the temporal location against the bandpass frequency of the wavelet. Notice the high-frequency periodic spiking observable in the scalogram. (Original in colour.) (From Addison, P. et al., *IEEE Engineering in Medicine and Biology Magazine*, 19, 104–109. Copyright 2000, IEEE.)

is much more compact in extent in frequency than that found for other traces where no atrial pulsing was apparent. Furthermore, there is some evidence of 'pulsing' in this band between 1 and 2 Hz in the scalogram. This is confirmed in Figure 6.16d, where the location of a zero wavelet phase is plotted over a short range of the bandpass frequencies, between 1.1 and 1.5 Hz. The phase plot exhibits a strikingly regular pattern with the zero phase lines aligning well with the atrial pulsing of the pressure trace in Figure 6.16a.

The information within the VF waveform can be extracted using wavelet techniques and used to predict whether a patient will respond to defibrillation shock. Watson et al. (2004, 2005) provided such a cardioversion outcome prediction (COP) method, capturing the temporal behaviour of signal components using a wavelet entropy-like intermittency measure derived from the ECG scalogram modulus maxima. A low-oscillation Morlet wavelet

282 ■ The Illustrated Wavelet Transform Handbook

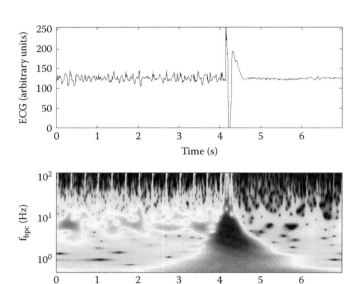

FIGURE 6.13 (See colour insert.) **Attempted defibrillation of human ventricular fibrillation. Top:** 7 seconds of human ECG exhibiting VF containing a defibrillation shock event. **Bottom:** Scalogram corresponding to the ECG signal. Notice the high-frequency spiking prior to the shock evident in the scalogram. (Original in colour.) (From Addison, P. et al., *IEEE Engineering in Medicine and Biology Magazine*, 21, 58–65, 2002b. Copyright 2002, IEEE.)

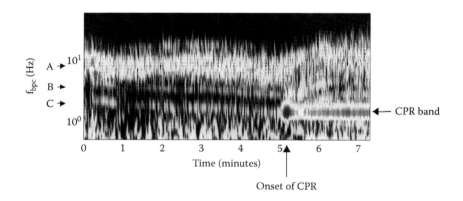

FIGURE 6.14 (See colour insert.) **The energy scalogram for the first 7 minutes of porcine ventricular fibrillation.** CPR is initiated at 5 minutes as indicated. (Original in colour.) (From Watson, J. N. et al., *Resuscitation*, 43, 121–127, 2000.)

(Addison et al., 2002a) was employed and the resulting metrics were input into a Bayesian classifier. The study was conducted on the ECGs of 878 out-of-hospital cardiac arrest (pre-shock) traces and it was shown that the wavelet-based method compared favourably with other non-wavelet techniques. Watson et al. (2006a) set out the specific steps required for the production of a robust wavelet-based technology for use within an external defibrillation device, dealing with data issues, cross-validation issues and the evaluation of the

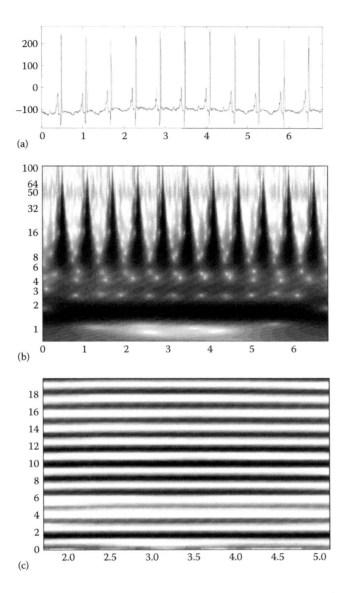

FIGURE 6.15 Wavelet scalogram versus STFT spectogram for normal sinus rhythm: (a) Original rhythmic ECG signal. **(b)** Morlet-based scalogram corresponding to (a). **(c)** Spectogram corresponding to (a) generated using a short-time Fourier transform with a 3.4 second Hanning window.

learning systems (the use of box plots, receiver operator characteristic [ROC] curves and error bars based on cross-validated system performance). The paper has wider implications for the development of rigorous techniques for this kind of wavelet-based analysis. The COP metric was subsequently employed by Box et al. (2008) to compare the relative efficacy of manual and automated (machine-based) chest compressions. Shandilya et al. (2012) have detailed an approach utilizing dual-tree complex wavelet transform features for predicting defibrillation success (including the mean, standard deviation, energy and entropy of the wavelet coefficients), where 90 pre-shock signals were used in the study.

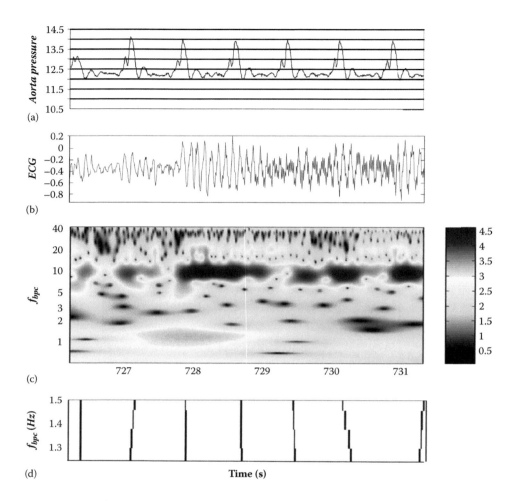

FIGURE 6.16 Simultaneous ECG and pressure recordings: (a) The aorta pressure trace, with (b) ECG, (c) corresponding wavelet energy plot obtained using the Morlet wavelet and (d) the zero phase lines of the Morlet wavelet transform. The plots correspond to the time period 726.23–731.31 seconds after the initiation of VF. (Original in colour.) (From Addison, P. et al., *IEEE Engineering in Medicine and Biology Magazine*, 21, 58–65, 2002b. Copyright 2002, IEEE.)

They demonstrated an improvement in the performance of the classifier when additional features from an end-tidal carbon dioxide (ETCO$_2$) signal were combined with those of the ECG, where the area under the curve (AUC) for the receiver operator characteristic (ROC) curve was boosted from 85.0% to 93.3%. The results were markedly better than another popular approach: the amplitude spectrum analysis (AMSA) technique. A further demonstration of the method has been provided by Shandilya et al. (2013) using a larger data set. A wavelet technique for the prediction of shock outcomes for implanted defibrillators has been reported by Bajaj et al. (2005), a near-real-time wavelet-based method for shock outcome prediction has been developed by Umapathy et al. (2009) and Rasooli et al. (2015) have described a wavelet-based scheme which includes blind source separation for the analysis of pre-shock VF waveforms. The superior nature of wavelet-based and nonlinear methods for determining shock outcomes over more traditional time domain and frequency domain-only techniques is discussed fully in the review by He et al. (2013).

Figure 6.17a shows an ECG trace from a patient who has experienced an out-of-hospital cardiac arrest. An episode of large amplitude chest compression artefact can clearly be seen (from around 15 to 40 seconds) corresponding to a period of cardiopulmonary resuscitation (CPR). This dominant artefact obscures the underlying heart rhythm in the ECG. A key aim of current research is to be able to shock the patient back into a normal (sinus) rhythm without stopping CPR in order to determine the underlying heart rhythm. Prior to the CPR signal in Figure 6.17a, the ECG exhibits asystole (no cardiac activity and not a shockable rhythm) and subsequent to the CPR region a VF waveform can be seen (which is shockable). The corresponding wavelet transform is given below the signal in Figure 6.17b. CPR artifact dominates the signal and makes it very difficult for traditional techniques to determine the underlying rhythm: in particular, whether the underlying rhythm is a shockable rhythm or not. In a method described by Watson et al. (2006b), the modulus maxima lines of the scalogram were computed and the magnitudes of these extracted across a high characteristic frequency band in the scalogram. These values were then reordered from lowest to highest to give the plot in Figure 6.17c. The largest values in the plot correspond to the large amplitude CPR artefact; however, the overall shape of the curve is dependent upon the rhythm underlying the CPR components; for example, VF results in a smoothly monotonic drop in coefficient values, whereas asystole results in a steep drop. The morphology of the reordered coefficient plots may be characterized by a metric (e.g. the amplitude or slope of the curve at a specific point) over time. Figure 6.17d contains a plot of the characteristic measure extracted from the reordered coefficient curve plotted against time for the CPR signal shown in Figure 6.17a. A distinctly increasing change in the value of the measure can be seen throughout the CPR and it was proposed that this may be compared with a threshold value in order to determine whether a shockable or non-shockable rhythm is present 'beneath' the chest compression artifact.

Atrial fibrillation (AF) is an arrhythmia associated with the asynchronous contraction of the atrial muscle fibres. It is the most prevalent cardiac arrhythmia in the Western world and is associated with significant morbidity. Figure 6.18 shows the wavelet transform decomposition of a 10 second segment of ECG from a patient with atrial fibrillation. Below the trace is a scalogram plot obtained using a Morlet-based wavelet transform. This yields high temporal resolution in the wavelet domain, but generates a very large data set. The modulus maxima of the scalogram provides a compact form of this information and is plotted below the scalogram in the figure where the QRS modulus maxima lines can be seen to dominate the plot. Watson et al. (2007) extended a previous method (Watson et al., 2001) by partitioning the maxima lines into two sets: (1) those related to QRS complexes alone and (2) the other components of the ECG. They generated two energy spectra from these modulus maxima lines from which they derived spectral metrics. (These partitioned modulus maxima spectra are shown in Figure 6.19.) The metrics from the non-QRS set of maxima were used in conjunction with temporal (entropy) markers derived from maxima lines in the 20–30 Hz region of the scalogram to partition successful from unsuccessful dc cardioversions in a population of AF patients. Figure 6.20 contains the plot of the entropy

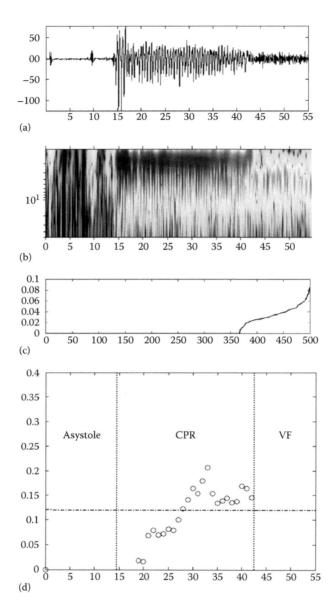

FIGURE 6.17 (a) The ECG trace showing asystole followed by a central region of CPR resulting in a VF waveform at the end of the trace. (Horizontal axis = time in seconds. Vertical axis = ECG amplitude in arbitrary units.) (b) The wavelet transform of the signal in (a). (Horizontal axis = time in seconds. Vertical axis = Characteristic wavelet frequency in hertz.) (c) Coefficient reordering curve obtained from the modulus maxima coefficients extracted at a selected characteristic frequency level. (Vertical axis – coefficient amplitude. Horizontal axis – reordered coefficient number.) (d) Wavelet characteristic measure values during the CPR episode. The shockable/non-shockable rhythm threshold is shown by the dashed line. Shockable rhythms are identified as those where the characteristic measure falls above the line. (Horizontal axis = time in seconds. Vertical axis = Marker value in arbitrary units.) (From Watson, J. N. et al., *MEDISIP 3rd International Conference: Advances in Medical Signals and Information Processing*, 128–131, Glasgow, Scotland, 2006b.)

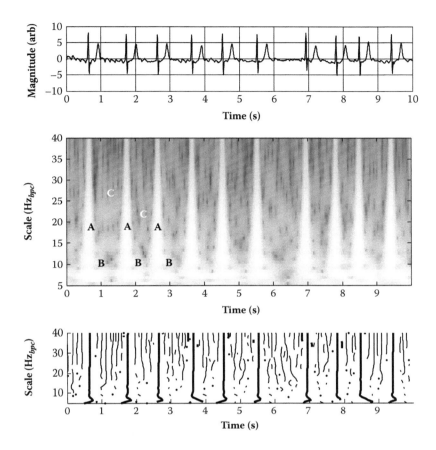

FIGURE 6.18 **Shows the ECG (Surface ECG: II) of a patient from the sample set.** Beneath is the wavelet scalogram with sample QRS, T and atrial activity labelled A, B and C, respectively. Bottom is a modulus maxima plot of the same scalogram with QRS features bold for clarity. (From Watson, J. N. et al., *Computers in Biology and Medicine*, 37, 517–523, 2007.)

metric against the peak amplitude. The decision boundary plotted in the figure corresponds to the 76% specificity and 100% sensitivity of predicting successful defibrillation in these patients. The technique provides the effective partitioning of more subtle 'background' ECG features from the dominant QRS complexes, both temporally and spectrally, in order to extract clinically useful information from them. The reader may also want to refer to the previous work by Watson et al. (2001) in this area, where the inverse wavelet transform of the partitioned sets of maxima lines were used to produce time–frequency filtered versions of the AF signal.

Alcaraz and Rieta (2012a) demonstrated that a wavelet entropy measure was the highest signal predictor of the termination of spontaneous paroxysmal AF and also in the determination of the termination of AF using electrocardioversion treatment. The authors considered a number of wavelet functions in their work: Haar, Daubechies, Coiflet, biorthogonal, reverse biorthogonal and symlets. Alcaraz and Reita (2012b) employed a central tendency measure of first difference scatter plots applied to wavelet coefficient vectors,

288 ■ The Illustrated Wavelet Transform Handbook

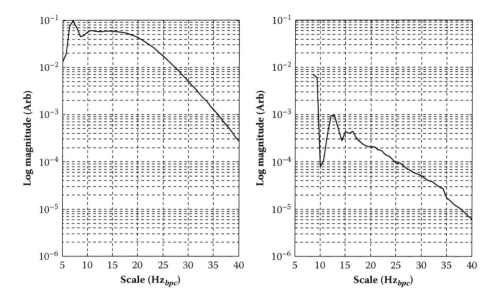

FIGURE 6.19 **Wavelet spectra:** **(a)** Wavelet spectrum for the QRS features. **(b)** Wavelet spectrum for the other components of the ECG signal. (From Watson, J. N. et al., *Computers in Biology and Medicine*, 37, 517–523, 2007.)

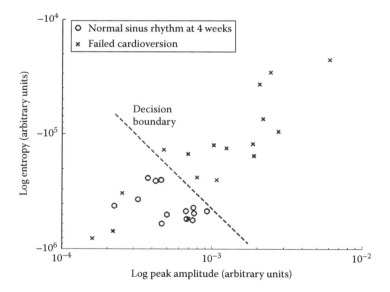

FIGURE 6.20 **Scatter plot of variables 'entropy' and 'peak amplitude':** **(a)** A linear decision boundary is positioned to obtain 100% sensitivity for the identification of those patients whose cardioversion resulted in at least 1 month of normal sinus rhythm. (From Watson, J. N. et al., *Computers in Biology and Medicine*, 37, 517–523, 2007.)

finding it a promising tool for the characterization of AF signals. Again, these authors considered a number of wavelet functions in their analysis. The same group (Ródenas et al., 2015) has described a wavelet entropy method for the automatic detection of AF episodes from a single lead electrocardiogram which can detect episodes as brief as five beats in length. Romero et al. (2011) have used both wavelet and Fourier methods to investigate the dominant frequency distribution in AF after pulmonary vein isolation and Zhao et al. (2013b) have demonstrated the use of wavelets in tracking wavefronts in non-contact catheter signals in the left atrium. Sovilj et al. (2010) have described an approach for identifying patients undergoing coronary artery bypass grafting (CABG) with the highest risk of developing AF. They employed an ECG detector based on a dyadic wavelet decomposition and also used wavelet-based features, among others, as input to their prediction model.

6.2.5 ECG Data Compression

ECGs are collected both over long periods of time and at high resolution. This creates substantial volumes of data. Data compression seeks to reduce the number of bits of information required to transmit or store digitized ECGs without significant loss of signal information. An early paper by Crowe et al. (1992) suggested the wavelet transform for compressing both ECG and heart rate variability data sets. Using discrete orthonormal wavelet transforms and Daubechies D10 wavelets, Chen et al. (1993) compressed ECG data sets resulting in compression ratios up to 22.9:1 while retaining clinically acceptable signal quality. More recently, compressed sensing was compared with a 'state-of-the-art' digital wavelet transform-based compression solution for wireless body sensor networks (WBSNs) by Mamaghanian et al. (2011), who found that the wavelets gave superior performance. This work has been extended by Zhang et al. (2013a) for use in the wireless telemonitoring of the fetal ECG. Other work in this area includes that conducted by Gurkan (2012), who used variable length classified vector sets, Chouakri et al. (2013), who employed wavelet transforms within an algorithm comprising higher-order statistics and Huffman coding and Polania et al. (2011), who employed a matching pursuit compressive sensing method based on Daubechies wavelets.

6.2.6 Hardware Implementation

The implementation of wavelet-based ECG algorithms on hardware requires smart techniques to deliver low-power, low-cost, lightweight and intelligent solutions, working in real time and often involving wireless links. Stojanović et al. (2011) have presented the design for a real-time hardware-based QRS detector for incorporation on an FPGA chip. The system employs the integer Haar wavelet transform at its core. Figure 6.21 contains a schematic of the on-chip architecture for QRS detection from Stojanović et al. (2013). Details of microprocessor design with a Mexican hat CWT at its core are provided by Cheng et al. (2012) and the porting of wavelet-based delineation of the ECG waveform to a wearable embedded sensor platform with limited storage and processing has been described by Boichat et al. (2009). Rizzi et al. (2008) generated a fast parallelized algorithm for the real-time signal processing of the ECG and Lestussi et al. (2011) compared three methods for online processing of the ECG with an offline technique. Using quadratic spline wavelets and incorporating

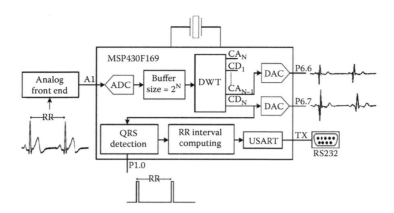

FIGURE 6.21 MC architecture for QRS detection. (From Stojanović, R. et al., *Computer Science and Information Systems*, 10(1), 503–523, 2013.)

a modulus maxima recognition stage, Ieong et al. (2012) produced a 0.83 μW detection processor with high sensitivity and specificities for beat detection using the Massachusetts Institute of Technology and Boston's Beth Israel Hospital (MIT BIH) arrhythmia database. Custom instructions to expedite wavelet processing on hardware have been detailed by Milosevic et al. (2011). Di Marco and Chiari (2011) have described a wavelet-based ECG delineation method based on 32 bit integer linear algebra. They reported results comparable with other approaches for identifying a number of ECG fiducial points: P onset, P peak, P offset, QRS onset, QRS offset, T peak and T Offset. Smartphone applications have been considered by Weng et al. (2013), Borodin et al. (2012) and Jannah et al. (2013). A further discussion of emerging analog biomedical signal processing applications of the wavelet transform can be found in the detailed contribution by Akansu et al. (2010).

6.3 NEUROELECTRIC WAVEFORMS

The electroencephalogram (EEG) signal is obtained from a set of electrodes which are usually placed on the scalp. In some cases, however, specially designed subdural electrodes are surgically implanted below the skull to monitor electrical activity obscured by the bone. The electrical potentials picked up by the electrodes – the EEG signal – reflects brain electrical activity owing to both intrinsic dynamics and responses to external stimuli. Experimental external stimuli can take the form of evoked potentials (EPs), which are well defined sensory inputs such as sounds, flashes, smells or touches, or or event-related potentials (ERPs) where experiments are set up to probe higher cognitive functions such as those associated with memory function or mechanical response. A good place to begin this section is with the comprehensive introduction to the wavelet analysis of neuroelectric waveforms given by Samar et al. (1999). This provides a concept-driven (with a minimal involvement of mathematics) account of the use of wavelet techniques in the examination of EEG and event-related potential waveforms. This is complemented by the more recent review by Pavlov et al. (2012) which considers the use of wavelet analysis for neurodynamic problems across a range of distinct scales of activity: from the microscopic cell level, via small neural assemblages, to the analysis of the EEG.

6.3.1 Evoked Potentials and Event-Related Potentials

Early work by Ademoglu et al. (1997) investigated the transient response of EEG signals to a set of brief visual stimuli. The authors studied a specific class of evoked potential (EP) due to visual stimuli – pattern-reversal visual evoked potentials, or PRVEPs – in an attempt to aid the clinical diagnosis of dementia. Figure 6.22a contains signal plots of 24 normal PRVEPs and 16 from patients with dementia. Using a quadratic B-spline wavelet (Figure 6.22b), the researchers decomposed each PRVEP signal containing 512 data points into six coefficient scales ($m = 1$–6) plus a residual component. That is, the multiresolution analysis was halted at scale index $m = 6$. Thus, the transform vector comprises eight approximation coefficients which contain the information from scale indices 7 to 9. The authors found that the residual scale coefficients had consistent sign changes for the normal cases (top of Figure 6.22c). Reconstructing the waveforms using only these residual scale coefficients produces identifiable, overlapping waveforms for the normal cases (middle of Figure 6.22c). The lower plot of Figure 6.22c shows the synthesized signals, reconstructed from the residual scale, for the PRVEPs obtained from patients with dementia where no regular pattern is evident.

A five-level Coiflet wavelet decomposition of the EEG was used by Zhang et al. (2006) to probe the auditory brainstem response (ABR) signal contained in the early part of the auditory evoked potential (AEP). AEPs consist of a series of waves that represent processes of the transduction, transmission and processing of auditory information from the cochlea to the brain stem. The ABR component reflects the neurophysiological processes within the brain in response to the auditory stimulus that is not affected by arousal, attention, drowsiness or drugs. The researchers found that the wavelet approach was suited to the detection of this low-amplitude component of the signal. The wavelet coefficients were input into a Bayesian classifier and subjected to cross-validation. The authors found that the method led to a significant reduction in the repetitions required to pick out the ABR from the signal. In other work, Hu et al. (2011) have developed a wavelet filtering technique using a binary weighting matrix which was multiplied by the Morlet wavelet transform of the signal after it has been processed by probabilistic independent component analysis. They used this two-step method to process somatosensory evoked potentials (SEPs) from multichannel EEGs in order to detect single-trial SEPs. This is often difficult to do in practice, frequently requiring large numbers of repeated trials due to very low signal-to-noise ratios (SNRs).

Manganotti et al. (2013) have recently detailed a preliminary study of sleeping volunteers undergoing transcranial magnetic stimulation (TMS). Following on from a previous study, (Manganotti et al., 2012), they employed a Morlet wavelet decomposition of the signal at four separate EEG bands – delta(1–4 Hz); theta (4–8 Hz); alpha (7–12 Hz); and beta (15–22 Hz) – to produce averaged waveforms for each band. The method is succinctly illustrated in Figure 6.23. The authors demonstrated a reciprocal synchronization/desynchronization effect on the slow and fast oscillatory activity in response to TMS after sleep deprivation and sleep states. Garcia et al. (2011) also used a Morlet-based wavelet approach to analyze TMS-evoked potentials at four separate EEG bands, which they defined as theta (4–6 Hz); alpha (8–12 Hz); beta (15–28 Hz); and gamma (30–40 Hz). From their work they

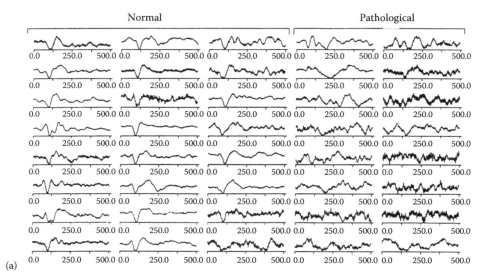
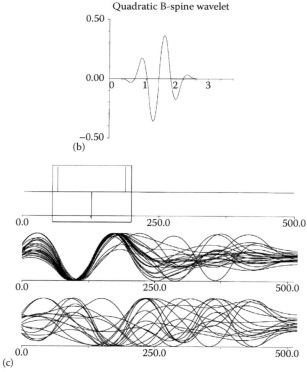

FIGURE 6.22 **An analysis of pattern-reversal visual evoked potentials (PRVEP) by spline wavelets: (a)** A subset of the 24 normal and 16 pathological PRVEPs considered in the study. **(b)** The quadratic spline wavelet used in the study. **(c)** Coefficient sign pattern and waveform reconstructions for PRVEPS. The coefficient sign pattern in the residual scale for a consistent (N70-P100-N130) complex, the superimposed waveforms are the synthesized delta-theta activity of the 24 normal (showing this pattern) and 16 pathological (lacking this pattern) PRVEPs. (From Ademoglu, A. et al., *IEEE Transactions on Biomedical Engineering*, 44(9), 881–890. Copyright 1997, IEEE.)

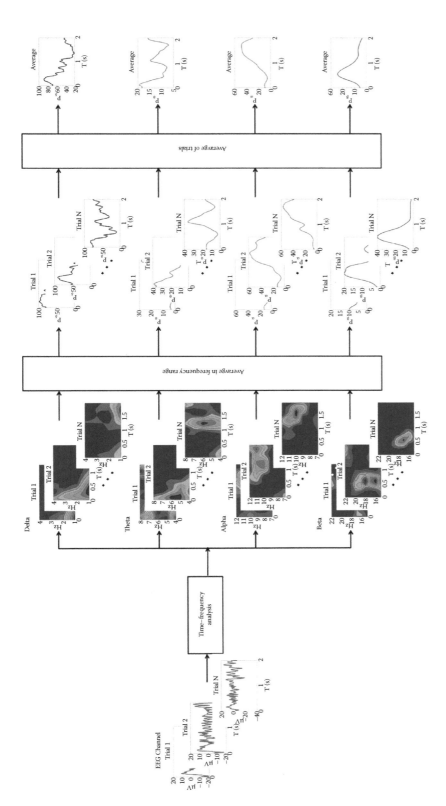

FIGURE 6.23 Schematic representation of the different steps of the analysis at the single-subject level. For each EEG channel trials with intervals greater than 2.2 s were selected for the analysis. Time–frequency analysis was performed on each trial with a continuous Morlet wavelet transform in the main frequency bands delta (1–4 Hz), theta (4–8 Hz), alpha (7–12 Hz) and beta (15–22 Hz). The power was then averaged in the frequency ranges of interest, producing a power time course for each trial. Finally, profiles for each trial were averaged and normalized to the baseline value. This procedure was applied for each condition (wake state, sleep deprivation and sleep). (Original in colour.) (From Manganotti, P. et al., *Frontiers in Human Neuroscience*, 7, 767, 2013.)

concluded that, rather than a technique for stimulating lesions, TMS generates natural brain signals and engages functional networks.

An early investigation of auditory evoked potentials (AEP's) as a measure of the depth of anaesthesia in dogs was detailed by Huang et al. (1999). They developed an automated monitoring system to control the delivery of intravenous anaesthetic based on a neural network classification of significant wavelet coefficients determined from the decomposed signal using Daubechies D20 wavelets. In later work, Schwilden et al. (2005) and Jeleazcov et al. (2006) characterized acoustic and somatosensory evoked potentials using a number of wavelet-based coefficient metrics to monitor the depth of anaesthesia in humans. A number of other research workers have examined evoked potentials in the EEG. For example, Wang et al. (2012a) used Morlet wavelet processing of the EEG combined with parallel factor analysis (PARAFAC) to quantify the differences in patients with chronic pain from controls using somatosensory evoked potential recordings. Their results indicated that frontal cortical activation and a lower response frequency were characteristics of evoked EEG responses in chronic pain subjects. The researchers also considered the effect of aging on the results. Combaz et al. (2009) implemented a feature extraction procedure for the EEG using a least -square support vector machine in order to classify characters observed by the patient on a screen during a brain–computer interface study. Kaspar et al. (2010) have described the Morlet–based examination of steady state, visually evoked potentials generated by object recognition at various flicker rates. Zhang et al. (2011a) have compared a local polynomial model (LPM) with the continuous Morlet wavelet transform of event-related EEGs. Thie et al. (2012) used both Gaussian and Coiflet wavelets in a method to separate multifocal visual evoked potentials from noisy signals obtaining 94% classification accuracy. Additionally, the discrete wavelet transform has been utilized by Ramirez-Cortes et al. (2010) as part of a neuro-fuzzy algorithm to detect P-300 rhythms.

The t-continuous wavelet transform (t-CWT) was employed by Steppacher et al. (2013) to assess higher-order cortical information processing in patients with unresponsive wakefulness syndrome. (The t-CWT technique, developed by Bostanov and Kotchoubey, 2004, 2006, is based on the Student's t-statistic and the CWT.) In the work, the wavelet transforms between two trial conditions were compared using the Student t-statistic plotted as a timescale map as shown in Figure 6.24. The figure shows one patient who exhibited a clear N-400 component and another who did not. The authors found that such an assessment of the N-400 ERP could help to evaluate a patient's long-term prognosis.

A study has been reported by Busch et al. (2009) which found that monitoring the wavelet phase of EEG activity in the theta and alpha bands allowed prediction of whether each of a series of identical visual stimuli would be observed or not ('hits' or 'misses'). By considering the phase of the wavelet transform values at each specific location in the time–frequency plane across trials, they were able to quantify an inter-trial coherence (ITC) metric. ITC represents the complex average of unit length complex vectors and provides a measure of phase locking across trials. It was incorporated within a phase bifurcation index (Φ) in order to differentiate the phase distributions associated with hits or misses. (The formulae for ITC and Φ are given in the Busch et al. paper.) The authors concluded that the visual detection threshold varies over time with the phase of the ongoing spontaneous

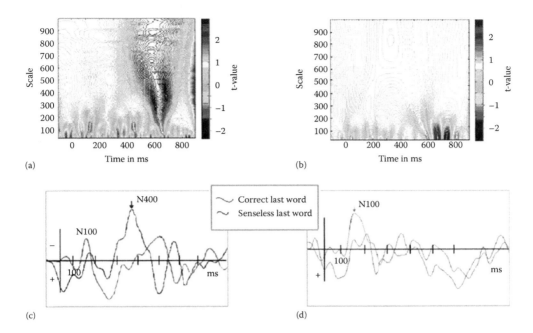

FIGURE 6.24 Examples of ERP detection methods with the N400 paradigm: **(a, b)** Results of the t-continuous wavelet transform at electrode Pz, represented by the t-tested difference scalogram of two unresponsive wakefulness syndrome patients. **(c, d)** Visually inspected ERP averages of the same two patients at the same electrode. The patient on the left shows a clear N400 evident from both analyses; in the patient on the right, no N400 is detectable. (Original in colour.) (From Steppacher, I. et al., *Annals of Neurology*, 73(5), 594–602, 2013.)

brain oscillations. Figure 6.25 contains a plot of the phase bifurcation index computed from the time–frequency plane. This work was extended by Busch and VanRullen (2010) to investigate how the connection between ongoing EEG oscillations and perception is modulated by selective attention to reveal fluctuations in 'sustained attention'.

A combined Morlet wavelet decomposition and multiple linear regression technique has been utilized by Hu et al. (2010) to enhance SNR in order to automatically detect ERPs in single trials. Their algorithm involved setting wavelet coefficients less than a predefined threshold to zero, then performing the inverse transform. (The wavelet filtering method was performed in a manner similar to their work on EPs described earlier in this section [Hu et al., 2011]). Huart et al. (2012) have studied chemosensory ERPs by injecting odors into airstreams within the nostril using an olfactometer. They discovered that across-trial averaging in the Morlet wavelet–based time–frequency plane markedly enhanced the SNR of the elicited responses and revealed EEG activity which could not be identified using conventional time-domain techniques. Cahn et al. (2013) investigated ERPs during Vipassana meditation using ICA and Morlet wavelets. Their findings suggested that this meditative state evoked enhanced perceptual clarity and decreased autonomic reactivity. A couple of papers by Cong et al. (2010, 2013) have briefly described their derivation of time–frequency representations of the ERP from which multi-domain features were extracted using tensor factorization.

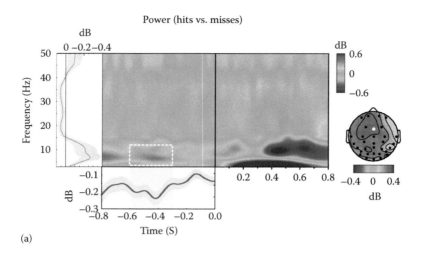

(a)

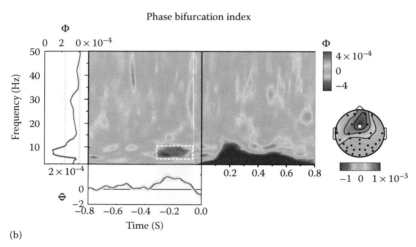

(b)

FIGURE 6.25 (See colour insert.) **Raw effects of oscillatory power and phase: (a)** Difference in spectral power in decibels between hits and misses, averaged across channels and subjects. Negative values indicate stronger power for misses. Left inset shows power difference averaged across time points in the prestimulus window (shaded areas SEM). The main difference is found in the 6–12 Hz frequency range. Bottom inset: Power difference averaged across frequencies in this range, with a maximally negative difference in the prestimulus time range between −600 and −300 ms (grey shaded area). The topography shows the distribution of the power difference from 6 to 12 Hz and from −600 to −300 ms preceding stimulus onset. **(b)** Phase bifurcation index (Φ), averaged across all channels and subjects. Positive values indicate that phase distributions are locked to different phase angles for hits and misses (e.g. in the prestimulus time range), while negative values indicate that only one condition is phase locked (e.g. phase locking exclusively for hits in the ERP time range). Left inset shows Φ averaged across all time points in the pre-stimulus window (vertical lines represent the 95% confidence interval). Bottom inset: Φ averaged across frequencies between 6 and 10 Hz (shaded area SE). Phase bifurcation is strongest from −300 to −50 ms preceding stimulus onset (grey shaded area). The topography shows the distribution of Φ from 6 to 10 Hz and from −300 to −50 ms preceding stimulus onset. (Original in colour.) (From Busch, N. A. et al., *Journal of Neuroscience*, 29(24), 7869–7876, 2009.)

A number of other authors have mentioned the use of wavelets within their processing methodology for ERPs, including Kuncheva and Rodriguez (2013) for ERP classification; McDonald et al. (2010) in a study of spatiotemporal profiles of word processing using intercranial EEG, functional magnetic resonance imaging (fMRI) and magnetoencephalography (MEG); Naue et al. (2011) in an examination of auditory ERPs; Gandhi et al. (2012) in their exploration of EEG responses to facial contrast-chimeras; Brenner et al. (2009) and Hall et al. (2011a) in their investigations of schizophrenia; Kiiski et al. (2012) in their analysis of low-frequency event-related EEG activity in multiple sclerosis patients; and Paiva et al. (2016) in their study of intensity dependence of the N1 and P2 waves of the auditory ERP.

6.3.2 Epileptic Seizures and Epileptogenic Foci

Various wavelet transforms were compared by Schiff et al. (1994) in early work to characterize epileptic seizures exhibited in the EEG. Figure 6.26 shows an EEG signal (Figure 6.26a) with its associated transform plot generated using a continuous Mexican hat transform calculated at each time step (Figure 6.26b). Schiff and coworkers used a variety of wavelet techniques from the continuous Mexican hat method shown in Figure 6.26b, to a discrete B-spline wavelet critically sampled using cubic B-spline wavelets and a multiresolution framework (not shown). They found that the use of spline techniques to speed up computation did not impair feature extraction from the signal. In addition, they pointed out that, since they do not require a fixed length data window, wavelet transforms provide an improved approach for spike detection in the data compared with windowed Fourier analysis. A number of advanced Morlet wavelet–based computational tools have been exploited by Li et al. (2007) to explore the relationship between epileptic EEGs acquired at different regions of the hippocampus. One of their specimen signal pairs is shown in Figure 6.27. They employed wavelet power representations, co-scalograms, cross-wavelet local correlation in phase and intensity, phase synchronization and wavelet bicoherence (WBC) plots to illustrate the relationship between signals. Figure 6.28 contains their combined plot of the wavelet amplitude and phase coherence for the signals in Figure 6.27. They concluded that these tools can successfully analyze and quantify temporal interactions between neuronal oscillators. More recently, Song et al. (2016) have constructed an automatic seizure

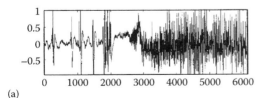
(a)
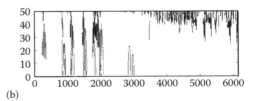
(b)

FIGURE 6.26 Wavelet transforms for electroencephlographic spike and seizure localization: (a) EEG recorded at 200 Hz from subdural electrode overlying frontal lobe seizure focus. (b) Continuous Mexican hat wavelet transform redundantly calculated in standard fashion. Contour lines are shown at values determined from ±1.96 S.D. of the *surrogate data* wavelet coefficients at each scale. For all the plots the abscissas are sample values in units of 1/200 s (30 s traces), while ordinates are arbitrary units. (From Schiff, S. J. et al., *Electroencephalography and Clinical Neurophysiology*, 91, 442–455, 1994.)

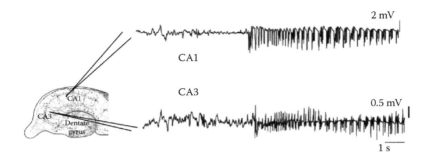

FIGURE 6.27 Neuronal activity recorded from the CA1 and CA3 regions of the hippocampus during the transition from pre-seizure to seizure states. (The duration of this recording is 12 s, the sampling frequency is 2500 Hz.) (Original in colour.) (From Li, X. et al., *Journal of Neuroscience Methods*, 160, 178–185, 2007.)

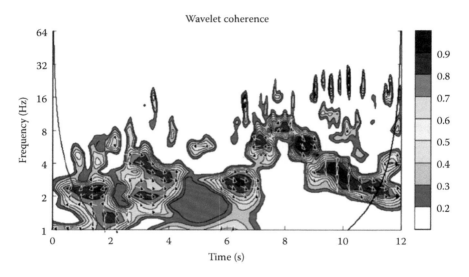

FIGURE 6.28 Wavelet coherence of two neuronal oscillations from the CA3 and CA1 regions. The cross-spectrum phase difference is shown by the arrow's direction (in phase: an arrow pointing right, means CA3 leads CA1; antiphase: an arrow pointing left, means CA1 leads CA3). The amplitude and phase information in the EEG signal are integrated to describe the first-order relationship between the EEG signals. The thick black line near to the darkest tone (black) is the significant level of 99% with re-sampling of AFFT. (Original in colour.) (From Li, X. et al., *Journal of Neuroscience Methods*, 160, 178–185, 2007.)

detection algorithm based on a Mahalanobis similarity-based feature extraction method and the DWT-decomposed signal. In other work, absence epilepsy has been the subject of study by Zeng et al. (2016), who developed a seizure detection technique based on the compressibility of the EEG which employed continuous wavelet transforms.

Epileptic events in the EEG from newborns have been interrogated by Mitra et al. (2009) using wavelet scale–based features in work to develop a reliable neonatal seizure detection system. Sakkalis et al. (2010) have developed a number of Morlet wavelet–based EEG 'biomarkers' to indicate the presence of minor neurophysiological signs in children with controlled epilepsy who showed no clinical or electrophysiological signs of brain dysfunction. They found that differences could be detected during a control task (rest), but not on a more demanding mathematical task. DWTs were applied in an explorative study of epilepsy classification using relevance vector machines (RVMs) by Lima et al. (2009). In other work, wavelet energy and entropy measures were computed from Morse wavelet transform modulus maxima in preictal and interictal EEGs by Gadhoumi et al. (2012) in order to discriminate between the two states in patients with mesial temporal lobe epilepsy.

A number of groups have used artificial neural networks (ANNs) and wavelets to process epileptic EEGs, including Arab et al. (2010), who classified healthy, petit mal and clonic epilepsy with around 80% accuracy using a two-stage neural network classifier; Übeyli (2009), who also classified healthy, pre-seizure and seizure EEGs using a combined neural network comprising three sets of networks trained using the Levenberg–Marquardt algorithm with different targets; and Guo et al. (2009, 2010), who have described an approach for epileptic seizure detection using wavelet-based energy and approximate entropy (ApEn) measures derived from multiwavelet transforms using a set of wavelet functions. (The same group later provided details of a seizure detection method employing the DWT and genetic programming [Guo et al., 2011].)

6.3.3 Sleep Studies

The EEG is an integral part of polysomnography – the comprehensive recording of multiple sleep parameters (brain function [EEG], heart rate [ECG], muscle activity [EMG], respiratory effort [piezo-bands], oxygen saturation [SpO_2], etc.) used to provide a recording of physiological changes occurring during sleep in order to diagnose sleep disorders. There have been numerous attempts to employ the wavelet transform in the analysis of the EEG during sleep. For example, Morlet-based wavelet phase synchronization measurements were employed by Mezeiová and Palus (2012) to assess the coupling between EEG signals from different scalp locations during various stages of sleep in healthy adults. Sleigh et al. (2010) have used wavelet coscalograms and coherence of electrocorticogram (ECoG) signals in their study of the slow-wave sleep pattern to rapid eye movement (REM) transitions in rats. Piantoni et al. (2013) demonstrated gamma activity associated with the upslope and peak of slow- wave structures in child EEGs during sleep using a Morlet wavelet decomposition. Le Van Quyen et al. (2010) examined gamma patterns in microelectrode recordings in the human cortex during sleep utilizing Morlet wavelet transforms, and produced a number of comprehensive plots to illustrate their work. The detection of sleep hypopnea (mild cessation of breathing) was tackled using a discrete Daubechies wavelet technique by Übeyli et al. (2010) and topographic EEG brain mapping during obstructive apnea (complete cessation of breathing) using Morlet-based power across discrete frequency bands has been described by Belo et al. (2011).

The development of complex wavelets derived from the EEG itself has been described by Sitnikova et al. (2009) in their study of sleep spindles and spike-wave discharge (SWD) components in rat EEGs. They were able to detect all SWDs as components having high wavelet power in the 30–50 Hz band using a standard Morlet wavelet decomposition. However, this standard method did not perform as well in identifying sleep spindles, exhibiting a 55%–67% identification rate. (Sleep spindles are associated with stage 2 of non-rapid eye movement [NREM] sleep and comprise brief bursts of fast activity, around 12–14 Hz, that rapidly increase in amplitude and then rapidly decay.) Using a generic 'spindle wavelet', the construction of which they describe, this was increased to 87%. The remaining sleep spindles could be recognized using a second spindle wavelet; however, this had to be selected individually for each trace. Figure 6.29 shows examples of the spindle wavelets and Figure 6.30 illustrates their application to the sleep EEG, comparing them with the standard Morlet wavelet. Further work in this area by this group is presented in Grubov et al. (2012) and Nazimov et al. (2013). More recently, Sitnikova et al. (2016) have investigated the sleep characteristics of rats with absence epilepsy. Their study made use of the Morlet-based ridges on the transform surface to characterize the nature of oscillatory events in the EEG. An algorithm for the accurate automatic detection of sleep spindles using the Morlet wavelet has been developed by Tsanas and Clifford (2015). They found that the algorithm outperformed six other contemporary sleep spindle detection approaches. Spindles in the EEG and MEG have also been investigated using Morlet wavelets by Dehghani et al. (2011) using topographic brain mappings. In the vein of transform modification, it is also worth

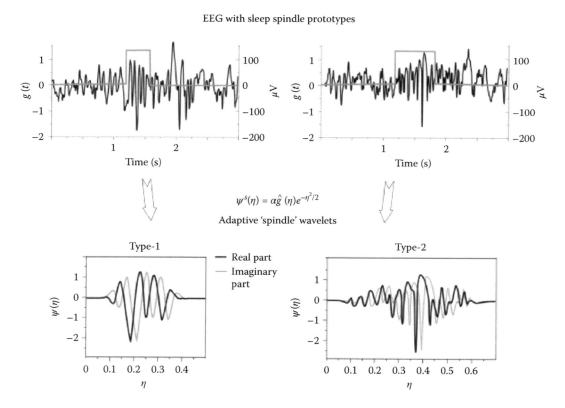

FIGURE 6.29 **The algorithm used for constructing adoptive 'spindle' wavelets.** Sleep spindle prototypes are selected in native EEG, $g(t)$, converted into the complex form and normalized with Gaussian function (see text for details). (From Sitnikova, E. et al., *Journal of Neuroscience Methods*, 180, 304–316, 2009.)

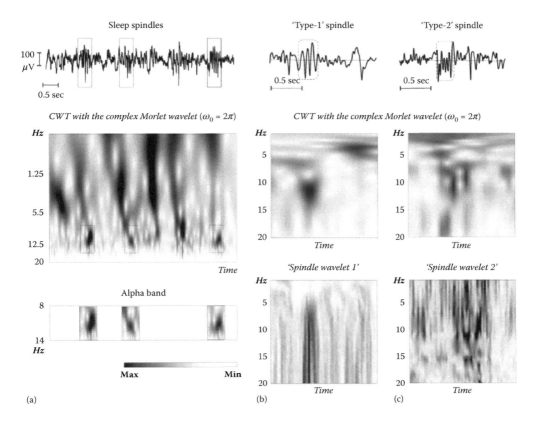

FIGURE 6.30 CWT in EEG with sleep spindles (in squares): (a) The complex Morlet perfectly represents time–frequency characteristics of sleep spindles. In wavelet spectrum, a hallmark of spindles events was an increased energy in alpha band (8–14 Hz). **(b, c)** In the Morlet-based wavelet spectrum (top plates), 'type 2' sleep spindle demonstrates high energy in frequencies <16 Hz, but in 'type 1' spindle these frequencies are hardly present. The CWT with two adoptive 'spindle wavelets' (bottom plates) poorly represents the time–frequency structure of EEG signal. Despite good performance in localizing sleep spindles, 'spindle wavelets' are not suitable for time–frequency analysis (they are not capable for extracting frequency-domain information). (From Sitnikova, E. et al., *Journal of Neuroscience Methods*, 180, 304–316, 2009.)

consulting Inoue et al. (2005), who, in an earlier publication, described a modified wavelet transform (MWT) decomposition and its use in detecting discrete sleep stages. In their method, the number of oscillations within the Gaussian window may be tailored to the scale of interest. They described the transform as having a character midway between a wavelet transform and a short-time Fourier transform.

Brief mention of wavelet methods employed for sleep analysis can also be found in Marzano et al. (2011), who investigated dream recall phenomena using Morlet wavelets; Pizza et al. (2011), who explored slow eye movement distributions; Spoormaker et al. (2010), who used maximal overlap discrete wavelet transforms to probe the large-scale functional brain networking during REM sleep; Fulda et al. (2011), who probed the rapid eye movement in rodents using the Mexican hat modulus maxima; and Zorick and Mandelkern

(2013), who scrutinized the multifractal nature of the sleep and awake EEG also using wavelet transform Modulus maxima techniques.

6.3.4 Other Areas

A number of researchers have used wavelets to understand the effects of drugs on the EEG, including Bonhomme et al. (2008), who mapped the dose-dependent relative regional cerebral activity changes associated with chlondine infusion in healthy volunteers; Berke (2009), who studied the effect of direct and indirect dopamine agonists on rat EEGs; and Pang et al. (2009), who found the unexpected abrupt cessation of type 2 theta oscillations in an EEG from mice during halothane exposure due to the administration of lidocaine. All three studies utilized Morlet wavelet decompositions.

The general problem of denoising the EEG has been considered by a number of researchers. Safieddine et al. (2012) considered a number of schemes for the removal of muscle artifact from the EEG and found that their wavelet technique offered better denoising for less noisy data sets. Ghandeharion and Erfanian (2010) employed wavelets combined with independent component analysis and a variety of signal measures (mutual information, projection strength, correlation, Kurtosis) in a fully automated algorithm to suppress ocular artifact in the EEG. Other wavelet-based blind source separation methods for signal denoising have been described by Vazquez et al. (2012) and Zima et al. (2012).

Evoked power and phase locking information from a Morlet transform of the EEG signal was analyzed by Hall et al. (2011b), whose results supported the use of auditory gamma-band responses as putative endophenotypes for schizophrenia. Catarino et al. (2013) examined task-related functional activity in autism spectrum conditions (ASCs) using Morlet-based wavelet transform coherence and found that interhemispheric coherence was reduced in people with ASC. Human emotions have been classified using wavelet-based EEG analysis by Murugappan et al. (2011) and Rizon (2010). Perceptual switching (of a Necker cube) has been tackled by Ozaki et al. (2012) and Nakatani et al. (2011) using complex continuous wavelets. Van De Ville et al. (2010) have performed a wavelet-based fractal analysis of the EEG to reveal scale-free dynamics during rest. In their method, they embedded the microstate sequences from the EEG within a random walk sequence prior to the analysis using Daubechies wavelets. They found strong statistical evidence that the dynamics were indeed scale-free over six dyadic scales corresponding with a 256 ms to 16 s range. Finally, a recent study concerning the characterization of tonic cold pain from EEG data by Hadjileontiadis (2015) has employed wavelet higher-order spectral (WHOS) features: the wavelet bispectrum (WBS), bicoherence and instantaneous bispectrum (refer to Chapter 2, Section 2.22). The experimental results revealed the potential of WHOS-based features to capture the transition between 'relaxed' and 'painful' states.

6.4 PHOTOPLETHYSMOGRAM

The photoplethysmogram (PPG or 'pleth') is used in pulse oximetry and takes the form of a smooth pulsatile signal indicating the change in light absorbance at a peripheral vascular site (e.g. finger, toe, earlobe, forehead) corresponding to a localized increase in blood volume with each heartbeat pulse. A schematic of the signal is provided in Figure 6.31a. Over recent years, it has been increasingly recognized as a carrier of a tremendous amount of

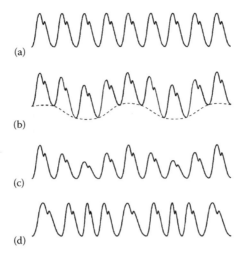

FIGURE 6.31 Modulations of the PPG due to respiration (modulation through two complete respiratory cycles shown): **(a)** PPG showing unmodulated cardiac pulse waveforms. **(b)** Baseline modulation (cardiac pulses riding on top of baseline shown dashed). **(c)** Amplitude modulation (cardiac pulse amplitude varying over respiratory cycle). **(d)** 'Respiratory sinus arrhythmia' (RSA) (pulse period varying over respiratory cycle). (From Addison, P. S. et al., *Journal of Clinical Monitoring and Computing*, 26, 45–51, 2012a.)

physiological information, including heart rate and oxygen saturation (its primary function in pulse oximetry) along with characteristics that correlate to respiratory rate (RR), respiratory effort, blood pressure, hypovolemia, vasomotion, arousal, age-related arterial stiffening and so on. For many years, manufacturers developed signal processing schemes to improve the determination of pulse rate and oxygen saturation (SpO_2) from the PPG. These schemes have, in general, attempted to remove other signal components in order to clean up the pulse signal component prior to analysis. However, these 'other components' are increasingly being exploited, and not least for the extraction of respiratory information.

6.4.1 Respiratory Modulations, Respiratory Rate and Respiratory Effort

Respiratory rate (RR) is well known to be a clinically important parameter owing to the fact that it provides information pertaining to many aspects of a patient's respiratory status. However, the most common technique for monitoring RR in practice is intermittent, manual observation. It is perhaps, one of the most important vital signs still to be mostly measured manually. It was suggested by Leonard et al. (2003) that respiratory rate may be measured from a standard pulse oximeter using wavelet transform-based methods after they identified phase cycling in the wavelet transform phase plot corresponding to breathing (region B in Figure 6.32). The extraction of individual breaths using phase information was later carried out by the same group (Leonard et al., 2004a) for a cohort of 22 healthy subjects. The determination of respiratory rate in children was then examined using this methodology by Leonard et al. (2006) and in chest clinic patients by Clifton et al. (2007). Figure 6.33a contains a scalogram corresponding to an 18 minute signal acquired during a bronchoscopy procedure. The patient was conscious during the first 200 seconds before

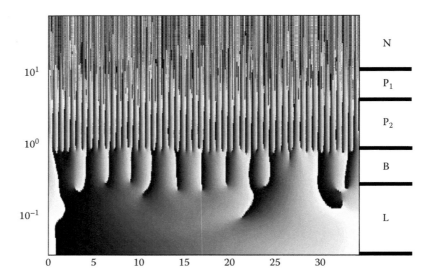

FIGURE 6.32 Wavelet plot showing an easily identified LF feature in region B (breathing) and easily separated from noise (N) and pulse features (P1 and P2). Horizontal axis time (s). Vertical axis frequency (Hz). (From Leonard, P. et al., *Emergency Medicine Journal*, 20, 524–525, 2003.)

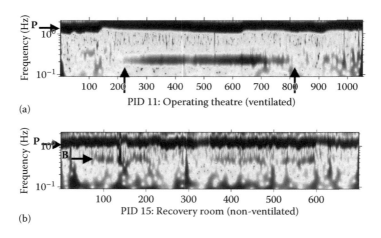

FIGURE 6.33 Wavelet–transform time–frequency plots from a selection of patients: (a) PID 11: operating theatre (ventilated). (b) PID 15: recovery room (non-ventilated). (From Clifton, D. et al., *Journal of Clinical Monitoring and Computing*, 21(1), 55–61, 2007.)

being anaesthetized from 200 to 800 seconds; thereafter regaining consciousness. The pulse band is marked 'P' in the figure. A very obvious breathing band can be seen to appear in the plot during the period of anaesthesia (between the two marker arrows). This corresponds to the period during the procedure where the patient was mechanically ventilated. This kind of dominant band feature is typical of all mechanically ventilated patient signals. (Note that the band corresponds to the baseline modulation in Figure 6.31b – see also the following paragraph). Either side of the ventilated region in Figure 6.33a, the patient was breathing for herself. Figure 6.33b shows 11 minutes of a scalogram from a non-ventilated patient in the recovery room. A breathing band can be observed at around 0.44 Hz (26 bpm) marked by a 'B' in the plot. However, this 'natural' breathing band is intermittent in nature and exhibits slight variability in frequency in contrast to the dominant, fixed frequency band observed during mechanical ventilation shown in Figure 6.33a. This variability is to be expected from natural respiration. Figure 6.34 contains a 3-D plot of a scalogram corresponding to a neonatal PPG (Addison, 2004). It can be seen from the figure that the relatively regular pulse band (P) contrasts markedly with the erratic respiration (B) typical of these infants.

The PPG actually contains three main respiratory modulations – baseline, amplitude and frequency – stemming from related physiological origins as described in Addison et al. (2012a). These are depicted in Figure 6.31b–d and are often non-stationary in nature. Wavelet-based decoupling of these non-stationary PPG components has been described by Addison and Watson (2004a) in the context of extracting respiratory information contained within the signal. In the method, named secondary wavelet feature decoupling (SWFD) and described in Chapter 2, Section 2.16, the baseline modulation is extracted directly from the original transform of the signal. However, in this transform the pulse band also contains respiratory information in the form of both amplitude and frequency modulations. These can be extracted by computing the pulse band maxima with respect to scale and projecting

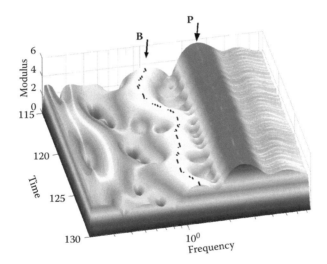

FIGURE 6.34 (See colour insert.) **A wavelet transform of a pulse oximeter signal from a premature baby.** Data were taken over a 15-second period using a wavelet of different frequencies. The 'P' band corresponds to the baby's regular pulse rate. The 'B' ridge corresponds to the baby's breathing rate, while the alternate black and white markings that lie along the peak of the ridge indicate the inspiration and expiration of breath, respectively. This breathing ridge is determined automatically by an algorithm that searches the transform surface for maxima and decides which of these correspond to respiration. The decision criteria are based on a careful study of the link between wavelet features and patient respiration. A complex wavelet was used in the analysis, which means that the vertical axis is actually the square of the modulus of the transform. (Original in colour.) (From Addison, P., *Physics World*, 17(3), 35–39, 2004.)

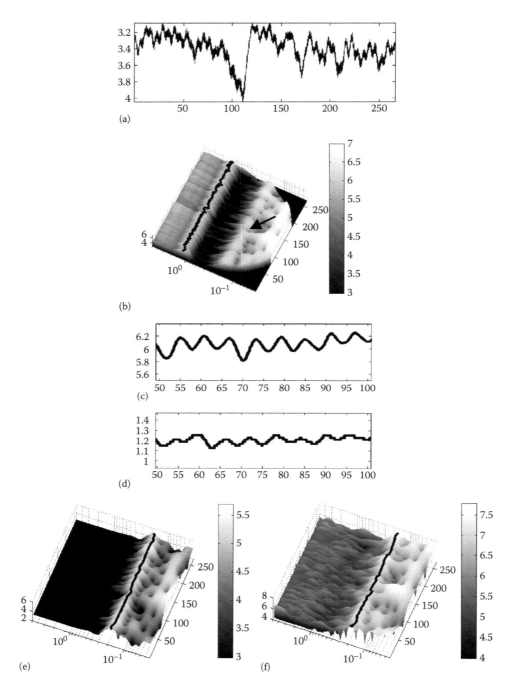

FIGURE 6.35 (a) Original signal and (b) 3-D view of scalogram with pulse band ridge locus path superimposed. (c) Blow up of RAP signal. (d) Blow up of RFP signal. (e) RAP signal scalogram in 3-D with ridge shown. (f) RFP signal scalogram in 3-D with ridge shown. (From Addison, P. S., and Watson, J. N., *International Journal of Wavelets, Multiresolution and Information Processing*, 2(1), 43–57, 2004a. Copyright 2005, World Scientific.)

horizontally to give a ridge amplitude perturbation (RAP) signal and vertically to give a ridge frequency perturbation (RFP) signal. These may then be wavelet-transformed to produce their own breathing bands. The technique is illustrated in Figure 6.35. Figure 6.35a contains a photoplethysmogram with its corresponding scalogram given in Figure 6.35b. The pulse band ridge is drawn across the surface as a thick black line. In addition, a region of artifact is indicated by an arrow. The horizontal projection of part of the pulse ridge is shown in Figure 6.35c. This is the RAP signal. The corresponding RFP signal, obtained using a vertical projection of the ridge, is shown in Figure 6.35d. We see obvious respiratory modulations present in the RAP and RFP signals. The secondary scalograms of the RAP and RFP signals are shown in Figure 6.35e,f, both with ridges at the respiration frequency. In this way, by extracting modulation information from the original pulse band ridge in wavelet space, non-stationary modulations of non-stationary, non-sinusoidal signal components may be extracted. In this case, the modulations represent respiratory activity but the technique has general applicability to other signals and other modulation components.

A fully automated wavelet-based algorithm for the determination of respiratory rate from the pulse oximeter was reported by Addison et al. (2012a). The resulting parameter, RR_{oxi}, was tested on a cohort of 139 healthy volunteers and excellent agreement was found between RR_{oxi} and a gold standard reference from an $ETCO_2$ device. The measures exhibited a mean difference and standard deviation of −0.23 and 1.14 breaths per minute (brpm), respectively. This work was followed up by a 63-patient study of the performance of the RR_{oxi} measure on the general care floor (Addison et al., 2015), where again excellent agreement was found between RR_{oxi} and RR_{ETCO2} (mean/s.d. = -0.48/1.77 brpm). The study demonstrated that the novel Morlet-based algorithm was a potentially viable approach for monitoring respiratory rate in this patient group. A number of related papers cover the performance of the algorithm: in the post-anaesthesia care unit (Mestek et al., 2012a), during coached breathing (Mestek et al., 2012b), in obese subjects (Mestek et al., 2012c), in chronic obstructive pulmonary disease patients (Mestek et al., 2012d), in congestive heart failure patients (Mestek et al., 2012e), during cold room hypoxia (Addison et al., 2012b) and in patients with high respiratory rates (Mestek et al., 2013).

In a closely related area, Addison and Watson (2010) detailed a method for the wavelet-based extraction of respiratory effort information from the photoplethysmogram where a change in energy of the breathing band on the transform surface is shown to be associated with a change in the effort to breathe caused by, for example, breathing against an increased resistance. More recently, it has been shown that a novel flexible probe arrangement may be used to enhance the respiratory components in the PPG at a non-traditional chest site (Addison et al., 2012c).

6.4.2 Oxygen Saturation

The measurement of oxygen saturation, SpO_2, is the primary purpose of the pulse oximeter device and is determined using two PPG signals: one red and one infrared. Light at these wavelengths is absorbed differentially by oxygenated and deoxygenated haemoglobin. The red and infrared lights are switched on and off very quickly in sequence at the probe and, because infrared is invisible to the naked eye, the probe appears to the human eye to glow red continuously at the attachment site (e.g. fingertip). In practice, a ratio is calculated from the amplitudes of the cardiac pulse components of the red and infrared PPGs. This ratio is related to the amount of oxygenated and deoxygenated haemoglobin as they absorb each wavelength of light differently. Hence, a value of SpO_2 may be determined via a previously developed calibration curve or look-up table. Over the last few decades, many temporal and spectral algorithmic schemes have been developed to determine SpO_2 from the red and infrared PPGs, particularly during periods of degraded signals. However, two novel wavelet-based methods for determining SpO_2 were developed by Addison and Watson (2004b, 2005) – both employing Morlet wavelet decompositions, but in different modalities. In the first method, illustrated in Figure 6.36, the scalogram of the red signal is divided by the infrared scalogram, shown in Figure 6.36a, thus forming the wavelet ratio surface of Figure 6.36b. This surface represents the ratio of all the components in the two signals expressed in the time–frequency plane. The question now is: where in the time–frequency plane is the best place to determine the ratio for SpO_2? As we are concerned with the pulse component of the signal, this has an easy answer: the ridge of the pulse band in one of the original transforms is determined and its time–frequency coordinates are projected onto the wavelet ratio surface. This projection corresponds to the optimal time–frequency path for the determination of the ratio from the pulse components and is shown as 'P' in Figure 6.36b. In this way, the ratio is determined through time and used to determine a current SpO_2 value. In the second method, the real part of the transform is computed for both PPGs, as shown in Figure 6.37a. Both of these real scalograms are then scanned across at a number of scales (Figure 6.37b). Hence, at each scale, two wavelet components are obtained for the red and infrared signals. These components are plotted against each other (in a similar manner to generating a Lissajous figure from two sinusoidal signals). If there is no noise, a closed Lissajous-like figure should be observed as all the components of the two signals move in phase. For regions where signal noise is present, the Lissajous opens up and becomes noisy. By scanning over a continuous range of scales and plotting the Lissajous for each, we essentially construct a 3-D Lissajous-like object. A 3-D view and an end view of this construction is given in Figure 6.37c,d, respectively. By examining the 3-D object at each scale, we can search for the least noisy (most closed up) Lissajous component (Figure 6.37e) and use this to determine the proper ratio of the two signals (the ratio is actually the slope of the Lissajous).

Running wavelet archetyping (RWA) was developed to enhance the pulsatile information within the photoplethysmographic signal (Addison, 2015a). The technique, which is fully described in Chapter 2, Section 2.18, is particularly useful for noisy signals and/or signals with low-amplitude cardiac pulse components, for example, due to low perfusion at the sensor site. Figure 6.38 contains the wavelet transform of a PPG before and after RWA was applied. The original PPG was of relatively poor quality, as often seen in practice, and the pulse band is barely visible across the original scalogram. The pulsatile component can be more clearly observed in the RWA-processed signal. In fact, the pulse band in the archetype scalogram is relatively continuous. This kind of enhancement facilitates the extraction of the pulse rate in the time–frequency domain through, for example, ridge detection methods.

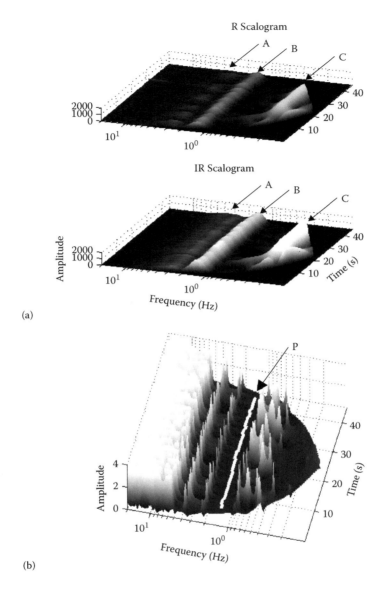

FIGURE 6.36 Wavelet ratio surface method: (a) The red and infrared scalograms corresponding to 45 seconds of PPG signals. Region A contains high-frequency noise and secondary pulse signals, region B represents the pulse signals and region C includes the breathing signal and general movement artefact. **(b)** The wavelet ratio surface derived from the division of the red by the infrared scalograms shown in (a). (From Addison, P. S., and Watson, J. N., *Medical Engineering and Physics*, 27(3), 245–248, 2005.)

6.4.3 The Video Photoplethysmogram (Video-PPG)

The extraction of physiological information from photoplethysmographic signals obtained from video image streams (video-PPGs) is a new and exciting area of physiological monitoring. A few groups have formulated wavelet-based approaches to process these new signals in order to derive heart rate, including Bousefsaf et al. (2013), Zaunseder et al. (2014)

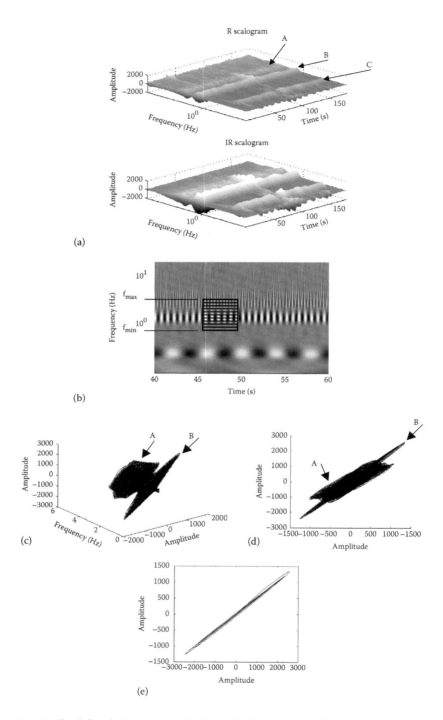

FIGURE 6.37 Method for deriving wavelet-based 3-D Lissajous: (**a**) Scalograms of the red and infrared pleths. (**b**) Schematic of the sliding window used to obtain the wavelet components for the 3-D Lissajous. (**c**) 3-D view of wavelet-based 3-D Lissajous. (**d**) End of view of (a). (**e**) End of view of selected component. Regions A, B and C are the same as in previous figure. (From Addison, P. S., and Watson, J. N., *Measurement Science and Technology*, 15, L1–L4, 2004b.)

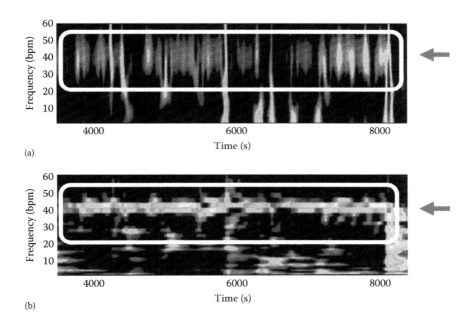

FIGURE 6.38 (a) Wavelet transform scalogram of a photoplethysmogram. (b) Running wavelet archetype corresponding to (a). The arrow points to the position of the pulse band. (Note that the frequency here is in beats-per-minute 'bpm.' Location of pulse band marked by an arrow in plots.) (From Addison, P. S., *Electronics Letters*, 51, 1153–1155, 2015a.)

and Huang and Dung (2016). A novel modular wavelet–based approach to the extraction of both heart rate and oxygen saturation from video-PPGs has recently been developed by the author (Addison, 2016). The method is shown schematically in Figure 6.39 and comprises extracting video-PPGs, performing the transforms, undertaking running wavelet archetyping to enhance the pulse information, extraction of the pulse ridge time–frequency information (and thus, a heart rate [HR$_{[vid]}$] signal), creation of a wavelet ratio surface, projection of the pulse ridge onto the ratio surface to determine the ratios from which a saturation trending signal is derived and calibrating this signal to provide an absolute saturation signal (S$_{vid}$O$_2$). This kind of modular continuous wavelet transform-based approach can be generalized to include any number of physiological parameters, and is advocated by the author as a powerful methodology to deal with noisy, non-stationary biosignals in general (Addison, 2016). The technique was applied to a video sequence from a standard, commercially available RGB (red, green, blue) video camera. An R video-PPG signal is shown in Figure 6.40a with its corresponding wavelet modulus plot in Figure 6.40b. A distinct pulse band can be seen across the transform modulus plot at around 85 beats per minute (bpm). The corresponding processed RWA scalogram is shown in Figure 6.40c, where the smoothing of the band relative to the original scalogram in Figure 6.40b is evident. The pulse rate may be derived by extracting the RWA ridge frequency over time. This is shown in Figure 6.41a. Apart from localized edge effects, the video and pulse oximeter heart rates stay within 2 bpm of each other during the period of investigation (Figure 6.41b). The projection of the pulse rate onto the ratio surface is shown in Figure 6.41c and the subsequently

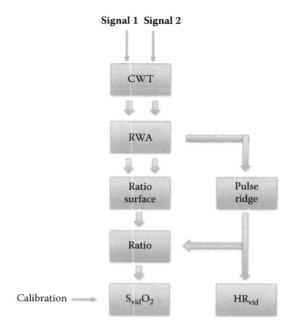

FIGURE 6.39 Flow diagram of method for determining oxygen saturation and heart rate from video signals. Two CWTs are computed, one for each input signal, and two RWAs are computed. These are then combined to form a single ratio surface. (From figure 1 of Addison, P., *Healthcare Technology Letters*, 3, 2, 111–115, 2016.)

derived oxygen saturation curve is shown in Figure 6.41d. Note that the $S_{vid}O_2$ appears to fall more rapidly than the pulse oximeter oxygen saturation, S_pO_2. This is a manifestation of the difference in filtering characteristics of the two signals: one is wavelet-based and is applied to a video signal, and the other uses traditional pulse oximeter algorithmic components and is applied to a photoplethysmogram acquired at the finger.

6.4.4 Other Areas

There are now many papers concerning the use of wavelet analysis for the extraction of a wide range of physiological information from the PPG. For example, Yacin et al. (2011) performed a reconstruction of the gastric myoelectric activity (GMA) slow wave from a finger PPG using a radial basis function neural network (RBFNN) and a Daubechies wavelet decomposition. Watson et al. (2003) and Leonard et al. (2004b) have developed an illness severity indicator to identify children with signs of circulatory compromise due to significant bacterial illnesses based on a Morlet wavelet–based decomposition. The method employed a Bayesian classifier which separated ill from healthy data with a high sensitivity and specificity. Dehkordi et al. (2015) recently devised a technique based on synchrosqueezing for the extraction of respiratory-induced intensity variations in the PPG baseline. Ju et al. (2013) have explored the possibility of using a smartphone to detect the PPG and use it in the calculation of cardiovascular parameters which employs a symlet wavelet–based approach. A symlet wavelet was also used by Khandoker et al. (2013) to investigate the relationship between the PPG and oesophageal pressures (P_{es}) in sleep-disordered breathing.

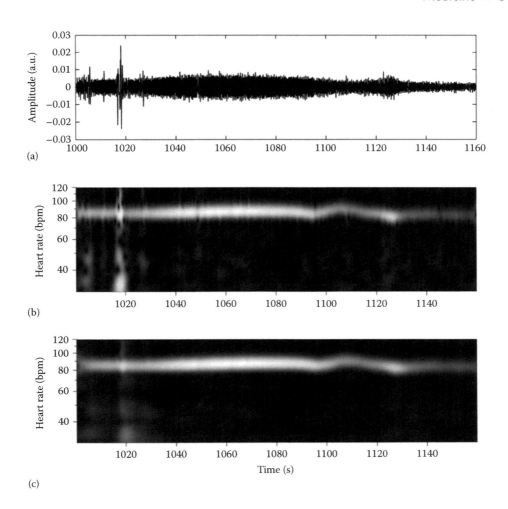

FIGURE 6.40 R signal with corresponding scalogram and associated running wavelet archetype: **(a)** R signal. **(b)** Transform modulus plot. **(c)** Corresponding RWA scalogram. (From figure 2 of Addison, P., *Healthcare Technology Letters*, 3, 2, 111–115, 2016.)

They compared the ability of three PPG-derived signals to correlate with changes in P_{es}: pulse interval time series, pulse amplitude time series and a level 8 symlet wavelet decomposition of the signal. They found that the highest correlation for obstructive sleep apnea ($r = .80$) was obtained for the wavelet decomposition-based metric. However, none of the methods performed well for central apneas. (Note that obstructive apneas generate enhanced modulations as the patient struggles to breathe against a closed airway, whereas during central events no effort is made to breathe and modulations are therefore expected to cease.) The feasibility of using wavelet analysis to quantify pulse arrival times (PATs) at the finger was considered in an investigative study by Allen et al. (2013). (PAT is the transit time for the pulse wave to travel from the heart to the peripheral site on the body where the pulse oximeter probe is attached. It may be measured from a characteristic, or fiducial, point on the ECG to a point on the PPG pulse.) Using Morlet wavelet decompositions, they calculated the PAT from the wavelet phase shift between ECG and PPG signals

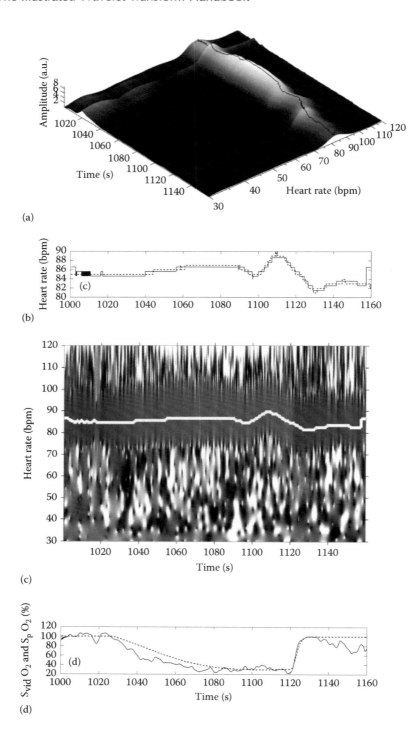

FIGURE 6.41 **Heart rate and oxygen saturation from video: (a)** Ridge found from the RWA pulse band maxima. **(b)** Dashed line: Pulse oximeter heart rate HR_p. Full line: Video HR_{vid}. **(c)** Pulse band projected onto ratio surface. **(d)** Dashed line: Pulse oximeter SpO_2. Full line: Video $Sv_{id}O_2$. (From figures 4 and 5 of Addison, P., *Healthcare Technology Letters*, 3(2), 111–115, 2016)

at the dominant frequency and related it to microvascular blood flow responses in terms of a normalized measure of the deviation in PAT during a deep inspiratory gasp maneuver.

6.5 PATHOLOGICAL SOUNDS, ULTRASOUNDS AND VIBRATIONS

This section deals with the wavelet analysis of sound signals of clinical relevance, including cardiovascular sounds, lung sounds, swallowing sounds, snoring, speech and the acoustic response of the ear.

6.5.1 Cardiovascular System

The turbulent sounds generated by femoral artery stenosis in dogs was analyzed by Akay et al. (1994) using both wavelet transforms and short-time Fourier transforms. They compared signals from the unblocked case (0% occlusion) with 72% and 85% occlusion and found an increase in power associated with the higher frequencies for the occlusion cases as seen in their wavelet power plots. They also found that the power corresponding to the first two wavelet frequency bandwidths (100–250 Hz) associated with arterial stenosis increased significantly after the injection of the vasodilator drug papaverine. In later work, Yuenyong et al. (2011) detailed a wavelet technique for automatic heart sound analysis without segmenting the signal. They considered two types of abnormal heart sounds from an electronic stethoscope: extra heart sounds (e.g. S_3) and murmurs. A Morlet wavelet was employed to produce an envelope signal from which three features were extracted (number of peaks, average distance between peaks and signal energy). These were added to another 32 additional features extracted using a DWT. Principle component analysis and bootstrapping was then applied before classification using an artificial neural network. The method was found to exhibit a high degree of robustness to noise.

A number of groups have incorporated wavelet techniques in targeting heart murmurs specifically. Safara et al. (2012), for example, used a wavelet packet entropy approach for heart murmur classification. They examined a number of potential wavelet functions for the task (Meyer, symlet, Coiflet and Daubechies), choosing the Daubechies wavelet as it exhibited the lowest error. Kumar et al. (2010) made use of a number of signal-derived features, including wavelet-based ones, to classify seven different murmur types, and Delgado-Trejos et al. (2009) considered wavelet analysis among a number of approaches, including time–frequency transforms (wavelet, STFT, Wigner–Ville), perceptual and fractal methods, for the automatic detection of murmurs in the phonocardiogram (PCG). An adaptive wavelet thresholding scheme was utilized by Zhao et al. (2013a) to remove noise from the PCG signal prior to analysis using the Hilbert–Hwang transform and empirical mode decomposition to detect coronary artery disease from diastolic murmur sounds; Ning and Atanasov (2010) found that the Daubechies db6 wavelet decomposition combined with autoregressive modelling was able to separate heart sounds from murmurs in most cases, except when a continuous large amplitude murmur was present; and a Morlet-based method incorporating a hidden Markov model was employed by Zhong et al. (2013) to classify heart murmur signal components. Finally, it is worth consulting the detailed set of wavelet transform scalograms and corresponding short-time Fourier transform spectrograms for auscultatory heart sounds in the paper by Nogata et al. (2012). The authors

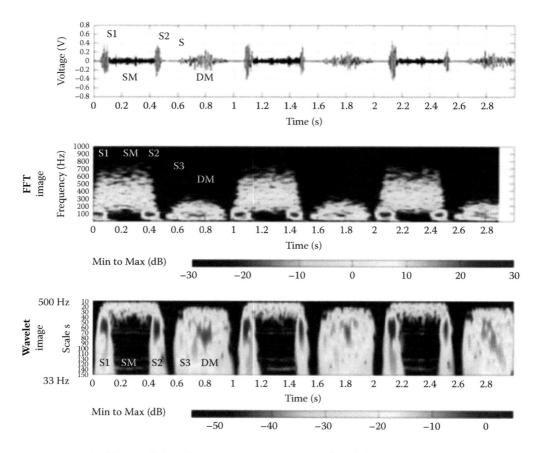

FIGURE 6.42 Audiovisual-based recognition images analyzed for VSD: 3rd degree. (Original in colour.) (From Nogata, F. et al., *Global Journal of Technology and Optimization*, 3, 43–48, 2012.)

suggested using such plots in an 'audiovisual' recognition method for auscultatory heart sounds. They produced images covering a wide variety of heart conditions including varying degrees of mitral regurgitation, mitral stenosis, ventricular septal defect, hypertrophic obstructive cardiomyopathy, dilated cardiomyopathy and sounds associated with high blood pressure. Figure 6.42 contains an example of one of their plots for the murmurs caused by a third-degree ventricular septal defect. Three murmurs are present in the signal and the associated scalogram – S_3, SM and DM – along with the normal valve closure sounds – S_1 and S_2. They combined their plots to form an image-based recognition map of the various heart diseases considered in the study. Nogata et al. (2015) provide a similar analysis for breathing sounds.

A wavelet-based technique for the detection of PCG S_1 and S_2 sounds acquired by a microphone placed within a patient's clothes has been described by Addison and Watson (2003). In such a configuration, very little high-frequency information pertaining to these signal components is available due to the very poor coupling between the microphone and skin. However, the method was able to reconstruct the envelopes corresponding to the S_1 and S_2 components from which a heart rate could be determined. It was demonstrated that respiratory rate could also be estimated using the approach. See also Liu et al. (2012a), who used a wavelet threshold shrinkage method to denoise stethoscope signals from four electronic stethoscopes embedded in an auscultation vest. After considering eight mother wavelets, Ergen et al. (2012) reported that the Morlet wavelet was the most reliable in terms of the non-invasive detection of heart abnormalities in the PCG. Balasubramaniam and Nedumaran (2010) have reported on the evaluation of a real-time execution performance of two systems of phonocardiograms – a PC-based system and a DSP-based system – where Morlet wavelet decomposition was employed. In another study, the separation of phonocardiographic and ballistocardiographic components has been accomplished by Zazula et al. (2012) using a Morlet wavelet–based technique applied to an optical interferometry signal.

Chan et al. (1997) and Lui et al. (1998) utilized the wavelet transform to detect venous air embolism (VAE) during surgery. Due to its life-threatening nature, the fast detection of VAE is essential to ensure prompt clinical treatment. They used a quadratic spline wavelet to analyze Doppler heart sounds (DHS) from dogs. A trace of the heart sound signal is plotted at the top of Figure 6.43. The arrow indicates the point at which 0.02 mL of air was injected. Wavelet transforms of the time series are plotted below the original signal for scale indices $m = 1$, 2 and 3. (Note that the authors use j for scale index.) A distinct increase in the wavelet coefficients at the first scale is obvious from the plot. A quantitative assessment was made by plotting the normalized power of single heartbeats at each scale. Figure 6.44 shows the normalized power plotted against heartbeat for 22 beats. The largest peak corresponds to the first scale and shows its usefulness as an indicator for venous air embolism. The authors also found a relationship between the sum of the normalized power

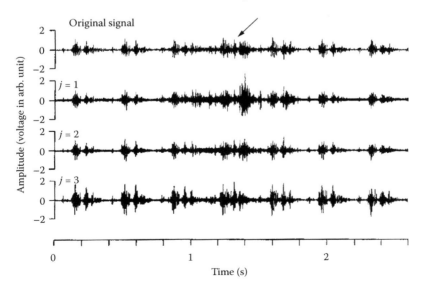

FIGURE 6.43 A typical DHS signal containing seven heartbeats and its WT at different scales ($j = 1$ to 3). With a 0.02 mL of air injected, the embolic heartbeat (marked by the arrow) was confirmed by an experienced anaesthetist by listening to the DHS signal. (From Chan, B. C. et al., *IEEE Transactions on Biomedical Engineering*, 44(4), 237–246. Copyright 1997, IEEE.)

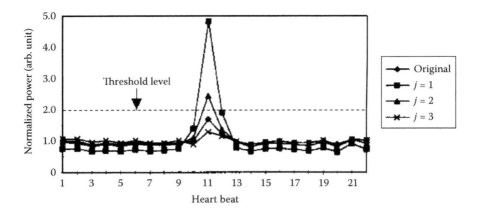

FIGURE 6.44 Normalized power of individual heartbeats obtained after WT of a DHS containing the time segment in Figure 6.43. The embolic heartbeat is identified by an increase in power (for $j = 1$ and 2) above a threshold level (two times the mean power level of the control signal before air injection). (From Chan, B. C. et al., *IEEE Transactions on Biomedical Engineering*, 44(4), 237–246. Copyright 1997, IEEE.)

of the heartbeats above a set threshold, which they called the 'cumulative embolic power' (CEP), and the volume of air injected. They stated that such a relationship could prove important to the anaesthetist, allowing him or her to act only when a clinically significant volume of air embolism is present. The authors concluded that the wavelet transform of the heart sound signal can provide both fast detection of VEA and an accurate estimation of the embolic air present. Later work by Das et al. (2009) employed a Daubechies wavelet packet decomposition in a neural network-based diagnosis of valvular heart disease. See also Hanbay (2009), who have described a study which utilized both wavelet packet decomposition and fast Fourier transforms (FFTs) to extract features for an expert system to diagnose valvular heart disease.

6.5.2 Lung Sounds, Swallowing, Snoring and Speech

In an early paper by Sankur et al. (1996), a wavelet-based detector was employed to discriminate crackles in pathological respiratory sounds. They used a Daubechies D6 wavelet, which has a similar shape to the crackle waveform, and found their method to be superior to two existing crackle detection schemes. Wavelet analysis was exploited by Hadjileontiadis and Panas (1997) to separate discontinuous adventitious sounds (DAS) from vesicular sounds (VS) in pulmonary acoustic signals. Adventitious sounds were divided into two categories: continuous (wheezes and rhonchi) and discontinuous (crackles and squawks), and indicated an underlying physiological malfunction. The algorithm of Hadjileontiadis and Panas combines multiresolution analysis with hard thresholding to provide a technique to partition DAS from VS. The nonlinear nature of wheeze signals was examined later by Taplidou and Hadjileontiadis (2007) using wavelet bicoherence. Figure 6.45 contains a plot of an asthmatic breath signal exhibiting a prolonged exhalation phase, along with the corresponding Morlet wavelet transform magnitude. The researchers explored a

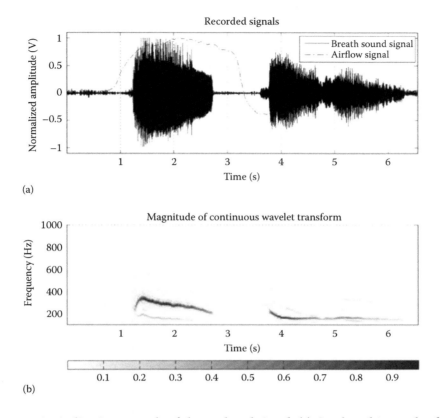

FIGURE 6.45 **An indicative example of the analyzed signal: (a)** One breathing cycle of a breath sound recording from an asthmatic patient with two dominant wheezes (one inspiratory and one expiratory); **(b)** time–frequency (TF) representation of the analyzed signal using the continuous wavelet transform. (From Taplidou, S. A., and Hadjileontiadis, L. J., *Computers in Biology and Medicine*, 37, 563–570, 2007.)

variety of coherence-based metrics including wavelet bispectrum (WBS), wavelet bicoherence (WBC), summed wavelet bicoherence (SWBC) and an evolutionary wavelet bicoherence (EWBC) (all of which they define in the paper). Figure 6.46 contains the evolutionary squared wavelet bicoherence which indicates the time evolution of the nonlinearities in the breath sound signal. From the figure, the authors concluded that phase coupling occurred in both breathing phases, only at the time instances where wheezing exists. Further work in this area concerning wavelet-based higher-order spectral features has been described by Taplidou and Hadjileontiadis (2010).

Hardware involving 14 microphones attached to the chest wall and a wavelet decomposition scheme to detect adventitious sounds from their signals has been detailed by Sen and Kahya (2005). More recently, Sakai et al. (2012) have provided a technique for extracting pulmonary sounds from low-quality auscultation signals via a sparse representation using a Daubechies wavelet. The method was shown to be highly robust against random noise and signal quantization. Input metrics from a number of different wavelets were utilized by Hashemi et al. (2011) to classify monophonic and polyphonic wheezing sounds. They

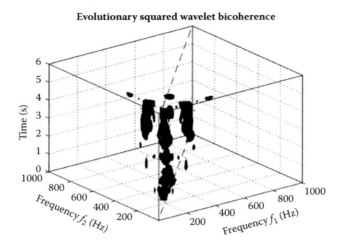

FIGURE 6.46 Evolutionary squared wavelet bicoherence for isosurface $b^2_w(f_1, f_2) = 0.2$. (From Taplidou, S. A., and Hadjileontiadis, L. J., *Computers in Biology and Medicine*, 37, 563–570, 2007.)

found biorthogonal wavelets offered better accuracy but that a feature set selected from among all of the wavelet decompositions performed best when selected. Bahoura (2009) used a number of feature extraction techniques, including wavelet transforms, in combination with a number of classification methods (vector quantization, Gaussian mixture models and artificial neural networks) to discriminate between wheeze and normal classes of respiratory sounds. Kandaswamy et al. (2004) applied wavelet-based neural networks to classify lung sounds into six categories: normal, wheeze, crackle, squawk, stridor and rhonchus. Interestingly, Yan et al. (2012) have reported on a Daubechies D4 wavepacket analysis of auscultation signals used in traditional Chinese medicine where it was found that they could differentiate between healthy subjects and those with deficiencies.

A number of groups have investigated the process of swallowing using wavelet methods. Sejdić et al. (2010) have proposed a denoising algorithm for dual-axis swallowing accelerometry signals. The algorithm employed a novel denoising approach comprising a search for an optimal denoising threshold within a reduced wavelet subspace and compared favourably with previous methods when applied to dry, wet and chin tuck swallows. They found Meyer wavelets performed best for the task. The same group Steele et al. (2013) again used Meyer wavelet decompositions to detect thin-liquid aspiration using dual-axis accelerometry signals from the front of the neck. Jestrovic et al. (2013) used the same technique when considering the effect of viscosity (of the swallowed substance) on the resulting swallowing sounds. Sazonov et al. (2010) proposed automated swallowing detection techniques based on the mel-scale Fourier spectrum, Coiflet-based wavelet packets and support vector machines. They found that the results suggested that the methods exhibited a high efficiency in separating swallowing sounds from artifacts originating from respiration, intrinsic speech, head movements, food ingestion and ambient noise.

The low-frequency modulations present in vowel sounds produced by Parkinson's disease patients has been the subject of investigation by Cnockaert et al. (2008) using a Morlet wavelet signal decomposition. They discovered differences in the modulations between male and female speakers and found that the modulation frequency was statistically higher and the ratio of the modulation energies in the (3–7 Hz) and (7–15 Hz) bands is significantly lower for Parkinsonian speakers of both genders when compared with healthy controls. More recent work by others includes that of Vilda et al. (2011), who performed a wavelet analysis of glottal features for use in voice biometry, Carvalho et al. (2011), who employed wavelet transforms and artificial neural networks to identify voice disorders and Basak et al. (2011), who described a wavelet-based wireless health monitoring system based on wearable sensors and vocabulary-based acoustic signal processing. This latter system collects and quantifies meaningful information on events and processes such as coughing, sneezing, activity level and sleep to diagnose symptoms and monitor healthy habits. Finally, time–frequency characteristics of the snoring signal have been investigated using discrete wavelets by Zhang et al. (2013b), who suggested the use of the technique to distinguish between benign and hypoxia-related snoring.

6.5.3 Acoustic Response

Otoacoustic emissions (OAEs) are acoustic signals emitted by the cochlea, either occurring spontaneously or in response to an acoustic stimulus, and reflect the active processes that are involved in the transduction of mechanical energy into electrical energy. Their form is related to the status of the cochlea and can be used to monitor cochlear functionality in patients exposed to prolonged noise and/or ototoxic agents. Tognola et al. (1998) have studied the acoustic response of the cochlea to acoustic stimuli of brief duration – specifically clicks of about 100 μs duration. The time–frequency properties of these click-evoked otoacoustic emissions (CEOAEs) have a close relationship with cochlear mechanisms. The authors compared various time–frequency analysis techniques, including the short-time Fourier transform, wavelet transforms, the Wigner distribution and two smoothed Wigner-Ville distributions: the pseudo-smoothed Wigner distribution and the Choi–Williams distribution. They found that, although there was no optimal method in an absolute sense, the wavelet transform offered the best compromise between time–frequency resolution and the attenuation of interference terms. Two examples of CEOAEs are shown in Figure 6.47: one for an adult and one for a full-term neonate. The otoacoustic emission response of the neonate exhibits a typical sustained burst-like behaviour up to 20 ms whereas the adult OAE shows clear frequency dispersion (i.e. reduction in high-frequency components in time). The wavelet scalograms corresponding to the CEOAEs in Figure 6.47 are shown in Figure 6.48a,b. The frequency dispersion of the adult signal is evident in the scalogram plot where low-frequency components have a longer duration in the scalogram plot and reach maximal amplitude at longer latencies than the high-frequency components. Figure 6.48c shows a scalogram for an adult suffering from noise-induced hearing loss. This hearing impaired patient had hearing loss greater than 30 dB above a frequency of 2.5 kHz. The lack of OAE response at frequencies above this 2.5 kHz threshold is evident in the scalogram plot associated with this patient.

A marked improvement in the detectability of neonatal CEOAEs in terms of SNRs in the lower-frequency ranges and whole wave reproducibility has been demonstrated by

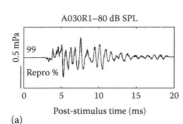
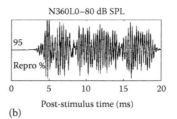

FIGURE 6.47 Click-evoked otoacoustic emissions (CEOAEs) from (a) a normal hearing adult and (b) a full-term baby. To reduce the influence of the stimulus artefact, responses have been windowed 2.5/20 ms post-stimulus time. In each row, two replicate recordings from the same ear (A and B replicate recordings in ILO equipment) are superimposed. Numbers on the left of each panel are the reproducibility values (in percentage points) between the two replicates. (From Tognola, G. et al., *IEEE Transactions on Biomedical Engineering*, 45(6), 686–697. Copyright 1998, IEEE.)

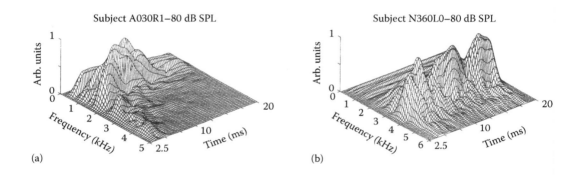
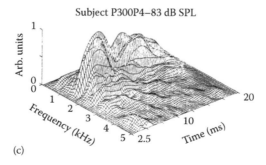

FIGURE 6.48 Time–frequency energy densities of CEOAEs: (a) Time–frequency distribution (energy density, normalized arbitrary units) of a CEOAE at 80 dB SPL of subject A030R1 (normal hearing adult). (b) Time–frequency distribution (energy density, normalized arbitrary units) of a CEOAE at 80 dB SPL of subject N360L0 (full-term neonate). (c) Time–frequency distribution (energy density, normalized arbitrary units) of a CEOAE at 83-dB SPL of subject P300P4 (suffering from noise-induced hearing loss). (From Tognola, G. et al., *IEEE Transactions on Biomedical Engineering*, 45(6), 686–697. Copyright 1998, IEEE.)

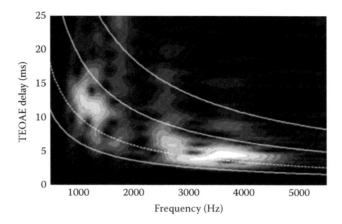

FIGURE 6.49 Time–frequency distribution of a TEOAE response to a 60 dB click stimulus. The main part of the response is concentrated along a quasi-hyperbolic stripe defined by the two lower solid lines. As in the SFOAE case, within this stripe, a further separation between two components of different latency is attempted (dashed line). A faint double latency component is also visible in a parallel stripe, attributable to double intra-cochlear reflection. (Reproduced with permission from Moleti, A. et al., *Journal of the Acoustical Society of America*, 132(4), 2455–2467. Copyright 2012, Acoustical Society of America.)

Zhang et al. (2011b). They employed a mask function in their wavelet-based scheme to suppress part of the time–frequency information in order to remove unwanted noise and preserve desired signal components. They stated that the technique has the potential to enhance recorded CEOAE responses, reducing false positive rates and hence reducing referral rates in neonatal screening programmes. Moleti et al. (2012) have examined transient evoked OAEs (TEOAEs), distortion product OAEs (DPOAEs) and stimulus frequency OAEs (SFOAEs). Figure 6.49 shows the wavelet transform plot for TEOAE response to a 60 dB click stimulus together with an expected hyperbolically delineated region in time and frequency superimposed on the plot. They concluded that their wavelet approach improves the SNR of any OAE response and may be applied to effectively separate the different components of the various OAEs studied. Geven et al. (2012) analyzed CEAOEs in a group of tinnitus patients comparing them to healthy controls. The influence of contralateral acoustic stimulation (CAS) was investigated in both ears of the subjects using a Gaussian wavelet-based analysis of the signals. They found that the suppression by CAS of CEAOEs in tinnitus patients was comparable with that in control subjects; that is, that there was no significant difference in suppression found between the two groups. Wavelet methods were also used in work with tinnitus patients by Paglialonga et al. (2011a) and, in a separate study, in Williams syndrome patients by Paglialonga et al. (2011b).

6.6 BLOOD FLOW AND BLOOD PRESSURE

Laser Doppler flowmetry (LDF) is a non-invasive technique for the measurement of cutaneous microcirculation. It was employed by Bračič and Stefanovska (1998) in an attempt to shed light on the cardiovasular control mechanisms using wavelet-based spectral methods.

They measured the peripheral blood flow in human skin over 20 minute periods using LDF. Subsequent decomposition of the acquired signals using the Morlet-based wavelet transform to form energy density scalograms revealed five characteristic frequency peaks. These local maxima in the wavelet-based energy spectrum, they hypothesized, can be attributed to the heart rate (1 Hz), respiratory activity (0.3 Hz), blood pressure regulation (0.1 Hz), neurogenic (0.04 Hz) and metabolic activity (0.01 Hz). They proposed a variety of statistical measures to characterize their wavelet-based power spectra and used them to reveal differences in the dynamics of the blood flow between two distinct groups: a control group of healthy young subjects and a group of athletes. They demonstrated that the increased blood flow in the trained subjects resulted from both the greater stroke volume and increased compliance of the peripheral vessels. In a related article, Kvernmo et al. (1998) compared the wavelet-based spectral analysis of these signals before and after exercise. In another related study, Kvernmo et al. (1999) used their methods to determine the effect of vasodilators (endothelium-dependent and endothelium–independent) on the oscillatory components present in these human cutaneous blood perfusion signals. A later paper by Jan et al. (2012) investigated the effect of local cooling on the metabolic and myogenic activities of the skin during prolonged loading periods. The study, involving rats, generated wavelet plots, shown in Figure 6.50, similar to the human plots in the Bracic and Stefanovska (1998) paper but with a shift to higher frequencies for the characteristic peaks. They defined the characteristic frequency ranges of interest for metabolic activity

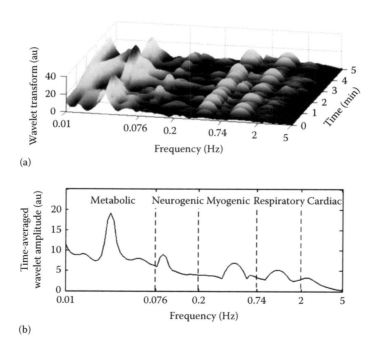

FIGURE 6.50 An example of (a) wavelet transforms of blood flow oscillations and (b) time-averaged wavelet amplitudes of blood flow oscillations. (Original in colour.) (From Jan, Y. K. et al., *Physiological Measurement*, 33(10), 1733–1745, 2012.)

(0.01–0.076 Hz), neurogenic (0.076–0.2 Hz), myogenic (0.2–0.74 Hz), respiratory activity (0.74–2 Hz) and heart rate (2–5 Hz). Their work demonstrated that local cooling provides a protective effect by reducing ischemia in weight-bearing soft tissues.

Humeau et al. (2000) have presented a model of laser Doppler flowmetry (LDF) signals produced when an arterial occlusion is removed. In such cases, the LDF signal increases then returns to its initial value. This phenomena is known as 'reactive hyperemia' and its study is important in the evaluation of the functional aspects of arterial blood flow. Humeau and coworkers obtained their model parameters from experimental LDF signals acquired from a healthy subject after a 2 minute vascular occlusion. These signals were denoised prior to parameter estimation using fourth-order Symmlett wavelets in a multi-resolution decomposition. Reconstruction of the denoised signal was performed only with those wavelet coefficients from scales which were noise free. Using a Morlet wavelet–based approach, Landsverk et al. (2007) demonstrated the effect of general anesthesia to reduce skin blood flow oscillations of the perfusion signal related to sympathetic, myogenic activity and the component modulated by the endothelium. More recently, Humeau et al. (2010) used a Mexican hat-based, wavelet transform modulus maxima analysis to determine the multifractal nature of LDF signals, first using synthetic signals of known multifractal properties and then on a set of 12 subjects. They found that the multifractal properties observed from both LDF and HRV fluctuations (indicative of the peripheral and central cardiovascular systems respectively) were close to the range of scales relevant to physiology. (This work was also mentioned earlier in Section 6.2.3 in the context of ECG-derived HRV.) In other work, Sun et al. (2012) detected vasomotor changes in skin microcirculation associated with severity of peripheral neuropathy in patients with type 2 diabetes by analyzing the wavelet spectrum split into three subintervals corresponding to endothelial-, neorugenic- and myogenic-related activities. Hsiu et al. (2011) has assessed the effects of acupuncture stimulation on local skin blood flows using Morlet mother wavelets and found that needling the Hegu acupoint significantly increased blood flows. Finally, Shiogai et al. (2010) have provided a comprehensive treatise concerning the nonlinear dynamics of cardiovascular aging revealed through wavelet analysis of LDF signals.

Doppler ultrasound uses reflected sound waves to examine the flow of blood through major arteries and veins. Doppler time series of middle cerebral artery flow velocities were explored by Sejdic et al. (2013) using a number of features, including a relative wavelet energy feature and a wavelet entropy feature. Exploiting a discrete Meyer wavelet, they demonstrated that these velocity signals can contribute valuable information regarding cerebral blood flow. Doppler ultrasonic blood flow signals from the wrist were investigated by Liu et al. (2012b) using features derived from a number of methods including Hilber–Huang, wavelet packet, wavelet transforms together with approximate entropy, time warp edit distance, auto-regressive model and fiducial point-based spatial features. These were all fed into a multiple kernel learning algorithm which showed promise in integrating multiple features in order to enhance classification performance when distinguishing between healthy patients and those with sugar diabetes, nephropathy and gastrointestinal diseases. Lockhart et al. (2011) and Hamilton et al. (2012) have undertaken analysis of pulsed Doppler blood velocity waveforms from the common carotid, ophthalmic and retinal arteries in a

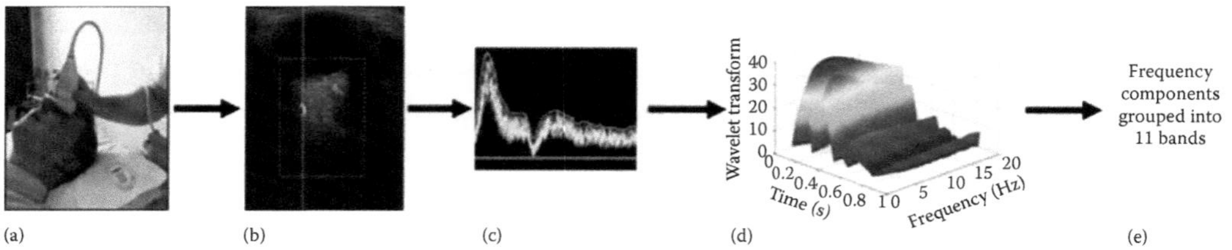

FIGURE 6.51 Steps undertaken during waveform analysis: (a) Position of ultrasound probe on closed eyelid. (b) Colour Doppler being used to locate ophthalmic artery. (c) Typical waveform from ophthalmic artery. (d) 3-D reconstruction of wavelet transform output following waveform analysis. (e) Wavelet transform frequency data is grouped into 11 frequency bands to facilitate analysis. (Original in colour.) (From Hamilton, P. K. et al., *British Journal of Diabetes and Vascular Disease*, 12, 40–47, 2012.)

cohort of 39 patients with type 1 diabetes. They decomposed the signal into 11 bands and showed statistically significant differences in some band amplitudes between the patients and healthy controls. Figure 6.51 shows the steps used in their acquisition and analysis of the signals. The same group (Hamilton et al. 2011) had less success when attempting to use flow-mediated dilation of the brachial artery as a reproducible indicator of microvascular function in type 1 diabetes patients. In other work on the nature of the Doppler ultrasound velocity waveforms, Agnew et al. (2015) have shown that wavelet entropy can detect subtle changes in the waveform structure in response to nitric-oxide mediated changes in arteriolar smooth muscle tone.

Cerebral autoregulation is the protective mechanism in the brain which regulates blood flow as blood pressures fluctuate. Its study lends itself to wavelet-based methods which compare relationships between signals (Addison, 2015b). Papademetriou et al. (2012) utilized wavelet cross-correlation methods to investigate the relationship between mean arterial pressure (MAP) and oxyhaemoglobin concentration (HbO_2) with a view to shedding light upon the cerebral autoregulation mechanism. Statistically significant differences between near-infrared spectroscopy (NIRS) channels placed on the right and left scalp indicated that the right hemisphere of the brain was more susceptible to the disruption of cerebral oxygenation. See also Payne et al. (2011), who employed the Morlet wavelet decomposition to determine an instantaneous time-varying phase difference signal from a blood pressure and NIRS signal. From the difference signal, they derived circular mean phase and synchronization index parameters; the latter being an inverse circular statistical analogue of variance (refer back to Chapter 2, Section 2.21.4). These parameters were used in their analysis of the relationship between the two signals. Considering both experimental and theoretical models, the Payne group found that autoregulation response was strongly dependent upon carbon dioxide partial pressure but less so on arterial oxygen saturation. More recently, Cui et al. (2014) have described a Morlet-based wavelet coherence technique to assess the relationship between cerebral tissue oxygen haemoglobin concentrations using NIRS signals and arterial blood pressures. They found a difference in wavelet coherence between elderly and young subjects indicating an altered cerebral autoregulation caused by aging. The same group (Gao et al., 2015) found that sit-to-stand positional changes in young and elderly subject groups manifested as significantly different wavelet phase coherence within certain frequency bands and argued that this age-related difference may prove useful in identifying risk for dynamic cerebral autoregulation processes. Rowley et al. (2007) have used Morlet wavelet cross-correlation (WCC) to interrogate the synchronization between blood pressure and oxyhaemoglobin concentration levels. They found that WCC typically exhibits two peaks for healthy controls, one of which (at 0.33 Hz) typically disappears during a head-up maneuver and the other (at 0.10 Hz) shifts to higher wavelet scales. However, for patients with autonomic failure, the 0.33 Hz peak is not significant and the 0.1 Hz peak does not shift significantly during the maneuver. The use of wavelet transform approaches to the NIRS-based analysis of cerebral autoregulation has recently been comprehensively reviewed by the author (Addison, 2015b). And, in new work, the use of synchrosqueezing the cross-wavelet transform to identify stable phase coupling between blood pressure and NIRS signals has been suggested (Addison, 2015c). The technique also employs low-oscillation complex wavelets (Addison et al., 2002a). (These methods are detailed in Chapter 2, Sections 2.20, 2.21 and 2.11 respectively). Finally, in a non-NIRS approach to the investigation of autoregulation, the wavelet spectral energy of oscillations in intracranial blood pressure (ICP) in acute traumatic brain injury patients was evaluated by Kvandal et al. (2013). They also assessed wavelet phase coherence and phase shift between the ICP and arterial blood pressure (ABP) signals. They found that a phase shift between ICP and ABP in the interval 0.07–0.14 Hz indicated normal cerebrovascular reactivity while a phase shift in the 0.006–0.07 Hz region indicated altered reactivity.

Blood pressure is the pressure exerted by the circulating blood against the walls of the blood vessels. It usually refers to the arterial pressure in the upper arm, but other common blood pressures include central venous pressure, jugular venous pressure, pulmonary artery pressure and intracranial pressure. Morlet wavelet spectra combined with an entropy-based complexity measure were used by Pavlov et al. (2009) to study gender-related differences in the dynamic response of rat blood pressure signals to stress and nitrous oxide (NO) deficiencies. They found that female rats demonstrated more favourable patterns of cardiovascular responses to stress and more effective NO control of cardiovascular activity than males. Postolache et al. (2009) have used DWT processing together with both radial basis function and multilayer perceptron neural network methods to analyze simultaneous heart rate and blood pressure variation signals in their work to develop a rapid evaluation tool for the assessment of the cardiovascular autonomic nervous system control in rats. Rocha et al. (2010) employed a Haar wavelet within a neural network method to forecast blood pressure time series in order to predict hypotensive episodes. Their technique produced high sensitivities and specificities (94.7% and 93.6%) in predicting acute hypotensive events in a 50 record patient dataset. More recently, Garg et al. (2013) used Morlet-based wavelet coherence to identify the presence of coupling between blood pressure and calf muscle EMG signals. Figure 6.52 contains a plot of their wavelet transform coherence, together with a band coherence measure derived from averaging the output over scales in the corresponding frequency ranges. They found that the method effectively identified the presence of linear coupling between the EMG and blood pressure signals during quiet standing.

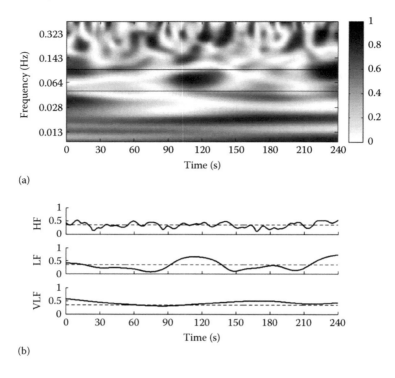

FIGURE 6.52 (a) Time–frequency map of coherence between EMG and SBP obtained from one participant using the wavelet transform coherence (WTC) analysis method. (b) Three plots of band coherence obtained from averaging over the corresponding frequency bands, with the solid line representing the coherence between EMG and SBP, and the dashed straight line representing the significance level for the corresponding frequency band. (From Garg, A. et al., *Biomedical Engineering Online*, 12(132), 1–14, 2013.)

6.7 MEDICAL IMAGING

There has been considerable research concerning the application of wavelets to the denoising, visual enhancement and compression of medical images. A wide variety of medical images, including magnetic resonance imaging (MRI), positron emission tomography (PET), computed tomography (CT), ultrasonic and optical images, have been considered by numerous research groups. This section provides only a very brief selection of examples of the application of wavelet approaches to the elucidation of medical images.

6.7.1 Optical Imaging

Laser speckle contrast imaging was employed by Holstein-Rathlou et al. (2011) to measure the blood flow dynamics of up to 100 nephrons simultaneously on the renal surface of anaesthetized rats. They found that synchronization may take place among nephrons not immediately adjacent on the kidney surface. A fractal analysis of the retinal vasculature was performed using Gabor wavelet transforms by Che Azemin et al. (2011). Figure 6.53 shows a retinal image probed using the Gabor wavelet to produce a surface of maximum response (also shown). The group went on to find a correlation between the reduction in the retinal

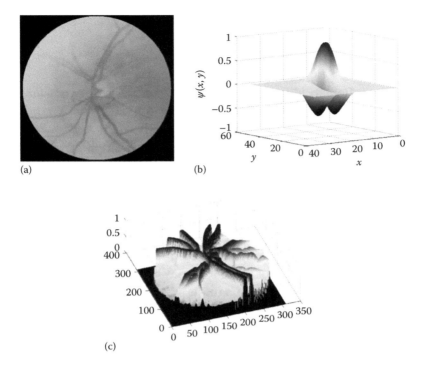

FIGURE 6.53 Steps for vessel enhancement: (a) Original retina image cropped at a 200 pixel radius from the optic disk centre. (b) Gabor wavelet at 0°orientation. (c) Surface of the maximum responses of the Gabor wavelet transform over 18 angles. (Original in colour.) (From Che Azemin, M. C. et al., *IEEE Transactions on Medical Imaging*, 30(2), 243–250. Copyright 2011, IEEE.)

vascular complexity with age. An automated tool for classifying human carcinoma cells in microscopic images using complex wavelet transform-based methods has been developed by Keskin et al. (2013). They found that their technique outperformed classical covariance-based schemes. The study of the diffuse optical imaging of the oxy- and deoxyhemoglobin levels in the brain using near-infrared light has been described by Abdelnour et al. (2009) where three-dimensional images of the oxygen levels in the brain are reconstructed using surface-based wavelets and prior knowledge of MRI structural images. The paper contains a number of 3-D image representations of the brain with superimposed blood oxygenation information.

A number of authors have turned their attention to medical thermography where highly detailed infrared images of the human body are used for diagnosis. For example, Etehadtavakol et al. (2013) have explored the use of both real and complex DWT-based texture features for the classification of breast thermograms, Cooke et al. (2011) briefly mentioned the use of a wavelet filter to reduce noise artifact in the thermographic imaging of traumatic injuries and Ghosh et al. (2010) have compared a genetic algorithm approach with the segmentation of thermographic hand images with a range of methods including a wavelet-based technique. In a slightly different vein, Abbas et al. (2011) have extracted respiratory time series from the thermographic images of neonates which they probed using continuous (Daubechies) wavelet analysis.

6.7.2 Ultrasonic Images

A detailed investigation to identify the optimal mother wavelet transform for use in ultrasound image denoising has been carried out by Adamo et al. (2013). This included Daubechies, Coiflet, Symmlett and biorthogonal wavelet families. The authors tested the wavelets on kidney, heart and liver images corrupted with multiplicative speckle noise and discovered that the orthogonal families generally outperformed the biorthogonal ones. Complex Daubechies wavelets were found by Khare et al. (2010) to be a clear improvement over other methods in the despeckling of ultrasound images of the hepatic vein, portal vein, gall bladder, kidney and liver. Tsiaparas et al. (2011) have compared multiresolution approaches for the classification of atherosclerositic tissue by employing a number of basis functions from different wavelet families. They found that the dominant texture feature exhibited directionality suggestive of a biomechanical factor at play in the form of plaque straining. A review of computer-aided diagnosis systems for the detection and classification of breast cancer, which includes the application of wavelet methods to this problem, has been conducted by Cheng et al. (2010). In another more recent review, Sudarshan et al. (2016) have provided a comprehensive account of the use of wavelet techniques for cancer diagnosis in ultrasound images.

The use of wavelet packet methods to process photoacoustic (PA) images (where acoustic responses are generated from tissue excited by absorbed light energy) has been described by Zalev and Kolios (2011). They used the technique to discriminate between normal and abnormal vasculature based on features generated from the wavelet packet analysis of its structural morphology. In other work, Hirasawa et al. (2013) have proposed an approach to quantifying optical absorption coefficients from PA signals using the Morlet-based CWT.

6.7.3 Computed Tomography, Magnetic Resonance Imaging and Other Radiographic Images

Early work by Zheng et al. (2000) tested the effect of the digital compression of CT images on the detection of coronary artery calcifications (CAC). They found that images compressed up to 20:1 using both JPEG and wavelet algorithms were acceptable for primary diagnosis of CAC by experienced radiologists. More recently, Wu and Lin (2013) have proposed a stationary wavelet transform method for automatically extracting parotid lesions in head and neck CT images; Chen et al. (2012b) used the stationary wavelet transform to suppress streak artifacts in thoracic low-dose CT images and found it less obscuring of the original low-frequency information which suffers from noise; and Arai et al. (2011) have described a technique for extracting line features from the multifidus muscle in order to determine its area. They found that Daubechies wavelets were superior to Haar wavelets in this regard. They also found a high correlation between the muscle area with visceral fat. In other work, a novel segmentation approach based on a wavelet density model has been described by Chang et al. (2013) for use in cone-beam computed tomography (CBCT).

Wavelet-based schemes for MRI analysis have found widespread use in enhancing diagnostics across a range of specialities in medicine. Saritha et al. (2013) combined wavelet-based spider entropy plots and neural networks to classify MRI brain images (normal, stroke, infectious disease, degenerative disease and tumor). They found a marked reduction in features that resulted from the method and they obtained a 100% accuracy detecting normal images and degenerative disease images from other classes of images. Li et al. (2012) have applied a Daubechies wavelet–based k-means clustering method to segment the renal compartments in a 3-D dynamic contrast-enhanced (DCE-)MRI of the human kidney (Figure 6.54). They advocated the method as a feasible tool for the automated perfusion and glomerular filtration rate quantification. An improved compressed sensing scheme for coronary MRI has been described by Akçakaya et al. (2011) which exploits the dependencies of wavelet coefficients as well as their sparsity. In other MRI work, Zhang et al. (2011c) have employed a wavelet-based neural network classification technique for MRI brain image classification; Hassanien and Kim (2012) used a hybrid approach including fuzzy sets, neural networks and wavelet-based feature extraction to diagnose breast cancer from MRI scans; Tafti et al. (2010) have briefly described a wavelet method for probing the fractal structure of phase contrast MRI flow field images; and Hackmack et al.

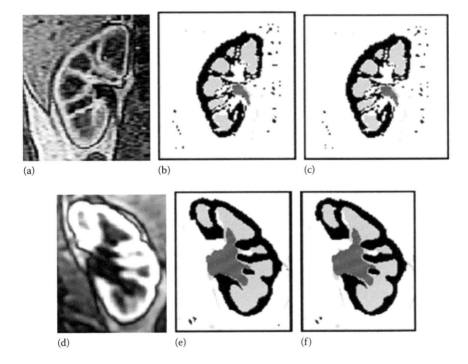

FIGURE 6.54 Segmentation result: (a, d) The original kidney image. (b, e) The segmentation result of proposed wavelet-based clustering. (c, f) The segmentation result of k-means. The black part depicts the renal cortex, the light grey part shows the renal medulla and the dark grey part represents the pelvis. Upper row data from data set 1, slice 10 of 20, acquired at 1.5 T; lower row middle slice of data set 3, acquired at 3.0 T. (From Li, S. et al., *Computerized Medical Imaging and Graphics*, 36, 108–118, 2012.)

(2012) have provided details of a new class of features generated by taking local scale information and directionality into account in wavelet coefficients calculated from MRI images. They used this approach to discriminate between normal healthy subjects and those with multiple sclerosis.

Functional MRI (fMRI) has also benefited from wavelet techniques. For example, Skidmore et al. (2011) used wavelet analysis and graph analytic techniques to probe functional connectivity between subcortical and cortical regions in a study of Parkinson fMRI data and Khalidov et al. (2011) employed novel 'activelets' sparse representation wavelet analysis to investigate the blood oxygen level dependent (BOLD) fMRI signals.

A number of groups have applied wavelet techniques in the processing of PET images. Stefan et al. (2012) improved the signal-to-noise ratio in PET scans using a redundant discrete wavelet transform and found that this provided sharper restorations than other reconstruction methods. Knešaurek et al. (2009) demonstrated that Daubechies-based wavelet temporal smoothing significantly improved the repeatability of myocardial blood flow (MBF) and coronary flow reserve (CFR) assessment in PET imaging. And, in other work, a combined wavelet and curvelet transform-based denoising technology for PET images has been developed by Le Pogam et al. (2013) where the curvelet transform accounts for the non-optimal processing of edge continuities of the wavelet transform.

6.8 OTHER APPLICATIONS IN MEDICINE

6.8.1 Electromyographic Signals

Electromyographic signals (EMG) represent the electrical activity of muscle during contraction. A number of authors have applied wavelet techniques in the investigation of muscle fatigue. Chowdhury et al. (2013), for example, have assessed neck and shoulder muscle fatigue from surface EMG (sEMG) during stacking exercises. They employed 10 common wavelet functions in a DWT method and found that wavelet coefficients corresponding to the 12–23 Hz frequency band demonstrated the highest sensitivity to fatigue due to the dynamic repetitive exertions. Other authors have also examined fatigue, including Yochum et al. (2012), who produced a wavelet-based fatigue index, Herrera et al. (2011), who examined a chaotic model of the EMG, Zhang et al. (2012a), who considered the influence of fatigue on surface EMG-based human computer interfaces and Wu et al. (2016), who produced a bacterial foraging, particle swarm optimization method for the diagnosis of fatigue status from the EMG that used input features which included those derived by wavelet pre-processing.

Morlet-based cross-wavelet power spectra have been developed as a tool for studying interference and rectified EMGs by Neto et al. (2010). They demonstrated that only the interference EMG could accurately capture the increase in oscillatory drive through changes in the beta and Piper band activity in the EMG with voluntary force. The same group (Pereira et al., 2010) used normalized Morlet wavelet spectra to investigate the EMG from agonist, antagonist and synergist muscles during a heel raise task sustained to failure. Other examples of the application of the wavelet transform to EMG analysis include Rafiee et al. (2011), who utilized a mother wavelet matrix incorporating 324 different wavelet functions to extract features for use in a forearm EMG signal for prosthetics; Fairley et al. (2011), who created an automated system for the assessment of the phasic EMG metric in Parkinson's disease patients; Stirling et al. (2011), who examined the underlying temporal rhythms (Piper rhythms indicative of central control) in muscles during running using nonlinearly scaled wavelet; and Vannozzi et al. (2010), whose Morlet-based automatic, user-independent algorithm detects onset and offset of surface EMG bursts in both simulated and patient data.

In order to improve the processing of the EMG, von Tscharner et al. (2011) developed an ECG suppression method which combines wavelets and independent component analysis which does not eliminate low-frequency EMG signal (as a standard Butterworth filter does) and does not require an ECG reference signal. Niegowski and Zivanovic (2016) have detailed an approach for the suppression of the ECG in the surface EMG using a wavelet-based unsupervised learning technique. In addition, the efficient storage and transmission of surface EMG has been tackled by Trabuco et al. (2014) using a dynamic bit allocation scheme to encode discrete wavelet coefficients and entropy coding to minimize redundancy and pack the data efficiently.

6.8.2 Posture, Gait and Activity

A number of researchers have focused on posture characterization, including Kirchner et al. (2012), who used Coiflet and biorthogonal wavelets in an analysis of the temporal pattern of body sway; Lockhart et al. (2013), who employed a Daubechies wavelet denoising method to elucidate postural events and transition durations in signals acquired by inertial measurement units; and and Mataar et al. (2013), who examined the use of the stabilogram as a behavioural biometric marker for the automatic recognition of individuals using discrete wavelet decomposition.

An analysis of accelerometer signals acquired from the ankles of healthy volunteers to identify specific gait events has been described by Khandelwal and Wickstrom (2014). Using a symlet wavelet–based technique, they identified 'toe off' and 'heel strike' events in the acceleration signals. Post-abdominal surgery gait changes have been examined by Atallah et al. (2013), who developed an approach based on DWT feature extraction from ear-worn activity signals. Lee and Lim (2012) have developed a Haar-based technique to discriminate Parkinson's disease patients from healthy controls using vertical ground reaction force signals from a series of foot-based sensors. They derived features from the wavelet coefficients and input them into a neural network with fuzzy membership functions. A 2-D Haar-based DWT approach for detecting human gender from moving images has been described by Arai and Andrie (2011) and Arai and Asmara (2014), attaining high classification accuracies. Xue et al. (2010) have also used image analysis of walking humans to classify their task-dependent gaits by employing a Haar-based decomposition of infrared images.

Activity classification was undertaken in chest-mounted triaxial accelerometer signals using a discrete Daubechies wavelet by Godfrey et al. (2011). Using a threshold applied to the negative peaks of the signal, episodes of walking were correctly identified with 100% accuracy. A technique to differentiate falling from other activities such as walking, sitting and door slamming has been developed by Yazar et al. (2013) using single tree complex wavelet transform analysis of vibration sensor signals. They found that the wavelet approach was superior to Fourier transform and mel-cepstrum feature extraction approaches. And in other work, an activity recognition programme has been developed by Mashita et al. (2012) for incorporation within smartphones, so that such devices can detect the activity context of the user and use this information to switch the content, interface or methods of display accordingly. They analyzed phone accelerometer signals using Gabor, Haar and Mexican hat wavelets and compared the results with an FFT-based technique. It was found that the Mexican hat wavelet-based methodology produced the best recognition rate.

6.8.3 Analysis of Multiple Biosignals

A number of authors have focused on the analysis of the relationship between biosignals, some of whom have already been cited earlier in this chapter, including the work on autoregulation and aging by Cui et al. (2014), autoregulation and acute traumatic brain injury by Kvandal et al. (2013) and coupling between blood pressure and calf muscle EMG signals by Garg et al. (2013) (Figure 6.52). Other examples of the use of wavelet methods to compare biosignals include the work of De Melis et al. (2007), who compared optical vibrometry signals with cardiophonograms and Bernjak et al. (2012), who have used wavelet phase coherence to investigate the coherently coupled skin and muscle sympathetic nerve oscillations over time by simultaneously measuring peroneal nerve skin and tibial nerve muscle sympathetic activities, along with the ECG, photoplethysmogram, arterial pressure, respiration, blood flow and temperature. This latter group found strong support for the existence of multiple central sympathetic neural oscillators in human subjects.

A number of authors have concentrated specifically on cardiorespiratory interactions using heart rate and respiratory signals. For example, Petrock et al. (2008) employed the cross-wavelet transform of the HRV and respiratory signal to quantify weak correlations in the cardiorespiratory system and Keissar et al. (2008) used Morlet-based wavelet coherence to elucidate the interaction between heart rate and respiratory signals. (See also the reference to Keissar et al., 2009a, in Section 6.2.3 and Figure 6.9.) More recently, in an attempt to provide a non-invasive tool for monitoring the response to noxious stimuli (nociception) during surgery, Brouse et al. (2011) examined the potential of wavelet transform cardiorespiratory coherence to detect movement events in pediatric patients undergoing general anaesthesia. Using a wavelet coherence threshold of 0.7, they obtained a 95% sensitivity in the ability to detect patient movement (a sign of patient nociception).

6.8.4 Miscellaneous

More and more areas of medicine are opening up to the wavelet method. The following are a few further examples from the literature:

- Li et al. (2011c) constructed a symlet wavelet-based denoising and dynamic peak-picking algorithm to count labelled cells within blood flow passing a cytometer in an *in vivo* study of the depletion kinetics of circulating metastatic cancer cells in mice.

- Leise et al. (2013) examined how wavelets may be employed in the analysis of actograms used in biological rhythm research.

- Papaioannou et al. (2012) applied wavelet energy and entropy features to investigate the temperature variability in three classes of patients (systemic inflammatory response syndrome [SIRS] sepsis and septic shock) finding significant statistical differences between infectious and non-infectious inflammatory states.

- Arvinti-Costache et al. (2011) detailed a wavelet-based method, employing both the Haar and Coiflet wavelets, to remove baseline drift in fetal magnetocardiograms. (The relative performance of Fourier and wavelet-based approaches for filtering these signals has also been described by Arvinti et al., 2011.)

- Klein et al. (2012) used a Meyer wavelet to assess the minute-to-minute urine flow variability which they proposed may be useful for the diagnosis of hypovolemia.

- Brazhe et al. (2008) performed Morlet wavelet decompositions in order to analyze the time-dependent frequencies of observed optical path differences in a laser interference microscopy study of nerve fibres.

- Obeid et al. (2010) extracted heartbeat information from microwave Doppler radar signals using a Daubechies wavelet technique which exhibited high accuracy in determining heart rate and heart rate variability.

- Tagluk et al. (2010) classified sleep apnea types into obstructive, central and mixed using a neural network whose inputs were DWT coefficients.

- Bhatawadekar et al. (2013) utilized Daubechies wavelets to remove artefacts from signals acquired during the forced oscillation techniques (FOT) used to determine mechanical properties of the respiratory system.

- Ponnui et al. (2012) developed an automated expert system for urolithiasis classification based on Daubechies wavelets which successfully de-trends the infrared spectra used in the technique.

- Pandey and Pandey (2007) applied Meyer wavelet–based denoising to cancel corrupting respiratory artifacts from impedance cardiography signals used to better determine an estimate of cardiac output.

- Podtaev et al. (2012) developed a scheme for extracting wave amplitude information from impedance cardiography signals and used it successfully as a discriminator on signals from patients with and without hypertension.

- Zhao and Davis (2009) demonstrated that wavelet analysis aided an ant colony algorithm by providing a lower dimensional search space, thus expediting the searching of pertinent features in mass spectral data used in ovarian cancer diagnostics.

- Shi et al. (2011) distinguished between mitochondrial and chloroplast proteins in a study of the prediction of protein locations at the subcellular level using a biorthogonal wavelet.

- Liu et al. (2013) described a method to extract features of colorectal cancer data where protein markers, used in the prediction of patient survival, were located based on optimized CWT-based features.

- Meeker et al. (2011) demonstrated, using a Morlet-based approach, that uncoupled suprachiasmatic nucleus (brain) cells are capable of a wide range of time-varying periodic behaviours where at least 80% of the cells exhibited some circadian rhythms.

CHAPTER 7

Fractals, Finance, Geophysics, Astronomy and Other Areas

7.1 INTRODUCTION

This final chapter covers a variety of subject areas in briefer detail than the preceding chapters. Most of the chapter is devoted to four main topics: fractals, finance, geophysics and astronomy. First, we will look at how we can use wavelet transforms to characterize the scaling properties of self-similar fractal and multifractal objects. After this, we will consider the emerging role of wavelet methods in financial analysis. The next section of the chapter is devoted to geophysics – where wavelet transform analysis began with the analysis of seismic signals. Following this, the use of the wavelet transform in astronomy is reviewed. Finally, in the last section, we will briefly explore a selection of other areas where wavelet analysis has made an impact that have not been covered within the rest of the book.

7.2 FRACTALS

Fractals are objects which display self-similarity over scales. These objects can be exactly self-similar, as in Figure 7.1a, where the exact form of the object is repeated at smaller and smaller scales, or they can be statistically self-similar (Figure 7.1b), where the statistical properties of the object are consistent across scales (Addison, 1997). Many natural fractals are statistically self-similar, for example, coastlines, cracking, tree branching, stock market indices, permeabilities in the subsurface, the distribution of galaxies and so on. Previous chapters have already touched upon some natural phenomena whose fractal properties have been interrogated using wavelet methods; for example, bubbly fluid flows and sediment transport phenomena (Chapter 4), engineering surface characterization and chaotic attractors (Chapter 5), heart rate variability, phonocardiogram murmur detection, laser Doppler flow signals and retinal vasculature images (Chapter 6). All natural fractals exhibit self-similarity only over a finite range of scales and hence, unlike their regular mathematical counterparts, their fractal description eventually breaks down. They do,

338 ■ The Illustrated Wavelet Transform Handbook

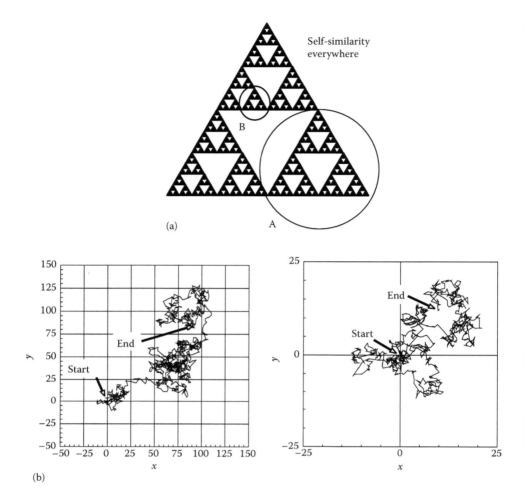

FIGURE 7.1 Exactly self-similar and statistically self-similar fractals: (a) The Sierpinski gasket: A fractal object which is exactly self-similar over all scales. (Each of the circles contains self-similar parts of the whole gasket at different scales.) (b) The two-dimensional trajectory of ordinary Brownian motion (The right-hand plot contains the first 1/16th of the trajectory of the left-hand plot blow up to maintain the same degree of resolution between each plot.) (From Addison, P. S., *Fractals and Chaos: An Illustrated Course*, CRC Press, Bristol, 1997.)

however, exhibit these fractal properties over a sufficiently large range of scales to allow fractal geometric methods to be usefully employed in their description. The property of self-similarity across scales makes wavelet transform analysis a natural candidate for the interrogation of such objects.

7.2.1 Exactly Self-Similar Fractals

Figure 7.2a shows another common exactly self-similar fractal, the triadic Cantor set. The construction method is shown in the plot where, at each step in the generation of the fractal set, the middle third is removed from the remaining line segments. The iteration process begins on the unit line and proceeds *ad infinitum* to construct the set. Figure 7.2b shows a

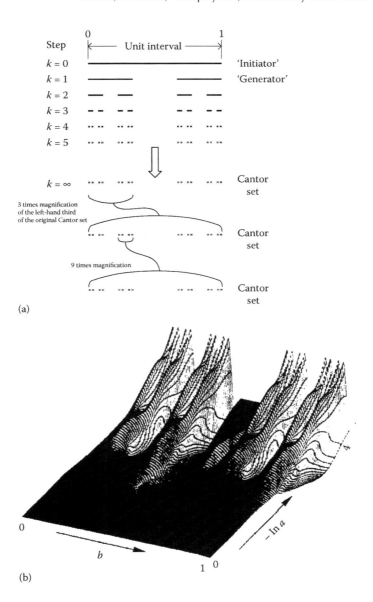

FIGURE 7.2 **Wavelet analysis of the triadic Cantor set:** (a) Construction of the triadic Cantor set. (From Addison, P. S., *Fractals and Chaos: An Illustrated Course*, CRC Press, Bristol, 1997.) (b) Transform plots for the Triadic Cantor Set: Note that $(\text{sgn}(T).|T(a,b)|^{1/2})$ is plotted against $\ln(a)$ and b. (From Arneodo, A. et al., *Wavelets*, 182–196. Springer, Berlin, 1989.)

3-D plot of a Mexican hat wavelet transform for the triadic Cantor set (Arneodo et al., 1989). The branching structure of the set is easily seen in the transform plot. Figure 7.3 contains a regular snowflake fractal generated in the plane together with transform plots at three a scales using the radial (2-D) Mexican hat wavelet (Argoul et al., 1989). Contour lines, set at an arbitrary value, show the construction rule of the fractal snowflake in Figure 7.4. As the scale of the wavelet tends to zero, the transform plot approximates more and more the

340 ■ The Illustrated Wavelet Transform Handbook

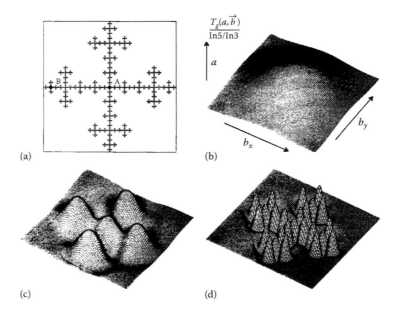

FIGURE 7.3 **The wavelet transform of a regular snowflake.** The snowflake is shown in **(a)**. The scale parameter a is successively divided by the same factor $l = 3$: $a = a^*$ **(b)**, $a = a^*/3$ **(c)**, $a = a^*/3^2$ **(d)**. $T(a,b)$ is expressed in $a^{\log(5)/\log(3)}$ units in order to reveal the self-similarity of the geometry of the snowflakes. (From Argoul, F. et al., *Physics Letters A*, 135(6), 327–336, 1989.)

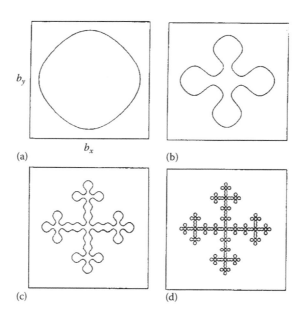

FIGURE 7.4 **Isocontour lines of the wavelet transform of the snowflake fractal.** The isocontour line $T(a,b)/a^{\log(5)/\log(3)} = k$ (k arbitrary chosen) for different values of the scale parameter: $a = a^*$ **(a)** $a = a^*/3$ **(b)**, $a = a^*/3^2$ **(c)**, $a = a^*/3^3$ **(d)**. (From Argoul, F. et al., *Physics Letters A*, 135(6), 327–336, 1989.)

snowflake itself. See also Antoine et al. (1997), who analyzed a fractal Koch curve in their paper concerning the characterization of shapes using the information gained from the maxima lines (both modulus maxima and ridges) of the continuous wavelet transform. The Sierpinski gasket, Cantor set, snowflake fractal and the Koch curve are exactly self-similar. However, very few examples of exact self-similarity are to be found in nature, (e.g. some fern shapes exhibit nearly exact self-similarity over a few scales). Most natural fractals exhibit stochastic self-similarity and consequently most research concerning fractal objects and processes in nature focuses on stochastic fractals.

7.2.2 Stochastic Fractals

Figure 7.5 shows a sequence of shots for a diffusion limited aggregation (DLA) cluster and the associated wavelet transforms. DLAs are essentially stochastic snowflakes *grown* using a numerical technique which allows particles to wander randomly about the plane (in a Brownian motion) until they encounter the DLA aggregate, sticking to it on impact (Addison, 1997). DLA has been used to model a variety of physical phenomena including bacterial colonies, viscous fingering and electrochemical deposition. See also an extension

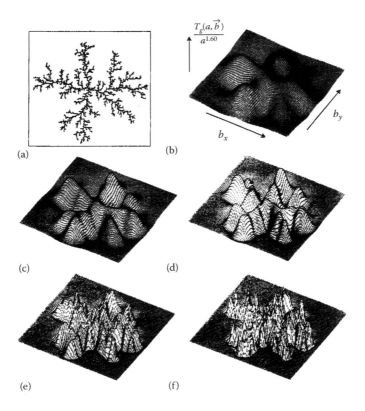

FIGURE 7.5 Wavelet transforms of a DLA cluster. The DLA cluster is shown in (a). The scale parameter a is successively divided by the same factor $\delta = 1.55$: $a = a^*$ (b), $a = a^*/\delta$ (c), $a = a^*/\delta^2$ (d), $a = a^*/\delta^3$ (e), $a = a^*/\delta^4$ (f). $T(a,b)$ is expressed in $a^{1.60}$ units in order to reveal the self-similarity of the geometry of the DLA clusters. (From Argoul, F. et al., *Physics Letters A*, 135(6), 327–336, 1989.)

to this type of work contained in the paper by Pei et al. (1995) and the later study of DLA on a linear substrate using a Haar wavelet by Postnikov et al. (2010).

One area where wavelet transform analysis has been used extensively is in the determination of the scaling properties of fractional Brownian motion (fBm) (and its derivative, fractional Gaussian noise, fGn). Over recent years, these correlated random functions, first proposed by Mandelbrot and Van Ness (1968) as a generalization of Brownian motion, have been suggested as models for a whole range of natural phenomena including DNA sequences, geometrical tolerancing in mechanical design, risk analysis, polymer models, landscape surfaces, image textures, the dynamics of nerve growth, crack profiles, permeability fields in porous media and non-Fickian diffusive processes (Addison and Ndumu, 1999). The *smoothness* of the fBm function increases with the Hurst exponent H, which can vary in the range $0 < H < 1$. Three examples of fBm, denoted $B_H(t)$, are shown in Figure 7.6 for different H's. Fractional Brownian motions are non-stationary random processes where the standard deviation, σ, of the fBm trace deviations ΔB_H (Figure 7.6c) taken over a sliding window of length, s, scales as

$$\sigma \propto s^H \tag{7.1}$$

$H > 0.5$ corresponds to persistent fBm where the trace has a tendency to persist in its progression in the direction in which it was moving, $H < 0.5$ corresponds to anti-persistent fBm where the trace has a tendency to turn back upon itself and $H = 0.5$ corresponds to regular Brownian motion where the trace is free to move in either direction from step to step (for more information see Mandelbrot, 1982, or Addison, 1997). The fractal dimension of an fBm trace function can be found from the Hurst exponent through the simple relationship:

$$D = 2 - H \tag{7.2}$$

There are many methods used in practice to determine either H or D for experimental data suspected of fBm scaling. One common method uses the Fourier power spectrum $P_F(f)$, which for an fBm, should scale as

$$P_F(f) \propto f^{-(2H+1)} \tag{7.3}$$

Hence, a logarithmic plot of power against frequency allows H to be determined from the slope of the spectrum. Wavelet power spectra, $P_W(f)$, both continuous and discrete, also exhibit this scaling and can therefore also be used to determine H (and hence D if required). Figure 7.7 shows a single realization of an fBm trace with $H = 0.6$ together with both its Fourier and wavelet power spectra (refer back to Chapter 2, Section 2.9). A continuous Mexican hat wavelet was used in the decomposition. The slope corresponding to −2.2 (i.e. $H = 0.6$) is plotted on the graph for comparison. We can see from the plot that the wavelet spectrum is much smoother due to the finite bandwidth of the spectral components associated with the wavelets. This can have an advantage when trying to abstract the fractal

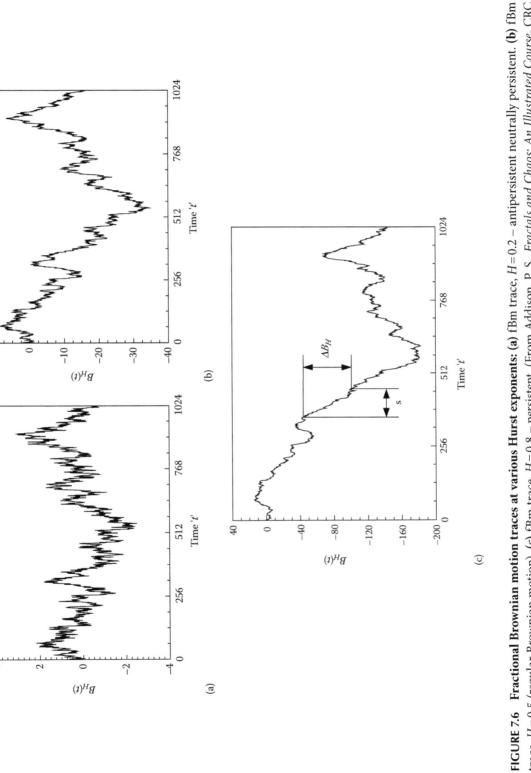

FIGURE 7.6 Fractional Brownian motion traces at various Hurst exponents: **(a)** fBm trace, $H = 0.5$ (regular Brownian motion). **(c)** fBm trace, $H = 0.8$ – persistent. (From Addison, P. S., *Fractals and Chaos: An Illustrated Course*, CRC Press, Bristol, 1997.)

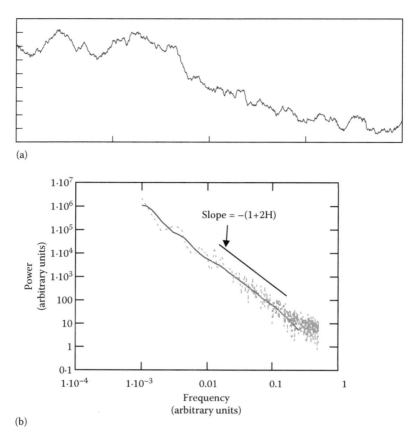

FIGURE 7.7 **Fourier and wavelet spectra of fractional Brownian motion:** (a) A single realization of an fBm trace function $H = 0.6$. (b) Fourier (dashed line) and wavelet (smooth line) power spectra of the trace in (a).

information from such data sets when only a limited number of data are available – too few to smooth the Fourier spectrum through ensemble averaging.

We can compute the fractal scaling characteristics of a data set suspected of exhibiting fBm scaling from its wavelet decomposition. The mean absolute discrete wavelet coefficient value scales as

$$\langle |T_{m,n}| \rangle_m \propto a_m^{\left(H+\frac{1}{2}\right)} \tag{7.4}$$

where

$$\langle |T_{m,n}| \rangle_m = \frac{\sum_{n=0}^{2^{M-m}-1} |T_{m,n}|}{2^{M-m}} \tag{7.5}$$

is the mean absolute value of the wavelet coefficients at scale a_m. Alternatively, we may use the scaling of the variance of discrete wavelet coefficients at scale index m, which as we saw in Chapter 4, Section 4.2.1, is given by

$$\left\langle T_{m,n}^2 \right\rangle_m = \frac{\sum_{n=0}^{2^{M-m}-1} (T_{m,n})^2}{2^{M-m}} \tag{7.6}$$

From Chapter 4, Equation 4.10b, we know that coefficient variance is simply related to the power spectrum as follows:

$$P_W(f_m) \propto \left\langle T_{m,n}^2 \right\rangle_m \tag{7.7}$$

Combining this expression with the power law relationship for fBm of index H given in Expression 7.3 and the fact that frequency f is inversely proportional to the wavelet scale a ($=2^m$), we obtain the scaling relationship:

$$\left\langle T_{m,n}^2 \right\rangle_m \propto a_m^{(2H+1)} \tag{7.8a}$$

In the literature, σ_m^2 is often used as a compact notation for the variance of the discrete wavelet coefficients at index m, hence, the relationship is written

$$\sigma_m^2 \propto a_m^{(2H+1)} \tag{7.8b}$$

If we take the square root of both sides of Expression 7.8b to get

$$\sigma_m \propto a_m^{H+\frac{1}{2}} \tag{7.9}$$

we can see that the scaling of the coefficient standard deviations is consistent with that given by Expression 7.4. We expect this, as both the mean absolute value of the coefficients and the standard deviation of the coefficients are first-order measures of spread. Furthermore, for an orthonormal multiresolution expansion using a dyadic grid the scale a is proportional to 2^m. We can, therefore, take base 2 logarithms of both sides of Expression 7.8b to get the often seen expression:

$$\log_2\left(\sigma_m^2\right) = (2H+1)m + \text{constant} \tag{7.10}$$

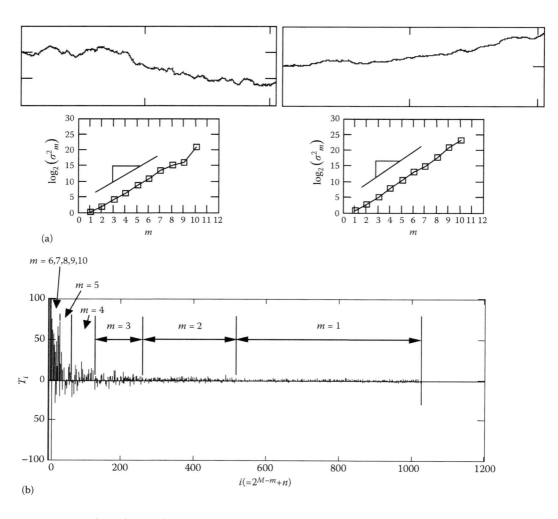

FIGURE 7.8 The relationship between the wavelet coefficient variance and Hurst exponent for fractional Brownian motion: (a) fBm traces ($H=0.6$ left/$H=0.8$ right) with their corresponding $\log_2(\sigma_m^2)$ against m plots. The slopes of $2H+1=2.2$ (left) and 2.6 (right) are shown on the plots above the data points. (Arbitrary axis units.) (b) The sequentially indexed wavelet coefficients for the $H=0.6$ fBm trace in (a). The coefficient values have been cut off at the largest scales as the coefficients at these scales dominate.

where the constant depends both on the wavelet used and the Hurst exponent. See, for example, Flandrin (1992), who defines the constant and gives it explicitly for the Haar wavelet. Figure 7.8a shows plots of $\log_2(\sigma_m^2)$ against m for two fBm's: one of index $H=0.6$ and the other $H=0.8$. These were computed using Haar wavelets. According to Expression 7.10, the plotted points should fall on lines of slopes 2.2 and 2.6, respectively. These gradients are superimposed on the plots. Good agreement is found between the experimental plots and the theoretical line. Remember that this is for a single realization of the fBm

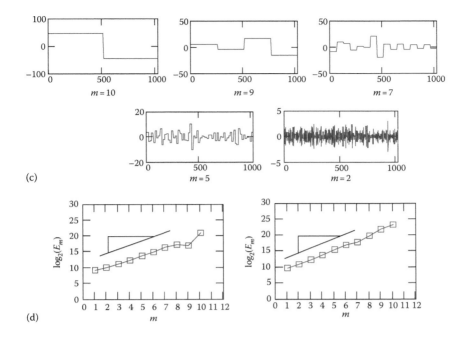

FIGURE 7.8 (CONTINUED) The relationship between the wavelet coefficient variance and Hurst exponent for fractional Brownian motion: **(c)** Reconstruction using only those coefficients at the indicated scale. Notice the reduction in the vertical axis scale with decreasing coefficient scales. **(d)** $\log_2(E_m)$ against m plots corresponding to the fBm traces given in (a) ($H = 0.6$ left/$H = 0.8$ right). The slopes of $2H + 1 = 1.2$ (left) and 1.6 (right) are shown on the plots above the data points.

trace and ensemble averaging over many traces will produce a much more accurate result. Figure 7.8b contains a plot of the coefficients in sequential format for the fBm of index $H = 0.6$ in Figure 7.8a. The increase in the variance with scale is evident from the plot. Figure 7.8c contains example signal reconstructions using only the coefficients at the levels specified.

Notice that from these arguments, we can also see that the wavelet scale–dependent energy:

$$E_m = \sum_{n=0}^{2^{M-m}-1} (T_{m,n})^2 \tag{7.11}$$

(Chapter 3, Equation 3.52) has the fractal scaling law:

$$\log_2(E_m) = 2Hm + \text{constant} \tag{7.12}$$

that is

$$E_m \propto a_m^{2H} \tag{7.13}$$

This is the same scaling as that given by Equation 7.1. This makes sense, as E_m is a measure of the scale-dependent variance of the signal. Figure 7.8d shows the $\log_2 (E_m)$ against m plot for the traces shown in Figure 7.8a. These plots have slopes of around 1.2 and 1.6, respectively, as we would expect. We have concentrated on discrete transform coefficients $T_{m,n}$ as they are prevalent in the fractal-wavelet literature. However, continuous transforms $T(a,b)$ obviously exhibit the same scaling law as Equation 7.13, that is

$$E(a) \propto a^{2H} \tag{7.14}$$

(refer back to Chapter 2, Equation 2.22) but now the a scale parameter is continuous (or at least for practical purposes a discretized approximation based on a non-dyadic grid) and the slope of the plot of $\log(E(a))$ against $\log(a)$ is used to find H.

There are many practical examples of the use of the wavelet-based fractal analysis in the literature. For example, light intensity time series from dusty plasmas have been examined by Safaai et al. (2013). Figure 7.9 shows the logarithmic plot of coefficient variance against scale for dusty signals at various pressures and a signal containing no

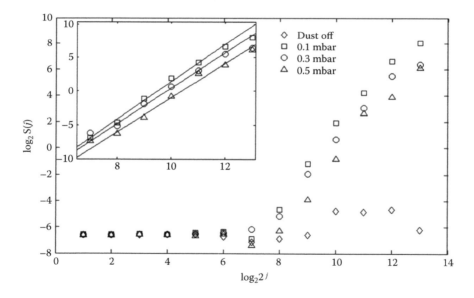

FIGURE 7.9 $\mathbf{Log_2}$-$\mathbf{log_2}$ **plot of** $S(j)$ **versus scale,** 2^j **for DLS time series with dust at 0.1, 0.3 and 0.5 mbar and without dust in the plasma.** The inset is marked by the linear regression lines fitted on the scalograms, where the slope gives the scaling exponent $\gamma = 2H + 1$. Scale 2^j is inversely proportional to frequency and j refers to the scale index. (Note that $S(j)$ is the equivalent of $\langle T_{m,n}^2 \rangle_m$ of Equations 7.7 and 7.8 in this text.) (Reproduced with permission from Safaai, S. S. et al., Fractal dynamics of light scattering intensity fluctuation in disordered dusty plasmas. *Physics of Plasmas*, 20, 103702-1–103702-8. Copyright 2013, American Institute of Physics.)

dust. The dust signals exhibit clear scaling for the higher scales, $j = 8$ to 13, corresponding to 0.026–1.6 seconds. The Hurst exponents were extracted from the signals and compared with approximate entropy metrics confirming the degree of irregularity of the signals.

A number of authors have demonstrated the usefulness of wavelet-based fractal analysis of financial data. Bayraktar et al. (2004), for example, employed a Daubechies wavelet-based approach to determine the Hurst exponent in S&P 500 financial index data. They found the method robust to non-stationarities including seasonal volatility. The Hurst parameter was calculated to be around 0.6 for most of the 1990s but dropped closer to 0.5 in the period 1997–2000, coincident with the growth of Internet trading among small investors. The effect of monetary policy on the fractal structure of macromonetary data was examined by Mulligan and Koppl (2011), who demonstrated that most time series considered were more anti-persistent prior to a structural break in monetary policy. This was interpreted as indicating a changed preference in favour of activist monetary policy. In later work, Aloui and Nguyen (2014) demonstrated long-range dependencies in stock returns from various Mediterranean countries indicated through their Hurst exponents extracted using a Morlet-based approach.

Numerous papers have appeared concerning the wavelet-based fractal analysis of biosignals. An integrated index for the detection of sudden cardiac death (SCD) was developed by Acharya et al. (2015), which included nonlinear features such as the fractal dimension and Hurst exponent extracted from the second-level Daubechies-based decomposition of the ECG. It was found that the resulting index was able to predict SCD with high accuracy from a few minutes of signal prior to the event. The epileptic electroencephalogram (EEG) was analyzed by Janjarasjitt and Loparo (2009), where the spectral exponent of the discrete wavelet coefficient variances was used to identify regions of signal corresponding to epileptic seizures. (See also their related work on the same topic using the electrocorticogram [ECoG] [Janjarasjitt and Loparo, 2014].) Other work on the EEG includes the denoising of schizophrenic EEG data using discreet wavelet transform (DWT) decomposition, singular spectrum analysis and the subsequent reconstruction of the coefficients prior to differentiation against healthy controls according to fractal dimension by Akar et al. (2015); the extraction of wavelet-based fractal features for active segment selection using a combined continuous wavelet transform (CWT) and DWT multiresolution analysis by Hsu et al. (2007); and the novel hybrid compression method using both fractal and wavelet coding for multichannel EEG by Saeedi et al. (2014). In other bioengineering work, the fractal structure of the retinal vasculature has been explored by Che Azemin et al. (2010) using a wavelet decomposition followed by a Fourier fractal dimension (FFD) measure. (This research was also referenced briefly in the optical imaging section of Chapter 6, Section 6.7.1 and Figure 6.53.)

In addition to those mentioned elsewhere in this section, there are numerous applications of wavelet-based fractal methods in image analysis, including the wavelet technique to embed fractal information for image steganography by Wu and Noonan (2012), the image watermarking scheme based on the lifting wavelet transform and singular value decomposition by Ghaderi et al. (2013), the genetic algorithm DWT method for fractal image compression by Wu (2014), the fractal-wavelet techniques for denoising sonar images by Wang et al. (2014), the wavelet maxima–based approach to analyzing lacunarity in cancer images by Bogdan (2010) and the texture analysis of images using Gabor wavelets combined with volumetric fractal dimension estimates carried out by Zuniga et al. (2014).

Many other areas have benefited from wavelet-based fractal analysis. In astronomy, Maclachlan et al. (2013) demonstrated that short and long gamma ray burst (GRB) time series exhibit different Hurst exponents, with the short GRBs tending to lower values in general. The Hurst exponent has also been computed for crystal growth images by Zhang et al. (2012b) and electrical power data by Zhai (2015). Tafti et al. (2009) have considered the wavelet analysis of multivariate fractional Brownian motion fields including the estimation of the corresponding Hurst exponents. Gaci et al. (2010) used wavelet techniques to analyze local regularity in strata heterogeneities of sonic borehole log data. Khoshelham et al. (2011) employed wavelet denoising of rock surface profiles prior to estimating their Hurst exponent and fractal dimension. (See also Dougan et al., 2000, who compared wavelet, Fourier, box counting and variable bandwidth methods for deriving H for crack profiles.) More recently, Jiang et al. (2015a) decomposed mode shapes using a Daubechies wavelet reconstruction method prior to calculating the fractal dimension of the signals in work to identify crack location in structural beam elements and Cantzos et al. (2015) have estimated the fractal characteristics of electromagnetic time series prior to earthquakes using both wavelet and rescaled range analysis.

7.2.3 Multifractals

Multifractal theory concerns itself with fractal objects which cannot be completely described using a single fractal dimension (monofractals). They have, in effect, an infinite number of dimension measures associated with them. This section presents a short summary of wavelet-based multifractal characterization.

The multifractal scaling of an object is characterized by

$$N_\varepsilon \propto \varepsilon^{-f(\alpha)} \qquad (7.15)$$

where N_ε is the number of boxes of length ε required to cover the object and $f(\alpha)$ is the dimension spectrum, which can be interpreted as the fractal dimension of the set of points with scaling index α (Hilborn, 1994). We can find the multifractal spectrum of a signal by partitioning it into N boxes of length ε. A probability density, $P(\varepsilon,i)$, of the signal in each box, labelled i, is calculated where $P(\varepsilon,i)$ is the fraction of the total mass of the object in each box. The qth order moments $M(\varepsilon,q)$ are then calculated as follows:

$$M(\varepsilon,q) = \sum_{i=1}^{N(\varepsilon)} P(\varepsilon,i)^q \qquad (7.16)$$

For a multifractal object, this moment function scales as

$$M(\varepsilon,q) \propto \varepsilon^{\tau(q)} \qquad (7.17)$$

From this scaling both α and the $f(\alpha)$ spectrum can be calculated from

$$\alpha(q) = \frac{d\tau(q)}{dq} \qquad (7.18)$$

and

$$f(\alpha) = q\alpha(q) - \tau(q) \qquad (7.19)$$

In the wavelet-based method for calculating the $f(\alpha)$ spectrum, the function of Equation 7.16 is replaced by the wavelet-based moment function:

$$M(a,q) = \sum_i |T(a,b_i)|^q \qquad (7.20)$$

where $|T(a,b_i)|$ is the ith wavelet transform modulus maxima found at scale a. By summing only over the modulus maxima, this incorporates the multiplicative structure of the singularity distribution within the calculation of the partition function (Muzy et al., 1991). For a multifractal object, this wavelet-based moment function scales as

$$M(a,q) \propto a^{\tau(q)} \qquad (7.21)$$

This relationship is then used to calculate α and the $f(\alpha)$ spectrum and hence characterize the multifractal object or process under investigation. Note that for negative q values, Equation 7.20 becomes unstable in the neighbourhood of points on maxima lines where the wavelet transform is close to zero. However, this can be remedied by replacing the value of the wavelet transform modulus at each maximum by the supremum value along the corresponding maxima line at all scales smaller than a. (Muzy et al., 1993, 1994; Argyris et al., 2015).

A modulus maxima method based on Gaussian derivative wavelets may be used for determining multifractal spectra as demonstrated by Haase and Lehle (1998), who illustrated the method on time series from a circle map and a turbulent velocity signal. The turbulence signal was acquired from within an axisymmetric jet and is plotted in Figure 7.10a. Figure 7.10b shows the energy spectrum of the signal showing the classic −5/3 power law of the inertial range of turbulence. Figure 7.10c contains the wavelet transform plot with its modulus maxima, shown in Figure 7.10d. A Mexican hat wavelet was employed in the decomposition. Finally, the $f(\alpha)$ spectrum, computed from the modulus maxima lines, is

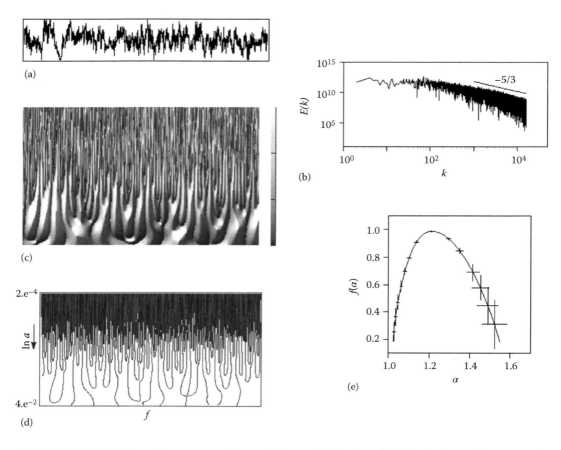

FIGURE 7.10 Multifractal spectrum of a turbulent velocity signal: (a) Turbulent velocity signal. (b) The energy spectrum of the signal in (a) showing the −5/3 law. (c) Wavelet transform of (a) using a Mexican hat wavelet. (d) Skeleton of maxima lines from (c). (e) The multifractal spectrum $f(\alpha)$ resulting from the wavelet modulus maxima method. (Figures provided by Dr Maria Haase, Institut für Höchstleistungsrechnen, Universität Stuttgart.)

shown in Figure 7.10e. Note that if the velocity signal were a monofractal, such as fBm, the multifractal spectrum would collapse to a single point. More recently Jing-Jing and Fei (2015) investigated the multifractal nature of turbulence in the urban canopy layer using Mexican hat wavelets and found a higher intermittency in this fluid flow than for wind tunnel flows at moderate Reynolds numbers. Other fluids-related work in this area includes the wavelet-based multifractal formalism used to characterize the 'intermittency' and 'roughness' associated with sediment transport phenomena by Singh et al. (2009) (already mentioned in Chapter 4, Section 4.6); the use of the third derivative of Gaussian wavelets to characterize the multifractal nature of bubble paths in a bubble column by Mosdorf et al. (2011) (also mentioned in Chapter 4, Section 4.6); and the calculation of the multifractal spectra associated with the turbulent flow of molten salt in a pipe, generated by Sona et al. (2014). Interestingly, Deliege and Nicolay (2014) employed a multifractal formalism to demonstrate that European air surface temperature signals were, in fact, monofractal in nature.

Wavelet-based multifractal analysis has now been employed in the elucidation of a wide variety of physical phenomena and papers have emerged from many fields of study. Benouioua et al. (2014) developed a non-intrusive diagnostic tool for the analysis of fuel cells based on the multifractal analysis of their voltage signals. In their work, which utilized the Mexican hat wavelet, they presented a nice illustration of the multifractal analysis of a monofractal and a multifractal (shown in Figure 7.11). The monofractal nature of the Weierstrass fractal is evident from the linear $\tau(q)$ versus q relationship, whereas the equivalent relationship for the multifractional Brownian motion signal is nonlinear, leading to a distinct multifractal spectrum exhibiting a concave arch. Figure 7.12 contains their analysis of voltage signals from a reference cell and cells in three types of poor operating conditions where it may be observed that the four curves are well separated and may be suitable as the basis of a diagnostic tool. (Note that they use the equivalent $D(h)-h$ instead of $f(\alpha)-\alpha$ multifractal spectral plots.)

The multifractal nature of eye movements has been probed by Hampson and Mallen (2011) using a Mexican hat–based analysis. In the study, they investigated the aberration dynamics of five subjects while they viewed a stationary target close to optical infinity. They found that the most frequently occurring Hölder exponent was 0.31+/−0.10, suggesting that the fluctuations are anti-persistent. Gerasimova et al. (2014) used computer-aided multifractal analysis of dynamic infrared imaging to detect breast cancers. They found that the multifractal temporal complexity of fluctuations in temperature from healthy breasts is lost when a malignant tumour is present. We have already come across the multifractal analysis of heart rate variability and Doppler flow signals by Humeau (2010) and the multifractal nature of the sleep and awake EEG by Zorick and Mandelkern (2013) in Chapter 6. Other medical examples include the work of Dick and Svyatogor (2015), who determined the degree of multifractality of the EEG during photic stimulation of patients with dyscirculatory encephalopathy; Wendt et al. (2014), who used a wavelet transform p-leader multifractal formalism to analyze the heart rate variability of survival probability in congestive heart failure (CHF) patients; Leonarduzzi et al. (2014), who also employed the wavelet p-leader technique to characterize fetal heart rate variability; and Pavlov et al. (2016), who studied the multifractal nature of cerebrovascular dynamics.

Examples of wavelet-based multifractal analysis from other areas include the studies by Golestani and Gras (2012), who demonstrated that the combination of predation and pressure associated with the distribution of food plays an important part in the emergence of multifractal phenomena in the simulations of ecosystems; Piñuela et al. (2007, 2010), who probed the multifractal nature of 2-D and 3-D images of soil samples; Pascoal and Monteiro (2014), who characterized a financial market index (PS120) using wavelet-based multifractal analysis among other methods; and Du et al. (2014), who extracted multifractal features and used them as input to a classifier for fault diagnosis in rotating machinery, where they found that the optimum performance was obtained using both multifractal and wavelet packet energy features in combination. Geophysical-specific applications include the analysis of well-log data for lithofacies segmentation (Ouadfeul and Aliouane, 2011), the detection of hydrocarbon fluids in seismograms (Khan and Fadzil, 2007) and the monitoring of infected vegetation in remotely sensed data (Chávez et al., 2010). For

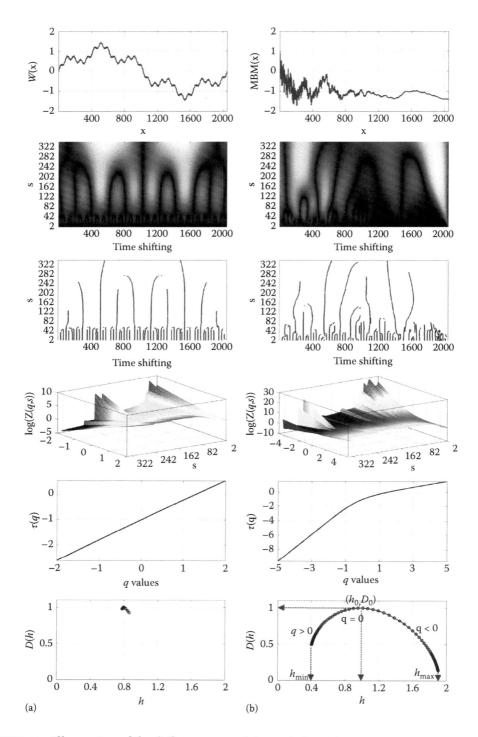

FIGURE 7.11 Illustration of the different steps of the multifractal spectrum computation based on WTMM of (**a**) Weierstrass fractal signal with $h = 0.7$ and $n = 50$ (left column) and (**b**) multi-fractional Brownian motion signal with Hölder function $h(x) = 0.1 + 0.8x$ and $\sigma = 1$ (right column). (From Benouioua, D. et al., *International Journal of Hydrogen Energy*, 39, 2236–2245, 2014.)

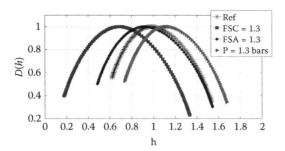

FIGURE 7.12 **Typical curves of singularity spectra.** (From Benouioua, D. et al., *International Journal of Hydrogen Energy*, 39, 2236–2245, 2014.)

further background information on multifractals in general, with reference to the wavelet method, the reader is referred to Muzy et al. (1994), Lopes and Betrouni (2009) and Mukli et al. (2015) and to Section 8.5.2 of the book by Argyris et al. (2015) which provides a lucid explanation of multifractal spectra and wavelet transform modulus maxima.

7.3 FINANCE

This is a very interesting application of wavelet analysis which concerns a pertinent problem (of which there is often a large financial incentive to solve!). The data sets (stock market indices, commodity prices, exchange rates, real estate prices, growth data, mortgages, etc.) are typically highly non-stationary, exhibit significant complexity and involve both (pseudo-) random processes and intermittent deterministic processes. A good example would be financial indices generated from, among other effects, a large number of small-scale (pseudo-random) share dealings combined with large-scale (deterministic) interest rate adjustments. When the first edition of this book was published, the application of wavelets to finance was very much in its infancy compared with other subject areas. However, this is certainly no longer the case with a large number of papers now appearing each year covering all aspects of financial data analysis (including stock market indices, commodity prices, currency exchange rates, inflation rates, etc.), and with wide-ranging investigations studying the links between various financial data streams. A useful overview of the application of wavelets in finance has been written by Chakrabarty et al. (2015a). They have provided a breakdown of the emerging discipline into four broad categories: horizon heterogeneities, denoising of financial time series, structural features and testing statistics and methodologies. Their review concentrates on the first category which is further broken down as per Figure 7.13. They concluded that agents, operating at different horizons, consider different information or interpret the same information differently, which can lead to certain trends in the time–frequency space. The wavelet 'primer' by Aguiar-Conraria and Soares (2011) is also worth consulting as it provides much of the wide-ranging theory used in the analysis of financial time series, and with a number of examples provided. More specific examples of applications in the field are detailed in the rest of this section.

Contagious heterogeneities in Asian stock markets (Chinese, South Korean, Taiwanese and Indian) have been investigated by Dewandaru et al. (2015) where wavelet analysis was

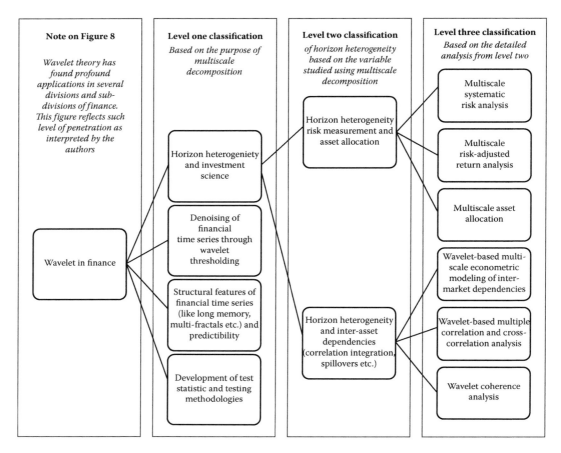

FIGURE 7.13 Wavelet applications in finance. (From Chakrabarty, A. et al., *Physica A: Statistical Mechanics and Its Applications*, 429, 45–61, 2015.)

used to examine the nature of contagion. In the study, stock market returns were decomposed into two parts: (1) a low-frequency part and (2) a high-frequency part, which can be linked to interdependence and contagion, respectively. Figure 7.14 shows two examples of their Morlet-based plots of wavelet squared–coherency between market pairs, where the wavelet phase difference is superimposed on the plots using directional arrows. These coherency plots allowed for the evaluation of episodes of market integration in the short term (due to some capital market imperfections such as regulation, taxes, transaction costs, etc.) as well as in the long term emanating from fundamental/economic linkages. The figure also contains the time series plots of phase for the frequency band corresponding to a 1–4 year period. They used their results to make an informed analysis of the relative stock market behaviours and the links between them. For example, they found that the India–Taiwan index pair (SENSEX–TAIEX) portrays a transition from low coherency to high coherency in the 2–4 year frequency band in 1993, with this 'co-movement' increasing gradually and persistently. The arrows indicate a more-or-less in-phase relationship with the phase difference time series indicating a slightly leading role of India over Taiwan. The India–South Korea markets (SENSEX–KOSPI), on the other hand, exhibit a

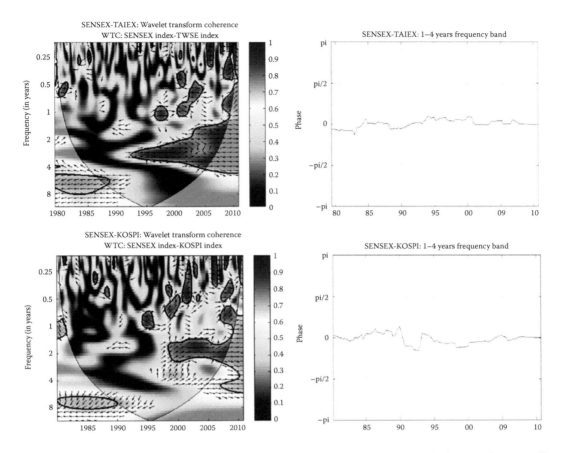

FIGURE 7.14 (See colour insert.) **Wavelet squared coherency and phase difference for overall stock markets.** The left column shows wavelet squared coherency and phase difference. The 5% significance level against red noise is shown as a thick contour. The relative phase relationship is indicated by arrows: Right: in-phase; Left: anti-phase; Down: first series leading second series by 90°; Up: second series leading first series by 90°. The right column shows wavelet phase-difference at frequency band 1–4 years. (Original figure in colour.) (From Dewandaru, G. et al., *Physica A: Statistical Mechanics and Its Applications*, 419, 241–259, 2015.)

transition to high coherence much later in 1998. The authors advocated the wavelet method as it provides an ability to observe the multi-horizon nature of co-movements, volatility and lead-lag relationships in the data. It is also worth consulting the earlier work by Rua and Nunes (2009) on the co-movement of various stock market returns in this regard. A Morlet-based wavelet coherence analysis was also employed by Albulescu et al. (2015) to assess contagion and dynamic correlations of the main European stock index futures markets. They found that stock index futures were highly correlated at all decomposition levels and that this correlation increases around episodes of financial distress. Figure 7.15 shows an example of the rolling wavelet correlation series for the Financial Times Stock Exchange 100 Futures Index (FTSEF), Deutscher Aktienindex 30 Futures Index (DAXF) and Crédit Agricole Consumer 40 Futures Index (CACF) at different financial cycle scales

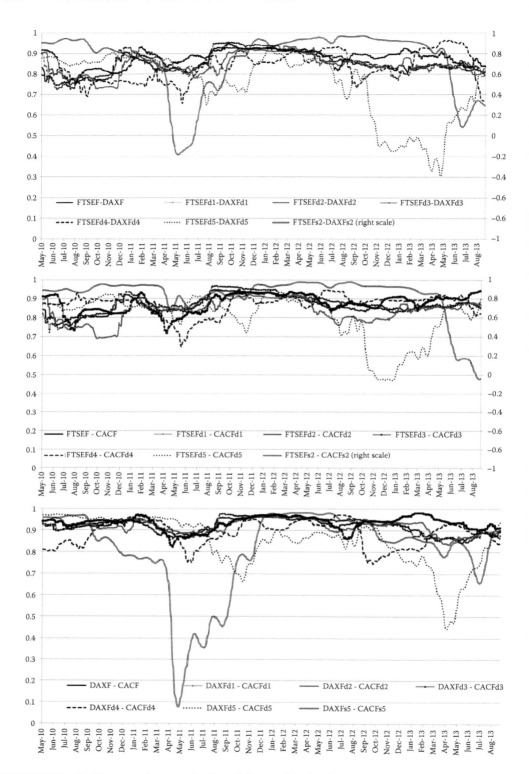

FIGURE 7.15 Rolling wavelet correlation of the stock index futures returns. (From Albulescu, C. T. et al., *Procedia Economics and Finance*, 20, 19–27, 2015.)

resulting from their analysis. Using a discrete wavelet transform approach, Berger (2015) investigated how information at different timescales in U.S. stocks could be used in portfolio management and demonstrated that contagion affects both short and long runs of asset returns. The modelling of regimes of volatility using a wavelet-hidden Markov model has been studied by Gencay et al. (2010). They discovered an 'asymmetric vertical dependence' of information flow, where low volatility at a long time horizon was followed by low volatility at a short-time horizon, whereas high volatility states at long time horizons were not necessarily followed by short-term volatility states. Fernández-Macho (2012) developed multiple correlation and cross-correlation metrics to quantify the multiscale behaviours of 11 main Eurozone stock market indices. Using their technique, they found an almost exact linear relationship for periods greater than 1 year.

Other work concerning the interrogation of stock market indices includes the multi-resolution analysis of the Chinese stock market by Huang et al. (2010); the testing of six different wavelet types for stock market index analysis by Machado et al. (2012); the heterogeneity of investment horizons in the Malaysian stock market by Najeeb et al. (2015); the stock market integration in Asian countries by Tiwari et al. (2013a); volatility spillover across sectorial indices in the Indian financial market by Chakrabarty et al. (2015b); the combination of wavelet transforms, multivariate adaptive regression splines and support vector regression for stock index forecasting by Kao et al. (2013); the co-movement dynamics between central and eastern European stock markets during European integration by Dajcman et al. (2012) and between central European stock markets by Barunik et al. (2011); the causal linkages between U.S. and East Asian stock markets by Kim (2010); the stock market modelling using local linear wavelet neural networks by Chen et al. (2005); the assessment of market risk in emerging markets by Rua and Nunes (2012); the wavelet coherence analysis of the interdependency between bond markets in Poland, the Czech republic, Hungary and Germany by Yang and Hamori (2015); the forecasting of market turning points by Bai et al. (2015); the assessment of risk contagion among international stock markets by Asgharian and Nossman (2011); and the short-time extrapolation of stock market indices by Gosse (2010).

A significant amount of work on stock indices has involved understanding their relationship with other financial signals. Some of the more recent investigations include the study of the relationship between oil and stock markets in the G7 countries by Khalfaoui et al. (2015) using maximal overlap discrete wavelet methods; between shocks in crude oil prices and the stock market in China by Huang et al. (2015) using a Haar-based multiresolution analysis; between interest rates and stock returns in Spain by Moya-Martinez et al. (2015), again using a Haar-based multiresolution analysis; between inflation and stock prices in India by Bhanja et al. (2012) using wavelet coherence; between inflation and stock returns in Pakistan by Tiwari et al. (2015), again using wavelet coherence analysis methods; between U.S. stock and real estate markets by Chou and Chen (2011) and, similarly between U.S. housing and stock markets by Li et al. (2015) using a Morlet-based partial coherency approach.

The nature of foreign exchange rate dynamics is another emerging topic of research in this area and includes the work of Tiwari et al. (2013b), who examined the asymmetric causality between oil prices and exchange rates using a wavelet transform framework. More recently, Bekiros and Marcellinio (2013) investigated the micro-foundations of across-scale causal heterogeneity on the basis of trader behaviour with different time horizons. They proposed an entropy-based method for the determination of the optimal decomposition level and found no indication of global causal behaviour dominating at all timescales. Andrieş et al. (2016) employed wavelet power, coherence and multiple correlation and cross-correlation analyses (of Fernández-Macho, 2012, cited previously) in their study of the behaviour of exchange rates in central and eastern European countries. Sandoval and Hernandez (2015) used wavelet-transformed order book volume dynamics in their investigation of foreign exchange trading strategies. In a study of major carry trade currencies and carry trade diversification opportunities, Orlov and Aijo (2015) examined the trade payoffs at five different timescales (overnight, 1 week, 1 month, 3 months, 1 year) using wavelet correlation techniques. They found that portfolio composition based on wavelet correlation of returns with dynamic rebalancing led to increased Sharpe ratios (a measure for calculating risk adjusted return) when compared with simply diversified portfolios and stock market proxy for most of the timescales considered. They concluded that wavelet correlation methods can find patterns in exchange rate movements which might be exploited by investors.

Energy prices have also attracted the attention of researchers. Wavelet analysis in this area includes the forecasting of short-term demand and gas prices using wavelet methods combined with adaptive machine learning–time series models by Nguyen and Nabney (2010); the use of wavelet coherence to uncover interesting dynamics in the relationship between crude oil, gasoline, heating oil and natural gas by Vacha and Barunik (2012); the analysis of the relationship between oil prices and exchange rates by Shahbaz et al. (2015); the multivariate wavelet denoising approach to estimating the portfolio value at risk for crude oil markets by He et al. (2012); the wavelet-based nonlinear autoregressive distributed lags model (W-NARDL) for assessing exchange-rate pass-through to crude oil prices by Jammazi et al. (2015); the 'grey correlation'–based wavelet analysis of crude oil price co-movements by Jia et al. (2015); the study of annual and biannual variations of electricity prices by Schlüter (2009); and the wavelet cross-bicoherence technique employed by Tonn et al. (2010) to detect frequencies of strong correlation between future prices of natural gas and oil.

The denoising of option prices using wavelets was carried out by Haven et al. (2012), where they found that the estimation of risk-neutral density functions and out-of-sample price forecasting is significantly improved after noise was removed using the wavelet method. Cui et al. (2015) investigated arbitrage trading of metal futures using a Mexican hat and neural network–based TGARCH (threshold generalized autoregressive conditional heteroskedasticity) model. Other areas where wavelets have been employed for financial signals include inflation (Ysusi, 2009; Tiwari et al., 2014a; Jiang et al., 2015b), land price cycles and trends (Hannonen, 2006), gold price time series (Lineesh et al., 2010), growth and export data (Dar et al., 2013), interest rates and growth data (Gallegati et al., 2013), money supply and the liquidity effect (Michis, 2011), the co-movement of real growth rates (Aloui et al., 2016) and mortgages and gross domestic product (González-Concepción

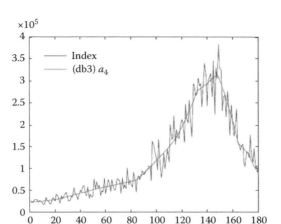

FIGURE 7.16 db3 filter of SCTfe Index. (Original in colour. The index is the more erratic signal in the plot.) (From González-Concepción, C. et al., *Journal of Applied Mathematics*, 2012, 917247, 1–17, 2012.)

et al., 2012). Figure 7.16 contains a plot of the Spanish index, SCTfe (household mortgages to gross domestic product [GDP] index for the Spanish province of Santa Cruz de Tenerife) from this last paper, together with its wavelet-filtered version using a Daubechies discrete wavelet filter. The reader is also referred to the end of Sections 7.2.2 and 7.2.3 where a few references are already provided for the wavelet-based fractal and multifractal analysis of financial signals.

7.4 GEOPHYSICS

We have already covered the application of wavelet analysis to geophysical flows in Chapter 4, including canopy flows, wind forecasting, wind-generated ocean surface waves, large-scale oceanic and atmospheric flow phenomena and associated biological processes. In this section, we cover some of the other areas of geophysics that have utilized wavelet transforms methods, including well logging, core analysis, seismology and topographic feature analysis. However, as with the previous section, the reader is also referred to the end of Sections 7.2.2 and 7.2.3 where a few references are provided for the wavelet-based fractal and multifractal interpretation of geophysical data.

7.4.1 Properties of Subsurface Media: Well Logging, Cores and Seismic Methods

A range of standard techniques have been developed for identifying stratigraphic changes in subsurface media by 'logging' the details of various measurements: resistivity, porosity, gamma radiation, density, sonic, nuclear magnetic resonance and so on. Numerous papers have now been published on the application of wavelet methods to the study of such 'well logging' data. For example, Perez-Muñoz et al. (2012) analyzed gamma ray logs using the Haar wavelet transform to identify electro-facies associations and also Morlet wavelet–based wavelet coherence to identify patterns in gamma ray logs of different wells. Karacan and Olea (2014) conducted work to infer strata separation intervals in longwall overburden by a method that relies only on commonly available drill hole gamma log signals. They used both Mexican hat and Morlet wavelets in their study to seek out singularities that may be the precursors for strata separations and associated methane flow paths when formations are exposed to mining stresses. The work aimed to identify potential methane inflows in mining operations from only well-log signals, thus improving mining safety. Henriques et al. (2015) attempted to improve the analysis of well-log data using wavelet cross-correlation. Signal pairs from gamma ray, sonic transmissivity and electrical resistivity logs from the same wells were first studied before analyzing the same log data signal types from different wells. In the latter investigation, they shifted the well-log data relative to each other from the two wells to gain maximum correlation of the data. This is shown in Figure 7.17 which contains, as an example, the sonic transmissivity (DT) signals from two wells shifted by 15.24 m relative to each other prior to wavelet cross-correlation and computation of a local correlation coefficient (a normalized weighted sum across scales). The local correlation index highlights the strong correlation across scales at around 1524 m (these strong correlations are not seen at zero displacement). The authors suspected that this strong cross-correlation from centimetre to meter scales was due to the regions crossed by the two wells being part of the same geological structure or interface.

Other studies of well-logs include that by Xuexu et al. (2013), who investigated Milankovitch cycles (i.e. long-term variations in the orbit of the earth that result in climate change over periods of hundreds of thousands of years) in gamma logging data; Perez-Muñoz et al. (2013), who performed an analysis involving a number of different mother wavelets applied to various types of log data; Davis and Christensen (2013), who, instead of a Mexican hat CWT, used a simple approximation that is piecewise-continuous in their study of layer selection of borehole conductivity logs; Heidary and Javaherian (2013), who employed principal component analysis and Coiflet wavelet transforms in the determination of reservoir fluid contacts; Pan et al. (2008), who examined both gamma ray logs and spontaneous potential logs using wavelet and Fourier transforms; and Saljooghi and Hezarkhani (2014), who studied the application of wavelet neural networks, based on a variety of wavelet types, to the prediction of porosity from well-log data.

Ge et al. (2015) developed an improved noise reduction algorithm for low-field nuclear magnetic resonance data used in the evaluation of pore structure in both numerical simulations and rock cores. The algorithm employed a hybrid thresholding function to avoid drawbacks from traditional soft and hard thresholding. In an attempt to monitor Pacific climate patterns (e.g. El Niño Southern Oscillation [ENSO] and Southern Annular Mode [SAM]), ice cores from coastal Antarctic locations were analyzed by Jones et al. (2014) using cross-wavelet coherence. They found that microclimatic differences, and possibly the post-depositional movement of snow, made climate reconstruction problematic. Soon et al. (2014) also scrutinized ice cores from East Antarctica using the Morlet-based cross-wavelet transform to determine the existence and nature of Holocene solar and climatic variations on centennial to millennial timescales. Ion concentration depth profiles of tropical ice cores from the eastern Bolivian Andes have been examined by Gay et al. (2014), using both Fourier and wavelet methods to establish relationships between ice depth and ice age,

Fractals, Finance, Geophysics, Astronomy and Other Areas ■ 363

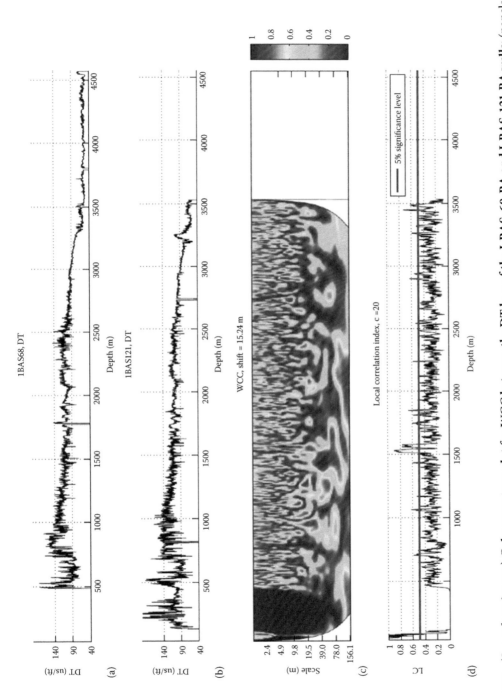

FIGURE 7.17 (See colour insert.) Colour contour plot for WCC between the DT logs of the I-BAS-68-BA and I-BAS-121-BA wells: (panels a and b) are shown in panel (c). Strong correlations near the depth position corresponding to 1524 m can be seen in all the scales, as shown by the dark vertical lines. This high degree of correlation was obtained when the first set of data was displaced by 100 units (15.24 m), relative to the second set. In panel (d), the local correlation index, where the x-axis represents the corresponding depths in the second well log. (Original figure in colour.) (From Henriques, M. V. C. et al., *Physica A: Statistical Mechanics and its Applications*, 417, 130–140, 2015.)

and a Morlet wavelet-based transform has been employed by Herb et al. (2015) to interogate age model data of a drill core from the NE Tibetan Plateau. In other work, Muñoz et al. (2009) revealed inter-annual, decadal, inter-decadal and centennial scales in the laminated structure of stalagmites by applying a wavelet decomposition to their thickness variations.

It is generally accepted that modern wavelet transform analysis really took off in the early 1980s with the Morlet wavelet, developed to aid in the interrogation of seismic signals (e.g. Goupillaud et al., 1984). Seismology may, therefore, be thought of as the birthplace of modern wavelet transform analysis. Since then, many other wavelets have been developed and used to examine these and many other signals in geophysics. Recently, Reine et al. (2009) employed the CWT and three other transforms to determine seismic attenuation measurements and found that the transforms with systematic varying windows (the S transform and CWT) allowed for more robust estimates of seismic attenuation than those with fixed windows (Fourier and Gabor transforms). Saunders et al. (2015) found that the CWT allowed for improved identification and mapping of thinner potential hydrocarbon reservoirs. De Matos et al. (2007) developed a semi-automated method based on wavelet transform modulus maxima line amplitudes (WTMMLAs) to identify seismic trace singularities in each geologically oriented segment which they used to build up a facies map using self-organized clustering methods. Modulus maxima were also utilized by Liu (2013) in a method to accurately determine arrival times of seismic P waves. Duchesne et al. (2006) incorporated Daubechies wavelets in a scheme to denoise computerized tomography (CT) scan data used to correlate sediment cores with very high-resolution seismic reflection sections. The authors chose a wavelet technique over Fourier-based smoothing to avoid losing localized information. Figure 7.18 contains a CT trace and its wavelet-filtered version from one of their core samples. Wavelets have formed the basis of many other techniques developed for use in seismic analysis, including the localization and denoising technique for in situ testing of soil and pavement properties by Golestani et al. (2013); the Morlet transform phase residues applied to map stratigraphic discontinuities by de Matos et al. (2010); the study of 2-D and 3-D seismic data to detect stratigraphic channels in oil fields by Shokrollahi and Riahi (2013) and hydrocarbon reservoirs by Saadatinejad and Hassani (2013); the entropy-based thresholding methods for denoising seismic signals by Beenamol et al. (2012); the parallel compression algorithm for non-cable seismographs by Zheng and Liu (2012); the seismic multiscale decomposition technique used for efficient selective filtering of seismic data by Leite et al. (2013); the use of the signal-to-noise ratio calculated by the wavelet modulus maxima-based method for the detection of sub-seismic fault footprints by Xu et al. (2015); the application of a combined empirical mode decomposition and wavelet transform scheme to characterize seismic frequency attenuation by Xue et al. (2014); and the modelling of wave dispersion characteristics using CWTs by Holschneider et al. (2005) and Kulesh et al. (2005).

Morlet-based synchrosqueezing was employed by Herrera et al. (2014) in order that individual components of seismic signals can be used to image the subsurface more accurately. One of their signals is shown in Figure 7.19a. The corresponding CWT, complete ensemble empirical mode decomposition (CEEMD) and synchrosqueezed transform (SST) are shown in Figures 19b–d. The authors went on to perform the analysis for a whole seismic

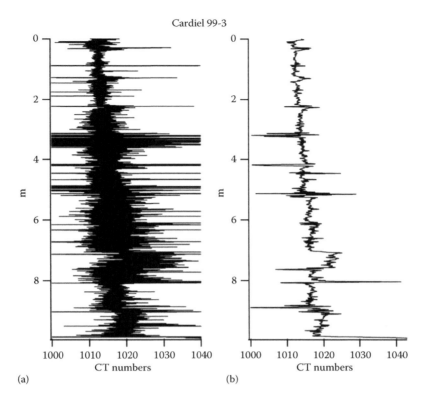

FIGURE 7.18 (a) Raw CT scan profile containing higher-frequency components masking lower-frequency components. (b) Application of the wavelet transform on the CT scan signal permitted to image properly lower frequencies contained within the signal. (From Duchesne, M. J. et al., *Journal of Geophysical Research: Solid Earth*, 111(B10103), 1–16, 2006.)

record, computing the cumulative spectral energy at 80% of total energy. These results are shown in Figure 7.20. They found that the SST exhibited the least speckle noise. In addition, the strong reflector at 0.9 s was better represented, which can be seen by comparing Figure 7.20c with Figure 7.20a,b. They stated that the SST could yield more favourable results as it has the ability to adapt the mother wavelet whereas the CEEMD does not require a basis to be specified. Examples of synchrosqueezing seismic data are also to be found in Chen et al. (2014b), who found the method useful in detecting deep-layer weak signal, which is usually smeared in seismic data.

The analysis of earthquake-generated seismic waves has been tackled by Chen et al. (2015) using the Morlet wavelet; Nicolis and Mateu (2015) using 2-D anisotropic wavelet entropy; Al-Hashimi et al. (2013) using DWTs to detect P- and S-phases in three-component seismic data; Fang et al. (2015) using wavelet-based sparsity-constrained inversion; Burjanek et al. (2010) using Morlet wavelets to investigate the variations in ambient vibrations of unstable mountain slopes; and Simons et al. (2011) using spherical wavelets to resolve a global tomographic model. See also the work on seismic-magnetic anomalies by Han et al. (2011) and the cross-wavelet analysis of infrasound and seismic

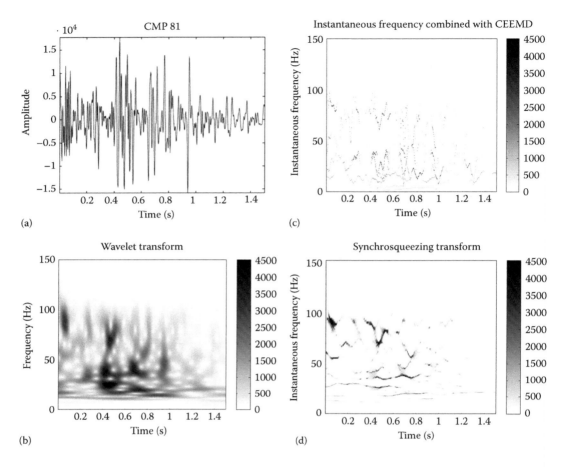

FIGURE 7.19 (a) Individual trace at common mid-point (CMP) 81. It crosses the channel at 0.42 s. CMP 81. Time–frequency representation from (b) CWT, (c) CEEMD and (d) SST. All of these show a decrease in frequency content over time, yet the CEEMD and SST results are the least smeared. (From Herrera, R. H. et al., *Geophysics*, 79(3), V55–V64, 2014.)

signals associated with Mount Etna explosive activity by Cannata et al. (2013). The reader is further referred to Duarte et al. (2014), who have provided a brief overview of recent advances in seismic signal processing, including the role of the wavelet transform in solving the inverse problem. There is also some mention of the analysis of the seismic (earthquake) loading of structures in a couple of sections in Chapter 5 of this book.

7.4.2 Remote Sensing

The interrogation of vegetation characteristics from remote sensing images is an active area of research. Martinez and Gilabert (2009) studied vegetation dynamics by decomposing normalized difference vegetation index (NVDI) signals. Figure 7.21 shows the intra- and inter-annual components of the NVDI signals collected for various vegetation covers generated using a Meyer wavelet multiresolution analysis. The authors found that the method provided relevant information concerning vegetation dynamics at the regional

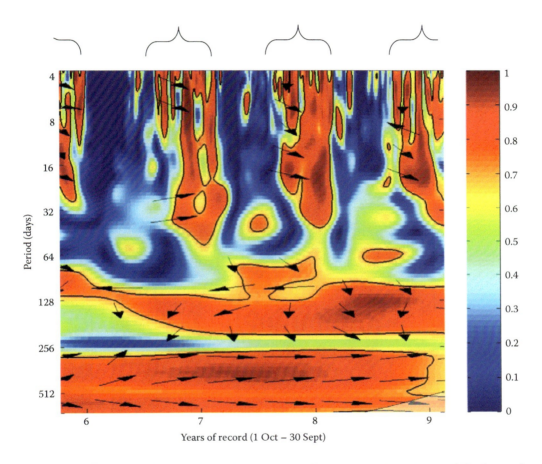

FIGURE 4.37 **Close-up of squared wavelet coherence between precipitation and discharge for Krycklan for ~3 years of record.** (Brackets have been added to the top of the plot to indicate where the high-coherence regions occur.) (From Carey, S. K. et al., *Water Resources Research*, 49(10), 6194–6207, 2013.)

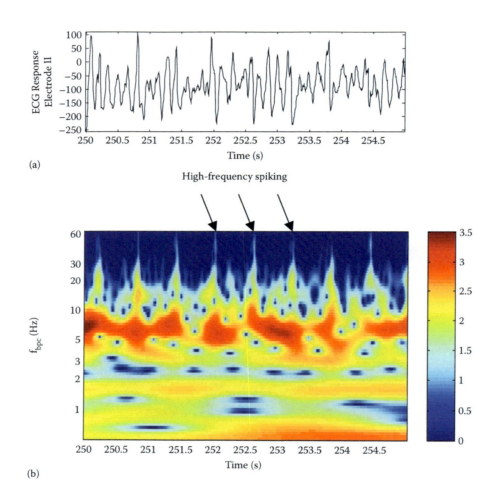

FIGURE 6.12 Wavelet transform of ECG exhibiting ventricular fibrillation: (a) Single channel ECG showing a region of VF. (b) The corresponding scalogram of the temporal location against the bandpass frequency of the wavelet. Notice the high-frequency periodic spiking observable in the scalogram. (From Addison, P. et al., *IEEE Engineering in Medicine and Biology Magazine*, 19, 104–109. Copyright 2000, IEEE.)

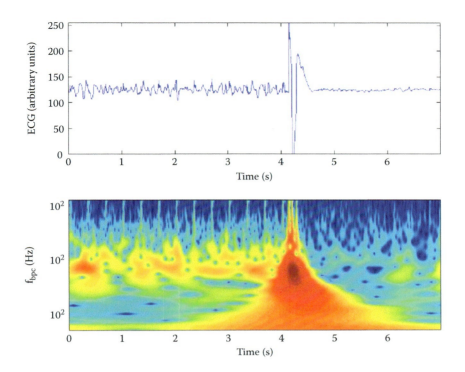

FIGURE 6.13 Attempted defibrillation of human ventricular fibrillation. **Top:** 7 seconds of human ECG exhibiting VF containing a defibrillation shock event. **Bottom:** Scalogram corresponding to the ECG signal. Notice the high-frequency spiking prior to the shock evident in the scalogram. (From Addison, P. et al., *IEEE Engineering in Medicine and Biology Magazine*, 21, 58–65, 2002b. Copyright 2002, IEEE.)

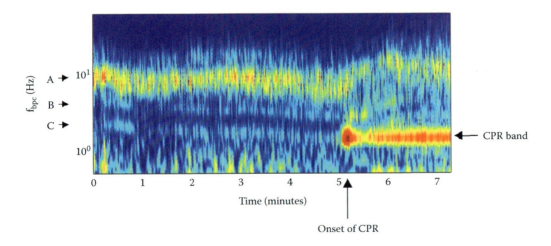

FIGURE 6.14 The energy scalogram for the first 7 minutes of porcine ventricular fibrillation. CPR is initiated at 5 minutes as indicated. (From Watson, J. N. et al., *Resuscitation*, 43, 121–127, 2000.)

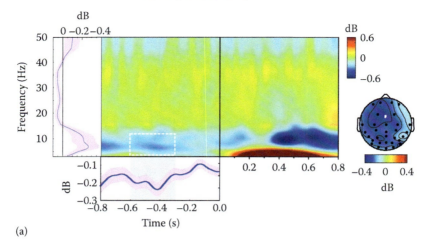

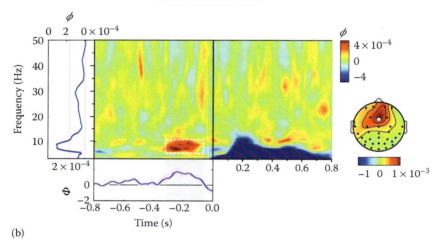

FIGURE 6.25 Raw effects of oscillatory power and phase: (a) Difference in spectral power in decibels between hits and misses, averaged across channels and subjects. Negative values indicate stronger power for misses. Left inset shows power difference averaged across time points in the prestimulus window (shaded areas SEM). The main difference is found in the 6–12 Hz frequency range. Bottom inset: Power difference averaged across frequencies in this range, with a maximally negative difference in the prestimulus time range between −600 and −300 ms (grey shaded area). The topography shows the distribution of the power difference from 6 to 12 Hz and from −600 to −300 ms preceding stimulus onset. (b) Phase bifurcation index (Φ), averaged across all channels and subjects. Positive values indicate that phase distributions are locked to different phase angles for hits and misses (e.g. in the prestimulus time range), while negative values indicate that only one condition is phase locked (e.g. phase locking exclusively for hits in the ERP time range). Left inset shows Φ averaged across all time points in the pre-stimulus window (vertical lines represent the 95% confidence interval). Bottom inset: Φ averaged across frequencies between 6 and 10 Hz (shaded area SE). Phase bifurcation is strongest from −300 to −50 ms preceding stimulus onset (grey shaded area). The topography shows the distribution of Φ from 6 to 10 Hz and from −300 to −50 ms preceding stimulus onset. (From Busch, N. A. et al., *Journal of Neuroscience*, 29(24), 7869–7876, 2009.)

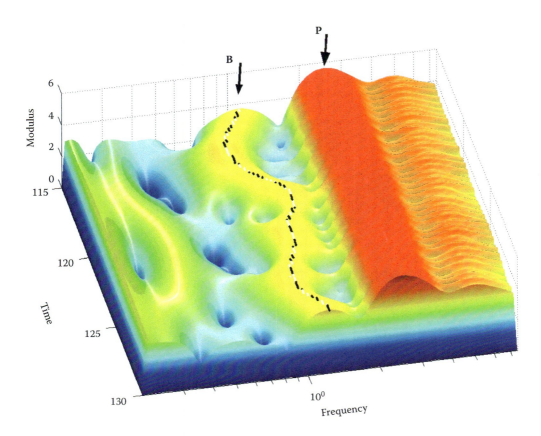

FIGURE 6.34 A wavelet transform of a pulse oximeter signal from a premature baby. Data were taken over a 15-second period using a wavelet of different frequencies. The 'P' band corresponds to the baby's regular pulse rate. The 'B' ridge corresponds to the baby's breathing rate, while the alternate black and white markings that lie along the peak of the ridge indicate the inspiration and expiration of breath, respectively. This breathing ridge is determined automatically by an algorithm that searches the transform surface for maxima and decides which of these correspond to respiration. The decision criteria are based on a careful study of the link between wavelet features and patient respiration. A complex wavelet was used in the analysis, which means that the vertical axis is actually the square of the modulus of the transform. (From Addison, P., *Physics World*, 17(3), 35–39, 2004.)

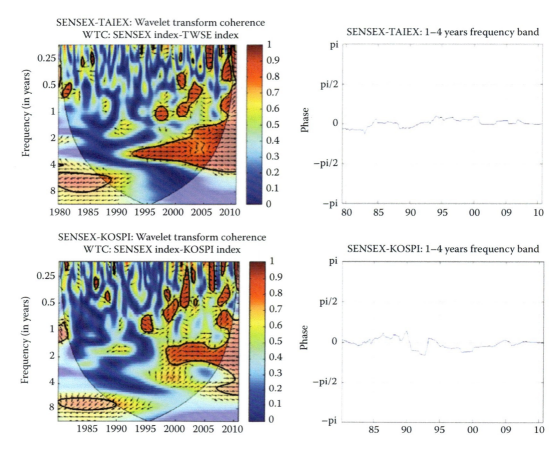

FIGURE 7.14 Wavelet squared coherency and phase difference for overall stock markets. The left column shows wavelet squared coherency and phase difference. The 5% significance level against red noise is shown as a thick contour. The relative phase relationship is indicated by arrows: Right: in-phase; Left: anti-phase; Down: first series leading second series by 90°; Up: second series leading first series by 90°. The right column shows wavelet phase-difference at frequency band 1–4 years. (From Dewandaru, G. et al., *Physica A: Statistical Mechanics and Its Applications*, 419, 241–259, 2015.)

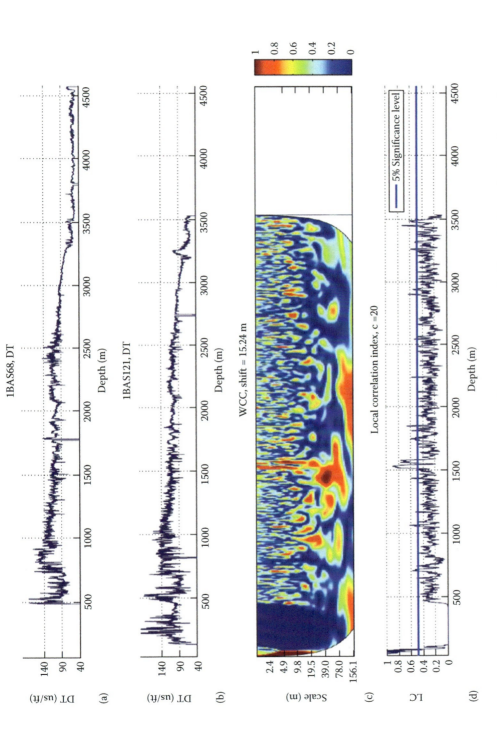

FIGURE 7.17 Colour contour plot for WCC between the DT logs of the I-BAS-68-BA and I-BAS-121-BA wells: (panels **a** and **b**) are shown in panel (**c**). Strong correlations near the depth position corresponding to 1524 m can be seen in all the scales, as shown by the dark vertical lines. This high degree of correlation was obtained when the first set of data was displaced by 100 units (15.24 m), relative to the second set. In panel (**d**), the local correlation index, where the x-axis represents the corresponding depths in the second well log. (From Henriques, M. V. C. et al., *Physica A: Statistical Mechanics and its Applications*, 417, 130–140, 2015.)

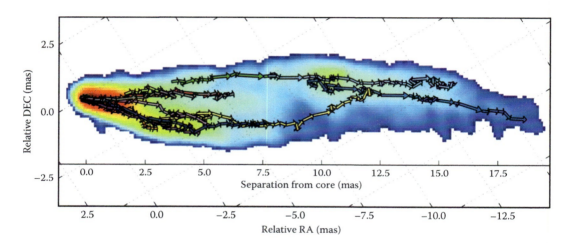

FIGURE 7.28 **Two-dimensional tracks of structural patterns detected by WISE at scales 3–4 of the SWD.** The tracks are overplotted on a stacked-epoch image of the jet rotated by an angle of 0.55 rad. Several generic 'flow lines' are clearly visible in the jet. These patterns are difficult to detect with the standard Gaussian model-fitting analysis. The image is rotated. (From Mertens, F., and Lobanov, A., *Astronomy and Astrophysics*, 574(A67), 1–14, 2015. Copyright ESO. Reproduced with permission.)

Fractals, Finance, Geophysics, Astronomy and Other Areas ■ 367

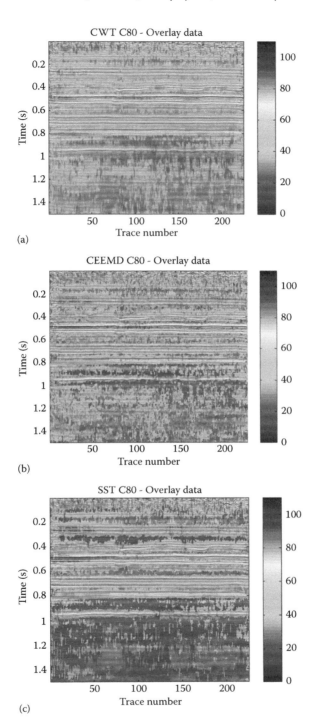

FIGURE 7.20 Characteristic frequencies for the vertical cross section. C80 attribute for **(a)** CWT, **(b)** CEEMD and **(c)** SST. CEEMD and SST show a sparser representation than the CWT. SST has even less speckle noise, and the strong reflector at 0.9 s is better represented. (From Herrera, R. H. et al., *Geophysics*, 79(3), V55–V64, 2014.)

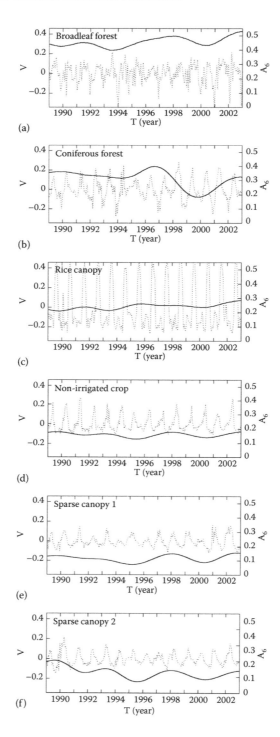

FIGURE 7.21 NDVI time series decomposition into the intra-annual component V (dashed line and left *y*-axis) and the inter-annual component A6 (bold line and right *y*-axis) for the selected vegetation canopies. (From Martínez, B., and Gilabert, M. A., *Remote Sensing of Environment*, 113, 1823–1842, 2009.)

scale, including mean and minimum NVDI, the amplitude of the phonological cycle, the timing of the maximum NVDI and the magnitude of the land cover change. The interannual components of the sparse canopies exhibited high inter-annual variability due to their sensitivity to precipitation whereas the rice canopy exhibited a much flatter curve due to it being properly irrigated. A technique for estimating the individual heights (HTs) and crown diameters (CDs) of conifer trees based on the wavelet transformation of Lidar data has been developed by Falkowski et al. (2006). Their method employed 2-D Mexican hat wavelets and is described schematically in Figure 7.22a The authors found that spatial wavelet analysis (SWA) performed slightly better than a variable window filter (VWF) when it came to crown diameter estimation, although no difference could be discerned between the two methods for determining heights. The scatter plots of the two method estimates against field-measured crown diameters are also shown in Figure 7.22b from

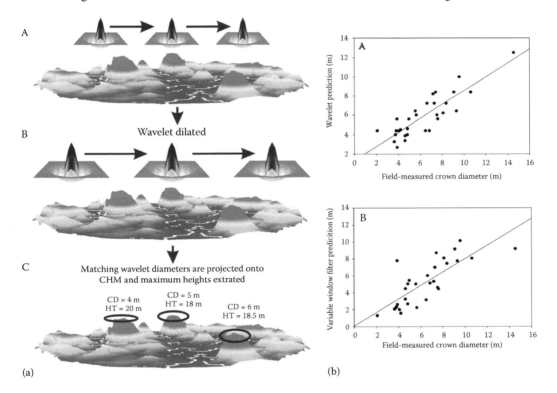

FIGURE 7.22 (a) A schematic diagram of the process used to determine tree crown diameters and tree heights in this study. (A) The smallest 2-D Mexican hat daughter wavelet is convolved with the lidar-derived canopy height model (CHM). (B) The 2-D daughter wavelets are then dilated in 0.1 m increments across a series of scales and convolved (at each scale) with the CHM. (C) The size of the 2-D daughter wavelet is retained when it best "matches" the size of features within the CHM (tree crowns). The height of each tree (HT) is then determined by extracting the maximum height from the CHM within each predicted crown diameter (CD). (b) Scatterplots of field-measured tree crown diameters versus tree crown diameters predicted by (A) spatial wavelet analysis (SWA) and (B) variable window filters (VWF). (From Falkowski, M. J. et al., *Canadian Journal of Remote Sensing*, 32(2), 153–161, 2006.)

which a correlation (R) value of 0.79 and RMSD of 1.66 m for VWF, compared with 0.86 and 1.35 m, respectively, for SWA, was calculated from the data. In other work, Cheng et al. (2011) developed a technique to determine leaf water content using a Mexican hat–based transform approach, which relies on using wavelet features which are strongly sensitive to changes in water content and insensitive to variation in leaf structure properties across species. More recently, Bell et al. (2015) analyzed a 28-year time series of giant kelp biomass derived from Landsat imagery using the Morlet wavelet transform.

Czuba et al. (2015) has developed a wavelet method for the detection of undocumented levees in remotely sensed data. These undocumented, and usually unmaintained, levees complicate flood forecasting, risk management and emergency response. The algorithm involved thresholding wavelet coefficients at various scales and threshold values to identify features consistent with levees. The authors checked the results against site visits and an alternative (hillshade) method. They suggested that a complimentary technique using both the wavelet and hillshade method may provide more reliable results more efficiently. Channels and flood plains were also the subject of investigation by McKean et al. (2009), who decomposed the channel thalweg (line of deepest river channel) height profile using a 6th order Gaussian wavelet transform. The transform spectral power was computed along this Lidar-derived profile and a distinct boundary was observed between the meandering channel with its undulating bedforms and the straight channel with a predominantly plane bedform. This is shown in Figure 7.23. More recently, Pan et al. (2015) employed a CWT-based method to assess the performance of a high-resolution airborne full-waveform Lidar for shallow river bathymetry. See also Booth et al. (2009), who performed a landslide inventory map using 2-D CWTs and Lescarmontier et al. (2012), who analyzed GPS variations in the Metz glacier ice tongue and found several dominant periods associated with waves propagating along and across the glacier.

A wavelet-based image registration algorithm for aligning remotely sensed images has been proposed by Hong and Zhang (2008). Their technique employs modulus maxima edge detection to determine feature points which were then used within a mapping scheme to rectify the image. Figure 7.24 illustrates the generation of feature points by their method. They found that the technique provided improved registration results by enabling semi-automation of high-volume data and reduced local distortions caused by local height variation. Other work of interest in the remote sensing arena includes an image fusion technique comprising a combined wavelet transform and sparse representation method (Cheng et al., 2015); a multi-temporal image change detection technique using undecimated wavelet transforms (Celik and Ma, 2011); a novel image classification method for remote sensing data sets using overcomplete wavelet transforms (Myint et al., 2015); a method for the extraction of roads from sensed images using quaternionic wavelets (Naouai et al., 2011); a wavelet fuzzy hybrid scheme for land cover classification (Shankar et al., 2011); a 2-D wavelet approach for identifying potential fields (Sailhac and Gibert, 2003); a wavelet ridge–based phase extraction method of extracting precession, obliquity and eccentricity characteristics of frequencies from geomagnetic intensity signals (Saracco et al., 2009); a lithosphere anisotropy mapping technique using 2-D wavelet coherence (Kirby and

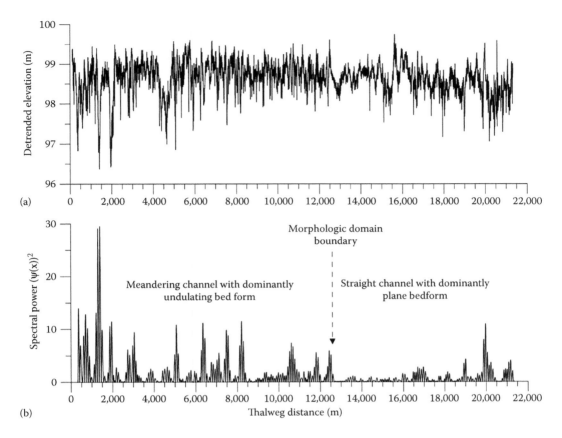

FIGURE 7.23 Gaussian (order 6) transform of a channel profile using a wavelet dilation of 100 m: **(a)** De-trended elevation profile along the channel thalweg. **(b)** Spatial distribution of spectral power in the thalweg profile. The transition from a meandering pool–riffle channel with undulating bed topography to a straight channel with plane bed topography is noted at the dashed arrow. (From McKean, J. et al., *Remote Sensing*, 1, 1065–1096, 2009.)

Swain, 2006); and a combined wavelet packet–grey model for feature extraction in hyperspectral data (Yin et al., 2013).

7.5 ASTRONOMY: SIGNALS AND IMAGES

Wavelet methods have made a significant impact in the study of astronomical signals and images, originating from both within our own solar system and beyond. In this section, we briefly review a number of articles covering a broad range of current topics in this field.

Much effort has gone into the wavelet analysis of solar phenomena. For example, Li et al. (2010) employed the cross-wavelet transform and wavelet coherence to investigate the phase synchrony of monthly solar flare indices of the northern and southern hemispheres which have 10.7- and 10.1-year cycles, respectively. They found that the cross-wavelet transform revealed fluctuating phase angles at frequencies corresponding to the Schwabe cycle. Biskri et al. (2010) used thresholded Morlet decompositions of solar coronal magnetic loops to extract their underlying structure. Figure 7.25 contains a plot of the maxima of an

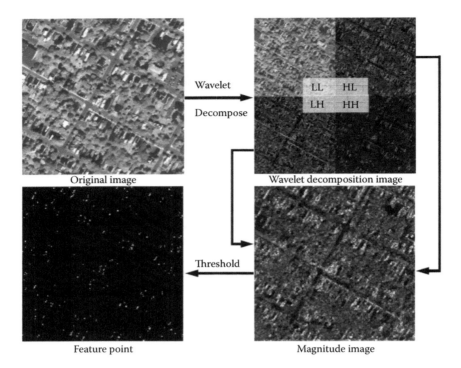

FIGURE 7.24 Workflow of feature point extraction based on wavelet decomposition (image size: 512×512 pixels, 1 pixel = 1 m). Width of image is 512 m. (From Hong, G., and Zhang, Y., *Computers and Geosciences*, 34, 1708–1720, 2008.)

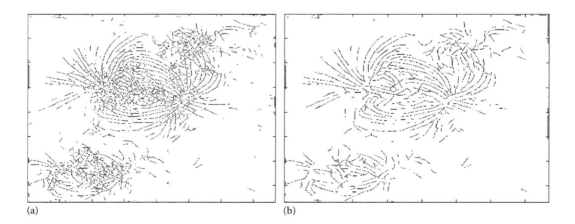

FIGURE 7.25 Segmentation of the transition region and coronal explorer (TRACE) image: (a) The noisy segmented image. (b) The same after the global and local thresholding. (From Biskri, S. et al., *Solar Physics*, 262, 373–385, 2010.)

image containing solar magnetic loops generated using the real part of a 2-D Morlet wavelet. The figure also contains the filtered image obtained after global and local thresholding. The authors found that the approach was both robust and fast for this image segmentation task. Ravindra and Javaraiah (2015) applied Morlet wavelets in their analysis of solar sunspot activity in cycle 23 and the rising phase of 24. They found that the wavelet power spectrum revealed several different periodicities in the north–south hemisphere asymmetries of sunspot area data. (See also the earlier investigation of the monthly averaged sunspot index using wavelet ridges by Polygiannakis et al., 2003.) Other studies of solar-related phenomena interrogated using the wavelet method include the work of Wintoft (2005), who investigated the coupling of the solar wind to the time difference horizontal geomagnetic field; Bloomfield et al. (2004), who applied wavelet phase coherence analysis to a quiet-sun magnetic element; Lundstedt et al. (2006), who explored solar activity using wavelet transform methods including skeletons of the group sunspot number and ^{14}C production rate time series; Vaquero et al. (2010), who considered the 155-day periodicity of solar flares; Johnson (2010), who studied solar magnetic activity using an edge-adapted wavelet transform and a renormalized admissible wavelet transform; McAteer et al. (2010), who employed Mexican hat wavelet analysis in their examination of solar flare activity; Kestener et al. (2010), who probed the multifractal properties of solar photospheric magnetic structures using modulus maxima methods; Bershadskii (2010), who performed wavelet regression de-trended fluctuation analysis of monthly subspot numbers; Deng et al. (2015), who studied the daily coronal index using Morlet wavelet analysis and empirical mode decomposition; and the work of Lopez-Montes et al. (2015), who performed a combined wavelet and fractal analysis of mid-latitude ionospheric disturbances associated with major geomagnetic storms occurring due to the arrival of coronal mass ejections. Finally, Poluianov and Usoskin (2014) performed a critical analysis of the hypothesis of the planetary tidal influence on solar activity using wavelet coherence between torque and heliosperic modulation potential. They showed no statistical significance and proposed that the hypothesis was not based on solid ground.

An in-depth examination of the radial structure of Saturn's rings has been conducted by Tiscareno et al. (2007) using Morlet wavelets with variable central frequencies (see also Addison et al., 2002a, in this regard). Figure 7.26 shows a close-up image of one of the sections of the ring structure used in the work along with a context image. The 'corduroy' pattern in the figure is caused by moonlet wakes excited by Pan (one of Saturn's moons). The wavelet analysis of a radial scan of the close-up image is shown in Figure 7.27. The upper wavelet plot, at a lower central frequency, is better at highlighting the density waves whereas the lower plot, at a relatively higher central frequency, clearly shows three frequency profiles associated with the moonlet wake structures. The authors found the wavelet method to be a powerful technique for identifying waves that are very weak or packed closely together, especially because the quasi-periodic radial structures have frequencies that vary with radius. More recently, Tiscareno et al. (2013) again employed wavelet analysis to interrogate the Iapetus −1:0 nodal bending wave, an extensive spiral wave in Saturn's rings whose variations in radial frequency indicate variations in the underlying ring surface density. Colwell et al. (2009) and Baillie et al. (2011) used a 'weighted wavelet Z' transform,

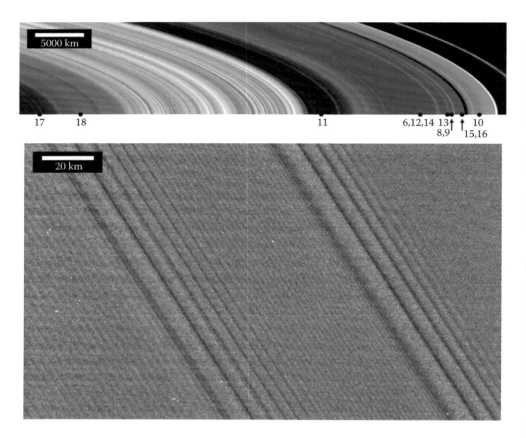

FIGURE 7.26 **Top:** Location within Saturn's main ring system of the lower figure marked by arrow at points 15, 16. **Bottom:** A portion of Cassini image N1467346329, showing a 'corduroy' pattern caused by moonlet wakes excited by Pan. Also visible are the Pandora 11:10 and Prometheus 15:14 density waves. (From Tiscareno, M. S. et al., *Icarus*, 189, 14–34, 2007.)

based on the Morlet wavelet, to add the wavelet transform of a number of individual noisy profiles. Hedman et al. (2016) further refined the method by accounting for the effect of the wave's driving frequency on the wavelet's complex phase while adding multiple noisy profiles, in order to boost the signal of the target wave. Additionally, Hedman et al. (2014) have explored overstabilities in Saturn's A ring, and Rein and Latter (2013) have employed wavelet transforms to elucidate results of simulations in viscous overstability in the rings.

Wavelet transforms have also been utilized to study many phenomena outside our solar system. Mertens and Lobanov (2015) have developed a wavelet-based image structure evaluation (WISE) method for multiscale decomposition segmentation and the tracking of structural patterns in astronomical images. Figure 7.28 contains the two-dimensional tracks detected in the parsec scale radio jet of the Quasar 3C 273, calculated using their WISE analysis technique. The tracks are superimposed on a single epoch image of the jet. The authors state that the method offers an effective and objective way to classify structural patterns in images of astronomical objects and to track their evolution traced by multiple observations of the same object. Wavelet decomposition was carried out by Krivonos et al.

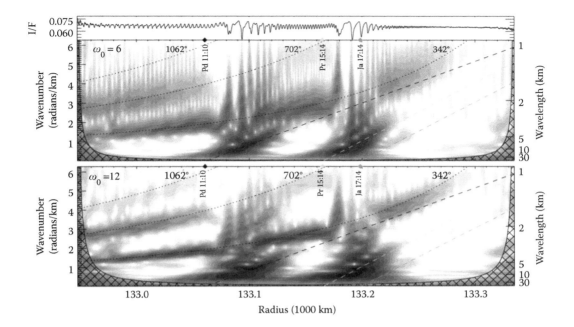

FIGURE 7.27 **Radial scan and two wavelet transforms from Cassini image N1467346329, showing wakes excited by Pan along with several density waves.** The upper wavelet plot uses a central frequency $\omega_0 = 6$; the lower plot uses $\omega_0 = 12$, resulting in increased resolution in the spectral (y) dimension at the expense of smearing in the radial (x) dimension. Density waves are clearer in the upper plot, including the strong Pandora 11:10 and Prometheus 15:14 waves, but also the weak third-order Janus 17:14. Dashed lines indicate model density wave traces, assuming a background surface density $\sigma_0 = 40$ g/cm². Moonlet wakes excited by Pan are clearer in the lower plot. The three dotted lines denote the frequency profiles of wakes that have travelled 342°, 702° or 1062° in synodic longitude since their last encounters with Pan. (From Tiscareno, M. S. et al., *Icarus*, 189, 14–34, 2007.)

(2014) during their study of x-ray emissions from a young densely packed massive star cluster and Finogeunov et al. (2015) applied wavelets in the reconstruction of x-ray emission maps. Siwak et al. (2014) performed a Morlet wavelet analysis of the light curve from T-Tauri type star TWHya. Figure 7.29 contains the resulting transform of the light curve showing a stable oscillation of about 4.18 days. See also Gazak et al. (2012), who have presented a graphical user interface software package for the analysis of exoplanet transit light curves, which incorporates a wavelet-based likelihood function which parametrizes both correlated and uncorrelated noise, allowing for accurate uncertainty measurement.

Leonard et al. (2015) performed wavelet multiresolution analysis of weak lensing mass reconstructions (see also Pires et al., 2010 in this regard) and a technique for the detection of dark matter's substructure using a multiresolution method for gravitational lensing has been employed by Bacon et al. (2010). Gupta and Kumar (2015) have examined an asymmetric triaxial galaxy computer model enclosed by a spherical dark halo component using wavelet transform ridges. Shaviv et al. (2014) analyzed oxygen isotopes in the fossil record to search for an imprint of the solar system's galactic motion, their findings suggested the

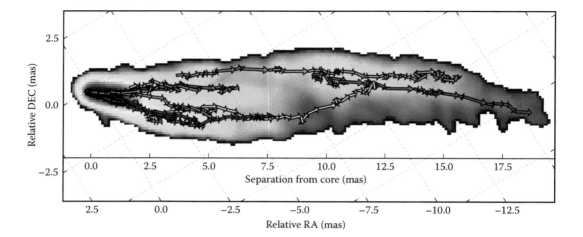

FIGURE 7.28 (See colour insert.) **Two-dimensional tracks of structural patterns detected by WISE at scales 3–4 of the SWD.** The tracks are overplotted on a stacked-epoch image of the jet rotated by an angle of 0.55 rad. Several generic 'flow lines' are clearly visible in the jet. These patterns are difficult to detect with the standard Gaussian model-fitting analysis. The image is rotated. (Original figure in colour.) (From Mertens, F., and Lobanov, A., *Astronomy and Astrophysics*, 574(A67), 1–14, 2015. Copyright ESO. Reproduced with permission.)

presence of a disk dark matter component. See also Hojjati et al. (2010), who studied the utility of wavelets to detect redshift evolution in the dark energy equation of state.

In other work, the cold spot in the cosmic microwave background has been extensively studied by Vielva (2010) using the kurtosis of the spherical Mexican hat wavelet coefficients and Paykari et al. (2012) have investigated the sparsity of the cosmic microwave background using wavelet and discrete cosine transforms. Redshift-space enhancement of line-of-sight baryon acoustic oscillations has been carried out by Tian et al. (2011) and, in later work, Everett et al. (2015) investigated baryonic acoustic oscillations in the matter power spectrum and concluded that the wavelet method has many advantages over the traditional Fourier technique. Batista et al. (2011) developed a technique to amplify the signal-to-noise ratio of cosmic ray maps using Mexican hat wavelet families. Young et al. (2012) employed a weighted wavelet-Z statistic in their analysis of pulse intensity modulations of a pulsar. Lachowicz and Done (2010) used both Morlet-based wavelet transforms and matching pursuits decompositions to study the lack of coherence associated with quasi-periodic oscillations observed in black hole binary systems. Masias et al. (2015) considered two wavelet methods – a stationary wavelet transform and a Mexican hat–based transform – in a combined multiscale distilled sensing scheme for source detection in long wavelength images. See also the earlier work of Masias et al. (2012), who reviewed source detection approaches in astronomical images where a large number of wavelet methods are listed and discussed.

It can be seen then, that wavelets are now commonplace in astronomy. Other work of interest includes an examination of the nature of galactic radio signals (Sukharev and Aller, 2014); an investigation of the rotational behaviour of Kepler stars with planets

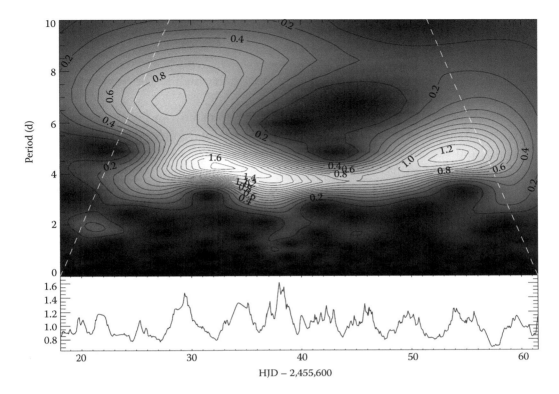

FIGURE 7.29 The Morlet-6 wavelet transform of the 2011 TW Hya microvariability and oscillations of stars telescope (MOST) data. The amplitudes of the transform are expressed by greyscale intensities and contours. Edge effects are present outside the white broken lines but they do not affect our conclusions. At the bottom, the MOST light curve (in mean star level flux units), resampled into a uniformly distributed time grid with 0.07047 d spacing, is shown. (From Siwak, M. et al., *Monthly Notices of the Royal Astronomical Society*, 444(1), 327–335, 2014. Used with permission from Oxford University Press and the authors.)

(Paz-Chinchon et al., 2015); an analysis of the hierarchically clustered young stellar populations in the Small Megallenic Cloud (Gouliermis et al., 2014); the search for sterile neutrinos (Mertens et al., 2015); the separation of galaxy clusters from the low-frequency radio sky (Wang et al., 2010); Faraday rotation measure synthesis (Frick et al., 2010); the mapping of particle acceleration in the cool core of galaxy clusters (Giacintucci et al., 2014); the enhanced cross-correlation of gravitational waves using a wavelet entropy filter (Terenzi and Sturani, 2009); the development of a radio de-convolution algorithm which employs an isotropic undecimated wavelet transform (Dabbech et al., 2015); the decomposition of de-reddened Hα emission data in a study relating dust, gas and the rate of star formation (Tabatabaei and Berkhujsen, 2010); the extraction of redshifted 21 cm sky survey signal from foreground noise (Gu et al., 2013); and the investigation of fractal nature of gamma ray bursts (Maclachlan et al., 2013 – already mentioned in Section 7.2.2). See also Romeo et al. (2004), who summarized wavelet add-on code for denoising N-body simulations.

7.6 OTHER AREAS

This final section briefly details some of those areas not covered elsewhere in the book including chemistry, quantum mechanics, biological phenomena and a few more unusual ones. These are not covered in great detail and only a very high-level overview is provided involving a limited number of examples.

A number of workers have employed the wavelet transform in the study of quantum mechanics. Ashmead (2012) has investigated the use of Morlet wavelets in quantum mechanics and has provided an explicit form of the inverse Morlet transform and the manifestly covariant form of the four-dimensional Morlet wavelet. Altaisky and Kaputkin (2013) and Brennen et al. (2014) have described the application of the continuous wavelet transform in quantum field theory, while Di Falco et al. (2012) have reported on their use of wavelets to study the lifetime statistics of a quantum chaotic resonator.

In chemistry, Harrop et al. (2002) applied a modified Morlet wavelet in an investigation of the structure of disordered glass materials. Later, Kopriva et al. (2009) used the Morlet wavelet to extract multiple component ^1H and ^{13}C NMR spectra from two mixtures. More recently, Cai et al. (2015) used Daubechies DB12 wavelets to separate peaks in near-infrared spectroscopy (NIRS) data. See also Hoang (2014), who has comprehensively reviewed wavelet-based spectral analysis in analytical chemistry. Examples of other applications in this area include that of Otal et al. (2015), who characterized the structure of Yb doped $Z_n O$ using Cauchy continuous wavelets; Rios et al. (2015), who investigated the corrosion of steel in crude oil using a wavelet analysis of electrochemical noise; Sohrabi and Zarkesh (2014), who improved the spectral resolution of components of the anti-HIV drug, employing Daubechies wavelets; and Wang et al. (2011), who demonstrated the enhanced appearance of spectroscopic features through the use of wavelet transforms for quantitative chemical imaging of element diffusion in heterogeneous media.

Mi et al. (2005) have described the use of Morlet and Mexican hat wavelets for the detection of ecological patterns illustrating their work with the analysis of diversity patterns in mixed oak forests. Zaugg et al. (2008) have developed a method for the automated identification of birds from their wing-flapping effects manifest in radar signals. More recently, Recknagel et al. (2013) used wavelet coherency to study ecological relationships, thresholds and timelags determining phytoplankton community dynamics in a lake; Pittiglio et al. (2013) employed wavelets to monitor the elephant response to spatial heterogeneity in a savannah landscape; and Rodriguez et al. (2014) have developed a monthly forecasting model for anchovy catches using a stationary wavelet transform to separate the raw time series into annual and inter-annual components. See also the paper by Cazelles et al. (2013) which considers the impact of statistical tests on the wavelet analysis of ecological and epidemiological time series.

Features related to congested traffic have been extracted using Mexican hat wavelet decomposition by Zheng et al. (2011) and used to identify active bottlenecks, transient traffic and traffic oscillations. Malegori and Ferrini (2012) have applied wavelet transforms to atomic force microscopy (AFM) where they examined the forces in a thermally driven cantilever and Cho et al. (2013) have demonstrated a significantly improved the signal-to-noise ratio (SNR) of the atomic force microscopy (AFM) cantilever signal using a wavelet-based time–frequency filter. More unusual applications of the wavelet transform analysis includes the examination of the rise and fall of individual states and empires by Yoroshenko et al. (2015), who utilized a variety of wavelet functions for this task. There is also the edge detection system for the game of cricket that has been developed by Rock et al. (2011) using wavelet transform features of a microphone signal (based behind the stumps) fed into a neural network. Finally, there is the identification of features in pork and turkey ham images for classification purposes undertaken by Jackman et al. (2010). This latter work employed a wavelet decomposition to efficiently and effectively condense the textural information required for use in a genetic algorithm to discriminate various grades of the foods.

References

Abbas, A. K., Heimann, K., Jergus, K., Orlikowsky, T., and Leonhardt, S. (2011). Neonatal non-contact respiratory monitoring based on real-time infrared thermography. *Biomedical Engineering Online*, 10(93), 1–17.

Abdelnour, F., Schmidt, B., and Huppert, T. J. (2009). Topographic localization of brain activation in diffuse optical imaging using spherical wavelets. *Physics in Medicine and Biology*, 54(20), 6383–6413.

Abdilghanie, A. M., and Diamessis, P. J. (2013). The internal gravity wave field emitted by a stably stratified turbulent wake. *Journal of Fluid Mechanics*, 720, 104–139.

Abi-Abdallah, D., Chauvet, E., Bouchet-Fakri, L., Bataillard, A., Briguet, A., and Fokapu, O. (2006). Reference signal extraction from corrupted ECG using wavelet decomposition for MRI sequence triggering: Application to small animals. *Biomedical Engineering Online*, 5(11), 1–12.

Abid, A. Z., Gdeisat, M. A., Burton, D. R., and Lalor, M. J. (2007). Ridge extraction algorithms for one-dimensional continuous wavelet transform: A comparison. *Journal of Physics: Conference Series*, 76, 012045, 1–7.

Acharya, U. R., Fujita, H., Sudarshan, V. K., Sree, V. S., Eugene, L. W. J., Ghista, D. N., and San Tan, R. (2015). An integrated index for detection of sudden cardiac death using discrete wavelet transform and nonlinear features. *Knowledge-Based Systems*, 83, 149–158.

Adamczak, S., and Makieła, W. (2011). Analyzing variations in roundness profile parameters during the wavelet decomposition process using the Matlab environment. *Metrology and Measurement Systems*, 18(1), 25–34.

Adamo, F., Andria, G., Attivissimo, F., Lanzolla, A. M. L., and Spadavecchia, M. (2013). A comparative study on mother wavelet selection in ultrasound image denoising. *Measurement*, 46, 2447–2456.

Adamowski, J., Adamowski, K., and Prokoph, A. (2013). Quantifying the spatial temporal variability of annual streamflow and meteorological changes in eastern Ontario and southwestern Quebec using wavelet analysis and GIS. *Journal of Hydrology*, 499, 27–40.

Addison, P. S. (1997). *Fractals and Chaos: An Illustrated Course*. Bristol, UK: CRC Press.

Addison, P. S. (1999). Wavelet analysis of the breakdown of a pulsed vortex flow. *Proceedings of the Institution of Mechanical Engineers, Part C: Journal of Mechanical Engineering Science*, 213(3), 217–229.

Addison, P. S. (2004). The little wave with the big future. *Physics World*, 17(3), 35–39.

Addison, P. S. (2005). Wavelet transforms and the ECG: A review. *Physiological Measurement*, 26, R155–R199.

Addison, P. S. (2014). Wavelet analysis of biosignals: From Pretty Pictures to Product. Abstracts of papers presented at the 2014 Meeting of the Society for Technology in Anesthesia (STA), January 15–18, 2014. Anesthesia and Analgesia Supplement, December 2014.

Addison, P. S. (2015a). Running wavelet archetype: Time–frequency ensemble averaging requiring no fiducial points. *Electronics Letters*, 51, 1153–1155.

Addison, P. S. (2015b). A review of wavelet transform time–frequency methods for NIRS-based analysis of cerebral autoregulation. *IEEE Reviews in Biomedical Engineering*, 8, 78–85.

Addison, P. S. (2015c). Identifying stable phase coupling associated with cerebral autoregulation using the synchrosqueezed cross-wavelet transform and low oscillation Morlet wavelets. *In Engineering in Medicine and Biology Society (EMBC), 2015 37th Annual International Conference of the IEEE* (pp. 5960–5963), August 25–29, 2015. Milan, Italy: IEEE.

Addison, P. S. (2016). Modular continuous wavelet processing of biosignals: Extracting heart rate and oxygen saturation from a video signal. *Healthcare Technology Letters 3*, 2, 111–115. In print.

Addison, P. S., McGonigle, S., and Watson, J. N. (2011). Systems and methods for signal rephasing using the wavelet transform. *U.S. Patent Application Number: 13/149,755*. Filing Date: May 31, 2011. Publication Date: December 6, 2012. (U.S. Patent 2012/0310051)

Addison, P. S., Morvidone, M., Watson, J. N., and Clifton, D. (2006). Wavelet transform reassignment and the use of low-oscillation complex wavelets. *Mechanical Systems and Signal Processing, 20*, 1429–1443.

Addison, P. S., Murray, K. B., and Watson, J. N. (2001). Wavelet transform analysis of open channel wake flows. *ASCE Journal of Engineering Mechanics, 127*(1), 58–70.

Addison, P. S., and Ndumu, A. S. (1999). Engineering applications of fractional Brownian motion: Self-affine and self-similar random processes. *Fractals, 7*, 151–157.

Addison, P. S., Walker, J., and Guido, R. C. (2009). Time-frequency analysis of biosignals. *IEEE Engineering in Medicine and Biology Magazine, 28*(5), 14–29.

Addison, P. S., and Watson, J. N. (2003). Analysis of acoustic medical signals. *WO/2003/055395 A1 Patent Application PCT/GB2002/005922*. Filing Date: December 24, 2002. Publication Date: July 10, 2003.

Addison, P. S., and Watson, J. N. (2003). Secondary wavelet feature decoupling (SWFD) and its use in detecting patient respiration from the photoplethysmogram. *Proceedings of the 25th Annual International Conference of the IEEE Engineering in Medicine and Biology Society, 3* (pp. 2602–2605). IEEE.

Addison, P. S., and Watson, J. N. (2004a). Secondary transform decoupling of shifted nonstationary signal modulation components: Application to photoplethysmography. *International Journal of Wavelets, Multiresolution and Information Processing, 2*(1), 43–57.

Addison, P. S., and Watson, J. N. (2004b). A novel time–frequency-based 3D Lissajous figure method and its application to the determination of oxygen saturation from the photoplethysmogram. *Measurement Science and Technology, 15*, L1–L4.

Addison, P. S., and Watson, J. N. (2005). Oxygen saturation determined using a novel wavelet ratio surface. *Medical Engineering and Physics, 27*(3), 245–248.

Addison, P. S., and Watson, J. N. (2010). Methods and apparatus for calibrating respiratory effort from photoplethysmography signals. *U.S. Patent Application Number: 12/771,792*. Filing Date: April 30, 2010. Publication Date: October 13, 2015. (U.S. Patent 2011/0270114) (Granted Patent Number: U.S. 9155493)

Addison, P. S., and Watson, J. N. (2015). Non-stationary feature relationship parameters for awareness monitoring. *U.S. Patent Application Number: 14/606,943*. Filing Date: January 27, 2015. Publication Date: July 30, 2015. (U.S. Patent 2015/0208940)

Addison, P. S., Watson, J. N., Clegg, G. R., Holzer, M., Sterz, F., and Robertson, C. E. (2000). Evaluating arrhythmias in ECG signals using wavelet transforms. *IEEE Engineering in Medicine and Biology Magazine, 19*, 104–109.

Addison, P. S., Watson, J. N., Clegg, G. R., Steen, P. A., and Robertson, C. E. (2002b). Finding coordinated atrial activity during ventricular fibrillation using wavelet decomposition. *IEEE Engineering in Medicine and Biology Magazine, 21*, 58–65.

Addison, P. S., Watson, J. N., and Feng, T. (2002a). Low-oscillation complex wavelets. *Journal of Sound and Vibration, 254*(4), 733–762.

Addison, P. S., Watson, J. N., Mestek, M. L., and Mecca, R. S. (2012a). Developing an algorithm for pulse oximetry derived respiratory rate (RRoxi): A healthy volunteer study. *Journal of Clinical Monitoring and Computing, 26*, 45–51.

Addison, P. S., Watson, J. N., Mestek, M. L., Ochs, J. P., Uribe, A. A., and Bergese, S. D. (2015). Pulse oximetry-derived respiratory rate in general care floor patients. *Journal of Clinical Monitoring and Computing*, 29(1), 113–120.

Addison, P. S., Watson, J. N., Mestek, M. L., and Wolstencroft, J. (2012c). Flexible pulse oximeter sensor design for monitoring respiratory modulations from the photoplethysmogram: A feasibility demonstration. *IAMPOV International Symposium* (Abstract 5, pp. 40–41), June 29–July 1, 2012. New Haven, CT.

Addison, P. S., Watson, J. N., Ochs, J. P., Neitenbach, A. M., and Mestek, M. L. (2012b). Continuous non-invasive respiratory rate derived from pulse oximetry during cold room hypoxia. *IAMPOV International Symposium* (Abstract 4, pp. 38–39), June 29–July 1, 2012. New Haven, CT.

Ademoglu, A., Micheli-Tzanakou, E., and Istefanopuos, Y. (1997). Analysis of pattern reversal visual evoked potentials (PRVEP's) by spline wavelets'. *IEEE Transactions on Biomedical Engineering*, 44(9), 881–890.

Agnew, C. E., Hamilton, P. K., McCann, A. J., McGivern, R. C., and McVeigh, G. E. (2015). Wavelet entropy of Doppler ultrasound blood velocity flow waveforms distinguishes nitric oxide-modulated states. *Ultrasound in Medicine and Biology*, 41(5), 1320–1327.

Aguiar-Conraria, L., Azevedo, N., and Soares, M. J. (2008). Using wavelets to decompose the time–frequency effects of monetary policy. *Physica A: Statistical Mechanics and its Applications*, 387, 2863–2878.

Aguiar-Conraria, L., and Soares, M. J. (2011). The continuous wavelet transform: A primer. NIPE Working Paper, 16, 1–43. NIPE: Núcleo de Investigaç00E3o em Políticas Económica, Universidade Do Minho.

Ahrens, M., Fischer, R., Dagen, M., Denkena, B., and Ortmaier, T. (2013). Abrasion monitoring and automatic chatter detection in cylindrical plunge grinding. *Procedia CIRP*, 8, 374–378.

Akansu, A. N., Serdijn, W. A., and Selesnick, I. W. (2010). Emerging applications of wavelets: A review. *Physical Communication*, 3, 1–18.

Akar, S. A., Kara, S., Latifoğlu, F., and Bilgiç, V. (2015). Investigation of the noise effect on fractal dimension of EEG in schizophrenia patients using wavelet and SSA-based approaches. *Biomedical Signal Processing and Control*, 18, 42–48.

Akay, M., Akay, Y. M., Welkowitz, W., and Lewkowicz, S. (1994). Investigating the effects of vasodilator drugs on the turbulent sound caused by femoral artery stenosis using short-term Fourier and wavelet transform methods. *IEEE Transactions on Biomedical Engineering*, 41(10), 921–928.

Akçakaya, M., Nam, S., Hu, P., Moghari, M. H., Ngo, L. H., Tarokh, V., Manning, W. J., and Nezafat, R. (2011). Compressed sensing with wavelet domain dependencies for coronary MRI: A retrospective study. *IEEE Transactions on Medical Imaging*, 30(5), 1090–1099.

Alam, M. M., and Sakamoto, H. (2005). Investigation of Strouhal frequencies of two staggered bluff bodies and detection of multistable flow by wavelets. *Journal of Fluids and Structures*, 20(3), 425–449.

Alam, M. M., Sakamoto, H., and Moriya, M. (2003). Reduction of fluid forces acting on a single circular cylinder and two circular cylinders by using tripping rods. *Journal of Fluids and Structures*, 18, 347–366.

Alam, M. M., Sakamoto, H., and Zhou, Y. (2006). Effect of a T-shaped plate on reduction in fluid forces on two tandem cylinders in a cross-flow. *Journal of Wind Engineering and Industrial Aerodynamics*, 94, 525–551.

Alam, M. M., and Zhou, Y. (2008). Strouhal numbers, forces and flow structures around two tandem cylinders of different diameters. *Journal of Fluids and Structures*, 24(4), 505–526.

Albulescu, C. T., Goyeau, D., and Tiwari, A. K. (2015). Contagion and dynamic correlation of the main European stock index futures markets: A time-frequency approach. *Procedia Economics and Finance*, 20, 19–27.

Alcântara, E. H., Stech, J. L., Lorenzzetti, J. A., and Novo, E. M. L. M. (2010). Cross wavelet, coherence and phase between water surface temperature and heat flux in a tropical hydroelectric reservoir. In *14th Workshop on Physical Processes in Natural Waters* (pp. 86–93), June 28–July 1, 2010. Reykjavik, Iceland: Faculty of Environmental Engineering, University of Iceland.

Alcaraz, R., and Rieta, J. J. *(2012a)*. Application of wavelet entropy to predict atrial fibrillation progression from the surface ECG. *Computational and Mathematical Methods in Medicine*, 2012, 245213, 1–9.

Alcaraz, R., and Rieta, J. J. (2012b). Central tendency measure and wavelet transform combined in the non-invasive analysis of atrial fibrillation recordings. *Biomedical Engineering Online*, *11*(46), 1–19.

Alhasan, A., White, D. J., and De Brabanterb, K. (2016). Continuous wavelet analysis of pavement profiles. *Automation in Construction*, *63*, 134–143.

Al-Hashmi, S., Rawlins, A., and Vernon, F. (2013). A wavelet transform method to detect P and S-Phases in three component seismic data. *Open Journal of Earthquake Research*, *2*, 1–20.

Allen, J., Di Maria, C., Mizeva, I., and Podtaev, S. (2013). Finger microvascular responses to deep inspiratory gasp assessed and quantified using wavelet analysis. *Physiological Measurement*, *34*, 769–779.

Almeida, R., Martinez, J. P., Rocha, A. P., and Laguna, P. (2009). Multilead ECG delineation using spatially projected leads from wavelet transform loops. *IEEE Transactions on Biomedical Engineering*, *56*(8), 1996–2005.

Aloui, C., Hkiri, B., and Nguyen, D. K. (2016). Real growth co-movements and business cycle synchronization in the GCC countries: Evidence from time-frequency analysis. *Economic Modelling*, *52*, 322–331.

Aloui, C., and Nguyen, D. K. (2014). On the detection of extreme movements and persistent behaviour in Mediterranean stock markets: A wavelet-based approach. *Applied Economics*, *46*(22), 2611–2622.

Al-Rousan, M., and Assaleh, K. (2011). A wavelet- and neural network-based voice system for a smart wheelchair control. *Journal of the Franklin Institute*, *348*, 90–100.

Altaisky, M. V., and Kaputkina, N. E. (2013). Continuous wavelet transform in quantum field theory. *Physical Review D*, *88*(2), 025015.

Amin, W., Davis, M. R., Thomas, G. A., and Holloway, D. S. (2013). Analysis of wave slam induced hull vibrations using continuous wavelet transforms. *Ocean Engineering*, *58*, 154–166.

An, M. S., Lim, H. Y., and Kang, D. S. (2013). *Design and realization of automatic fault diagnosis system of wind turbine based on LabVIEW*. In *The 2nd International Conference on Software Technology*. SoftTech, ASTL, *19* (pp. 175–178), Sandy Bay, Tasmania, Australia: Science and Engineering Research Support Society.

André, G., Marcos, M., and Daubord, C. (2013). Detection method of meteotsunami events and characterization of harbour oscillations in western Mediterranean. In *Coastal Dynamics 2013 Conference Proceedings* (pp. 83–92). June 24–28, 2013. Bordeaux, France: Bordeaux University.

Andrieş, A. M., Ihnatov, I., and Tiwari, A. K. (2016). Comovement of exchange rates: A wavelet analysis. *Emerging Markets Finance and Trade*, *52*(3), 574–588.

Antoine, J. P., Barachea, D., Cesar, R. M., and da Fontoura Costa, L. (1997). Shape characterization with the wavelet transform. *Signal Processing*, *62*, 265–290.

Antoine, J. P., Murenzi, R., Vandergheynst, P., and Ali, S. T. (2004). *Two-Dimensional Wavelets and Their Relatives*. Cambridge, UK: Cambridge University Press.

Arab, M. R., Suratgar, A. A., Martínez-Hernández, V. M., and Ashtiani, A. R. (2010). Electroencephalogram signals processing for the diagnosis of petit mal and grand mal epilepsies using an artificial neural network. *Journal of Applied Research and Technology*, *8*(1), 120–129.

Arai, K., and Andrie, R. (2011). Human gait gender classification using 2D discrete wavelet transforms energy. *IJCSNS International Journal of Computer Science and Network Security*, Vol II(12), 62–68.

Arai, K., and Asmara, R. A. (2014). Gender classification method based on gait energy motion derived from silhouette through wavelet analysis of human gait moving pictures. *International Journal of Information Technology and Computer Science (IJITCS)*, 6(3), 1–11.

Arai, K., Eguchi, Y., and Kitajima, Y. (2011). Extraction of line features from multifidus muscle of CT scanned images with morphologic filter together with wavelet multi resolution analysis. *International Journal of Advanced Computer Science and Applications. Special Issue on Artificial Intelligence*, 60–66.

Argoul, F., Arneodo, A., Elezgaray, J., Grasseau, G., and Murenzi, R. (1989). Wavelet transform of fractal aggregates. *Physics Letters A*, 135(6), 327–336.

Argyris, J., Faust, G., Haase, M., and Friedrich, R. (2015). *An Exploration of Dynamical Systems and Chaos: Completely Revised and Enlarged*, Second Edition (Section 8.5.2). Berlin: Springer.

Arneodo, A., Grasseau, G., and Holschneider, M. (1989). Wavelet transform analysis of invariant measures of some dynamical systems. In *Wavelets* (pp. 182–196). Berlin: Springer.

Arroyo, M. (2007). Wavelet analysis of pulse tests in soil samples. *Rivista Italiana di Italiana*, 30, 26–38.

Arumugam, S. S., Gurusamy, G., and Gopalasamy, S. (2009). Wavelet based detection of ventricular arrhythmias with neural network classifier. *Journal of Biomedical Science and Engineering*, 2(6), 439–444.

Arvinti, B., Isar, A., Stolz, R., and Costache, M. (2011). Performance of Fourier versus wavelet analysis for magnetocardiograms using a SQUID-acquisition system. In *Applied Computational Intelligence and Informatics (SACI), 2011 6th IEEE International Symposium on* (pp. 69–74). IEEE.

Arvinti-Costache, B., Costache, M., Stolz, R., Nafornita, C., Isar, A., and Toepfer, H. (2011). A wavelet based baseline drift correction method for fetal magnetocardiograms. In *New Circuits and Systems Conference (NEWCAS), 2011 IEEE 9th International* (pp. 109–112). IEEE.

Asgharian, H., and Nossman, M. (2011). Risk contagion among international stock markets. *Journal of International Money and Finance*, 30(1), 22–38.

Ashmead, J. (2012). Morlet wavelets in quantum mechanics. *Quanta*, 1, 58–70.

Atallah, L., Aziz, O., Gray, E., Lo, B., and Yang, G. Z. (2013). An ear-worn sensor for the detection of gait impairment after abdominal surgery. *Surgical Innovation*, 20(1), 86–94.

Atto, A. M., and Berthoumieu, Y. (2012). Wavelet packets of nonstationary random processes: Contributing factors for stationarity and decorrelation. *IEEE Transactions on Information Theory*, 58(1), 317–330.

Auger, F., and Flandrin, P. (1995). Improving the readability of time-frequency and time-scale representations by the reassignment method. *IEEE Transactions on Signal Processing*, 43(5), 1068–1089.

Auger, F., Flandrin, P., Lin, Y. T., McLaughlin, S., Meignen, S., Oberlin, T., and Wu, H. T. (2013). Time-frequency reassignment and synchrosqueezing: An overview. *IEEE Signal Processing Magazine*, 30(6), 32–41.

Awrejcewicz, J., Krysko, A. V., and Soldatov, V. (2009). On the wavelet transform application to a study of chaotic vibrations of the infinite length flexible panels driven longitudinally. *International Journal of Bifurcation and Chaos*, 19(10), 3347–3371.

Awrejcewicz, J., Krysko, A. V., Yakovleva, T. V., Zelenchuk, D. S., and Krysko, V. A. (2013). Chaotic synchronization of vibrations of a coupled mechanical system consisting of a plate and beams. *Latin American Journal of Solids and Structures*, 10, 163–174.

Awrejcewicz, J., Saltykova, O. A., Zhigalov, M. V., Hagedorn, P., and Krysko, V. A. (2011). Analysis of non-linear vibrations of single-layered Euler–Bernoulli beams using wavelets. *International Journal of Aerospace and Lightweight Structures*, 1(2), 203–219.

Baars, W. J., and Tinney, C. E. (2013). Transient wall pressures in an overexpanded and large area ratio nozzle. *Experiments in Fluids*, 54(2), 1–17.

Baars, W. J., Tinney, C. E., and Ruf, J. H. (2011). *Time-frequency analysis of rocket nozzle wall pressures during start-up transients*. Journal of Physics: Conference Series, 318(9), 092001, 1–10.

Bacon, D. J., Amara, A., and Read, J. I. (2010). Measuring dark matter substructure with galaxy-galaxy flexion statistics. *Monthly Notices of the Royal Astronomical Society*, 409, 389–395.

Bahoura, M. (2009). Pattern recognition methods applied to respiratory sounds classification into normal and wheeze classes. *Computers in Biology and Medicine*, 39, 824–843.

Bai, L., Yan, S., Zheng, X., and Chen, B. M. (2015). Market turning points forecasting using wavelet analysis. *Physica A: Statistical Mechanics and its Applications*. 437, 184–197.

Baillié, K., Colwell, J. E., Lissauer, J. J., Esposito, L. W., and Sremčević, M. (2011). Waves in Cassini UVIS stellar occultations: 2. The C ring. *Icarus*, 216, 292–308.

Bajaj, N., Leon, L. J., Vigmond, E., and Kimber, S. (2005). Fibrillation complexity as a predictor of successful defibrillation. In *27th Annual International Conference of the Engineering in Medicine and Biology Society, 2005* (pp. 768–771). IEEE.

Bakhshi, A. D., Ahmed, A., Gulfam, S., Khaqan, A., Yasin, A., Riaz, R., Alimgeer, K., Malik, S., Khan, S., and Dar, A. H. (2012). Detection of ECG T-wave alternans using maxima of continuous-time wavelet transform ridges. *Przegląd Elektrotechniczny*, 88(12b), 35–38.

Balakrishnan, S., Cacciola, M., Udpa, L., Rao, B. P., Jayakumar, T., and Raj, B. (2012). Development of image fusion methodology using discrete wavelet transform for eddy current images. *NDT & E International*, 51, 51–57.

Balasubramaniam, D., and Nedumaran, D. (2010). Efficient computation of phonocardiographic signal analysis in digital signal processor based system. *International Journal of Computer Theory and Engineering*, 2(4), 660–664.

Banerjee, S., and Mitra, M. (2012). An approach for ECG based cardiac abnormality detection through the scope of cross wavelet transform. In *2012 4th International Conference on Intelligent Human Computer Interaction (IHCI)* (pp. 1–6). IEEE.

Banfi, F., and Ferrini, G. (2012). Wavelet cross-correlation and phase analysis of a free cantilever subjected to band excitation. *Beilstein Journal of Nanotechnology*, 3, 294–300.

Bao, Y., Beck, J. L., and Li, H. (2010). Compressive sampling for accelerometer signals in structural health monitoring. *Structural Health Monitoring*, 10(3), 235–246.

Baruník, J., Vácha, L., and Krištoufek, L. (2011). Comovement of Central European stock markets using wavelet coherence: Evidence from high-frequency data. IES Working Paper No. 22/2011, 1–18.

Basak, A., Narasimhan, S., and Bhunia, S. (2011). KiMS: Kids' Health Monitoring System at daycare centers using wearable sensors and vocabulary-based acoustic signal processing. In *2011 13th IEEE International Conference on e-Health Networking Applications and Services (Healthcom)* (pp. 1–8). IEEE.

Batista, R. A., Kemp, E., and Daniel, B. (2011). Amplification of the signal-to-noise ratio in cosmic ray maps using the Mexican hat wavelet family. In *International Cosmic Ray Conference* (Vol.2, pp. 305–308). Beijing.

Bayraktar, E., Poor, H. V., and Sircar, K. R. (2004). Estimating the fractal dimension of the S & P 500 index using wavelet analysis. *International Journal of Theoretical and Applied Finance*, 7, 615–643.

Bayraktar, S., and Yilmaz, T. (2011). Experimental analysis of transverse jet using various decomposition techniques. *Journal of Mechanical Science and Technology*, 25(5), 1325–1333.

Beenamol, M., Prabavathy, S., and Mohanalin, J. (2012). Wavelet based seismic signal de-noising using Shannon and Tsallis entropy. *Computers and Mathematics with Applications*, 64, 3580–3593.

Behbahani, S., and Dabanloo, N. J. (2011). Detection of QRS complexes in the ECG signal using multiresolution wavelet and thresholding method. In *Computing in Cardiology*, 38, 805–808. IEEE.

Behera, A. K., Iyengar, A. S., and Panigrahi, P. K. (2014). Non-stationary dynamics in the bouncing ball: A wavelet perspective. *Chaos: An Interdisciplinary Journal of Nonlinear Science, 24*(4), 043107.

Bekiros, S., and Marcellino, M. (2013). The multiscale causal dynamics of foreign exchange markets. *Journal of International Money and Finance, 33*, 282–305.

Bell, T. W., Cavanaugh, K. C., and Siegel, D. A. (2015). Remote monitoring of giant kelp biomass and physiological condition: An evaluation of the potential for the Hyperspectral Infrared Imager (HyspIRI) mission. *Remote Sensing of Environment, 167*, 218–228.

Belo, D., Coito, A. L., Paiva, T., and Sanches, J. M. (2011). Topographic EEG brain mapping before, during and after obstructive sleep apnea episodes. In *Biomedical Imaging: From Nano to Macro, 2011 IEEE International Symposium on* (pp. 1860–1863). IEEE.

Belova, N. Y., Mihaylov, S. V., and Piryova, B. G. (2007). Wavelet transform: A better approach for the evaluation of instantaneous changes in heart rate variability. *Autonomic Neuroscience, 131*, 107–122.

Benedetto, J. J., Czaja, W., and Ehler, M. (2013). Wavelet packets for time-frequency analysis of multispectral imagery. *GEM-International Journal on Geomathematics, 4*(2), 137–154.

Benítez, R., Bolós, V. J., and Ramírez, M. E. (2010). A wavelet-based tool for studying non-periodicity. *Computers and Mathematics with Applications, 60*(3), 634–641.

Benkedjouh, T., Medjaher, K., Zerhouni, N., and Rechak, S. (2015). Health assessment and life prediction of cutting tools based on support vector regression. *Journal of Intelligent Manufacturing, 26*(2): 213–223.

Benouioua, D., Candusso, D., Harel, F., and Oukhellou, L. (2014). Fuel cell diagnosis method based on multifractal analysis of stack voltage signal. *International Journal of Hydrogen Energy, 39*, 2236–2245.

Berger, T. (2015). A wavelet based approach to measure and manage contagion at different time scales. *Physica A: Statistical Mechanics and its Applications, 436*, 338–350.

Berke, J. D. (2009). Fast oscillations in cortical-striatal networks switch frequency following rewarding events and stimulant drugs. *European Journal of Neuroscience, 30*(5), 848–859.

Bernardini, C., Benton, S. I., Chen, J. P., and Bons, J. P. (2014). Pulsed jets laminar separation control using instability exploitation. *AIAA Journal, 52*(1), 104–115.

Bernjak, A., Cui, J., Iwase, S., Mano, T., Stefanovska, A., and Eckberg, D. L. (2012). Human sympathetic outflows to skin and muscle target organs fluctuate concordantly over a wide range of time-varying frequencies. *Journal of Physiology, 590*(2), 363–375.

Bershadskii, A. (2010). Subharmonic and chaotic resonances in solar activity. *EPL (Europhysics Letters), 92*(5), 50012.

Bertoncini, C. A., Hinders, M. K., Thompson, D. O., and Chimenti, D. E. (2010). An ultrasonographic periodontal probe. *Review of Progress in Quantitative Nondestructive Evaluation, 29*, 1566–1573.

Beruvides, G., Quiza, R., del Toro, R., and Haber, R. E. (2013). Sensoring systems and signal analysis to monitor tool wear in microdrilling operations on a sintered tungsten–copper composite material. *Sensors and Actuators A: Physical, 199*, 165–175.

Bhanja, N., Dar, A. B., and Tiwari, A. K. (2012). Are stock prices hedge against inflation? A revisit over time and frequencies in India. *Central European Journal of Economic Modelling and Econometrics, 4*, 199–213.

Bhatawadekar, S. A., Leary, D., Chen, Y., Ohishi, J., Hernandez, P., Brown, T., McParland, C., and Maksym, G. N. (2013). A study of artifacts and their removal during forced oscillation of the respiratory system. *Annals of Biomedical Engineering, 41*(5), 990–1002.

Bilgin, S., Çolak, O. H., Polat, O., and Koklukaya, E. (2009). Determination of sympathovagal balance in ventricular tachiarrytmia patients with implanted cardioverter defibrillators using wavelet transform and MLPNN. *Digital Signal Processing, 19*, 330–339.

Biskri, S., Antoine, J. P., Inhester, B., and Mekideche, F. (2010). Extraction of solar coronal magnetic loops with the directional 2D Morlet wavelet transform. *Solar Physics, 262*, 373–385.

Blatter, C. (1998). *Wavelets: A Primer*. Natick, MA: AK Peters.

Bloomfield, D. S., McAteer, R. J., Lites, B. W., Judge, P. G., Mathioudakis, M., and Keenan, F. P. (2004). Wavelet phase coherence analysis: Application to a quiet-sun magnetic element. *Astrophysical Journal, 617*, 623.

Bogdan, A. (2010). Wavelet maxima based lacunarity texture analysis. In *2010 IEEE International Conference on Acoustics Speech and Signal Processing (ICASSP)* (pp. 1098–1101). IEEE

Boichat, N., Boichat, N., Atienza, D., and Khaled, N. (2009). Wavelet-based ECG delineation on a wearable embedded sensor platform. In *Wearable and Implantable Body Sensor Networks, 2009. BSN 2009. Sixth International Workshop on* (pp. 256–261). IEEE.

Boltežar, M., and Slavič, J. (2004). Enhancements to the continuous wavelet transform for damping identifications on short signals. *Mechanical Systems and Signal Processing, 18*, 1065–1076.

Bolzan, M. J. A., and Vieira, P. C. (2006). Wavelet analysis of the wind velocity and temperature variability in the Amazon forest. *Brazilian Journal of Physics, 36(4A)*, 1217–1222.

Bonhomme, V., Maquet, P., Phillips, C., Plenevaux, A., Hans, P., Luxen, A., Lamy, M., and Laureys, S. (2008). The effect of clonidine infusion on distribution of regional cerebral blood flow in volunteers. *Anesthesia and Analgesia, 106(3)*, 899–909.

Booth, A. M., Roering, J. J., and Perron, J. T. (2009). Automated landslide mapping using spectral analysis and high-resolution topographic data: Puget Sound lowlands, Washington, and Portland Hills, Oregon. *Geomorphology, 109*, 132–147.

Bordoloi, D. J., and Tiwari, R. (2014). Support vector machine based optimization of multi-fault classification of gears with evolutionary algorithms from time–frequency vibration data. *Measurement, 55*, 1–14.

Borodin, A., Pogorelov, A., and Zavyalova, Y. (2012). The cross-platform application for arrhythmia detection. In *Proceedings of the 12th Conference of Finnish-Russian University Cooperation in Telecommunications Program* (pp. 26–30).

Bostanov, V., and Kotchoubey, B. (2004). Recognition of affective prosody: Continuous wavelet measures of event-related brain potentials to emotional exclamations. *Psychophysiology, 41*, 259–268.

Bostanov, V., and Kotchoubey, B. (2006). The t-CWT: A new ERP detection and quantification method based on the continuous wavelet transform and Student's t-statistics. *Clinical Neurophysiology, 117*, 2627–2644.

Bousefsaf, F., Maaoui, C., and Pruski, A. (2013). Continuous wavelet filtering on webcam photoplethysmographic signals to remotely assess the instantaneous heart rate. *Biomedical Signal Processing and Control, 8*, 568–574.

Box, M. S., Watson, J. N., Addison, P. S., Clegg, G. R., and Robertson, C. E. (2008). Shock outcome prediction before and after CPR: A comparative study of manual and automated active compression–decompression CPR. *Resuscitation, 78(3)*, 265–274.

Bozhokin, S. V., and Suslova, I. M. (2013). Double wavelet transform of frequency-modulated nonstationary signal. *Technical Physics, 58(12)*, 1730–1736.

Bozhokin, S. V., and Suslova, I. B. (2014). Analysis of non-stationary HRV as a frequency modulated signal by double continuous wavelet transformation method. *Biomedical Signal Processing and Control, 10*, 34–40.

Bračič, M., and Stefanovska, A. (1998). Wavelet-based analysis of human blood-flow dynamics. *Bulletin of Mathematical Biology, 60*, 919–935.

Braga, C. C., Amanajás, J. C., Cerqueira, H. D. V., and Vitorino, M. I. (2014). The role of the tropical Atlantic and Pacific oceans SST in modulating the rainfall of Paraiba State, Brazil. *Revista Brasileira de Geofísica, 32(1)*, 97–107.

Brandner, P. A., Henderson, A. D., de Graaf, K. L., and Pearce, B. W. (2015b). Bubble breakup in a turbulent shear layer. *Journal of Physics: Conference Series, 656*, 012015, 1–4.

Brandner, P. A., Pearce, B. W., and De Graaf, K. L. (2015a). Cavitation about a jet in crossflow. *Journal of Fluid Mechanics, 768*, 141–174.

Brandner, P. A., Walker, G. J., Niekamp, P. N., and Anderson, B. (2010). An experimental investigation of cloud cavitation about a sphere. *Journal of Fluid Mechanics, 656*, 147–176.

Brazhe, A. R., Brazhe, N. A., Rodionova, N. N., Yusipovich, A. I., Ignatyev, P. S., Maksimov, G. V., Mosekilde, E., and Sosnovtseva, O. V. (2008). Non-invasive study of nerve fibres using laser interference microscopy. *Philosophical Transactions of the Royal Society of London A: Mathematical, Physical and Engineering Sciences, 366*, 3463–3481.

Brennen, G. K., Rohde, P., Sanders, B. C., and Singh, S. (2015). Multiscale quantum simulation of quantum field theory using wavelets. *Physical Review A, 92*(3), 032315.

Brenner, C. A., Kieffaber, P. D., Clementz, B. A., Johannesen, J. K., Shekhar, A., O'Donnell, B. F., and Hetrick, W. P. (2009). Event-related potential abnormalities in schizophrenia: A failure to "gate in" salient information? *Schizophrenia Research, 113*(2), 332, 1–14.

Briciu, A. E. (2014). Wavelet analysis of lunar semidiurnal tidal influence on selected inland rivers across the globe. *Scientific Reports, 4*(4193), 1–12.

Briggs, W. M., and Levine, R. A. (1997). Wavelets and field forecast verification. *Monthly Weather Review, 125*(6), 1329–1341.

Brol, S., and Grzesik, W. (2009). Continuous wavelet approach to surface profile characterization after finish turning of three different workpiece materials. *Advances in Manufacturing Science and Technology, 33*(1), 45–57.

Brouse, C. J., Karlen, W., Myers, D., Cooke, E., Stinson, J., Lim, J., Dumont, G. A., and Ansermino, J. M. (2011, August). Wavelet transform cardiorespiratory coherence detects patient movement during general anesthesia. In *Engineering in Medicine and Biology Society, EMBC, 2011 Annual International Conference of the IEEE* (pp. 6114–6117). IEEE.

Buchner, A. J., and Soria, J. (2013). Measurements of the three-dimensional topological evolution of a dynamic stall event using wavelet methods. *31st AIAA Applied Aerodynamics Conference 2013* (pp. 1736–1750), June 24–27, 2013. San Diego, CA: AAIA 2013–2818.

Burjánek, J., Gassner-Stamm, G., Poggi, V., Moore, J. R., and Fäh, D. (2010). Ambient vibration analysis of an unstable mountain slope. *Geophysical Journal International, 180*, 820–828.

Burke, M. J., and Nasor, M. (2012). ECG analysis using the Mexican-hat wavelet. In *Advances in Scientific Computing, Computational Intelligence and Applications*, Athens, *2001*, 26–31. WSES.

Burri, H., Chevalier, P., Arzi, M., Rubel, P., Kirkorian, G., and Touboul, P. (2006). Wavelet transform for analysis of heart rate variability preceding ventricular arrhythmias in patients with ischemic heart disease. *International Journal of Cardiology, 109*, 101–107.

Busch, N. A., Dubois, J., and VanRullen, R. (2009). The phase of ongoing EEG oscillations predicts visual perception. *Journal of Neuroscience, 29*(24), 7869–7876.

Busch, N. A., and VanRullen, R. (2010). Spontaneous EEG oscillations reveal periodic sampling of visual attention. *Proceedings of the National Academy of Sciences, 107*(37), 16048–16053.

Büssow, R. (2007). An algorithm for the continuous Morlet wavelet transform. *Mechanical Systems and Signal Processing, 21*(8), 2970–2979.

Cahn, B. R., Delorme, A., and Polich, J. (2013). Event-related delta, theta, alpha and gamma correlates to auditory oddball processing during Vipassana meditation. *Social Cognitive and Affective Neuroscience, 8*, 100–111.

Cai, C. B., Xu, L., Zhong, W., Tao, Y. Y., Wang, B., Yang, H. W., and Wen, M. Q. (2015). Studying a gas–solid multi-component adsorption process with near-infrared process analytical technique: Experimental setup, chemometrics, adsorption kinetics and mechanism. *Chemometrics and Intelligent Laboratory Systems, 144*, 80–86.

Camussi, R., Grilliat, J., Caputi-Gennaro, G., and Jacob, M. C. (2010). Experimental study of a tip leakage flow: Wavelet analysis of pressure fluctuations. *Journal of Fluid Mechanics, 660*, 87–113.

Camussi, R., Robert, G., and Jacob, M. C. (2008). Cross-wavelet analysis of wall pressure fluctuations beneath incompressible turbulent boundary layers. *Journal of Fluid Mechanics*, 617, 11–30.

Candès, E. J. (2003). What is.. a curvelet? *Notices of the American Mathematical Society*, 50(11), 1402–1403.

Candès, E. J., and Donoho, D. L. (1999). Curvelets: A surprisingly effective nonadaptive representation for objects with edges. *Proceedings of the International Conference on Curves and Surfaces, Volume 2. Curve and Surface Fitting* (pp. 105–120). July 1–7, 1999. Saint Malo, France. ADP011978.

Cannata, A., Montalto, P., and Patanè, D. (2013). Joint analysis of infrasound and seismic signals by cross wavelet transform: Detection of Mt. Etna explosive activity. *Natural Hazards and Earth System Sciences*, 13, 1669–1677.

Cantzos, D., Nikolopoulos, D., Petraki, E., Nomicos, C., Yannakopoulos, P. H., and Kottou, S. (2015). Identifying long-memory trends in pre-seismic MHz disturbances through support vector machines. *Journal of Earth Science and Climatic Change*, 6(3), 1–9.

Carey, S. K., Tetzlaff, D., Buttle, J., Laudon, H., McDonnell, J., McGuire, K., Seibert, J., Soulsby, C., and Shanley, J. (2013). Use of color maps and wavelet coherence to discern seasonal and interannual climate influences on streamflow variability in northern catchments. *Water Resources Research*, 49(10), 6194–6207.

Carmona, R., Hwang, W. L., and Torrésani, B. (1997). Characterization of signals by the ridges of their wavelet transforms. *IEEE Transactions on Signal Processing*, 45(10), 2586–2590.

Carvalho, R. T. S., Cavalcante, C. C., and Cortez, P. C. (2011). Wavelet transform and artificial neural networks applied to voice disorders identification. In *Nature and Biologically Inspired Computing (NaBIC), 2011 Third World Congress on* (pp. 371–376). IEEE.

Castillo, E., Morales, D. P., García, A., Martínez-Martí, F., Parrilla, L., and Palma, A. J. (2013). Noise suppression in ECG signals through efficient one-step wavelet processing techniques. *Journal of Applied Mathematics*, 763903, 1–13.

Catalao, J. P. S., Pousinho, H. M. I., and Mendes, V. M. F. (2011). Hybrid wavelet-PSO-ANFIS approach for short-term wind power forecasting in Portugal. *IEEE Transactions on Sustainable Energy*, 2(1), 50–59.

Catarino, A., Andrade, A., Churches, O., Wagner, A. P., Baron-Cohen, S., and Ring, H. (2013). Task-related functional connectivity in autism spectrum conditions: An EEG study using wavelet transform coherence. *Molecular Autism*, 4(1), 1–14.

Cavalieri, A. V., Jordan, P., Gervais, Y., Wei, M., and Freund, J. B. (2010). Intermittent sound generation and its control in a free-shear flow. *Physics of Fluids*, 22, 115113-1–115113-14.

Cazelles, B., Cazelles, K., and Chavez, M. (2013). Wavelet analysis in ecology and epidemiology: Impact of statistical tests. *Journal of the Royal Society Interface*, 11(91), 20130585, 1–10.

Celik, T., and Ma, K. K. (2011). Multitemporal image change detection using undecimated discrete wavelet transform and active contours. *IEEE Transactions on Geoscience and Remote Sensing*, 49(2), 706–716.

Chakrabarty, A., De, A., and Bandyopadhyay, G. (2015b). A wavelet-based MRA-EDCC-GARCH methodology for the detection of news and volatility spillover across sectoral indices: Evidence from the Indian financial market. *Global Business Review*, 16(1), 35–49.

Chakrabarty, A., De, A., Gunasekaran, A., and Dubey, R. (2015a). Investment horizon heterogeneity and wavelet: Overview and further research directions. *Physica A: Statistical Mechanics and its Applications*, 429, 45–61.

Chan, B. C., Chan, F. H., Lam, F. K., Lui, P. W., and Poon, P. W. (1997). Fast detection of venous air embolism in Doppler heart sound using the wavelet transform. *IEEE Transactions on Biomedical Engineering*, 44(4), 237–246.

Chang, R. C. H., Lin, C. H., Wei, M. F., Lin, K. H., and Chen, S. R. (2014). High-precision real-time premature ventricular contraction (PVC) detection system based on wavelet transform. *Journal of Signal Processing Systems*, 77(3), 289–296.

Chang, Y. B., Xia, J. J., Yuan, P., Kuo, T. H., Xiong, Z., Gateno, J., and Zhou, X. (2013). 3D segmentation of maxilla in cone-beam computed tomography imaging using base invariant wavelet active shape model on customized two-manifold topology. *Journal of X-Ray Science and Technology*, 21(2), 251–282.

Chatterjee, P., OBrien, E., Li, Y., and González, A. (2006). Wavelet domain analysis for identification of vehicle axles from bridge measurements. *Computers and Structures*, 84, 1792–1801.

Chaves, L. F., Calzada, J. E., Valderrama, A., and Saldaña, A. (2014). Cutaneous leishmaniasis and sand fly fluctuations are associated with El Niño in Panamá. *PLOS Neglected Tropical Diseases*, 8(10), e3210, 1–11.

Chávez, P., Yarlequé, C., Piro, O., Posadas, A., Mares, V., Loayza, H., Chuquillanqui, C., Zorogastúa P, Flexas, J., and Quiroz, R. (2010). Applying multifractal analysis to remotely sensed data for assessing PYVV infection in potato (*Solanum tuberosum* L.) crops. *Remote Sensing*, 2, 1197–1216.

Che Azemin, M., Kumar, D. K., Wong, T. Y., Kawasaki, R., Mitchell, P., and Wang, J. J. (2011). Robust methodology for fractal analysis of the retinal vasculature. *IEEE Transactions on Medical Imaging*, 30(2), 243–250.

Chen, D., Fan, J., and Zhang, F. (2013). Extraction the unbalance features of spindle system using wavelet transform and power spectral density. *Measurement*, 46, 1279–1290.

Chen, J., Itoh, N. S., and Hashimoto, T. (1993). ECG data compression by using wavelet transform. *IEICE Transactions on Information and Systems*, E76(12), 1454–1461.

Chen, J., Li, Z., Pan, J., Chen, G., Zi, Y., Yuan, J., Chen, B., and He, Z. (2016). Wavelet transform based on inner product in fault diagnosis of rotating machinery: A review. *Mechanical Systems and Signal Processing*, 70–71, 1–35.

Chen, K. C., Wang, J. H., Kim, K. H., Huang, W. G., Chang, K. H., Wang, J. C., and Leu, P. L. (2015). Morlet wavelet analysis of ML ≥3 earthquakes. *Terrestrial Atmospheric and Oceanic Sciences*, 26(2), Part I, 83–94.

Chen, W., Novak, M. D., Black, T. A., and Lee, X. (1997). Coherent eddies and temperature structure functions for three contrasting surfaces. Part I: Ramp model with finite microfront time. *Boundary-Layer Meteorology*, 84, 99–123.

Chen, X., Du, Z., Li, J., Li, X., and Zhang, H. (2014a). Compressed sensing based on dictionary learning for extracting impulse components. *Signal Processing*, 96, 94–109.

Chen, X., Hu, H., Zhang, J., and Zhou, Q. (2012a). An ECT system based on improved RBF network and adaptive wavelet image enhancement for solid/gas two-phase flow. *Chinese Journal of Chemical Engineering*, 20(2), 359–367.

Chen, Y., Dong, X., and Zhao, Y. (2005). Stock index modeling using EDA based local linear wavelet neural network. In *ICNN&B'05. International Conference on Neural Networks and Brain, 2005* (Vol.3, pp. 1646–1650). IEEE.

Chen, Y., Liu, T., Chen, X., Li, J., and Wang, E. (2014b). Time-frequency analysis of seismic data using synchrosqueezing wavelet transform. In *2014 SEG Annual Meeting* (pp. 303–312), October 26–31, 2014. Denver, CO: Society of Exploration Geophysicists.

Chen, Y., Yang, Z., Hu, Y., Yang, G., Zhu, Y., Li, Y., Luo, L., Chen, W. and Toumoulin, C. (2012b). Thoracic low-dose CT image processing using an artifact suppressed large-scale nonlocal means. *Physics in Medicine and Biology*, 57(9), 2667–2688.

Cheng, H. D., Shan, J., Ju, W., Guo, Y., and Zhang, L. (2010). Automated breast cancer detection and classification using ultrasound images: A survey. *Pattern Recognition*, 43, 299–317.

Cheng, J., Liu, H., Liu, T., Wang, F., and Li, H. (2015). Remote sensing image fusion via wavelet transform and sparse representation. *ISPRS Journal of Photogrammetry and Remote Sensing*, 104, 158–173.

Cheng, L. F., Chen, T. C., and Chen, L. G. (2012). Architecture design of the multi-functional wavelet-based ECG microprocessor for real-time detection of abnormal cardiac events. In *Engineering in Medicine and Biology Society (EMBC), 2012 Annual International Conference of the IEEE* (pp. 4466–4469). IEEE.

Cheng, T., Rivard, B., and Sanchez-Azofeifa, A. (2011). Spectroscopic determination of leaf water content using continuous wavelet analysis. *Remote Sensing of Environment*, 115(2), 659–670.

Cho, H., Felts, J. R., Yu, M. F., Bergman, L. A., Vakakis, A. F., and King, W. P. (2013). Improved atomic force microscope infrared spectroscopy for rapid nanometer-scale chemical identification. *Nanotechnology*, 24(44), 444007.

Chou, C. C., and Chen, S. L. (2011). Integrated or segmented? A wavelet transform analysis on relationship between stock and real estate markets. *Economics Bulletin*, 31(4), 3030–3040.

Chouakri, S. A., Djaafri, O., and Taleb-Ahmed, A. (2013). Wavelet transform and Huffman coding based electrocardiogram compression algorithm: Application to telecardiology. *Journal of Physics: Conference Series*, 454(012086), 1–16.

Chowdhury, S. K., Nimbarte, A. D., Jaridi, M., and Creese, R. C. (2013). Discrete wavelet transform analysis of surface electromyography for the fatigue assessment of neck and shoulder muscles. *Journal of Electromyography and Kinesiology*, 23, 995–1003.

Chuang, L. Z. H., Wu, L. C., and Wang, J. H. (2013). Continuous wavelet transform analysis of acceleration signals measured from a wave buoy. *Sensors*, 13(8), 10908–10930.

Chun, J., Ahn, K., Yoon, J. T., Suh, K. D., and Kim, M. (2013). Projection of extreme typhoon waves: Case study at Busan, *Korea Journal of Coastal Research*, Special Issue No. 65, 684–689.

Clemson, P., Stefanovska, A., Robnik, M., and Romanovski, V. G. (2012). Time series analysis of turbulent and non-autonomous systems. *AIP Conference Proceedings-American Institute of Physics*, 1468, 69–81.

Clifton, D., Addison, P. S., Stiles, M. K., Grubb, N., Watson, J. N., Clegg, G. R., and Robertson, C. E. (2003). Using wavelet transform reassignment techniques for ECG characterisation. *Computers in Cardiology*, 30, 581–584.

Clifton, D., Douglas, J. G., Addison, P. S., and Watson, J. N. (2007). Measurement of respiratory rate from the photoplethysmogram in chest clinic patients. *Journal of Clinical Monitoring and Computing*, 21(1), 55–61.

Cnockaert, L., Schoentgen, J., Auzou, P., Ozsancak, C., Defebvre, L., and Grenez, F. (2008). Low-frequency vocal modulations in vowels produced by Parkinsonian subjects. *Speech Communication*, 50(4), 288–300.

Coifman, R. R., and Donoho, D. L. (1995). Translation-invariant de-noising. In *Lecture Notes in Statistics 103* (pp. 125–150). New York: Springer.

Collineau, S., and Brunet, Y. (1993a). Detection of turbulent coherent motions in a forest canopy part I: wavelet analysis. *Boundary-Layer Meteorology*, 65, 357–379.

Collineau, S., and Brunet, Y. (1993b). Detection of turbulent coherent motions in a forest canopy part II: Time-scales and conditional averages. *Boundary-Layer Meteorology*, 66, 49–73.

Colwell, J. E., Cooney, J. H., Esposito, L. W., and Sremčević, M. (2009). Density waves in Cassini UVIS stellar occultations: 1. The Cassini division. *Icarus*, 200(2), 574–580.

Combaz, A., Manyakov, N. V., Chumerin, N., Suykens, J. A., and Hulle, M. (2009). Feature extraction and classification of EEG signals for rapid P300 mind spelling. In *Machine Learning and Applications, 2009. ICMLA'09. International Conference on* (pp. 386–391). IEEE.

Combet, F., Gelman, L., and LaPayne, G. (2012). Novel detection of local tooth damage in gears by the wavelet bicoherence. *Mechanical Systems and Signal Processing*, 26, 218–228.

Cong, F., Phan, A. H., Astikainen, P., Zhao, Q., Wu, Q., Hietanen, J. K., Ristaniemi, T., and Cichocki, A. (2013). Multi-domain feature extraction for small event-related potentials through non-negative multi-way array decomposition from low dense array EEG. *International Journal of Neural Systems*, 23(02), 1350006, 1–18.

Cong, F., Phan, A. H., Cichocki, A., Lyytinen, H., and Ristaniemi, T. (2010). Identical fits of nonnegative matrix/tensor factorization may correspond to different extracted event-related potentials. In *The 2010 International Joint Conference on Neural Networks (IJCNN)* (pp. 1–5). IEEE.

Cooke, W. H., Moralez, G., Barrera, C. R., and Cox, P. (2011). Digital infrared thermographic imaging for remote assessment of traumatic injury. *Journal of Applied Physiology*, 111, 1813–1818.

Courbebaisse, G., Bouffanais, R., Navarro, L., Leriche, E., and Deville, M. (2011). Time-scale joint representation of DNS and LES numerical data. *Computers and Fluids*, *43*(1), 38–45.

Crowe, J. A., Gibson, N. M., Woolfson, M. S., and Somekh, M. G. (1992). Wavelet transform as a potential tool for ECG analysis and compression. *Journal of Biomedical Engineering*, *14*, 268–272.

Cui, L., Huang, K., and Cai, H. J. (2015). Application of a TGARCH-wavelet neural network to arbitrage trading in the metal futures market in China. *Quantitative Finance*, *15*(2), 371–384.

Cui, R., Zhang, M., Li, Z., Xin, Q., Lu, L., Zhou, W., Han, Q., and Gao, Y. (2014). Wavelet coherence analysis of spontaneous oscillations in cerebral tissue oxyhemoglobin concentrations and arterial blood pressure in elderly subjects. *Microvascular Research*, *93*, 14–20.

Czuba, C., Williams, B. K., Westman, J., and LeClaire, K. (2015). *An Assessment of Two Methods for Identifying Undocumented Levees Using Remotely Sensed Data* (No. 2015–5009). Reston, VA: US Geological Survey.

Daamouche, A., Hamami, L., Alajlan, N., and Melgani, F. (2012). A wavelet optimization approach for ECG signal classification. *Biomedical Signal Processing and Control*, *7*(4), 342–349.

Dabbech, A., Ferrari, C., Mary, D., Slezak, E., Smirnov, O., and Kenyon, J. S. (2015). MORESANE: MOdel REconstruction by Synthesis-ANalysis Estimators. A sparse deconvolution algorithm for radio interferometric imaging. *Astronomy and Astrophysics*, *576(A7)*, 1–16.

Dajcman, S., Festic, M., and Kavkler, A. (2012). Comovement dynamics between central and eastern European and developed European stock markets during European integration and amid financial crises–A wavelet analysis. *Engineering Economics*, *23*(1), 22–32.

Dar, A. B., Bhanja, N., Samantaraya, A., and Tiwari, A. K. (2013). Export led growth or growth led export hypothesis in India: Evidence based on time-frequency approach. *Asian Economic and Financial Review*, *3*(7), 869–880.

Das, R., Turkoglu, I., and Sengur, A. (2009). Diagnosis of valvular heart disease through neural networks ensembles. *Computer Methods and Programs in Biomedicine*, *93*, 185–191.

Daubechies, I. (1992) *Ten Lectures on Wavelets*. CBMS-NSF Regional Conference Series in Applied Mathematics, SIAM, Philadelphia, PA.

Daubechies, I., Lu, J., and Wu, H. T. (2011). Synchrosqueezed wavelet transforms: An empirical mode decomposition-like tool. *Applied and Computational Harmonic Analysis*, *30*, 243–261.

Daubechies, I., and Maes, S. (1996). A nonlinear squeezing of the continuous wavelet transform based on auditory nerve models. In A. Aldroubi and M. Unser (Eds.), *Wavelets in Medicine and Biology* (pp. 527–546). Boca Raton, FL: CRC Press.

David, D. T., Kumar, S. P., Byju, P., Sarma, M. S. S., Suryanarayana, A., and Murty, V. S. N. (2011). Observational evidence of lower-frequency Yanai waves in the central equatorial Indian Ocean. *Journal of Geophysical Research: Oceans*, *116*, C06009, 1–17.

David, M., Hirsch, M., and Akselrod, S. (2006). Maturation of fetal cardiac autonomic control as expressed by fetal heart rate variability. *Computers in Cardiology*, *33*, 901–904.

Davis, A. C., and Christensen, N. B. (2013). Derivative analysis for layer selection of geophysical borehole logs. *Computers and Geosciences*, *60*, 34–40.

de Lannoy, G., De Decker, A., and Verleysen, M. (2008). A supervised wavelet transform algorithm for R spike detection in noisy ECGs. In *Biomedical Engineering Systems and Technologies* (pp. 256–264). Berlin: Springer.

de Matos, M. C., Davogusto, O., Zhang, K., and Marfurt, K. J. (2010). Continuous wavelet transform phase residues applied to detect stratigraphic discontinuities. In *2010 SEG Annual Meeting* (pp. 1494–1499), October 17–22, 2010. Denver, CO: Society of Exploration Geophysicists.

de Matos, M. C., Osorio, P. L., and Johann, P. R. (2007). Unsupervised seismic facies analysis using wavelet transform and self-organizing maps. *Geophysics*, *72*(1), P9–P21.

De Melis, M., Morbiducci, U., and Scalise, L. (2007, August). Identification of cardiac events by optical vibrocardiography: Comparison with phonocardiography. *In Engineering in Medicine and Biology Society, 2007. EMBS 2007. 29th Annual International Conference of the IEEE* (pp. 2956–2959). IEEE.

De Paula, A. V., and Möller, S. V. (2013). Finite mixture model applied in the analysis of a turbulent bistable flow on two parallel circular cylinders. *Nuclear Engineering and Design, 264,* 203–213.

Dehghani, N., Cash, S. S., and Halgren, E. (2011). Topographical frequency dynamics within EEG and MEG sleep spindles. *Clinical Neurophysiology, 122*(2), 229–235.

Dehkordi, P., Garde, A., Molavi, B., Petersen, C. L., Ansermino, J. M., and Dumont, G. A. (2015). Estimating instantaneous respiratory rate from the photoplethysmogram. In *37th Annual International Conference of the IEEE Engineering in Medicine and Biology Society (EMBC)* (pp. 6150–6153). IEEE.

Delgado-Trejos, E., Quiceno-Manrique, A. F., Godino-Llorente, J. I., Blanco-Velasco, M., and Castellanos-Dominguez, G. (2009). Digital auscultation analysis for heart murmur detection. *Annals of Biomedical Engineering, 37*(2), 337–353.

Deliège, A., and Nicolay, S. (2014). A wavelet leaders-based climate classification of European surface air temperature signals. In S. L. Copicentro Granada (Eds.), *Proceedings of the International Work-Conference on Ttime Series Analysis* (pp. 40–51).

Deng, L. H., Li, B., Xiang, Y. Y., and Dun, G. T. (2015). Multi-scale analysis of coronal Fe xiv emission: The role of mid-range periodicities in the sun–heliosphere connection. *Journal of Atmospheric and Solar-Terrestrial Physics, 122,* 18–25.

Dewandaru, G., Masih, R., and Masih, A. M. M. (2015). Why is no financial crisis a dress rehearsal for the next? Exploring contagious heterogeneities across major Asian stock markets. *Physica A: Statistical Mechanics and its Applications, 419,* 241–259.

Dey, D., Chatterjee, B., Chakravorti, S., and Munshi, S. (2010). Cross-wavelet transform as a new paradigm for feature extraction from noisy partial discharge pulses. *IEEE Transactions on Dielectrics and Electrical Insulation, 17*(1), 157–166.

Di Falco, A., Krauss, T. F., and Fratalocchi, A. (2012). Lifetime statistics of quantum chaos studied by a multiscale analysis. *Applied Physics Letters, 100*(18), 184101.

Di Marco, L. Y., and Chiari, L. (2011). A wavelet-based ECG delineation algorithm for 32-bit integer online processing. *Biomedical Engineering Online, 10*(23), 1–19.

Diab, M. O., Moslem, B., Khalil, M., and Marque, C. (2012). Classification of uterine EMG signals by using normalized wavelet packet energy. In *16th IEEE Mediterranean Electrotechnical Conference (MELECON)* (pp. 335–338). IEEE.

Dick, O. E., and Svyatogor, I. A. (2015). Wavelet and multifractal estimation of the intermittent photic stimulation response in the electroencephalogram of patients with dyscirculatory encephalopathy. *Neurocomputing, 165,* 361–374.

Dien, N. P. (2008). Damping identification using the wavelet-based demodulation method: Application to Gearbox signals. *Technische Mechanik, 28*(3–4), 324–333.

Do, M. N., and Vetterli, M. (2005). The contourlet transform: An efficient directional multiresolution image representation. *IEEE Transactions on Image Processing, 14*(12), 2091–2106.

Donoho, D. L., and Johnstone, J. M. (1994). Ideal spatial adaptation by wavelet shrinkage. *Biometrika, 81*(3), 425–455.

Donoho, D. L., and Johnstone, I. M. (1995). Adapting to unknown smoothness via wavelet shrinkage. *Journal of the American Statistical Association, 90*(432), 1200–1224.

Donoho, D. L., and Kutyniok, G. (2009). Geometric separation using a wavelet-shearlet dictionary. In *SAMPTA'09, International Conference on Sampling Theory and Applications* (pp. 95–99). Marseille Luminy, France: Centre International de Rencontres Mathématiques.

Dougan, L. T., Addison, P. S., and McKenzie, W. M. C. (2000). Fractal analysis of fracture: A comparison of dimension estimates. *Mechanics Research Communications, 27*(4), 383–392.

Dragon, R., Mörke, T., Rosenhahn, B., and Ostermann, J. (2011). Fingerprints for machines–characterization and optical identification of grinding imprints. In R. Mester and M. Felsberg (Eds.) *Pattern Recognition* (pp. 276–285). Berlin: Springer.

Du, S., Huang, D., and Lv, J. (2013). Recognition of concurrent control chart patterns using wavelet transform decomposition and multiclass support vector machines. *Computers and Industrial Engineering*, 66, 683–695.

Du, W., Tao, J., Li, Y., and Liu, C. (2014). Wavelet leaders multifractal features based fault diagnosis of rotating mechanism. *Mechanical Systems and Signal Processing*, 43, 57–75.

Duarte, L. T., Donno, D., Lopes, R. R., and Romano, J. M. T. (2014). Seismic signal processing: Some recent advances. In *2014 IEEE International Conference on Acoustics, Speech and Signal Processing (ICASSP)* (pp. 2362–2366). IEEE.

Duchesne, M. J., Long, B. F., Labrie, J., and Simpkin, P. G. (2006). On the use of computerized tomography scan analysis to determine the genesis of very high resolution seismic reflection facies. *Journal of Geophysical Research: Solid Earth*, 111(B10103), 1–16.

Dumont, J., Hernandez, A. I., and Carrault, G. (2010). Improving ECG beats delineation with an evolutionary optimization process. *IEEE Transactions on Biomedical Engineering*, 57(3), 607–615.

Dutta, S., Pal, S. K., Mukhopadhyay, S., and Sen, R. (2013). Application of digital image processing in tool condition monitoring: A review. *CIRP Journal of Manufacturing Science and Technology*, 6, 212–232.

Dziedziech, K., Staszewski, W. J., Basu, B., and Uhl, T. (2015b). Wavelet-based detection of abrupt changes in natural frequencies of time-variant systems. *Mechanical Systems and Signal Processing*, 64–65, 347–359.

Dziedziech, K., Staszewski, W. J., and Uhl, T. (2015a). Wavelet-based modal analysis for time-variant systems. *Mechanical Systems and Signal Processing*, 50, 323–337.

Elbarghathi, F., Tian, X., Tung Tran, V., Gu, F., and Ball, A. (2013). Multi-stages helical gearbox fault detection using vibration signal and Morlet wavelet transform adapted by information entropy difference. In *COMADEM 2013* (pp. 1–7), June 11–13, 2013. Helsinki, Finland.

Elsayed, M. A. (2006). Wavelet bicoherence analysis of wind–wave interaction. *Ocean Engineering*, 33, 458–470.

Endres, L. A. M., and Möller, S. V. (2009). Experimental study of the propagation of a far-field disturbance in the turbulent flow through square array tube banks. *Journal of the Brazilian Society of Mechanical Sciences and Engineering*, 31(3), 232–242.

Engels, T., Kolomenskiy, D., Schneider, K., and Sesterhenn, J. (2013). Two-dimensional simulation of the fluttering instability using a pseudospectral method with volume penalization. *Computers and Structures*, 122, 101–112.

Ergen, B., Tatar, Y., and Gulcur, H. O. (2012). Time-frequency analysis of phonocardiogram signals using wavelet transform: A comparative study. *Computer Methods in Biomechanics and Biomedical Engineering*, 15(4), 371–381.

Escamilla-Ambrosio, P. J., Liu, X., Lieven, N. A. J., and Ramirez-Cortes, J. M. (2011). ANFIS-2D wavelet transform approach to structural damage identification. In *2011 Annual Meeting of the North American Fuzzy Information Processing Society (NAFIPS)*, (pp. 1–6). IEEE.

Etehadtavakol, M., Ng, E. Y. K., Chandran, V., and Rabbani, H. (2013). Separable and non-separable discrete wavelet transform based texture features and image classification of breast thermograms. *Infrared Physics and Technology*, 61, 274–286.

Everett, S., Johnson, I., Murphy, J., and Tarpley, M. (2015). Detection of baryonic acoustic oscillations in the matter power spectrum. *DePaul Discoveries*, 4(1,4), 1–7.

Facco, P., Bezzo, F., Barolo, M., Mukherjee, R., and Romagnoli, J. A. (2009). Monitoring roughness and edge shape on semiconductors through multiresolution and multivariate image analysis. *AIChE Journal*, 55(5), 1147–1160.

Facco, P., Tomba, E., Roso, M., Modesti, M., Bezzo, F., and Barolo, M. (2010). Automatic characterization of nanofiber assemblies by image texture analysis. *Chemometrics and Intelligent Laboratory Systems*, 103, 66–75.

Fadili, J., and Starck, J. L. (2009). Curvelets and Ridgelets. In R. A. Meyers (Ed.), *Encyclopedia of Complexity and Systems Science* (pp. 1718–1738). New York: Springer.

Faezipour, M., Saeed, A., Bulusu, S. C., Nourani, M., Minn, H., and Tamil, L. (2010). A patient-adaptive profiling scheme for ECG beat classification. *IEEE Transactions on Information Technology in Biomedicine, 14*(5), 1153–1165.

Faezipour, M., Tiwari, T. M., Saeed, A., Nouranl, M., and Tamil, L. S. (2009). Wavelet based denoising and beat detection of ECG signal. *IEEE/NIH Life Science Systems and Applications Workshop. LiSSA 2009* (pp. 100–103), April 9–10, 2009. Bethesda, MD: IEEE.

Fairley, J. A., Georgoulas, G., Stylios, C. D., Vachtsevanos, G., Rye, D. B., and Bliwise, D. L. (2011). Phasic electromyographic metric detection based on wavelet analysis. In *2011 19th Mediterranean Conference on Control and Automation (MED)* (pp. 497–502). IEEE.

Falkowski, M. J., Smith, A. M., Hudak, A. T., Gessler, P. E., Vierling, L. A., and Crookston, N. L. (2006). Automated estimation of individual conifer tree height and crown diameter via two-dimensional spatial wavelet analysis of lidar data. *Canadian Journal of Remote Sensing, 32*(2), 153–161.

Fan, W., and Qiao, P. (2009). A 2-D continuous wavelet transform of mode shape data for damage detection of plate structures. *International Journal of Solids and Structures, 46*, 4379–4395.

Fang, H., Yao, H., Zhang, H., Huang, Y. C., and van der Hilst, R. D. (2015). Direct inversion of surface wave dispersion for three-dimensional shallow crustal structure based on ray tracing: Methodology and application. *Geophysical Journal International, 201*(3), 1251–1263.

Farge, M. (1992). Wavelet transforms and their applications to turbulence. *Annual Review of Fluid Mechanics, 24*(1), 395–458.

Farge, M., Kevlahan, N., Perrier, V., and Goirand, E. (1996). Wavelets and turbulence. *Proceedings of the IEEE, 84*(4), 639–669.

Farge, M., and Schneider, K. (2015). Wavelet transforms and their applications to MHD and plasma turbulence: A review. *Journal of Plasma Physics, 81*(6), 435810602, 1–43.

Farge, M., Schneider, K., Pannekoucke, O., and Van Yen, R. N. (2010). Multiscale representations: Fractals, self-similar random processes and wavelets. In H. J. S. Fernando (Ed.), *Handbook of Environmental Fluid Dynamics, Volume 2* (Chapter 23, pp. 311–332). Boca Raton, FL: CRC Press.

Feng, C., Lei, X., and Chun-Guang, L. (2012). Wavelet phase synchronization of Fractional-Order chaotic systems. *Chinese Physics Letters, 29*, 070501-1–070501-3.

Fernández-Macho, J. (2012). Wavelet multiple correlation and cross-correlation: A multiscale analysis of Eurozone stock markets. *Physica A: Statistical Mechanics and its Applications, 391*, 1097–1104.

Ferreres, E., Soler, M. R., and Terradellas, E. (2013). Analysis of turbulent exchange and coherent structures in the stable atmospheric boundary layer based on tower observations. *Dynamics of Atmospheres and Oceans, 64*, 62–78.

Ferzli, I., Chiprout, E., and Najm, F. N. (2010). Verification and codesign of the package and die power delivery system using wavelets. *IEEE Transactions on Computer-Aided Design of Integrated Circuits and Systems, 29*(1), 92–102.

Finoguenov, A., Tanaka, M., Cooper, M., Allevato, V., Cappelluti, N., Choi, A., Heymans, C., et al. (2015). Ultra-deep catalog of X-ray groups in the Extended Chandra Deep Field South. *Astronomy and Astrophysics, 576*(A130), 1–19.

Fisco, N. R., and Adeli, H. (2011). Smart structures: Part II—hybrid control systems and control strategies. *Scientia Iranica, 18*(3), 285–295.

Flandrin, P. (1992). Wavelet analysis and synthesis of fractional Brownian motion. *IEEE Transactions on Information Theory, 38*(2), 910–917.

Franco, C., Guméry, P. Y., Vuillerme, N., Fleury, A., and Fontecave-Jallon, J. (2012). Synchrosqueezing to investigate cardio-respiratory interactions within simulated volumetric signals. *Proceedings of the 20th European Signal Processing Conference (EUSIPCO)* (pp. 939–943), August 27–31, 2012. Bucharest, Romania: EUSIPCO.

Frick, P., Sokoloff, D., Stepanov, R., and Beck, R. (2010). Wavelet-based Faraday rotation measure synthesis. *Monthly Notices of the Royal Astronomical Society: Letters, 401*, L24–L28.

Fulda, S., Romanowski, C. P., Becker, A., Wetter, T. C., Kimura, M., and Fenzl, T. (2011). Rapid eye movements during sleep in mice: High trait-like stability qualifies rapid eye movement density for characterization of phenotypic variation in sleep patterns of rodents. *BMC Neuroscience*, 12(110), 1–13.

Gaci, S., Zaourar, N., Hamoudi, M., and Holschneider, M. (2010). Local regularity analysis of strata heterogeneities from sonic logs. *Nonlinear Processes in Geophysics*, 17, 455–466.

Gadaleta, M., and Giorgio, A. *(2012)*. A method for ventricular late potentials detection using time-frequency representation and wavelet denoising. *ISRN Cardiology*, 2012, 258769, 1–9.

Gadhoumi, K., Lina, J. M., and Gotman, J. (2012). Discriminating preictal and interictal states in patients with temporal lobe epilepsy using wavelet analysis of intracerebral EEG. *Clinical Neurophysiology*, 123(10), 1906–1916.

Gallegati, M., Ramsey, J. B., and Semmler, W. (2013). Time scale analysis of interest rate spreads and output using wavelets. *Axioms*, 2, 182–207.

Gandhi, M. S., Sathe, M. J., Joshi, J. B., and Vijayan, P. K. (2011). Two phase natural convection: CFD simulations and PIV measurement. *Chemical Engineering Science*, 66(14), 3152–3171.

Gandhi, T., Suresh, N., and Sinha, P. (2012). EEG responses to facial contrast-chimeras. *Journal of Integrative Neuroscience*, 11(2), 201–211.

Gao, J., Sultan, H., Hu, J., and Tung, W. W. (2010). Denoising nonlinear time series by adaptive filtering and wavelet shrinkage: A comparison. *IEEE Signal Processing Letters*, 17(3), 237–240.

Gao, L., Zai, F., Su, S., Wang, H., Chen, P., and Liu, L. (2011). Study and application of acoustic emission testing in fault diagnosis of low-speed heavy-duty gears. *Sensors*, 11, 599–611.

Gao, W., and Li, B. L. (1993). Wavelet analysis of coherent structures at the atmosphere–forest interface. *Journal of Applied Meteorology*, 32, 1717–1725.

Gao, Y., Zhang, M., Han, Q., Li, W., Xin, Q., Wang, Y., and Li, Z. (2015). Cerebral autoregulation in response to posture change in elderly subjects-assessment by wavelet phase coherence analysis of cerebral tissue oxyhemoglobin concentrations and arterial blood pressure signals. *Behavioural Brain Research*, 278, 330–336.

Garcia, J. O., Grossman, E. D., and Srinivasan, R. (2011). Evoked potentials in large-scale cortical networks elicited by TMS of the visual cortex. *Journal of Neurophysiology*, 106, 1734–1746.

García-Lorenzo, B., and Fuensalida, J. J. (2006). Processing of turbulent-layer wind speed with generalized SCIDAR through wavelet analysis. *Monthly Notices of the Royal Astronomical Society*, 372, 1483–1495.

Garg, A., Xu, D., and Blaber, A. P. (2013). Statistical validation of wavelet transform coherence method to assess the transfer of calf muscle activation to blood pressure during quiet standing. *Biomedical Engineering Online*, 12(132), 1–14.

Gay, M., De Angelis, M., and Lacoume, J. L. (2014). Dating a tropical ice core by time–frequency analysis of ion concentration depth profiles. *Climate of the Past Discussions*, 10, 1–15.

Gazak, J. Z., Johnson, J. A., Tonry, J., Dragomir, D., Eastman, J., Mann, A. W., and Agol, E. *(2012)*. Transit analysis package: An IDL graphical user interface for exoplanet transit photometry. *Advances in Astronomy*, 2012, 697967, 1–8.

Ge, X., Fan, Y., Li, J., Wang, Y., and Deng, S. (2015). Noise reduction of nuclear magnetic resonance (NMR) transversal data using improved wavelet transform and exponentially weighted moving average (EWMA). *Journal of Magnetic Resonance*, 251, 71–83.

Ge, Z. (2007). Significance tests for the wavelet power and the wavelet power spectrum. *Annales Geophysicae*, 25, 2259–2269.

Ge, Z. (2008). Significance tests for the wavelet cross spectrum and wavelet linear coherence. *Annales Geophysicae*, 26, 3819–3829.

Gençay, R., Gradojevic, N., Selçuk‖, F., and Whitcher, B. (2010). Asymmetry of information flow between volatilities across time scales. *Quantitative Finance*, 10(8), 895–915.

Geramifard, O., Xu, J. X., Zhou, J. H., and Li, X. (2012). A physically segmented hidden Markov model approach for continuous tool condition monitoring: Diagnostics and prognostics. *IEEE Transactions on Industrial Informatics*, 8(4), 964–973.

Gerasimova, E., Audit, B., Roux, S. G., Khalil, A., Gileva, O., Argoul, F., Naimark, O., and Arneodo, A. (2014). Wavelet-based multifractal analysis of dynamic infrared thermograms to assist in early breast cancer diagnosis. *Frontiers in Physiology*, 5(176), 1–11.

Geven, L. I., Wit, H. P., de Kleine, E., and van Dijk, P. (2012). Wavelet analysis demonstrates no abnormality in contralateral suppression of otoacoustic emissions in tinnitus patients. *Hearing Research*, 286, 30–40.

Ghaderi, K., Akhlaghian, F., and Moradi, P. (2013). A new robust semi-blind digital image watermarking approach based on LWT-SVD and fractal images. In *2013 21st Iranian Conference on Electrical Engineering (ICEE)* (pp. 1–5). IEEE.

Ghandeharion, H., and Erfanian, A. (2010). A fully automatic ocular artifact suppression from EEG data using higher order statistics: Improved performance by wavelet analysis. *Medical Engineering and Physics*, 32(7), 720–729.

Ghasemi, M., Ghaffari, A., SadAbadi, H., and Golbayani, H. (2010). QT interval measurement using RMED curve; a novel approach based on wavelet techniques. *Computer Methods in Biomechanics and Biomedical Engineering*, 13(6), 857–864. SPIE, San Jose, CA.

Ghosh, P., Mitchell, M., and Gold, J. (2010). Segmentation of thermographic images of hands using a genetic algorithm. In *Proceedings of SPIE-IS&T Electronic Imaging* (Vol. 7538, pp. 75380D-1–75380D-8). San Jose, California: SPIE.

Giacintucci, S., Markevitch, M., Brunetti, G., ZuHone, J. A., Venturi, T., Mazzotta, P., and Bourdin, H. (2014). Mapping the particle acceleration in the cool core of the galaxy cluster RX J1720. 1+2638. *Astrophysical Journal*, 795(1), 73.

Giri, B. K., Mitra, C., Panigrahi, P. K., and Iyengar, A. S. (2014). Multi-scale dynamics of glow discharge plasma through wavelets: Self-similar behavior to neutral turbulence and dissipation. *Chaos: An Interdisciplinary Journal of Nonlinear Science*, 24, 043135, 1–7.

Godfrey, A., Bourke, A. K., Olaighin, G. M., Van De Ven, P., and Nelson, J. (2011). Activity classification using a single chest mounted tri-axial accelerometer. *Medical Engineering and Physics*, 33, 1127–1135.

Gökdağ, H. (2013). A crack identification method for bridge type structures under vehicular load using wavelet transform and particle swarm optimization. *Advances in Acoustics and Vibration*, 634217, 1–10.

Golestani, A., and Gras, R. (2012). Identifying origin of self-similarity in EcoSim, an individual-based ecosystem simulation, using wavelet-based multifractal analysis. In *Proceedings of the World Congress on Engineering and Computer Science* (Vol. 2, pp. 1275–1282), October 24–26, 2012. San Francisco, CA: WCECS 2012.

Golestani, A., Kolbadi, S. M. S., and Heshmati, A. A. (2013). Localization and de-noising seismic signals on SASW measurement by wavelet transform. *Journal of Applied Geophysics*, 98, 124–133.

Gombarska, D., and Smetana, M. (2011). Wavelet based signal analysis of pulsed eddy current signals. *Przegląd Elektrotechniczny*, 87, 37–39.

González-Concepción, C., Gil-Fariña, M. C., and Pestano-Gabino, C. (2012). Using wavelets to understand the relationship between mortgages and gross domestic product in Spain. *Journal of Applied Mathematics*, 2012, 917247, 1–17.

Gosse, L. (2010). Analysis and short-time extrapolation of stock market indexes through projection onto discrete wavelet subspaces. *Nonlinear Analysis: Real World Applications*, 11(4), 3139–3154.

Gouliermis, D. A., Schmeja, S., Ossenkopf, V., Klessen, R. S., and Dolphin, A. E. (2014). Hierarchically clustered star formation in the magellanic clouds. In *The Labyrinth of Star Formation, Astrophysics and Space Science Proceedings* (Vol. 36. pp.447–451). Heidelberg: Springer International Publishing.

Goupillaud, P., Grossmann, A., and Morlet, J. (1984). Cycle-octave and related transforms in seismic signal analysis. *Geoexploration*, 23(1), 85–102.

Grassucci, D., Camussi, R., Kerhervé, F., Jordan, P., and Grizzi, S. (2010). Using wavelet transforms and linear stochastic estimation to study nearfield pressure and turbulent velocity signatures in free jets. *16th AIAA/CEAS Aeroacoustics Conference Stockholm* (AIAA 2010-3954, pp. 1–12), June 7–9, 2010. Sweden.

Gresil, M., Yu, L., Shen, Y., and Giurgiutiu, V. (2013). Predictive model of fatigue crack detection in thick bridge steel structures with piezoelectric wafer active sensors. *Smart Structures and Systems*, 12(2), 97–119.

Grinsted, A., Moore, J. C., and Jevrejeva, S. (2004). Application of the cross wavelet transform and wavelet coherence to geophysical time series. *Nonlinear Processes in Geophysics*, 11, 561–566.

Grizzi, S., and Camussi, R. (2012). Wavelet analysis of near-field pressure fluctuations generated by a subsonic jet. *Journal of Fluid Mechanics*, 698, 93–124.

Grubov, V. V., Sitnikova, E. Y., Koronovskii, A. A., Pavlov, A. N., and Hramov, A. E. (2012). Automatic extraction and analysis of oscillatory patterns on nonstationary EEG signals by means of wavelet transform and the empirical modes method. *Bulletin of the Russian Academy of Sciences: Physics*, 76(12), 1361–1364.

Gu, J., Xu, H., Wang, J., An, T., and Chen, W. (2013). The application of continuous wavelet transform based foreground subtraction method in 21 cm sky surveys. *Astrophysical Journal*, 773(38), 1–16.

Guharay, S. K., Thakur, G. S., Goodman, F. J., Rosen, S. L., and Houser, D. (2013). Analysis of nonstationary dynamics in the financial system. *Economics Letters*, 121, 454–457.

Guo, L., Rivero, D., and Pazos, A. (2010). Epileptic seizure detection using multiwavelet transform based approximate entropy and artificial neural networks. *Journal of Neuroscience Methods*, 193(1), 156–163.

Guo, L., Rivero, D., Dorado, J., Munteanu, C. R., and Pazos, A. (2011). Automatic feature extraction using genetic programming: An application to epileptic EEG classification. *Expert Systems with Applications*, 38, 10425–10436.

Guo, L., Rivero, D., Seoane, J. A., and Pazos, A. (2009). Classification of EEG signals using relative wavelet energy and artificial neural networks. In *Proceedings of the First ACM/SIGEVO Summit on Genetic and Evolutionary Computation* (pp. 177–184). New York: ACM.

Gupta, B. R., and Kumar, V. (2015). Time-frequency analysis of asymmetric triaxial galaxy model including effect of spherical dark halo component. *International Journal of Astronomy and Astrophysics*, 5, 106–115.

Gurkan, H. (2012). Compression of ECG signals using variable-length classified vector sets and wavelet transforms. *EURASIP Journal on Advances in Signal Processing*, 119, 1–17.

Gurley, K., and Kareem, A. (1999). Applications of wavelet transforms in earthquake, wind and ocean engineering. *Engineering Structures*, 21(2), 149–167.

Gurley, K., Kijewski, T., and Kareem, A. (2003). First-and higher-order correlation detection using wavelet transforms. *Journal of Engineering Mechanics*, 129, 188–201.

Gutknecht, E., Dadou, I., Charria, G., Cipollini, P., and Garcon, V. (2010). Spatial and temporal variability of the remotely sensed chlorophyll a signal associated with Rossby waves in the South Atlantic Ocean. *Journal of Geophysical Research: Oceans*, 115, C05004, 1–16.

Haase, M., and Lehle, B. (1998). Tracing the skeleton of wavelet transform maxima lines for the characterization of fractal distributions. In M. M. Novak (Ed.), *Fractals and Beyond* (pp. 241–250). Singapore: World Scientific.

Hachem, F. E., and Schleiss, A. J. (2012). Detection of local wall stiffness drop in steel-lined pressure tunnels and shafts of hydroelectric power plants using steep pressure wave excitation and wavelet decomposition. *Journal of Hydraulic Engineering*, 138(1), 35–45.

Hackmack, K., Paul, F., Weygandt, M., Allefeld, C., Haynes, J. D., and Alzheimer's Disease Neuroimaging Initiative. (2012). Multi-scale classification of disease using structural MRI and wavelet transform. *Neuroimage*, 62, 48–58.

Hadjileontiadis, L. (2015). EEG-based tonic cold pain characterization using wavelet higher-order spectral features. *IEEE Transactions on Biomedical Engineering, 62*(8), 1981–1991.

Hadjileontiadis, L. J., and Panas, S. M. (1997). Separation of discontinuous adventitious sounds from vesicular sounds using a wavelet-based filter. *IEEE Transactions on Biomedical Engineering, 44*(12), 1269–1281.

Hagelberg, C. R., Cooper, D. I., Winter, C. L., and Eichinger, W. E. (1998). Scale properties of microscale convection in the marine surface layer. *Journal of Geophysical Research: Atmospheres (1984–2012), 103*(D14), 16897–16907.

Hagelberg, C. R., and Gamage, N. K. K. (1994). Short term prediction of local wind conditions. *Boundary-Layer Meteorology, 70*, 217–246.

Hajj, M. R., Jordan, D. A., and Tieleman, H. W. (1998). Analysis of atmospheric wind and pressures on a low-rise building. *Journal of Fluids and Structures, 12*(5), 537–547.

Hajj, M. R., and Tieleman, H. W. (1996). Application of wavelet analysis to incident wind in relevance to wind loads on low-rise structures. *Journal of Fluids Engineering, 118*(4), 874–876.

Hall, M. H., Taylor, G., Salisbury, D. F., and Levy, D. L. (2011a). Sensory gating event-related potentials and oscillations in schizophrenia patients and their unaffected relatives. *Schizophrenia Bulletin, 37*(6), 1187–1199.

Hall, M. H., Taylor, G., Sham, P., Schulze, K., Rijsdijk, F., Picchioni, M., Toulopoulou, T., et al. (2011b). The early auditory gamma-band response is heritable and a putative endophenotype of schizophrenia. *Schizophrenia Bulletin, 37*(4), 778–787.

Hamdan, M. N., Jubran, B. A., Shabaneh, N. H., and Abu-Samak, M. (1996). Comparison of various basic wavelets for the analysis of flow-induced vibration of a cylinder in cross flow. *Journal of Fluids and Structures, 10*(6), 633–651.

Hamilton, P., Lockhart, C. J., McCann, A. J., Agnew, C. E., Harbinson, M. T., McClenaghan, V., Bleakley, C., McGivern, R. C., and McVeigh, G. (2011). Flow-mediated dilatation of the brachial artery is a poorly reproducible indicator of microvascular function in Type I diabetes mellitus. *QJM, 104*(7), 589–597.

Hamilton, P. K., Mccann, A. J., Agnew, C. E., Millar, A., Mcclenaghan, V. O., Mcgivern, R. C., and Mcveigh, G. (2012). Detecting early microvascular disease in type 1 diabetes: Wavelet transform analysis of Doppler blood velocity waveforms. *British Journal of Diabetes and Vascular Disease, 12*, 40–47.

Hampson, K. M., and Mallen, E. A. (2011). Multifractal nature of ocular aberration dynamics of the human eye. *Biomedical Optics Express, 2*(3), 464–477.

Han, L., Li, C. W., Guo, S. L., and Su, X. W. (2015). Feature extraction method of bearing AE signal based on improved FAST-ICA and wavelet packet energy. *Mechanical Systems and Signal Processing, 62*, 91–99.

Han, P., Hattori, K., Huang, Q., Hirano, T., Ishiguro, Y., Yoshino, C., and Febriani, F. (2011). Evaluation of ULF electromagnetic phenomena associated with the 2000 Izu Islands earthquake swarm by wavelet transform analysis. *Natural Hazards and Earth System Science, 11*, 965–970.

Han, T., Yang, B. S., Choi, W. H., and Kim, J. S. (2006). Fault diagnosis system of induction motors based on neural network and genetic algorithm using stator current signals. *International Journal of Rotating Machinery, 61690*, 1–13.

Hanbay, D. (2009). An expert system based on least square support vector machines for diagnosis of the valvular heart disease. *Expert Systems with Applications, 36*, 4232–4238.

Hanbay, D., Turkoglu, I., and Demir, Y. (2008). An expert system based on wavelet decomposition and neural network for modeling Chua's circuit. *Expert Systems with Applications, 34*, 2278–2283.

Hannonen, M. (2006). An analysis of trends and cycles of land prices using wavelet transforms. *International Journal of Strategic Property Management, 10*, 1–21.

Hardy, R. J., Best, J. L., Lane, S. N., and Carbonneau, P. E. (2009). Coherent flow structures in a depth-limited flow over a gravel surface: The role of near-bed turbulence and influence of Reynolds number. *Journal of Geophysical Research: Earth Surface, 114*, F01003, 1–18.

Hariharan, G., and Kannan, K. (2014). Review of wavelet methods for the solution of reaction–diffusion problems in science and engineering. *Applied Mathematical Modelling, 38*, 799–813.

Harrison, D. E., and Chiodi, A. M. (2013). Multi-decadal variability and trends in the El Niño-Southern Oscillation and tropical Pacific fisheries implications. *Deep Sea Research Part II: Topical Studies in Oceanography, 113*, 9–21.

Harrop, J. D., Taraskin, S. N., and Elliott, S. R. (2002). Instantaneous frequency and amplitude identification using wavelets: Application to glass structure. *Physical Review E, 66*, 026703, 1–9.

Hashemi, A., Arabalibiek, H., and Agin, K. (2011). Classification of wheeze sounds using wavelets and neural networks. In *International Conference on Biomedical Engineering and Technology* (Vol.11, pp. 127–131.). Singapore: IACSIT Press.

Hashizume, M., Chaves, L. F., Faruque, A. S. G., Yunus, M., Streatfield, K., and Moji, K. (2013). A differential effect of Indian Ocean dipole and El Niño on cholera dynamics in Bangladesh. *PLoS ONE, 8*(3), e60001, 1–11.

Hassanien, A. E., and Kim, T. H. (2012). Breast cancer MRI diagnosis approach using support vector machine and pulse coupled neural networks. *Journal of Applied Logic, 10*, 277–284.

Haven, E., Liu, X., and Shen, L. (2012). De-noising option prices with the wavelet method. *European Journal of Operational Research, 222*, 104–112.

He, K., Lai, K. K., and Xiang, G. (2012). Portfolio value at risk estimate for crude oil markets: A multivariate wavelet denoising approach. *Energies, 5*, 1018–1043.

He, M., Chen, B., Gong, Y., Wang, K., and Li, Y. (2013). Prediction of defibrillation outcome by ventricular fibrillation waveform analysis: A clinical review. *Journal of Clinical and Experimental Cardiology, S10*, 1–8.

Hedman, M. M., and Nicholson, P. D. (2016). The B-ring's surface mass density from hidden density waves: Less than meets the eye? Icarus. In print. Available online January 22, 2016.

Hedman, M. M., Nicholson, P. D., and Salo, H. (2014). Exploring overstabilities in Saturn's A ring using two stellar occultations. *Astronomical Journal, 148*(15), 1–9.

Heidary, M., and Javaherian, A. (2013). Wavelet analysis in determination of reservoir fluid contacts. *Computers and Geosciences, 52*, 60–67.

Henriques, M. V. C., Leite, F. E. A., Andrade, R. F. S., Andrade, J. S., Lucena, L. S., and Neto, M. L. (2015). Improving the analysis of well-logs by wavelet cross-correlation. *Physica A: Statistical Mechanics and its Applications, 417*, 130–140.

Herb, C., Appel, E., Voigt, S., Koutsodendris, A., Pross, J., Zhang, W., and Fang, X. (2015). Orbitally tuned age model for the late Pliocene–Pleistocene lacustrine succession of drill core SG-1 from the western Qaidam Basin (NE Tibetan Plateau). *Geophysical Journal International, 200*, 35–51.

Herrera, R. H., Han, J., and van der Baan, M. (2014). Applications of the synchrosqueezing transform in seismic time-frequency analysis. *Geophysics, 79*(3), V55–V64.

Herrera, V., Romero, J. F., and Amestegui, M. (2011). Detection and alert of muscle fatigue considering a Surface Electromyography Chaotic Model. In *Journal of Physics: Conference Series, 285*(1), 012046, 1–8. Bristol, UK: IOP Publishing.

Hess-Nielsen, N., and Wickerhauser, M. V. (1996). Wavelets and time-frequency analysis, *Proceedings of the IEEE, 84*(4), 523–540.

Hester, D., and González, A. (2012). A wavelet-based damage detection algorithm based on bridge acceleration response to a vehicle. *Mechanical Systems and Signal Processing, 28*, 145–166.

Higuchi, H., Lewalle, J., and Crane, P. (1994). On the structure of a two-dimensional wake behind a pair of flat plates. *Physics of Fluids (1994–present), 6*(1), 297–305.

Hilborn, R. C. (1994). *Chaos and Nonlinear Dynamics* (Vol. 2). New York: Oxford University Press.

Hirasawa, T., Fujita, M., Okawa, S., Kushibiki, T., and Ishihara, M. (2013). Improvement in quantifying optical absorption coefficients based on continuous wavelet-transform by correcting distortions in temporal photoacoustic waveforms. In *Proceedings of SPIE 8581, Photons Plus Ultrasound: Imaging and Sensing 2013* (Vol. 8581, pp. 85814J-1–85814J-7.). Bellingham, Washington, DC: International Society for Optics and Photonics.

Hoang, V. D. (2014). Wavelet-based spectral analysis. *Trends in Analytical Chemistry, 62*, 144–153.

Hojjati, A., Pogosian, L., and Zhao, G. B. (2010). Detecting features in the dark energy equation of state: A wavelet approach. *Journal of Cosmology and Astroparticle Physics, 2010*(04), JACAP04(2010)007.

Holschneider, M., Diallo, M. S., Kulesh, M., Ohrnberger, M., Lück, E., and Scherbaum, F. (2005). Characterization of dispersive surface waves using continuous wavelet transforms. *Geophysical Journal International, 163*, 463–478.

Holstein-Rathlou, N. H., Sosnovtseva, O. V., Pavlov, A. N., Cupples, W. A., Sorensen, C. M., and Marsh, D. J. (2011). Nephron blood flow dynamics measured by laser speckle contrast imaging. *American Journal of Physiology-Renal Physiology, 300*(2), F319–F329.

Hong, G., and Zhang, Y. (2008). Wavelet-based image registration technique for high-resolution remote sensing images. *Computers and Geosciences, 34*, 1708–1720.

Hramov, A. E., and Koronovskii, A. A. (2005). Time scale synchronization of chaotic oscillators. *Physica D: Nonlinear Phenomena, 206*(3), 252–264.

Hsiu, H., Hsu, W. C., Hsu, C. L., and Huang, S. M. (2011). Assessing the effects of acupuncture by comparing needling the hegu acupoint and needling nearby nonacupoints by spectral analysis of microcirculatory laser Doppler signals. *Evidence-Based Complementary and Alternative Medicine, 435928*, 1–9.

Hsu, C. F. (2013). Adaptive neural complementary sliding-mode control via functional-linked wavelet neural network. *Engineering Applications of Artificial Intelligence, 26*, 1221–1229.

Hsu, W. Y., Lin, C. C., Ju, M. S., and Sun, Y. N. (2007). Wavelet-based fractal features with active segment selection: Application to single-trial EEG data. *Journal of Neuroscience Methods, 163*, 145–160.

Hu, L., Mouraux, A., Hu, Y., and Iannetti, G. D. (2010). A novel approach for enhancing the signal-to-noise ratio and detecting automatically event-related potentials (ERPs) in single trials. *Neuroimage, 50*, 99–111.

Hu, L., Zhang, Z. G., Hung, Y. S., Luk, K. D. K., Iannetti, G. D., and Hu, Y. (2011). Single-trial detection of somatosensory evoked potentials by probabilistic independent component analysis and wavelet filtering. *Clinical Neurophysiology, 122*, 1429–1439.

Huang, J. W., Lu, Y. Y., Nayak, A., and Roy, R. J. (1999). Depth of anesthesia estimation and control. *IEEE Transactions on Biomedical Engineering, 46*(1), 71–81.

Huang, L., Wang, C., and Qin, S. (2010). A multiresolution wavelet based analysis of the Chinese stock market. In *2010 Third International Conference on Business Intelligence and Financial Engineering (BIFE)* (pp. 305–309). IEEE.

Huang, L., Xu, Y. L., and Liao, H. (2014). Nonlinear aerodynamic forces on thin flat plate: Numerical study. *Journal of Fluids and Structures, 44*, 182–194.

Huang, R. Y., and Dung, L. R. (2016). Measurement of heart rate variability using off-the-shelf smart phones. *Biomedical Engineering Online, 15*(11), 1–16.

Huang, S., An, H., Gao, X., and Huang, X. (2015). Identifying the multiscale impacts of crude oil price shocks on the stock market in China at the sector level. *Physica A: Statistical Mechanics and its Applications, 434*, 13–24.

Huart, C., Legrain, V., Hummel, T., Rombaux, P., and Mouraux, A. (2012). Time-frequency analysis of chemosensory event-related potentials to characterize the cortical representation of odors in humans. *PLoS One, 7*(3), e33221, 1–11.

Hubbard, B. B. (1996). *The World According to Wavelets: The Story of a Mathematical Technique in the Making*. Wellesley, MA: Ak Peters.

Humeau, A., Buard, B., Mahé, G., Chapeau-Blondeau, F., Rousseau, D., and Abraham, P. (2010). Multifractal analysis of heart rate variability and laser Doppler flowmetry fluctuations: Comparison of results from different numerical methods. *Physics in Medicine and Biology*, 55, 6279–6297.

Humeau, M. A., Saumet, J. L., and L'huillier, J. P. (2000). Simplified model of laser Doppler signals during reactive hyperaemia. *Medical and Biological Engineering and Computing*, 38, 80–87.

Iatsenko, D., Bernjak, A., Stankovski, T., Shiogai, Y., Owen-Lynch, P. J., Clarkson, P. B. M., McClintock, P. V. E., and Stefanovska, A. (2013). Evolution of cardiorespiratory interactions with age. *Philosophical Transactions of the Royal Society of London A: Mathematical, Physical and Engineering Sciences*, 371, 20110622, 1–18.

Ieong, C. I., Mak, P. I., Lam, C. P., Dong, C., Vai, M. I., Mak, P. U., Pun, S. H., Wan, F., and Martins, R. P. (2012) A 0.83-μW QRS detection processor using quadratic spline wavelet transform for wireless ECG acquisition in 0.35-μm CMOS. *IEEE Transactions on Biomedical Circuits and Systems*, 6(6), 586–595.

Immanuel, J. J. R., Prabhu, V., Christopheraj, V. J., Sugumar, D., and Vanathi, P. T. (2012). Separation of maternal and fetal ECG signals from the mixed source signal using FASTICA. *Procedia Engineering*, 30, 356–363.

Ince, T., Kiranyaz, S., and Gabbouj, M. (2009). A generic and robust system for automated patient-specific classification of ECG signals. *IEEE Transactions on Biomedical Engineering*, 56(5), 1415–1426.

Indrusiak, M. L. S., and Möller, S. V. (2011). Wavelet analysis of unsteady flows: Application on the determination of the Strouhal number of the transient wake behind a single cylinder. *Experimental Thermal and Fluid Science*, 35(2), 319–327.

Inoue, K., Tsujihata, T., Kumamaru, K., and Matsuoka, S. (2005). Feature extraction of human sleep EEG based on a peak frequency analysis. In *Proceedings of the 16th IFAC World Congress*, Prague (pp. 177–182). Amsterdam: Elsevier.

Iungo, G. V., and Lombardi, E. (2011). Time-frequency analysis of the dynamics of different vorticity structures generated from a finite-length triangular prism. *Journal of Wind Engineering and Industrial Aerodynamics*, 99(6), 711–717.

Jackman, P., Sun, D. W., Allen, P., Valous, N. A., Mendoza, F., and Ward, P. (2010). Identification of important image features for pork and turkey ham classification using colour and wavelet texture features and genetic selection. *Meat Science*, 84, 711–717.

Jacobitz, F. G., Schneider, K., Bos, W. J., and Farge, M. (2010). On the structure and dynamics of sheared and rotating turbulence: Anisotropy properties and geometrical scale-dependent statistics. *Physics of Fluids*, 22(8), 085101-1–085101-13.

Jacobitz, F. G., Schneider, K., Bos, W. J., and Farge, M. (2012). On helical multiscale characterization of homogeneous turbulence. *Journal of Turbulence*, 13(35), 1–16.

Jacques, L., Duval, L., Chaux, C., and Peyré, G. (2011). A panorama on multiscale geometric representations, intertwining spatial, directional and frequency selectivity. *Signal Processing*, 91, 2699–2730.

Jafari, M. G., and Chambers, J. A. (2005). Fetal electrocardiogram extraction by sequential source separation in the wavelet domain. *IEEE Transactions on Biomedical Engineering*, 52(3), 390–400.

Jammazi, R., Lahiani, A., and Nguyen, D. K. (2015). A wavelet-based nonlinear ARDL model for assessing the exchange rate pass-through to crude oil prices. *Journal of International Financial Markets, Institutions and Money*, 34, 173–187.

Jan, Y. K., Lee, B., Liao, F., and Foreman, R. D. (2012). Local cooling reduces skin ischemia under surface pressure in rats: An assessment by wavelet analysis of laser Doppler blood flow oscillations. *Physiological Measurement*, 33(10), 1733–1745.

Jänicke, H., Böttinger, M., Mikolajewicz, U., and Scheuermann, G. (2009). Visual exploration of climate variability changes using wavelet analysis. *IEEE Transactions on Visualization and Computer Graphics*, 15(6), 1375–1382.

Janjarasjitt, S., and Loparo, K. A. (2009). Wavelet-based fractal analysis of the epileptic EEG signal. In *ISPACS 2009. International Symposium on Intelligent Signal Processing and Communication Systems, 2009* (pp. 127–130). IEEE.

Janjarasjitt, S., and Loparo, K. A. (2014). Characteristic of spectral exponent of epileptic ECoG data corresponding to levels in wavelet-based fractal analysis. In *Proceedings of the International MultiConference of Engineers and Computer Scientists* (pp. 140–143), March 12–14, 2014. Hong Kong: IAENG.

Jannah, N., Hadjiloucas, S., Hwang, F., and Galvão, R. K. H. (2013). Smart-phone based electrocardiogram wavelet decomposition and neural network classification. In *Journal of Physics: Conference Series*, 450, 012019, 1–7. Bristol, UK: IOP Publishing.

Jansen, M. (2001). *Noise Reduction by Wavelet Thresholding. Lecture Notes in Statistics 161*. New York: Springer.

Janusek, D., Kania, M., Zaczek, R., Zavala-Fernandez, H., Zbiec, A., Opolski, G., and Maniewski, R. (2011). Application of wavelet based denoising for T-wave alternans analysis in high resolution ECG maps. *Measurement Science Review*, 11(6), 181–184.

Jaunet, V., Collin, E., and Bonnet, J. P. (2012). Wavelet series method for reconstruction and spectral estimation of laser Doppler velocimetry data. *Experiments in Fluids*, 52(1), 225–233.

Javadi, M., Ebrahimpour, R., Sajedin, A., Faridi, S., and Zakernejad, S. (2011). Improving ECG classification accuracy using an ensemble of neural network modules. *PLoS One*, 6(10), e24286, 1–13.

Jedliński, Ł., and Jonak, J. (2015). Early fault detection in gearboxes based on support vector machines and multilayer perceptron with a continuous wavelet transform. *Applied Soft Computing*, 30, 636–641.

Jeleazcov, C., Schneider, G., Daunderer, M., Scheller, B., Schüttler, J., and Schwilden, H. (2006). The discriminant power of simultaneous monitoring of spontaneous electroencephalogram and evoked potentials as a predictor of different clinical states of general anesthesia. *Anesthesia and Analgesia*, 103(4), 894–901.

Jemielniak, K., and Kossakowska, J. (2010). Tool wear monitoring based on wavelet transform of raw acoustic emission signal. *Advances in Manufacturing Science and Technology*, 34(3), 5–17.

Jemielniak, K., Urbański, T., Kossakowska, J., and Bombiński, S. (2012). Tool condition monitoring based on numerous signal features. *International Journal of Advanced Manufacturing Technology*, 59(1–4), 73–81.

Jena, D. P., Panigrahi, S. N., and Kumar, R. (2013). Gear fault identification and localization using analytic wavelet transform of vibration signal. *Measurement*, 46, 1115–1124.

Jena, D., Singh, M., and Kumar, R. (2012). Radial ball bearing inner race defect width measurement using analytical wavelet transform of acoustic and vibration signal. *Measurement Science Review*, 12(4), 141–148.

Jeong, M. K., Lu, J. C., and Wang, N. (2006). Wavelet-based SPC procedure for complicated functional data. *International Journal of Production Research*, 44(4), 729–744.

Jestrovic, I., Dudik, J. M., Luan, B., Coyle, J. L., and Sejdic, E. (2013). The effects of increased fluid viscosity on swallowing sounds in healthy adults. *Biomedical Engineering Online*, 12(1), 90.

Jia, X., An, H., Fang, W., Sun, X., and Huang, X. (2015). How do correlations of crude oil prices co-move? A grey correlation-based wavelet perspective. *Energy Economics*, 49, 588–598.

Jiang, C., Chang, T., and Li, X. L. (2015b). Money growth and inflation in China: New evidence from a wavelet analysis. *International Review of Economics and Finance*, 35, 249–261.

Jiang, X. J., and Whitehouse, D. J. (2012). Technological shifts in surface metrology. *CIRP Annals-Manufacturing Technology*, 61, 815–836.

Jiang, Y. Y., Li, B., Zhang, Z. S., and Chen, X. F. (2015a). Identification of crack location in beam structures using wavelet transform and fractal dimension. *Shock and Vibration*, *501*(832763), 1–10.

Jing-Jing, X. U., and Fei, H. U. (2015). Multifractal characteristics of intermittent turbulence in the urban canopy layer. *Atmospheric and Oceanic Science Letters*, *8*(2), 72–77.

Johnson, B., and Lind, R. (2010). Characterizing wing rock with variations in size and configuration of vertical tail. *Journal of Aircraft*, *47*, 567–576.

Johnson, R. W. (2010). Edge adapted wavelets, solar magnetic activity, and climate change. *Astrophysics and Space Science*, *326*, 181–189.

Jones, T. R., White, J. W. C., and Popp, T. (2014). Siple Dome shallow ice cores: A study in coastal dome microclimatology. *Climate of the Past*, *10*, 1253–1267.

Ju, B., Qian, Y. T., and Ye, H. J. (2013). Wavelet based measurement on photoplethysmography by smartphone imaging. *Applied Mechanics and Materials*, *380*, 773–777.

Jubran, B. A., Hamdan, M. N., and Shabaneh, N. H. (1998a). Wavelet and chaos analysis of flow induced vibration of a single cylinder in cross-flow. *International Journal of Engineering Science*, *36*, 843–864.

Jubran, B. A., Hamdan, M. N., Shabanneh, N. H., and Szepessy, S. (1998b). Wavelet and chaos analysis of irregularities of vortex shedding. *Mechanics Research Communications*, *25*(5), 583–591.

Kaewkongka, T., Au, Y. J., Rakowski, R., and Jones, B. E. (2001). Continuous wavelet transform and neural network for condition monitoring of rotodynamic machinery. In *Proceedings of the 18th IEEE Instrumentation and Measurement Technology Conference. IMTC 2001* (pp. 3, 1962–1966), May 21–23, 2001. IEEE: Budapest, Hungary.

Kaihatu, J., Devery, D., Erwin, R., and Goertz, J. (2012). The interaction between short ocean swell and transient long waves: Dissipative and nonlinear effects. *Coastal Engineering Proceedings*, *33*, 1–11.

Kailas, S. V., and Narasimha, R. (1999). The eduction of structures from flow imagery using wavelets Part I. The mixing layer. *Experiments in Fluids*, *27*(2), 167–174.

Kandaswamy, A., Kumar, C. S., Ramanathan, R. P., Jayaraman, S., and Malmurugan, N. (2004). Neural classification of lung sounds using wavelet coefficients. *Computers in Biology and Medicine*, *34*, 523–537.

Kang, Y., Belušić, D., and Smith-Miles, K. (2014). A note on the relationship between turbulent coherent structures and phase correlation. *Chaos: An Interdisciplinary Journal of Nonlinear Science*, *24*, 023114-1–023114-6.

Kankar, P. K., Sharma, S. C., and Harsha, S. P. (2011). Fault diagnosis of ball bearings using continuous wavelet transform. *Applied Soft Computing*, *11*(2), 2300–2312.

Kao, L. J., Chiu, C. C., Lu, C. J., and Chang, C. H. (2013). A hybrid approach by integrating wavelet-based feature extraction with MARS and SVR for stock index forecasting. *Decision Support Systems*, *54*, 1228–1244.

Karacan, C. Ö., and Olea, R. A. (2014). Inference of strata separation and gas emission paths in longwall overburden using continuous wavelet transform of well logs and geostatistical simulation. *Journal of Applied Geophysics*, *105*, 147–158.

Karamperidou, C., Engel, V., Lall, U., Stabenau, E., and Smith III, T. J. (2013). Implications of multiscale sea level and climate variability for coastal resources. *Regional Environmental Change*, *13(S1)*, S91–S100.

Kareem, A., and Kijewski, T. (2002). Time-frequency analysis of wind effects on structures. *Journal of Wind Engineering and Industrial Aerodynamics*, *90*(12), 1435–1452.

Kareem, A., and Wu, T. (2013). Wind-induced effects on bluff bodies in turbulent flows: Nonstationary, non-Gaussian and nonlinear features. *Journal of Wind Engineering and Industrial Aerodynamics*, *122*, 21–37.

Karel, J. M., Haddad, S. A., Hiseni, S., Westra, R. L., Serdijn, W., and Peeters, R. L. (2012). Implementing wavelets in continuous-time analog circuits with dynamic range optimization. *IEEE Transactions on Circuits and Systems I: Regular Papers*, *59*(2), 229–242.

Karl, T., Misztal, P. K., Jonsson, H. H., Shertz, S., Goldstein, A. H., and Guenther, A. B. (2013). Airborne flux measurements of BVOCs above Californian oak forests: Experimental investigation of surface and entrainment fluxes, OH densities, and damköhler numbers. *Journal of the Atmospheric Sciences*, 70, 3277–3287.

Karvounis, E. C., Tsipouras, M. G., and Fotiadis, D. I. (2009). Detection of fetal heart rate through 3-D phase space analysis from multivariate abdominal recordings. *IEEE Transactions on Biomedical Engineering*, 56(5), 1394–1406.

Kaspar, K., Hassler, U., Martens, U., Trujillo-Barreto, N., and Gruber, T. (2010). Steady-state visually evoked potential correlates of object recognition. *Brain Research*, 1343, 112–121.

Katul, G. G., Geron, C. D., Hsieh, C. I., Vidakovic, B., and Guenther, A. B. (1998). Active turbulence and scalar transport near the forest–atmosphere interface. *Journal of Applied Meteorology*, 37(12), 1533–1546.

Katul, G. G., and Parlange, M. B. (1995). Analysis of land surface heat fluxes using the orthonormal wavelet approach. *Water Resources Research*, 31, 2743–2749.

Katul, G. G., Parlange, M. B., and Chu, C. R. (1994). Intermittency, local isotropy, and non-Gaussian statistics in atmospheric surface layer turbulence. *Physics of Fluids (1994–present)*, 6(7), 2480–2492.

Katul, G., and Vidakovic, B. (1996). The partitioning of attached and detached eddy motion in the atmospheric surface layer using Lorentz wavelet filtering. *Boundary-Layer Meteorology*, 77(2), 153–172.

Katul, G., and Vidakovic, B. (1998). Identification of low-dimensional energy containing/flux transporting eddy motion in the atmospheric surface layer using wavelet thresholding methods. *Journal of the Atmospheric Sciences*, 55(3), 377–389.

Katunin, A., and Przystałka, P. (2014). Damage assessment in composite plates using fractional wavelet transform of modal shapes with optimized selection of spatial wavelets. *Engineering Applications of Artificial Intelligence*, 30, 73–85.

Kayhan, S., and Ercelebi, E. (2011). ECG denoising on bivariate shrinkage function exploiting interscale dependency of wavelet coefficients. *Turkish Journal of Electrical Engineering and Computer Sciences*, 19(3), 495–511.

Kedadouche, M., Thomas, M., Tahan, A., and Guilbault, R. (2015). Monitoring gears by vibration measurements: Lempel-Ziv complexity and approximate entropy as diagnostic tools. In *AVE2014: 4ième Colloque Analyse Vibratoire Expérimentale / Experimental Vibration Analysis. MATEC Web of Conferences*, (Vol. 20, pp. 07001-1–07001-7). London: EDP Sciences.

Keenan, D. (2008). Detection and correction of ectopic beats for HRV analysis applying discrete wavelet transform. *International Journal of Medical, Health, Biomedical, Bioengineering and Pharmaceutical Engineering*, 2(10), 358–364.

Keissar, K., Davrath, L. R., and Akselrod, S. (2008). Wavelet transform coherence estimates in cardiovascular analysis: Error analysis and feasibility study. *Computers in Cardiology*, 35, 461–464.

Keissar, K., Davrath, L. R., and Akselrod, S. (2009a). Coherence analysis between respiration and heart rate variability using continuous wavelet transform. *Philosophical Transactions of the Royal Society of London A: Mathematical, Physical and Engineering Sciences*, 367, 1393–1406.

Keissar, K., Gilad, O., and Akselrod, S. (2009b). Modified wavelet bicoherence as a diagnostic tool for very high frequency peaks in cardiovascular signals of normal and heart transplant subjects. *Computers in Cardiology*, 36, 677–680.

Kellnerova, R., Kukacka, L., Jurcakova, K., Uruba, V., and Janour, Z. (2011). *Comparison of wavelet analysis with velocity derivatives for detection of shear layer and vortices inside a turbulent boundary layer. Journal of Physics: Conference Series, 318(6), 062012, 1–10.*

Kenwright, D. A., Bahraminasab, A., Stefanovska, A., and McClintock, P. V. (2008). The effect of low-frequency oscillations on cardio-respiratory synchronization. *European Physical Journal B*, 65(3), 425–433.

Kerhervé, F., Jordan, P., Cavalieri, A. V. G., Delville, J., Bogey, C., and Juvé, D. (2012). Educing the source mechanism associated with downstream radiation in subsonic jets. *Journal of Fluid Mechanics, 710,* 606–640.

Keskin, F., Suhre, A., Kose, K., Ersahin, T., Cetin, A. E., and Cetin-Atalay, R. (2013). Image classification of human carcinoma cells using complex wavelet-based covariance descriptors. *PLoS One, 8(e52807),* 1–10.

Kestener, P., Conlon, P. A., Khalil, A., Fennell, L., McAteer, R. T. J., Gallagher, P. T., and Arneodo, A. (2010). Characterizing complexity in solar magnetogram data using a wavelet-based segmentation method. *Astrophysical Journal, 717,* 995–1005.

Keylock, C. J. (2010). Characterizing the structure of nonlinear systems using gradual wavelet reconstruction. *Nonlinear Processes in Geophysics, 17*(6), 615–632.

Keylock, C. J., Tokyay, T. E., and Constantinescu, G. (2011). A method for characterising the sensitivity of turbulent flow fields to the structure of inlet turbulence. *Journal of Turbulence, 12*(45), 1–30.

Khalfaoui, R., Boutahar, M., and Boubaker, H. (2015). Analyzing volatility spillovers and hedging between oil and stock markets: Evidence from wavelet analysis. *Energy Economics, 49,* 540–549.

Khalidov, I., Fadili, J., Lazeyras, F., Van De Ville, D., and Unser, M. (2011). Activelets: Wavelets for sparse representation of hemodynamic responses. *Signal Processing, 91*(12), 2810–2821.

Khan, M. M., and Fadzil, M. H. A. (2007). Singularity spectrum of hydrocarbon fluids in synthetic seismograms. In *International Conference on Intelligent and Advanced Systems, 2007. ICIAS 2007* (pp. 1236–1239). IEEE.

Khandelwal, S., and Wickström, N. (2014). Identification of gait events using expert knowledge and continuous wavelet transform analysis. In *7th International Conference on Bio-Inspired Systems and Signal Processing (BIOSIGNALS 2014)* (pp. 197–204), March 3–6, 2014. Angers, France: SciTePress.

Khandoker, A. H., Karmakar, C. K., and Palaniswami, M. (2009). Automated recognition of patients with obstructive sleep apnoea using wavelet-based features of electrocardiogram recordings. *Computers in Biology and Medicine, 39,* 88–96.

Khandoker, A. H., Karmakar, C. K., Penzel, T., Glos, M., and Palaniswami, M. (2013). Investigating relative respiratory effort signals during mixed sleep apnea using photoplethysmogram. *Annals of Biomedical Engineering, 41*(10), 2229–2236.

Khare, A., Khare, M., Jeong, Y., Kim, H., and Jeon, M. (2010). Despeckling of medical ultrasound images using Daubechies complex wavelet transform. *Signal Processing, 90,* 428–439.

Khazaee, M., Ahmadi, H., Omid, M., Moosavian, A., and Khazaee, M. (2014). Classifier fusion of vibration and acoustic signals for fault diagnosis and classification of planetary gears based on Dempster–Shafer evidence theory. *Proceedings of the Institution of Mechanical Engineers, Part E: Journal of Process Mechanical Engineering, 228,* 21–32.

Kheder, G., Kachouri, A., Taleb, R., Ben Messaoud, M., and Samet, M. (2009). Feature extraction by wavelet transforms to analyze the heart rate variability during two meditation techniques. *Advances in Numerical Methods, 11,* 379–387.

Khoshelham, K., Altundag, D., Ngan-Tillard, D., and Menenti, M. (2011). Influence of range measurement noise on roughness characterization of rock surfaces using terrestrial laser scanning. *International Journal of Rock Mechanics and Mining Sciences, 48,* 1215–1223.

Khujadze, G., Schneider, K., Oberlack, M., and Farge, M. (2011). Coherent vorticity extraction in turbulent boundary layers using orthogonal wavelets. *Journal of Physics: Conference Series, 318,* 022011, 1–10.

Kiiski, H., Reilly, R. B., Lonergan, R., Kelly, S., O'Brien, M. C., Kinsella, K., Bramham, J., et al. (2012). Only low frequency event-related EEG activity is compromised in multiple sclerosis: Insights from an independent component clustering analysis. *PLoS One, 7*(9), e45536, 1–12.

Kijewski-Correa, T., and Bentz, A. (2011). Wind-induced vibrations of buildings: Role of transient events. *Proceedings of the ICE—Structures and Buildings, 164*(4), 273–284.

Kim, H. (2010). Dynamic causal linkages between the US stock market and the stock markets of the East Asian economies. *Royal Institute of Technology Centre of Excellence for Science and Innovation Studies (CESIS), 236*, 1–23.

Kirby, J. F., and Swain, C. J. (2006). Mapping the mechanical anisotropy of the lithosphere using a 2D wavelet coherence, and its application to Australia. *Physics of the Earth and Planetary Interiors, 158*, 122–138.

Kirchner, M., Schubert, P., Schmidtbleicher, D., and Haas, C. T. (2012). Evaluation of the temporal structure of postural sway fluctuations based on a comprehensive set of analysis tools. *Physica A: Statistical Mechanics and its Applications, 391*, 4692–4703.

Klein, Y., Grinstein, M., Cohn, S. M., Silverman, J., Klein, M., Kashtan, H., and Shamir, M. Y. (2012). Minute-to-minute urine flow rate variability: A new renal physiology variable. *Anesthesia and Analgesia, 115*(4), 843–847.

Knešaurek, K., Machac, J., and Zhang, Z. (2009). Repeatability of regional myocardial blood flow calculation in 82 Rb PET imaging. *BMC Medical Physics, 9*(2), 1–9.

Kodera, K., Gendrin, R., and Villedary, C. (1978). Analysis of time-varying signals with small BT values. *Acoustics, IEEE Transactions on Speech and Signal Processing, 26*, 64–76.

Koenig, M., Cavalieri, A. V. G., Jordan, P., Delville, J., Gervais, Y., and Papamoschou, D. (2011). Farfield filtering of subsonic jet noise: Mach and Temperature effects. *AIAA/CEAS Aeroacoustics Conference (32nd AIAA Aeroacoustics Conference)* (pp. 1–19), June 5–8, 2011. Portland, OR: AIAA 2011-2926.

Koenig, M., Cavalieri, A., Jordan, P., Delville, J., Gervais, Y., Papamoschou, D., Samimy M, and Lele, S. (2010). Farfield filtering and source imaging for the study of jet noise. *16th AIAA/CEAS Aeroacoustics Conference* (pp. 1–24), June 7–9, 2010. Stockholm, Sweden: AIAA 2010-3779.

Kopriva, I., Jerić, I., and Smrečki, V. (2009). Extraction of multiple pure component 1 H and 13 C NMR spectra from two mixtures: Novel solution obtained by sparse component analysis-based blind decomposition. *Analytica Chimica Acta, 653*, 143–153.

Koronovskii, A. A., and Khramov, A. E. (2002). Wavelet bicoherence analysis as a method for investigating coherent structures in an electron beam with an overcritical current. *Plasma Physics Reports, 28*(8), 666–681.

Korürek, M., and Nizam, A. (2010). Clustering MIT–BIH arrhythmias with Ant Colony Optimization using time domain and PCA compressed wavelet coefficients. *Digital Signal Processing, 20*, 1050–1060.

Krajewski, J., Golz, M., Schnieder, S., Schnupp, T., Heinze, C., and Sommer, D. (2010). Detecting fatigue from steering behaviour applying continuous wavelet transform. In *Proceedings of the 7th International Conference on Methods and Techniques in Behavioral Research* (pp. 326–329), August 24–27, 2010. The Netherlands: ACM.

Kreitzer, P. J., Hanchak, M., and Byrd, L. (2012). Horizontal two phase flow regime identification: Comparison of pressure signature, electrical capacitance tomography (ECT) and high speed visualization. In *ASME 2012 International Mechanical Engineering Congress and Exposition* (pp. 1281–1291), November 9–15, 2012. New York: American Society of Mechanical Engineers.

Krishna, B., and YR Satyaji, R. (2011). Time series modeling of river flow using wavelet neural networks. *Journal of Water Resource and Protection, 3*, 50–59.

Krivonos, R. A., Tomsick, J. A., Bauer, F. E., Baganoff, F. K., Barriere, N. M., Bodaghee, A., Boggs, S. E., et al. (2014). First hard X-ray detection of the non-thermal emission around the Arches cluster: Morphology and spectral studies with NuSTAR. *Astrophysical Journal, 781*(107), 1–11.

Kulesh, M., Holschneider, M., Diallo, M. S., Xie, Q., and Scherbaum, F. (2005). Modeling of wave dispersion using continuous wavelet transforms. *Pure and Applied Geophysics, 162*, 843–855.

Kulkarni, J. R., Sadani, L. K., and Murthy, B. S. (1999). Wavelet analysis of intermittent turbulent transport in the atmospheric surface layer over a monsoon trough region. *Boundary-Layer Meteorology, 90*(2), 217–239.

Kumar, D., Carvalho, P., Antunes, M., Paiva, R. P., and Henriques, J. (2010). Heart murmur classification with feature selection. In *Engineering in Medicine and Biology Society (EMBC), 2010 Annual International Conference of the IEEE* (pp. 4566–4569). IEEE.

Kumar, P., and Foufoula-Georgiou, E. (1997). Wavelet analysis for geophysical applications. *Reviews of Geophysics*, 35(4), 385–412.

Kuncheva, L. I., and Rodríguez, J. J. (2013). Interval feature extraction for classification of event-related potentials (ERP) in EEG data analysis. *Progress in Artificial Intelligence*, 2(1), 65–72.

Kunpeng, Z., San, W. Y., and Soon, H. G. (2009). Wavelet analysis of sensor signals for tool condition monitoring: A review and some new results. *International Journal of Machine Tools and Manufacture*, 49, 537–553.

Kutyniok, G., Lemvig, J., and Lim, W. Q. (2012). Compactly supported shearlets. In M. Neamtu and L.L. Schumaker (Eds.) *Approximation Theory XIII, San Antonio, TX, 2010* (pp. 163–186). New York: Springer.

Kutyniok, G., and Sauer, T. (2007). From wavelets to shearlets and back again. In M. Neamtu and L.L. Schumaker (Eds.) *Approximation Theory XII, San Antonio, TX, 2007* (pp. 201–209). Nashville, TN: Nashboro Press

Kvandal, P., Sheppard, L., Landsverk, S. A., Stefanovska, A., and Kirkeboen, K. A. (2013). Impaired cerebrovascular reactivity after acute traumatic brain injury can be detected by wavelet phase coherence analysis of the intracranial and arterial blood pressure signals. *Journal of Clinical Monitoring and Computing*, 27, 375–383.

Kvernmo, H. D., Stefanovska, A., Bracic, M., Kirkebøen, K. A., and Kvernebo, K. (1998). Spectral analysis of the laser Doppler perfusion signal in human skin before and after exercise. *Microvascular Research*, 56, 173–182.

Kvernmo, H. D., Stefanovska, A., Kirkebøen, K. A., and Kvernebo, K. (1999). Oscillations in the human cutaneous blood perfusion signal modified by endothelium-dependent and endothelium-independent vasodilators. *Microvascular Research*, 57, 298–309.

Lachowicz, P., and Done, C. (2010). Quasi-periodic oscillations under wavelet microscope: The application of Matching Pursuit algorithm. *Astronomy and Astrophysics*, 515(A65), 1–11.

Landsverk, S. A., Kvandal, P., Bernjak, A., Stefanovska, A., and Kirkeboen, K. A. (2007). The effects of general anesthesia on human skin microcirculation evaluated by wavelet transform. *Anesthesia and Analgesia*, 105(4), 1012–1019.

Lane, S. N. (2007). Assessment of rainfall-runoff models based upon wavelet analysis. *Hydrological Processes*, 21(5), 586–607.

Latka, M., Turalska, M., Glaubic-Latka, M., Kolodziej, W., Latka, D., and West, B. J. (2005). Phase dynamics in cerebral autoregulation. *American Journal of Physiology-Heart and Circulatory Physiology*, 289(5), H2272–H2279.

Le Pogam, A., Hanzouli, H., Hatt, M., Le Rest, C. C., and Visvikis, D. (2013). Denoising of PET images by combining wavelets and curvelets for improved preservation of resolution and quantitation. *Medical Image Analysis*, 17(8), 877–891.

Le, T. H., and Nguyen, D. A. (2008). Temporo-spectral coherent structure of turbulence and pressure using Fourier and wavelet transforms. *AJSTD*, 25(2), 405–417.

Le, T. H., Tamura, Y., and Matsumoto, M. (2010). Spanwise pressure coherence on prisms based on spectral POD and wavelet transform tools. In *Proceedings of the 5th International Symposium on Computational Wind Engineering (CWE2010)* (pp. 23–27), May 23–27, 2010, Tokyo, Japan: IAWE.

Le Van Quyen, M., Staba, R., Bragin, A., Dickson, C., Valderrama, M., Fried, I., and Engel, J. (2010). Large-scale microelectrode recordings of high-frequency gamma oscillations in human cortex during sleep. *Journal of Neuroscience*, 30(23), 7770-7782.

Lee, B. C., Kao, C. C., and Doong, D. J. (2011). An analysis of the characteristics of freak waves using the wavelet transform. Terrestrial. *Atmospheric and Oceanic Sciences*, 22(3), 359–370.

Lee, C. I., Pakhomov, E., Atkinson, A., and Siegel, V. *(2010)*. Long-term relationships between the marine environment, krill and salps in the Southern Ocean. *Journal of Marine Biology*, 2010, 410129, 1–18.

Lee, I., and Sung, H. J. (2001). Characteristics of wall pressure fluctuations in separated flows over a backward-facing step: Part II. Unsteady wavelet analysis. *Experiments in Fluids*, 30(3), 273–282.

Lee, S. H., and Lim, J. S. (2012). Parkinson's disease classification using gait characteristics and wavelet-based feature extraction. *Expert Systems with Applications*, 39, 7338–7344.

Lee, S. Y., Rus, G., and Park, T. (2007). Detection of stiffness degradation in laminated composite plates by filtered noisy impact testing. *Computational Mechanics*, 41(1), 1–15.

Legarreta, I. R., Addison, P. S., Grubb, N. R., Clegg, G. R., Robertson, C. E., and Watson, J. N. (2005b). Analysis of ventricular late potentials prior to the onset of ventricular tachyarrhythmias: End of QRS point detector. *Computers in Cardiology*, 32, 471–474.

Legarreta, I. R., Addison, P. S., Reed, M. J., Grubb, N., Clegg, G. R., Robertson, C. E., and Watson, J. N. (2005a). Continuous wavelet transform modulus maxima analysis of the electrocardiogram: Beat characterisation and beat-to-beat measurement. *International Journal of Wavelets, Multiresolution and Information Processing*, 3(1), 19–42.

Legarreta, I. R., Reed, M. J., Addison, P. S., Grubb, N. R., Clegg, G. R., Robertson, C. E., and Watson, J. N. (2004a). Measurement of heart rate variability during recurrent episodes of ventricular tachyarrhythmia in one patient using wavelet transform analysis. *Computers in Cardiology*, 31, 469–472.

Legarreta, I. R., Reed, M. J., Addison, P. S., Grubb, N. R., Clegg, O. R., Robertson, C. E., and Watson, J. N. (2004b). Can heart rate variability analysis predict the acute onset of ventricular tachyarrhythmias? *Computers in Cardiology*, 31, 201–204.

Leise, T. L., Indic, P., Paul, M. J., and Schwartz, W. J. (2013). Wavelet meets actogram. *Journal of Biological Rhythms*, 28(1), 62–68.

Leite, F. E. A., Henriques, M. V. C., and Gurgel, V. C. (2013). Efficient selective filtering of seismic data using multiscale decomposition. *Nonlinear Processes in Geophysics*, 20, 207–211.

Leonard, A., Lanusse, F., and Starck, J. L. (2015). Weak lensing reconstructions in 2D and 3D: Implications for cluster studies. *Monthly Notices of the Royal Astronomical Society*, 449(1), 1146–1157.

Leonard, P. A., Clifton, D., Addison, P. S., Watson, J. N., and Beattie, T. (2006). An automated algorithm for determining respiratory rate by photoplethysmogram in children. *Acta Paediatrica*, 95, 1124–1128.

Leonard, P., Beattie, T. F., Addison, P. S., and Watson, J. N. (2003). Standard pulse oximeters can be used to monitor respiratory rate. *Emergency Medicine Journal*, 20, 524–525.

Leonard, P., Beattie, T. F., Addison, P. S., and Watson, J. N. (2004b). Wavelet analysis of pulse oximeter waveform permits identification of unwell children. *Emergency Medicine Journal*, 21, 59–60.

Leonard, P., Grubb, N. R., Addison, P. S., Clifton, D., and Watson, J. N. (2004a). An algorithm for the detection of individual breaths from the pulse oximeter waveform. *Journal of Clinical Monitoring and Computing*, 18(5), 309–312.

Leonarduzzi, R., Spilka, J., Wendt, H., Jaffard, S., Torres, M. E., Abry, P., and Doret, M. (2014). p-leader based classification of first stage intrapartum fetal HRV. In *Proceedings of the VI Congreso Latinoamericano de Ingeniería Biomédica (CLAIB), Paraná, Entre Ríos, Argentina* (IFBME Proceedings, Vol. 49, pp. 504–507), October 29–31, 2014.

Lepik, Ü. (2012). Exploring vibrations of cracked beams by the Haar wavelet method. *Estonian Journal of Engineering*, 18, 58–75.

Lescarmontier, L., Legrésy, B., Coleman, R., Perosanz, F., Mayet, C., and Testut, L. (2012). Vibrations of Mertz glacier ice tongue, East Antarctica. *Journal of Glaciology*, 58(210), 665–676.

Lestussi, F., Persia, L. D., and Milone, D. (2011). Comparison of on-line wavelet analysis and reconstruction: With application to ECG. *5th International Conference on Bioinformatics and Biomedical Engineering, (iCBBE) 2011* (pp. 1–4). IEEE.

Lewalle, J. (2010). Single-scale wavelet representation of turbulence dynamics: Formulation and Navier–Stokes regularity. *Physica D: Nonlinear Phenomena*, 239, 1232–1235.

Li, C., and Liang, M. (2012). Time-frequency signal analysis for gearbox fault diagnosis using a generalized synchrosqueezing transform. *Mechanical Systems and Signal Processing*, 26, 205–217.

Li, H. W., Zhou, Y. L., Hou, Y. D., Sun, B., and Yang, Y. (2014b). Flow pattern map and time-frequency spectrum characteristics of nitrogen–water two-phase flow in small vertical upward noncircular channels. *Experimental Thermal and Fluid Science*, 54, 47–60.

Li, H., Yi, T., Gu, M., and Huo, L. (2009). Evaluation of earthquake-induced structural damages by wavelet transform. *Progress in Natural Science*, 19, 461–470.

Li, H., Zhang, Y., and Zheng, H. (2011a). Application of Hermitian wavelet to crack fault detection in gearbox. *Mechanical Systems and Signal Processing*, 25, 1353–1363.

Li, K. J., Gao, P. X., Zhan, L. S., Shi, X. J., and Zhu, W. W. (2010). Relative phase analyses of long-term hemispheric solar flare activity. *Monthly Notices of the Royal Astronomical Society*, 401, 342–346.

Li, L., Li, K., Liu, C. C., and Liu, C. Y. (2011d). Comparison of detrending methods in spectral analysis of heartRate variability. *Research Journal of Applied Sciences, Engineering and Technology*, 3(9), 1014–1021.

Li, S., Zöllner, F. G., Merrem, A. D., Peng, Y., Roervik, J., Lundervold, A., and Schad, L. R. (2012). Wavelet-based segmentation of renal compartments in DCE-MRI of human kidney: Initial results in patients and healthy volunteers. *Computerized Medical Imaging and Graphics*, 36, 108–118.

Li, X. L., Chang, T., Miller, S. M., Balcilar, M., and Gupta, R. (2015). The co-movement and causality between the US housing and stock markets in the time and frequency domains. *International Review of Economics and Finance*, 38, 220–233.

Li, X., Tso, S. K., and Wang, J. (2000). Real-time tool condition monitoring using wavelet transforms and fuzzy techniques. *IEEE Transactions on Systems, Man, and Cybernetics, Part C: Applications and Reviews*, 30(3), 352–357.

Li, X., Yao, X., Fox, J., and Jefferys, J. G. (2007). Interaction dynamics of neuronal oscillations analysed using wavelet transforms. *Journal of Neuroscience Methods*, 160, 178–185.

Li, Y., Guo, J., Wang, C., Fan, Z., Liu, G., Wang, C., Gu, Z., Damm, D., Mosig, A., and Wei, X. (2011c). Circulation times of prostate cancer and hepatocellular carcinoma cells by in vivo flow cytometry. *Cytometry Part A*, 79, 848–854.

Li, Y., Wang, X., Lin, J., and Shi, S. (2014a). A wavelet bicoherence-based quadratic nonlinearity feature for translational axis condition monitoring. *Sensors*, 14, 2071–2088.

Li, Z., Yan, X., Yuan, C., Peng, Z., and Li, L. (2011b). Virtual prototype and experimental research on gear multi-fault diagnosis using wavelet-autoregressive model and principal component analysis method. *Mechanical Systems and Signal Processing*, 25, 2589–2607.

Liebling, M., Bernhard, T. F., Bachmann, A. H., Froehly, L., Lasser, T., and Unser, M. (2005). Continuous wavelet transform ridge extraction for spectral interferometry imaging. In *Biomedical Optics 2005* (pp. 397–402). San Jose, CA: International Society for Optics and Photonics.

Lilly, J. M., and Gascard, J. C. (2006). Wavelet ridge diagnosis of time-varying elliptical signals with application to an oceanic eddy. *Nonlinear Processes in Geophysics*, 13, 467–483.

Lilly, J. M., and Olhede, S. C. (2009). Wavelet ridge estimation of jointly modulated multivariate oscillations. *Conference Record of the 43rd Asilomar Conference on Signals, Systems and Computers* (pp. 452–456), November 1–9, 2009. Pacific Grove, CA.

Lilly, J. M., and Olhede, S. C. (2010). On the analytic wavelet transform. *IEEE Transactions on Information Theory*, 56(8), 4135–4156.

Lilly, J. M., and Olhede, S. C. (2012). Analysis of modulated multivariate oscillations. *IEEE Transactions on Signal Processing*, 60(2), 600–612.

Lilly, J. M., Scott, R. K., and Olhede, S. C. (2011). Extracting waves and vortices from Lagrangian trajectories. *Geophysical Research Letters*, 38, L23605, 1–5.

Lim, H. J., Sohn, H., DeSimio, M. P., and Brown, K. (2014). Reference-free fatigue crack detection using nonlinear ultrasonic modulation under various temperature and loading conditions. *Mechanical Systems and Signal Processing*, 45, 468–478.

Lim, H. S., Liu, J. J., Han, J. H., and Lee, J. M. (2012). Abrasion diagnosis and assessment of marine engine using the wavelet transform. In *2012 12th International Conference on Control, Automation and Systems (ICCAS)* (pp. 1661–1665). IEEE.

Lim, M. H., and Leong, M. S. (2013). Detection of early faults in rotating machinery based on wavelet analysis. *Advances in Mechanical Engineering*, 625863, 1–8.

Lima, C. A., Coelho, A. L., and Chagas, S. (2009). Automatic EEG signal classification for epilepsy diagnosis with Relevance Vector Machines. *Expert Systems with Applications*, 36(6), 10054–10059.

Lin, C. H. (2014). A novel hybrid recurrent wavelet neural network control of permanent magnet synchronous motor drive for electric scooter. *Turkish Journal of Electrical Engineering and Computer Sciences*, 22, 1056–1075.

Lineesh, M. C., Minu, K. K., and John, C. J. (2010). Analysis of nonstationary nonlinear economic time series of gold price: A comparative study. *International Mathematical Forum*, 5(34), 1673–1683.

Litak, G., Kecik, K., and Rusinek, R. (2013). Cutting force response in milling of Inconel: Analysis by wavelet and Hilbert-Huang transforms. *Latin American Journal of Solids and Structures*, 10, 133–140.

Litak, G., and Rusinek, R. (2011). 615. Vibrations in stainless steel turning: Multifractal and wavelet approaches. *Vibroengineering*, 13(1), 102–108.

Liu, G., and Luan, Y. (2014). Identification of protein coding regions in the eukaryotic DNA sequences based on Marple algorithm and wavelet packets transform. *Abstract and Applied Analysis*, 2014, 402567, 1–14.

Liu, H., Huang, W., Wang, S., and Zhu, Z. (2014). Adaptive spectral kurtosis filtering based on Morlet wavelet and its application for signal transients detection. *Signal Processing*, 96, 118–124.

Liu, J., Wang, H., Liu, W., and Zhang, J. (2012a). Autonomous detection and classification of congenital heart disease using an auscultation vest. *Journal of Computational Information Systems*, 8(2), 485–492.

Liu, L., Zuo, W., Zhang, D., Li, N., and Zhang, H. (2012b). Combination of heterogeneous features for wrist pulse blood flow signal diagnosis via multiple kernel learning. *IEEE Transactions on Information Technology in Biomedicine*, 16(4), 599–607.

Liu, P. C. (1994). Wavelet spectrum analysis and ocean wind waves. In E. Foufoula-Georgiou and P. Kumar (Eds.), *Wavelets in Geophysics* (pp. 151–166). New York: Academic Press.

Liu, P. C. (2000a). Wave grouping characteristics in nearshore Great Lakes. *Ocean Engineering*, 27, 1221–1230.

Liu, P. C. (2000b). Is the wind wave frequency spectrum outdated. *Ocean Engineering*, 27, 577–588.

Liu, P. C., and Babanin, A. V. (2004). Using wavelet spectrum analysis to resolve breaking events in the wind wave time series. *Annales Geophysicae*, 22, 3335–3345.

Liu, X. (2013). Time-arrival location of seismic P-wave based on wavelet transform modulus maxima. *Journal of Multimedia*, 8(1), 32–39.

Liu, Y., Aickelin, U., Feyereisl, J., and Durrant, L. G. (2013). Wavelet feature extraction and genetic algorithm for biomarker detection in colorectal cancer data. *Knowledge-Based Systems*, 37, 502–514.

Liu, Y. Z., Kang, W., and Sung, H. J. (2005). Assessment of the organization of a turbulent separated and reattaching flow by measuring wall pressure fluctuations. *Experiments in Fluids*, 38(4), 485–493.

Lockhart, C. J., McCann, A., Agnew, C. A., Hamilton, P. K., Quinn, C. E., McClenaghan, V., Patterson, C., McGivern, R. C., Harbinson, M. T., and McVeigh, G. E. (2011). Impaired microvascular properties in uncomplicated type 1 diabetes identified by Doppler ultrasound of the ocular circulation. *Diabetes and Vascular Disease Research*, 8(3), 211–220.

Lockhart, T. E., Soangra, R., Zhang, J., and Wu, X. (2013). Wavelet based automated postural event detection and activity classification with single IMU (TEMPO). *Biomedical Sciences Instrumentation*, 49, 224–233.

Lopes, R., and Betrouni, N. (2009). Fractal and multifractal analysis: A review. *Medical Image Analysis*, 13, 634–649.

Lopez-Montes, R., Pérez-Enríquez, R., Araujo-Pradere, E. A., and Cruz-Abeyro, J. A. L. (2015). Fractal and wavelet analysis evaluation of the mid latitude ionospheric disturbances associated with major geomagnetic storms. *Advances in Space Research*, 55(2), 586–596.

Lovejoy, S., and Schertzer, D. (2012). Haar wavelets, fluctuations and structure functions: Convenient choices for geophysics. *Nonlinear Processes in Geophysics*, 19(5), 513–527.

Low, K. R., Berger, Z. P., Lewalle, J., El-Hadidi, B., and Glauser, M. N. (2011). Correlations and wavelet based analysis of near-field and far-field pressure of a controlled high speed jet. *41st Fluid Dynamics Conference and Exhibit* (pp. 1–10), June 27–30, 2011. Honolulu, Hawaii: AIAA 2011-4020.

Lu, C. H., and Fitzjarrald, D. R. (1994). Seasonal and diurnal variations of coherent structures over a deciduous forest. *Boundary-Layer Meteorology*, 69, 43–69.

Lu, W. B., Li, P., Chen, M., Zhou, C. B., and Shu, D. Q. (2011). Comparison of vibrations induced by excavation of deep-buried cavern and open pit with method of bench blasting. *Journal of Central South University of Technology*, 18, 1709–1718.

Lu, W., Wei-Hua, C., and Feng-Chen, L. (2014). Large-eddy simulations of a forced homogeneous isotropic turbulence with polymer additives. *Chinese Physics B*, 23(3), 034701-1–034701-13.

Lui, P. W., Chan, B. C., Chan, F. H., Poon, P. W., Wang, H., and Lam, F. K. (1998). Wavelet analysis of embolic heart sound detected by precordial Doppler ultrasound during continuous venous air embolism in dogs. *Anesthesia and Analgesia*, 86, 325–331.

Lundstedt, H., Liszka, L., Lundin, R., and Muscheler, R. (2006). Long-term solar activity explored with wavelet methods. *Annales Geophysicae*, 24(2), 769–778.

Ma, H., Yu, T., Han, Q., Zhang, Y., Wen, B., and Xuelian, C. (2009a). Time-frequency features of two types of coupled rub-impact faults in rotor systems. *Journal of Sound and Vibration*, 321, 1109–1128.

Ma, J., Hussaini, M. Y., Vasilyev, O. V., and Le Dimet, F. X. (2009b). Multiscale geometric analysis of turbulence by curvelets. *Physics of Fluids*, 21, 075104-1–075104-19.

Ma, J., and Plonka, G. (2010). The curvelet transform. *IEEE Signal Processing Magazine*, 27(2), 118–133.

Ma, Y., Dong, G., and Ma, X. (2011). Separation of obliquely incident and reflected irregular waves by the Morlet wavelet transform. *Coastal Engineering*, 58, 761–766.

Ma, Y., Dong, G., Ma, X., and Wang, G. (2010). A new method for separation of 2D incident and reflected waves by the Morlet wavelet transform. *Coastal Engineering*, 57, 597–603.

Machado, J. T., Duarte, F. B., and Duarte, G. M. *(2012)*. Analysis of stock market indices with multidimensional scaling and wavelets. *Mathematical Problems in Engineering*, 2012, 819503, 1–15.

MacLachlan, G. A., Shenoy, A., Sonbas, E., Coyne, R., Dhuga, K. S., Eskandarian, A., Maximon, L. C., and Parke, W. C. (2013). The Hurst exponent of Fermi gamma-ray bursts. *Monthly Notices of the Royal Astronomical Society*, 436(4), 2907–2914

Madeiro, J. P., Cortez, P. C., Marques, J. A., Seisdedos, C. R., and Sobrinho, C. R. (2012). An innovative approach of QRS segmentation based on first-derivative, Hilbert and Wavelet Transforms. *Medical Engineering and Physics*, 34, 1236–1246.

Magini, M., Mocaiber, I., De Oliveira, L., Barbosa, W. L. D. O., Pereira, M. G., and Machado-Pinheiro, W. (2012). The role of basal HRV assessed through wavelet transform in the prediction of anxiety and affect levels: A case study. *Journal of Biomedical Graphics and Computing*, 2(1), 133.

Mahdavi, S. H., and Abdul Razak, H. (2015). A comparative study on optimal structural dynamics using wavelet functions. *Mathematical Problems in Engineering, 956793*, 1–10.

Makris, N., and Kampas, G. (2013). Estimating the "effective period" of bilinear systems with linearization methods, wavelet and time-domain analyses: From inelastic displacements to modal identification. *Soil Dynamics and Earthquake Engineering, 45*, 80–88.

Malegori, G., and Ferrini, G. (2012). Wavelet transforms in dynamic atomic force spectroscopy. In V. Belitto (Ed.), *Atomic Force Microscopy: Imaging, Measuring and Manipulating Surfaces at the Atomic Scale* (Chapter 5, pp. 71–98). INTECH Open Access Publisher. Rijeka, Croatia.

Mallat, S. G. (1989). A theory for multiresolution signal decomposition: The wavelet representation. *IEEE Transactions on Pattern Analysis and Machine Intelligence, 11*(7), 674–693.

Mallat, S. (2009). *A Wavelet Tour of Signal Processing: The Sparse Way*. Burlington, MA: Academic Press.

Mallat, S., and Hwang, W. L. (1992). Singularity detection and processing with wavelets. *IEEE Transactions on Information Theory, 38*(2), 617–643.

Mallat, S. G., and Zhang, Z. (1993). Matching pursuits with time-frequency dictionaries. *IEEE Transactions on Signal Processing, 41*(12), 3397–3415.

Mallat, S., and Zhong, S. (1992). Characterization of signals from multiscale edges. *IEEE Transactions on Pattern Analysis and Machine Intelligence, 14*(7), 710–732.

Mamaghanian, H., Khaled, N., Atienza, D., and Vandergheynst, P. (2011). Compressed sensing for real-time energy-efficient ECG compression on wireless body sensor nodes. *IEEE Transactions on Biomedical Engineering, 58*(9), 2456–2466.

Mandelbrot, B. B. (1982). *The Fractal Geometry of Nature*. Francisco, CA: W.H. Freeman.

Mandelbrot, B. B., and Van Ness, J. W. (1968). Fractional Brownian motions, fractional noises and applications. *SIAM Review, 10*(4), 422–437.

Manganotti, P., Formaggio, E., Del Felice, A., Storti, S. F., Zamboni, A., Bertoldo, A., Fiaschi, A. and Toffolo, G. M. (2013). Time-frequency analysis of short-lasting modulation of EEG induced by TMS during wake, sleep deprivation and sleep. *Frontiers in Human Neuroscience, 7*(767), 1–12.

Manganotti, P., Formaggio, E., Storti, S. F., De Massari, D., Zamboni, A., Bertoldo, A., Fiaschi, A. and Toffolo, G. M. (2012). Time-frequency analysis of short-lasting modulation of EEG induced by intracortical and transcallosal paired TMS over motor areas. *Journal of Neurophysiology, 107*, 2475–2484.

Maraun, D., and Kurths, J. (2004). Cross wavelet analysis: Significance testing and pitfalls. *Nonlinear Processes in Geophysics, 11*(4), 505–514.

Maraun, D., Kurths, J., and Holschneider, M. (2007). Nonstationary Gaussian processes in wavelet domain: Synthesis, estimation, and significance testing. *Physical Review E, 75*(1), 016707.

Markovic, D., and Koch, M. (2014). Long-term variations and temporal scaling of hydroclimatic time series with focus on the German part of the Elbe River Basin. *Hydrological Processes, 28*(4), 2202–2211.

Márquez, F. P. G., Tobias, A. M., Pérez, J. M. P., and Papaelias, M. (2012). Condition monitoring of wind turbines: Techniques and methods. *Renewable Energy, 46*, 169–178.

Martínez, B., and Gilabert, M. A. (2009). Vegetation dynamics from NDVI time series analysis using the wavelet transform. *Remote Sensing of Environment, 113*, 1823–1842.

Marzano, C., Ferrara, M., Mauro, F., Moroni, F., Gorgoni, M., Tempesta, D., Cipolli, C., and De Gennaro, L. (2011). Recalling and forgetting dreams: Theta and alpha oscillations during sleep predict subsequent dream recall. *Journal of Neuroscience, 31*(18), 6674–6683.

Mashita, T., Shimatani, K., Iwata, M., Miyamoto, H., Komaki, D., Hara, T., Kiyokawa, K., Takemura, H., and Nishio, S. (2012). Human activity recognition for a content search system considering situations of smartphone users. In *2012 IEEE Virtual Reality Short Papers and Posters (VRW)*, (pp. 1–2). IEEE.

Masias, M., Freixenet, J., Lladó, X., and Peracaula, M. (2012). A review of source detection approaches in astronomical images. *Monthly Notices of the Royal Astronomical Society*, 422(2), 1674–1689.

Masias, M., Lladó, X., Peracaula, M., and Freixenet, J. (2015). Multiscale distilled sensing: Astronomical source detection in long wavelength images. *Astronomy and Computing*, 9, 10–19.

Mataar, D., Fournier, R., Lachiri, Z., and Nait-Ali, A. (2013). Biometric application and classification of individuals using postural parameters. *International Journal of Computers and Technology*, 7(2), 580–593.

McAteer, R. J., Gallagher, P. T., and Conlon, P. A. (2010). Turbulence, complexity, and solar flares. *Advances in Space Research*, 45(9), 1067–1074.

McDonald, C. R., Thesen, T., Carlson, C., Blumberg, M., Girard, H. M., Trongnetrpunya, A., Sherfey, J. S., et al. (2010). Multimodal imaging of repetition priming: Using fMRI, MEG, and intracranial EEG to reveal spatiotemporal profiles of word processing. *Neuroimage*, 53(2), 707–717.

McKean, J., Nagel, D., Tonina, D., Bailey, P., Wright, C. W., Bohn, C., and Nayegandhi, A. (2009). Remote sensing of channels and riparian zones with a narrow-beam aquatic-terrestrial LIDAR. *Remote Sensing*, 1, 1065–1096.

Meeker, K., Harang, R., Webb, A. B., Welsh, D. K., Doyle, F. J., Bonnet, G., Herzog, E. D., and Petzold, L. R. (2011). Wavelet measurement suggests cause of period instability in mammalian circadian neurons. *Journal of Biological Rhythms*, 26(4), 353–362.

Mehr, A. D., Kahya, E., and Özger, M. (2014). A gene–wavelet model for long lead time drought forecasting. *Journal of Hydrology*, 517, 691–699.

Mendez, M. O., Corthout, J., Van Huffel, S., Matteucci, M., Penzel, T., Cerutti, S., and Bianchi, A. M. (2010). Automatic screening of obstructive sleep apnea from the ECG based on empirical mode decomposition and wavelet analysis. *Physiological Measurement*, 31, 273–289.

Meneveau, C. (1991a). Analysis of turbulence in the orthonormal wavelet representation. *Journal of Fluid Mechanics*, 232, 469–520.

Meneveau, C. (1991b). Dual spectra and mixed energy cascade of turbulence in the wavelet representation. *Physical Review Letters*, 66(11), 1450–1453.

Mengistu, S. G., Creed, I. F., Kulperger, R. J., and Quick, C. G. (2013). Russian nesting dolls effect–Using wavelet analysis to reveal non-stationary and nested stationary signals in water yield from catchments on a northern forested landscape. *Hydrological Processes*, 27, 669–686.

Merry, R., van de Molengraft, R., and Steinbuch, M. (2006). Removing non-repetitive disturbances in iterative learning control by wavelet filtering. In *Proceedings of the 2006 American Control Conference* (pp. 226–231), June 14–16, 2006. Minneapolis, MN: IEEE.

Mertens, F., and Lobanov, A. (2015). Wavelet-based decomposition and analysis of structural patterns in astronomical images. *Astronomy and Astrophysics*, 574(A67), 1–14.

Mertens, S., Dolde, K., Korzeczek, M., Glueck, F., Groh, S., Martin, R. D., Poon, A. W. P., and Steidl, M. (2015). Wavelet approach to search for sterile neutrinos in tritium β-decay spectra. *Physical Review D*, 91(4), 042005, 1–10.

Mestek, M. L., Addison, P. S., Kinney, A. R., Kelley, S. D. (2012c). Accuracy of continuous non-invasive respiratory rate derived from pulse oximetry in obese subjects. *Anesthesiology 2012: The American Society of Anesthesiologists Annual Meeting 2012*, October 13–17, in Washington, DC. Abstract A561.

Mestek, M. L., Addison, P. S., Neitenbach, A. M., Bergese, S. D., and Kelley, S. D. (2012a) Accuracy of continuous non-invasive respiratory rate derived from pulse oximetry in the post-anesthesia care unit. *ANESTHESIOLOGY 2012: The American Society of Anesthesiologists Annual Meeting 2012*, October 13–17, in Washington, DC. Abstract A094.

Mestek, M., Addison, P., Neitenbach, A. M., Bergese, S., and Kelley, S. (2012d). Accuracy of continuous noninvasive respiratory rate derived from pulse oximetry in chronic obstructive pulmonary disease patients. *CHEST Journal*, 142, 671A.

Mestek, M., Addison, P., Neitenbach, A. M., Bergese, S., and Kelley, S. (2012e). Accuracy of continuous noninvasive respiratory rate derived from pulse oximetry in congestive heart failure patients. *CHEST Journal*, *142*, 113A.

Mestek, M. L., Ochs, J. P., Addison, P. S., Neitenbach, A. M., Bergese, S. D., and Kelley, S. D. (2013). Accuracy of continuous non-invasive respiratory rate derived from pulse oximetry in patients with high respiratory rates. In *Anesthesia and Analgesia. Supplement: 'Abstracts of Papers Presented at the 2013 Annual Meeting of the Society for Technology in Anesthesia (STA), January 9–12, 2013'*, p. 49. Hagerstown, MD: Lippincott Williams & Wilkins.

Mestek, M. L., Watson, J. N., Ochs, J. P., Neitenbach, A. M., and Addison, P. S. (2012b) Accuracy of continuous noninvasive respiratory rate derived from pulse oximetry during coached breathing. *IAMPOV International Symposium* (Abstract 3, pp. 36–37), June 29–July 1, 2012. IAMPOV.

Meyers, S. D., Kelly, B. G., and O'Brien, J. J. (1993). An introduction to wavelet analysis in oceanography and meteorology: With application to the dispersion of Yanai waves. *Monthly Weather Review*, *121*(10), 2858–2866.

Mezeiová, K., and Paluš, M. (2012). Comparison of coherence and phase synchronization of the human sleep electroencephalogram. *Clinical Neurophysiology*, *123*(9), 1821–1830.

Michis, A. (2011). *Multiscale Analysis of the Liquidity Effect* Central bank of Cyprus. Working Paper Series, October 2011 (Working Paper No. 2011-5), 1–15.

Mi, X., Ren, H., Ouyang, Z., Wei, W., and Ma, K. (2005). The use of the Mexican Hat and the Morlet wavelets for detection of ecological patterns. *Plant Ecology*, *179*, 1–19.

Milosevic, M., Jovanov, E., and Milenković, A. (2011). Rapid processor customization for design optimization: A case study of ECG R-peak detection. In *Biomedical Circuits and Systems Conference (BioCAS), 2011 IEEE* (pp. 209–212). IEEE.

Mitra, J., Glover, J. R., Ktonas, P. Y., Kumar, A. T., Mukherjee, A., Karayiannis, N. B., Frost, J. D., Hrachovy, R. A., and Mizrahi, E. M. (2009). A multi-stage system for the automated detection of epileptic seizures in neonatal EEG. *Journal of Clinical Neurophysiology: Official Publication of the American Electroencephalographic Society*, *26*(4), 218.

Mochimaru, F., Fujimoto, Y., and Ishikawa, Y. (2002). Detecting the fetal electrocardiogram by wavelet theory-based methods. *Progress in Biomedical Research*, *7*, 185–193.

Moleti, A., Longo, F., and Sisto, R. (2012). Time-frequency domain filtering of evoked otoacoustic emissions. *Journal of the Acoustical Society of America*, *132*(4), 2455–2467.

Molini, A., Katul, G. G., and Porporato, A. (2010). Causality across rainfall time scales revealed by continuous wavelet transforms. *Journal of Geophysical Research: Atmospheres (1984–2012)*, *115*, D14123, 1–16.

Moriyama, O., Kuroiwa, N., Kanda, M., and Matsushita, M. (1998). Statistics and structure of granular flow through a vertical pipe. *Journal of the Physical Society of Japan*, *67*(5), 1603–1615.

Morris, A., Gozlan, R. E., Hassani, H., Andreou, D., Couppié, P., and Guégan, J. F. (2014). Complex temporal climate signals drive the emergence of human water-borne disease. *Emerging Microbes and Infections*, *3*, e56, 1–9.

Mosdorf, R., Wyszkowski, T., and Dąbrowski, K. (2011). Multifractal properties of large bubble paths in a single bubble column. *Archives of Thermodynamics*, *32*(1), 3–20.

Mota-Valtierra, G. C., Franco-Gasca, L. A., Herrera-Ruiz, G., and Macias-Bobadilla, G. (2011). ANN based tool condition monitoring system for CNC milling machines. *Ingeniería Investigación y Tecnología*, *12*(4), 461–468.

Mouri, H., Kubotani, H., Fujitani, T., Niino, H., and Takaoka, M. (1999). Wavelet analyses of velocities in laboratory isotropic turbulence. *Journal of Fluid Mechanics*, *389*, 229–254.

Moya-Martínez, P., Ferrer-Lapeña, R., and Escribano-Sotos, F. (2015). Interest rate changes and stock returns in Spain: A wavelet analysis. *BRQ Business Research Quarterly*, *18*, 95–110.

Mu, D., Chen, D., Fan, J., Wang, X., and Zhang, F. (2012). Carriage error identification based on cross-correlation analysis and wavelet transformation. *Sensors*, *12*, 9551–9565.

Mukli, P., Nagy, Z., and Eke, A. (2015). Multifractal formalism by enforcing the universal behavior of scaling functions. *Physica A: Statistical Mechanics and its Applications, 417,* 150–167.

Mulligan, R. F., and Koppl, R. (2011). Monetary policy regimes in macroeconomic data: An application of fractal analysis. *Quarterly Review of Economics and Finance, 51,* 201–211.

Muñoz, A., Sen, A. K., Sancho, C., and Genty, D. (2009). Wavelet analysis of Late Holocene stalagmite records from Ortigosa caves in Northern Spain. *Journal of Cave and Karst Studies. 71*(1), 63–72.

Murguía, J. S., and Campos-Cantón, E. (2006). Wavelet analysis of chaotic time series. *Revista mexicana de física, 52*(2), 155–162.

Murugappan, M., Nagarajan, R., and Yaacob, S. (2011). Combining spatial filtering and wavelet transform for classifying human emotions using EEG signals. *Journal of Medical and Biological Engineering, 31*(1), 45–51.

Muzy, J. F., Bacry, E., and Arneodo, A. (1991). Wavelets and multifractal formalism for singular signals: Application to turbulence data. *Physical Review Letters, 67*(25), 3515–3518.

Muzy, J. F., Bacry, E., and Arneodo, A. (1993). Multifractal formalism for fractal signals: The structure-function approach versus the wavelet-transform modulus-maxima method. *Physical Review E, 47*(2), 875–884.

Muzy, J. F., Bacry, E., and Arneodo, A. (1994). The multifractal formalism revisited with wavelets. *International Journal of Bifurcation and Chaos, 4*(02), 245–302.

Myint, S. W., Zhu, T., and Zheng, B. (2015). A novel image classification algorithm using overcomplete wavelet transforms. *IEEE Geoscience and Remote Sensing Letters, 12*(6), 1232–1236.

Najeeb, S. F., Bacha, O., and Masih, M. (2015). Does heterogeneity in investment horizons affect portfolio diversification? Some insights using M-GARCH-DCC and wavelet correlation analysis. *Emerging Markets Finance and Trade, 51,* 188–208.

Nakatani, H., Orlandi, N., and van Leeuwen, C. (2011). Precisely timed oculomotor and parietal EEG activity in perceptual switching. *Cognitive Neurodynamics, 5,* 399–409.

Nan, J. (2011). Wavelet analysis to detect multi-scale coherent eddy structures and intermittency in turbulent boundary layer. In J. C. Lerner and U. Boldes (Eds.), *Wind Tunnels and Experimental Fluid Dynamics Research* (Chapter 25, pp. 509–534). Published: July 27, 2011. Rijeka, Croatia: InTech.

Naouai, M., Hamouda, A., Akkari, A., and Weber, C. (2011). New approach for road extraction from high resolution remotely sensed images using the quaternionic wavelet. *In Pattern Recognition and Image Analysis, 6669,* 452–459.

Nason, G. P., and Silverman, B. W. (1995). The stationary wavelet transform and some statistical applications. *In Lecture Notes in Statistics 103* (pp. 281–299). New York: Springer.

Naue, N., Rach, S., Strüber, D., Huster, R. J., Zaehle, T., Körner, U., and Herrmann, C. S. (2011). Auditory event-related response in visual cortex modulates subsequent visual responses in humans. *Journal of Neuroscience, 31*(21), 7729–7736.

Nazimov, A. I., Pavlov, A. N., Nazimova, A. A., Grubov, V. V., Koronovskii, A. A., Sitnikova, E., and Hramov, A. E. (2013). Serial identification of EEG patterns using adaptive wavelet-based analysis. *European Physical Journal Special Topics, 222,* 2713–2722.

Nejadmalayeri, A., Vezolainen, A., De Stefano, G., and Vasilyev, O. V. (2014). Fully adaptive turbulence simulations based on Lagrangian spatio-temporally varying wavelet thresholding. *Journal of Fluid Mechanics, 749,* 794–817.

Nemes, A., Zhao, J., Lo Jacono, D., and Sheridan, J. (2012). The interaction between flow-induced vibration mechanisms of a square cylinder with varying angles of attack. *Journal of Fluid Mechanics, 710,* 102–130.

Neto, O. P., Baweja, H. S., and Christou, E. A. (2010). Increased voluntary drive is associated with changes in common oscillations from 13 to 60 Hz of interference but not rectified electromyography. *Muscle and Nerve, 42*(3), 348–354.

Neto, O. P., Pinheiro, A. O., Pereira, V. L., Pereira, R., Baltatu, O. C., and Campos, L. A. (2016). Morlet wavelet transforms of heart rate variability for autonomic nervous system activity. *Applied and Computational Harmonic Analysis, 40,* 200–206.

Neupauer, R. M., and Powell, K. L. (2005). A fully-anisotropic Morlet wavelet to identify dominant orientations in a porous medium. *Computers and Geosciences, 31*, 465–471.

Newland, D. E. (1993). *An Introduction to Random Vibrations, Spectral and Wavelet Analysis*, 3rd Edition, New York: Dover.

Ng, E. K., and Chan, J. C. (2012). Geophysical applications of partial wavelet coherence and multiple wavelet coherence. *Journal of Atmospheric and Oceanic Technology, 29*, 1845–1853.

Nguyen, H. T., and Nabney, I. T. (2010). Short-term electricity demand and gas price forecasts using wavelet transforms and adaptive models. *Energy, 35*, 3674–3685.

Nguyen, V. T., Euh, D. J., and Song, C. H. (2010). An application of the wavelet analysis technique for the objective discrimination of two-phase flow patterns. *International Journal of Multiphase Flow, 36*, 755–768.

Nicolis, O., and Mateu, J. (2015). 2D anisotropic wavelet entropy *with an application to earthquakes in Chile. Entropy, 17*, 4155–4172.

Niegowski, M., and Zivanovic, M. (2016). Wavelet-based unsupervised learning method for electrocardiogram suppression in surface electromyograms. *Medical Engineering and Physics, 38*(3), 248–256.

Ning, J., and Atanasov, N. (2010). Delineation of systolic and diastolic heart murmurs via wavelet transform and autoregressive modeling. *International Journal of Bioelectromagnetism, 12*(3), 114–120.

Ni, S. H., Isenhower, W. M., and Huang, Y. H. (2012a). Continuous wavelet transform technique for low-strain integrity testing of deep drilled shafts. *Journal of GeoEngineering, 7*(3), 97–105.

Niu, J., and Sivakumar, B. (2013). Scale-dependent synthetic streamflow generation using a continuous wavelet transform. *Journal of Hydrology, 496*, 71–78.

Ni, Y. Q., Xia, H. W., Wong, K. Y., and Ko, J. M. (2012b). In-service condition assessment of bridge deck using long-term monitoring data of strain response. *Journal of Bridge Engineering, 17*, 876–885.

Noël, J. P., Renson, L., and Kerschen, G. (2014). Dynamics of a strongly nonlinear spacecraft structure Part I: Experimental identification. In *13th European Conference on Spacecraft Structures, Materials and Environmental Testing* (pp. 1–7), April 1–4, 2014. Braunschweig, Germany.

Nogata, F., Yokota, Y., Kawamura, Y., Morita, H., and Uno, Y. (2015). Audio-visual recognition of auscultatory breathing sounds using Fourier and wavelet analyses. *Asian Journal of Computer and Information Systems, 3*(4), 96–105.

Nogata, F., Yokota, Y., Kawanura, Y., Morita, H., Uno, Y., and Walsh, W. R. (2012). Audio-visual based recognition of auscultatory heart sounds with Fourier and wavelet analyses. *Global Journal of Technology and Optimization, 3*, 43–48.

Nolan, G., Ringwood, J. V., and Holmes, B. (2007). Short term wave energy variability off the west coast of Ireland. In *Proceedings of the 7th European Wave and Tidal Energy Conference* (pp. 1–10), September 11–13, 2007. Porto, Portugal.

Nordbo, A., and Katul, G. (2013). A wavelet-based correction method for eddy-covariance high-frequency losses in scalar concentration measurements. *Boundary-Layer Meteorology, 146*, 81–102.

Noriega, M., Martínez, J. P., Laguna, P., Bailón, R., and Almeida, R. (2012). Respiration effect on wavelet-based ECG T-wave end delineation strategies. *IEEE Transactions on Biomedical Engineering, 59*(7), 1818–1828.

Noskov, V., Denisov, S., Stepanov, R., and Frick, P. (2012). Turbulent viscosity and turbulent magnetic diffusivity in a decaying spin-down flow of liquid sodium. *Physical Review E, 85*, 016303-1–016303-9.

Nourani, V., Hosseini Baghanam, A., Adamowski, J., and Kisi, O. (2014). Applications of hybrid wavelet–Artificial Intelligence models in hydrology: A review. *Journal of Hydrology, 514*, 358–377.

Nyander, A., Addison, P. S., McEwan, I., and Pender, G. (2003). Analysis of river bed surface roughnesses using 2D wavelet transform-based methods. *Arabian Journal for Science and Engineering, 28*(1; PART C), 107–122.

Obeid, D., Sadek, S., Zaharia, G., and El Zein, G. (2010). Touch-less heartbeat detection and measurement-based cardiopulmonary modeling. In *2010 Annual International Conference of the IEEE Engineering in Medicine and Biology Society (EMBC)*, (pp. 658–661). IEEE.

Okamoto, N., Yoshimatsu, K., Schneider, K., Farge, M., and Kaneda, Y. (2011). Coherent vorticity simulation of three-dimensional forced homogeneous isotropic turbulence. *Multiscale Modeling and Simulation*, 9(3), 1144–1161.

Okonkwo, C., Demoz, B., and Tesfai, S. *(2014)*. Characterization of West African jet streams and their association to ENSO events and rainfall in ERA-interim 1979–2011. *Advances in Meteorology*, 2014, 405617, 1–12.

Olkkonen, J. (Ed.). (2011) *Discrete Wavelet Transforms: Theory and Applications*. Rijeka, Croatia: InTech.

Oltean, M., Picheral, J., Lahalle, E., Hamdan, H., and Griffaton, J. (2013). Compression methods for mechanical vibration signals: Application to the plane engines. *Mechanical Systems and Signal Processing*, 41, 313–327.

Orlov, V., and Äijö, J. (2015). Benefits of wavelet-based carry trade diversification. *Research in International Business and Finance*, 34, 17–32.

Otal, E. H., Sileo, E., Aguirre, M. H., Fabregas, I. O., and Kim, M. (2015). Structural characterization and EXAFS wavelet analysis of Yb doped ZnO by wet chemistry route. *Journal of Alloys and Compounds*, 622, 115–120.

Ouadfeul, S., and Aliouane, L. (2011). Multifractal analysis revisited by the continuous wavelet transform applied in lithofacies segmentation from well-logs data. *International Journal of Applied Physics and Mathematics*, 1(1), 10–18.

Ozaki, T. J., Sato, N., Kitajo, K., Someya, Y., Anami, K., Mizuhara, H., Ogawa, S., and Yamaguchi, Y. (2012). Traveling EEG slow oscillation along the dorsal attention network initiates spontaneous perceptual switching. *Cognitive Neurodynamics*, 6, 185–198.

Özbay, Y. (2009). A new approach to detection of ECG arrhythmias: Complex discrete wavelet transform based complex valued artificial neural network. *Journal of Medical Systems*, 33, 435–445.

Paglialonga, A., Barozzi, S., Brambilla, D., Soi, D., Cesarani, A., Gagliardi, C., Comiotto, E., Spreafico, E., and Tognola, G. (2011b). Cochlear active mechanisms in young normal-hearing subjects affected by Williams syndrome: Time–frequency analysis of otoacoustic emissions. *Hearing Research*, 272, 157–167.

Paglialonga, A., Fiocchi, S., Del Bo, L., Ravazzani, P., and Tognola, G. (2011a). Quantitative analysis of cochlear active mechanisms in tinnitus subjects with normal hearing sensitivity: Time-frequency analysis of transient evoked otoacoustic emissions and contralateral suppression. *Auris Nasus Larynx*, 38, 33–40.

Paiva, T. O., Almeida, P. R., Ferreira-Santos, F., Vieira, J. B., Silveira, C., Chaves, P. L., Barbosa, F., and Marques-Teixeira, J. (2016). Similar sound intensity dependence of the N1 and P2 components of the auditory ERP: Averaged and single trial evidence. *Clinical Neurophysiology*, 127, 499–508.

Pakrashi, V., O'Connor, A., and Basu, B. (2010). A bridge–vehicle interaction based experimental investigation of damage evolution. *Structural Health Monitoring*, 9(4), 285–296.

Pal, S., Heyns, P. S., Freyer, B. H., Theron, N. J., and Pal, S. K. (2011). Tool wear monitoring and selection of optimum cutting conditions with progressive tool wear effect and input uncertainties. *Journal of Intelligent Manufacturing*, 22(4), 491–504.

Palmer, S., and Hall, W. (2012). Surface evaluation of carbon fibre composites using wavelet texture analysis. *Composites Part B: Engineering*, 43, 621–626.

Pandey, V. K., and Pandey, P. C. (2007). Wavelet based cancellation of respiratory artifacts in impedance cardiography. In *2007 15th International Conference on Digital Signal Processing* (pp. 191–194). IEEE.

Pang, D. S., Robledo, C. J., Carr, D. R., Gent, T. C., Vyssotski, A. L., Caley, A., Zecharia, A. Y., Wisden, W., Brickley, S. G., and Franks, N. P. (2009). An unexpected role for TASK-3 potassium channels in network oscillations with implications for sleep mechanisms and anesthetic action. *Proceedings of the National Academy of Sciences*, 106(41), 17546–17551.

Pan, S. Y., Hsieh, B. Z., Lu, M. T., and Lin, Z. S. (2008). Identification of stratigraphic formation interfaces using wavelet and Fourier transforms. *Computers and Geosciences*, 34, 77–92.

Pan, Z., Glennie, C., Hartzell, P., Fernandez-Diaz, J. C., Legleiter, C., and Overstreet, B. (2015). Performance assessment of high resolution airborne full waveform LiDAR for shallow river bathymetry. *Remote Sensing*, 7, 5133–5159.

Papademetriou, M. D., Tachtsidis, I., Elliot, M. J., Hoskote, A., and Elwell, C. E. (2012). Multichannel near infrared spectroscopy indicates regional variations in cerebral autoregulation in infants supported on extracorporeal membrane oxygenation. *Journal of Biomedical Optics*, 17(6), 067008-1–067008-9.

Papaioannou, V. E., Chouvarda, I. G., Maglaveras, N. K., and Pneumatikos, I. A. (2012). Temperature variability analysis using wavelets and multiscale entropy in patients with systemic inflammatory response syndrome, sepsis, and septic shock. *Critical Care*, 16(RS1), 1–15.

Park, C., Tang, J., and Ding, Y. (2010). Aggressive data reduction for damage detection in structural health monitoring. *Structural Health Monitoring*, 9(1), 59–74.

Pascoal, R., and Monteiro, A. M. (2014). Market efficiency, roughness and long memory in PSI20 index returns: Wavelet and entropy analysis. *Entropy*, 16, 2768–2788.

Pasquini, A. I., Lecomte, K. L., and Depetris, P. J. (2013). The Manso Glacier drainage system in the northern Patagonian Andes: An overview of its main hydrological characteristics. *Hydrological Processes*, 27, 217–224.

Pattiaratchi, C., and Wijeratne, E. M. S. (2014). Observations of meteorological tsunamis along the south-west Australian coast. *Natural Hazards*, 74, 281–303.

Pavlov, A. N., Abdurashitov, A. S., Sindeeva, O. A., Sindeev, S. S., Pavlova, O. N., Shihalov, G. M., and Semyachkina-Glushkovskaya, O. V. (2016). Characterizing cerebrovascular dynamics with the wavelet-based multifractal formalism. *Physica A: Statistical Mechanics and its Applications*, 442, 149–155.

Pavlov, A. N., Anisimov, A. A., Semyachkina-Glushkovskaya, O. V., Matasova, E. G., and Kurths, J. (2009). Analysis of blood pressure dynamics in male and female rats using the continuous wavelet transform. *Physiological Measurement*, 30, 707–717.

Pavlov, A. N., Hramov, A. E. E., Koronovskii, A. A., Sitnikova, E. Y., Makarov, V. A., and Ovchinnikov, A. A. (2012). Wavelet analysis in neurodynamics. *Physics-Uspekhi*, 55(9), 845–875.

Paykari, P., Starck, J. L., and Fadili, M. J. (2012). True CMB Power Spectrum Estimation. *Astronomical Data Analysis, 7th Conference*, ADA Online Proceedings (pp. 1–10), May 14–18, 2012. Cargese, Corsica.

Payne, S. J., Mohammad, J., Tisdall, M. M., and Tachtsidis, I. (2011). Effects of arterial blood gas levels on cerebral blood flow and oxygen transport. *Biomedical Optics Express*, 2(4), 966–979.

Paz-Chinchón, F., Leão, I. C., Bravo, J. P., de Freitas, D. B., Lopes, C. F., Alves, S., Catelan, M., Canto Martins, B. L., and De Medeiros, J. R. (2015). The rotational behavior of Kepler stars with planets. *Astrophysical Journal*, 803(2), 69.

Pei, S. C., Tseng, C. C., and Lin, C. Y. (1995). Wavelet transform and scale space filtering of fractal images. *IEEE Transactions on Image Processing*, 4(5), 682–687.

Peng, Z., Chu, F., and He, Y. (2002). Vibration signal analysis and feature extraction based on reassigned wavelet scalogram. *Journal of Sound and Vibration*, 253(5), 1087–1100.

Peng, Z. K., Jackson, M. R., Rongong, J. A., Chu, F. L., and Parkin, R. M. (2009). On the energy leakage of discrete wavelet transform. *Mechanical Systems and Signal Processing*, 23(2), 330–343.

Peng, Z. K., Meng, G., and Chu, F. L. (2011). Improved wavelet reassigned scalograms and application for modal parameter estimation. *Shock and Vibration*, 18, 299–316.

Pereira, R., Schettino, L., Machado, M., da Silva, P. A. V., and Neto, O. P. (2010). Task failure during standing heel raises is associated with increased power from 13 to 50 Hz in the activation of triceps surae. *European Journal of Applied Physiology*, 110(2), 255–265.

Perez-Munoz, T., Velasco-Hernandez, J. X., Altamira-Areyan, A., Velasquillo-Martinez, L. G., and Hernandez-Martinez, E. (2012) Multiscale coherence in the analysis of gamma rays in well characterization. *ORADM 2012 Workshop Proceedings. 3 Mathematical Modeling for Decision Making*, 3(2), 83–97.

Perez-Muñoz, T., Velasco-Hernandez, J., and Hernandez-Martinez, E. (2013). Wavelet transform analysis for lithological characteristics identification in siliciclastic oil fields. *Journal of Applied Geophysics*, 98, 298–308.

Perez-Ramirez, C. A., Amezquita-Sanchez, J. P., Adeli, H., Valtierra-Rodriguez, M., Camarena-Martinez, D., and Romero-Troncoso, R. J. (2016). New methodology for modal parameters identification of smart civil structures using ambient vibrations and synchrosqueezed wavelet transform. *Engineering Applications of Artificial Intelligence*, 48, 1–12.

Petrock, A. M., Donnelly, D. L., and Rosenberg, M. L. (2008). Quantifying cardio-pulmonary correlations using the cross-wavelet transform: Validating a correlative method. In *Engineering in Medicine and Biology Society, 2008. EMBS 2008. 30th Annual International Conference of the IEEE* (pp. 2940–2943). IEEE.

Piantoni, G., Astill, R. G., Raymann, R. J., Vis, J. C., Coppens, J. E., and Van Someren, E. J. (2013). Modulation of gamma and spindle-range power by slow oscillations in scalp sleep EEG of children. *International Journal of Psychophysiology*, 89(2), 252–258.

Pichot, V., Bourin, E., Roche, F., Garet, M., Gaspoz, J. M., Duverney, D., Antoniadis, A., Lacour, J. R., and Barthélémy, J. C. (2002). Quantification of cumulated physical fatigue at the workplace. *Pflügers Archive: European Journal of Physiology*, 445(2), 267–272.

Piñuela, J., Alvarez, A., Andina, D., Heck, R. J., and Tarquis, A. M. (2010). Quantifying a soil pore distribution from 3D images: Multifractal spectrum through wavelet approach. *Geoderma*, 155, 203–210.

Piñuela, J. A., Andina, D., McInnes, K. J., and Tarquis, A. M. (2007). Wavelet analysis in a structured clay soil using 2-D images. *Nonlinear Processes in Geophysics*, 14, 425–434.

Pires, S., and Starck, J. L. (2010). Light on dark matter with weak gravitational lensing. *IEEE Signal Processing Magazine*, 27(1), 76–85.

Pittiglio, C., Skidmore, A. K., van Gils, H. A., and Prins, H. H. (2013). Elephant response to spatial heterogeneity in a savanna landscape of northern Tanzania. *Ecography*, 36, 819–831.

Pizza, F., Fabbri, M., Magosso, E., Ursino, M., Provini, F., Ferri, R., and Montagna, P. (2011). Slow eye movements distribution during nocturnal sleep. *Clinical Neurophysiology*, 122, 1556–1561.

Podtaev, S., Stepanov, R., Dumler, A., Chugainov, S., and Tziberkin, K. (2012). Wavelet analysis of the impedance cardiogram waveforms. In *Journal of Physics: Conference Series*, 407(1), 012003. IOP Publishing.

Polania, L. F., Carrillo, R. E., Blanco-Velasco, M., and Barner, K. E. (2011). Compressed sensing based method for ECG compression. In *2011 IEEE International Conference on Acoustics, Speech and Signal Processing (ICASSP)* (pp. 761–764). IEEE.

Poluianov, S., and Usoskin, I. (2014). Critical analysis of a hypothesis of the planetary tidal influence on solar activity. *Solar Physics*, 289(6), 2333–2342.

Polygiannakis, J., Preka-Papadema, P., and Moussas, X. (2003). On signal–noise decomposition of time-series using the continuous wavelet transform: Application to sunspot index. *Monthly Notices of the Royal Astronomical Society*, 343, 725–734.

Ponnui, J., Tanthanuch, S., Phukpattaranont, P., and Wongkittisuksa, B. (2012). Automated expert system for urolithiasis classification from infrared spectrogram. *The 10th International PSU engineering Conference (IPEC-10)* (pp. 1–5), May 14–15, 2012. Hat Yai, Thailand.

Ponomarenko, V. I., Prokhorov, M. D., Bespyatov, A. B., Bodrov, M. B., and Gridnev, V. I. (2005). Deriving main rhythms of the human cardiovascular system from the heartbeat time series and detecting their synchronization. *Chaos, Solitons and Fractals, 23*, 1429–1438.

Postnikov, E. B. (2007). On precision of wavelet phase synchronization of chaotic systems. *Journal of Experimental and Theoretical Physics, 105*(3), 652–654.

Postnikov, E. B., and Lebedeva, E. A. (2010). Decomposition of strong nonlinear oscillations via modified continuous wavelet transform. *Physical Review E, 82*, 057201-1–057201-4.

Postnikov, E. B., Ryabov, A. B., and Loskutov, A. (2010). Analysis of patterns formed by two-component diffusion limited aggregation. *Physical Review E, 82*(5), 051403, 1–7.

Postolache, G., Carvalho, L. S., Postolache, O., Girão, P., and Rocha, I. (2009). HRV and BPV neural network model with wavelet based algorithm calibration. *Measurement, 42*, 805–814.

Putra, T. E., Abdullah, S., Schramm, D., Nuawi, M. Z., and Bruckmann, T. (2014). FCM-based optimisation to enhance the Morlet wavelet ability for compressing suspension strain data. *Procedia Materials Science, 3*, 288–294.

Qiu, J., Paw, U. K. T., and Shaw, R. H. (1995). Pseudo-wavelet analysis of turbulence patterns in three vegetation layers. *Boundary-Layer Meteorology, 72*, 177–204.

Quiroz, R., Yarlequé, C., Posadas, A., Mares, V., and Immerzeel, W. W. (2011). Improving daily rainfall estimation from NDVI using a wavelet transform. *Environmental Modelling and Software, 26*, 201–209.

Rafiee, J., Rafiee, M. A., and Tse, P. W. (2010). Application of mother wavelet functions for automatic gear and bearing fault diagnosis. *Expert Systems with Applications, 37*, 4568–4579.

Rafiee, J., Rafiee, M. A., Yavari, F., and Schoen, M. P. (2011). Feature extraction of forearm EMG signals for prosthetics. *Expert Systems with Applications, 38*(4), 4058–4067.

Rafiee, J., and Tse, P. W. (2009). Use of autocorrelation of wavelet coefficients for fault diagnosis. *Mechanical Systems and Signal Processing, 23*, 1554–1572.

Rahmati, A., Adhami, R., and Dimassi, M. (2015). Real-time electrical variables estimation based on recursive wavelet transform. *International Journal of Electrical Power and Energy Systems, 68*, 170–179.

Rajaee, T., Mirbagheri, S. A., Nourani, V., and Alikhani, A. (2010). Prediction of daily suspended sediment load using wavelet and neurofuzzy combined model. *International Journal of Environmental Science and Technology, 7*(1), 93–110.

Rajeev, P., and Wijesundara, K. K. (2014). Energy-based damage index for concentrically braced steel structure using continuous wavelet transform. *Journal of Constructional Steel Research, 103*, 241–250.

Ramírez-Cortes, J. M., Alarcon-Aquino, V., Rosas-Cholula, G., Gomez-Gil, P., and Escamilla-Ambrosio, J. (2010). P-300 rhythm detection using ANFIS algorithm and wavelet feature extraction in EEG signals. In *Proceedings of the World Congress on Engineering and Computer Science* (Vol.1, pp. 963–968). San Francisco, CA: International Association of Engineers.

Rasooli, M., Foomany, F. H., Balasundaram, K., Masse, S., Zamiri, N., Ramadeen, A., Hu, X., et al. (2015). Analysis of electrocardiogram pre-shock waveforms during ventricular fibrillation. *Biomedical Signal Processing and Control, 21*, 26–33.

Ravindra, B., and Javaraiah, J. (2015). Hemispheric asymmetry of sunspot area in solar cycle 23 and rising phase of solar cycle 24: Comparison of three data sets. *New Astronomy, 39*, 55–63.

Razali, S. M., Zhou, T., Rinoshika, A., and Cheng, L. (2010). Wavelet analysis of the turbulent wake generated by an inclined circular cylinder. *Journal of Turbulence, 11*(15), 1–25.

Recknagel, F., Ostrovsky, I., Cao, H., Zohary, T., and Zhang, X. (2013). Ecological relationships, thresholds and time-lags determining phytoplankton community dynamics of Lake Kinneret, Israel elucidated by evolutionary computation and wavelets. *Ecological Modelling, 255*, 70–86.

Reed, M. J., Robertson, C. E., and Addison, P. S. (2005). Heart rate variability measurements and the prediction of ventricular arrhythmias. *Quarterly Journal of Medicine, 98*, 87–95.

Rein, H., and Latter, H. N. (2013). Large-scale N-body simulations of the viscous overstability in Saturn's rings. *Monthly Notices of the Royal Astronomical Society, 431*(1), 145–158.

Reine, C., van der Baan, M., and Clark, R. (2009). The robustness of seismic attenuation measurements using fixed- and variable-window time-frequency transforms. *Geophysics, 74*(2), WA123–WA135.

Remick, K., Vakakis, A., Bergman, L., McFarland, D. M., Quinn, D. D., and Sapsis, T. P. (2014). Sustained high-frequency dynamic instability of a nonlinear system of coupled oscillators forced by single or repeated impulses: Theoretical and experimental results. *Journal of Vibration and Acoustics, 136*, 011013-1–011013-15.

Ren, Z. Y., Gao, C., Han, G., Ding, S., and Lin, J. (2014). DT-CWT robust filtering algorithm for the extraction of reference and waviness from 3-D nano scalar surfaces. *Measurement Science Review, 14*(2), 87–93.

Rios, E. C., Zimer, A. M., Mendes, P. C., Freitas, M. B., de Castro, E. V., Mascaro, L. H., and Pereira, E. C. (2015). Corrosion of AISI 1020 steel in crude oil studied by the electrochemical noise measurements. *Fuel, 150*, 325–333.

Risien, C. M., Reason, C. J. C., Shillington, F. A., and Chelton, D. B. (2004). Variability in satellite winds over the Benguela upwelling system during 1999–2000. *Journal of Geophysical Research: Oceans, 109*, C03010, 1–15.

Rizon, M. (2010). Discrete wavelet transform based classification of human emotions using electroencephalogram signals. *American Journal of Applied Sciences, 7*(7), 878–885.

Rizzi, M., D'Aloia, M., and Castagnolo, B. (2008). Fast parallelized algorithm for ECG analysis. *WSEAS Transactions on Biology and Biomedicine, 5*, 210–219.

Rizzo, P., Cammarata, M., Bartoli, I., di Scalea, F. L., Salamone, S., Coccia, S., and Phillips, R. (2010). Ultrasonic guided waves-based monitoring of rail head: Laboratory and field tests. *Advances in Civil Engineering, 291293*, 1–13.

Rizzo, P., Cammarata, M., Dutta, D., Sohn, H., and Harries, K. (2009). An unsupervised learning algorithm for fatigue crack detection in waveguides. *Smart Materials and Structures, 18*, 025016, 1–11.

Rocha, T., Paredes, S., Carvalho, P., Henriques, J., and Harris, M. (2010). Wavelet based time series forecast with application to acute hypotensive episodes prediction. *In Engineering in Medicine and Biology Society (EMBC), 2010 Annual International Conference of the IEEE* (pp. 2403–2406). IEEE.

Roche, F., Pichot, V., Sforza, E., Duverney, D., Costes, F., Garet, M., and Barthélémy, J. C. (2003). Predicting sleep apnoea syndrome from heart period: A time-frequency wavelet analysis. *European Respiratory Journal, 22*, 937–942.

Rock, R., Als, A., Gibbs, P., and Hunte, C. (2011). The 5th umpire: Cricket's edge detection system. In *CSC'11-8th International Conference on Scientific Computing*, July 18–21, 2011. Las Vegas, Nevada.

Ródenas, J., García, M., Alcaraz, R., and Rieta, J. J. (2015). *Wavelet entropy automatically detects episodes of atrial fibrillation from single-lead electrocardiograms.* Entropy, 17, 6179–6199.

Rodríguez, N., Cubillos, C., and Rubio, J. M. *(2014).* Multi-step-ahead forecasting model for monthly anchovy catches based on wavelet analysis. *Journal of Applied Mathematics, 2014*, 798464, 1–8.

Romeo, A. B., Horellou, C., and Bergh, J. (2004). A wavelet add-on code for new-generation N-body simulations and data de-noising (JOFILUREN). *Monthly Notices of the Royal Astronomical Society, 354*, 1208–1222.

Romero, I., Fleck, E., and Kriatselis, C. (2011). Frequency analysis of atrial fibrillation surface and intracardiac electrograms during pulmonary vein isolation. *Europace, 13*, 1340–1345.

Romero, I., Grubb, N. R., Clegg, G. R., Robertson, C. E., Addison, P. S., and Watson, J. N. (2008). T-wave alternans found in preventricular tachyarrhythmias in CCU patients using a wavelet transform-based methodology. *IEEE Transactions on Biomedical Engineering, 55*(11), 2658–2665.

Rong-Yi, Y., and Xiao-Jing, H. (2011). Phase space reconstruction of chaotic dynamical system based on wavelet decomposition. *Chinese Physics B, 20*(2), 020505-1–020505-5.

Rooijakkers, M. J., Rabotti, C., Oei, S. G., and Mischi, M. (2012). Low-complexity R-peak detection for ambulatory fetal monitoring. *Physiological Measurement, 33,* 1135–1150.

Rowley, A. B., Payne, S. J., Tachtsidis, I., Ebden, M. J., Whiteley, J. P., Gavaghan, D. J., Tarassenko, L., Smith, M., Elwell, C. E., and Delpy, D. T. (2007). Synchronization between arterial blood pressure and cerebral oxyhaemoglobin concentration investigated by wavelet cross-correlation. *Physiological Measurement, 28,* 161–173.

Rua, A., and Nunes, L. C. (2009). International co-movement of stock market returns: A wavelet analysis. *Journal of Empirical Finance, 16,* 632–639.

Rua, A., and Nunes, L. C. (2012). A wavelet-based assessment of market risk: The emerging markets case. *Quarterly Review of Economics and Finance, 52*(1), 84–92.

Rucka, M. (2011). Damage detection in beams using wavelet transform on higher vibration modes. *Journal of Theoretical and Applied Mechanics, 49*(2), 399–417.

Rus, G., Lee, S. Y., Chang, S. Y., and Wooh, S. C. (2006). Optimized damage detection of steel plates from noisy impact test. *International Journal for Numerical Methods in Engineering, 68,* 707–727.

Ruzzene, M., Fasana, A., Garibaldi, L., and Piombo, B. (1997). Natural frequencies and dampings identification using wavelet transform: Application to real data. *Mechanical Systems and Signal Processing, 11*(2), 207–218.

Saadatinejad, M. R., and Hassani, H. (2013). Application of wavelet transform for evaluation of hydrocarbon reservoirs: Example from Iranian oil fields in the north of the Persian Gulf. *Nonlinear Processes in Geophysics, 20,* 231–238.

Saeedi, J., Faez, K., and Moradi, M. H. (2014). Hybrid fractal-wavelet method for multi-channel EEG signal compression. *Circuits, Systems, and Signal Processing, 33*(8), 2583–2604.

Safaai, S. S., Muniandy, S. V., Chew, W. X., Asgari, H., Yap, S. L., and Wong, C. S. (2013). Fractal dynamics of light scattering intensity fluctuation in disordered dusty plasmas. *Physics of Plasmas, 20,* 103702-1–103702-8.

Safara, F., Doraisamy, S., Azman, A., Jantan, A., and Ranga, S. (2012). Wavelet packet entropy for heart murmurs classification. *Advances in Bioinformatics, 327269,* 1–6

Safieddine, D., Kachenoura, A., Albera, L., Birot, G., Karfoul, A., Pasnicu, A., Biraben, A., Wendling, F., Senhadji, L., and Merlet, I. (2012). Removal of muscle artifact from EEG data: Comparison between stochastic (ICA and CCA) and deterministic (EMD and wavelet-based) approaches. *EURASIP Journal on Advances in Signal Processing, 127,* 1–15.

Sahambi, J. S., Tandon, S. M., and Bhatt, R. K. P. (1997a). Using wavelet transforms for ECG characterization: An on-line digital signal processing system. *IEEE Engineering in Medicine and Biology, 16*(1), 77–83.

Sahambi, J. S., Tandon, S. N., and Bhatt, R. K. P. (1997b). Quantitative analysis of errors due to power-line interference and base-line drift in detection of onsets and offsets in ECG using wavelets. *Medical and Biological Engineering and Computing, 35*(6), 747–751.

Sailhac, P., and Gibert, D. (2003). Identification of sources of potential fields with the continuous wavelet transform: Two-dimensional wavelets and multipolar approximations. *Journal of Geophysical Research: Solid Earth, 108*(B5), 2262.

Sakai, T., Satomoto, H., Kiyasu, S., and Miyahara, S. (2012). Sparse representation-based extraction of pulmonary sound components from low-quality auscultation signals. In *2012 IEEE International Conference on Acoustics, Speech and Signal Processing (ICASSP)* (pp. 509–512). IEEE.

Sakkalis, V., Cassar, T., Zervakis, M., Giurcaneanu, C. D., Bigan, C., Micheloyannis, S., Camilleri, K. P., Fabri, S. G., Karakonstantaki, E., and Michalopoulos, K. (2010). A decision support framework for the discrimination of children with controlled epilepsy based on EEG analysis. *Journal of Neuroengineering and Rehabilitation, 7/1/124, 24,* 1–14.

Saljooghi, B. S., and Hezarkhani, A. (2014). Comparison of WAVENET and ANN for predicting the porosity obtained from well log data. *Journal of Petroleum Science and Engineering, 123,* 172–182.

Salpeter, N., and Hassan, Y. (2012). Large eddy simulations of jet flow interaction within staggered rod bundles. *Nuclear Engineering and Design, 251,* 92–101.

Samar, V. J., Bopardikar, A., Rao, R., and Swartz, K. (1999). Wavelet analysis of neuroelectric waveforms: A conceptual tutorial. *Brain and Language, 66,* 7–60.

Sandoval, J., and Hernández, G. (2015). Computational visual analysis of the order book dynamics for creating high-frequency foreign exchange trading strategies. *Procedia Computer Science, 51,* 1593–1602.

Sankur, B., Güler, E. Ç., and Kahya, Y. P. (1996). Multiresolution biological transient extraction applied to respiratory crackles. *Computers in Biology and Medicine, 26*(1), 25–39.

Saracco, G., Thouveny, N., Bourlès, D. L., and Carcaillet, J. T. (2009). Extraction of non-continuous orbital frequencies from noisy insolation data and from palaeoproxy records of geomagnetic intensity using the phase of continuous wavelet transforms. *Geophysical Journal International, 176,* 767–781.

Saritha, M., Joseph, K. P., and Mathew, A. T. (2013). Classification of MRI brain images using combined wavelet entropy based spider web plots and probabilistic neural network. *Pattern Recognition Letters, 34,* 2151–2156.

Sarkar, T. K., Su, C., Adve, R., Salazar-Palma, M., Garcia-Castillo, L. and Boix, R. R. (1998). A tutorial on wavelets from an electrical engineering perspective, part 1: discrete wavelet techniques. *IEEE Antennas and Propagation Magazine, 40*(5), 49–70.

Sarma, B., Chauhan, S. S., Wharton, A. M., and Iyengar, A. N. S. (2013). Continuous wavelet transform analysis for self-similarity properties of turbulence in magnetized DC glow discharge plasma. *Journal of Plasma Physics, 79*(05), 885–891.

Sartori, C. A. F., and Sevegnani, F. X. (2010). Fault classification and detection by wavelet-based magnetic signature recognition. *IEEE Transactions on Magnetics, 46*(8), 2880–2883.

Sasikala, P., and Wahidabanu, R. S. D. (2010). Robust R peak and QRS detection in electrocardiogram using wavelet transform. *International Journal of Advanced Computer Science and Applications, 1,* 48–53.

Sathe, M. J., Thaker, I. H., Strand, T. E., and Joshi, J. B. (2010). Advanced PIV/LIF and shadowgraphy system to visualize flow structure in two-phase bubbly flows. *Chemical Engineering Science, 65,* 2431–2442.

Saunders, M., Geiger, L., Negri, D., Stein, J. A., Sansal, T. A., and Springman, J. (2015). Improved stratigraphic interpretation using broadband processing–Sergipe Basin, Brazil. *First Break, 33,* 87–93.

Saxena, S. C., Kumar, V., and Hamde, S. T. (2002). Feature extraction from ECG signals using wavelet transforms for disease diagnostics. *International Journal of Systems Science, 33*(13), 1073–1085.

Sazonov, E. S., Makeyev, O., Schuckers, S., Lopez-Meyer, P., Melanson, E. L., and Neuman, M. R. (2010). Automatic detection of swallowing events by acoustical means for applications of monitoring of ingestive behavior. *IEEE Transactions on Biomedical Engineering, 57*(3), 626–633.

Schiff, S. J., Aldroubi, A., Unser, M., and Sato, S. (1994). Fast wavelet transformation of EEG. *Electroencephalography and Clinical Neurophysiology, 91,* 442–455.

Schlüter, S. (2009). *A Two-Factor Model for Electricity Prices with Dynamic Volatility.* IWQW discussion paper series, (No. 04/2009), 1–17.

Schneider, K., and Vasilyev, O. V. (2009). Wavelet methods in computational fluid dynamics*. *Annual Review of Fluid Mechanics, 42,* 473–503.

Schram, C., Rambaud, P., and Riethmuller, M. L. (2004). Wavelet based eddy structure eduction from a backward facing step flow investigated using particle image velocimetry. *Experiments in Fluids, 36*(2), 233–245.

Schulte, J. A., Duffy, C., and Najjar, R. G. (2014). Geometric and topological approaches to significance testing in wavelet analysis. *Nonlinear Processes in Geophysics Discussions, 1*, 1331–1363.

Schwilden, H., Kochs, E., Daunderer, M., Jeleazcov, C., Scheller, B., Schneider, G., Schüttler, J., Schwender, D., Stockmanns, G., and Pöppel, E. (2005). Concurrent recording of AEP, SSEP and EEG parameters during anaesthesia: A factor analysis. *British Journal of Anaesthesia, 95*(2), 197–206.

Seena, A., and Jin Sung, H. (2011). Wavelet spatial scaling for educing dynamic structures in turbulent open cavity flows. *Journal of Fluids and Structures, 27*(7), 962–975.

Segreto, T., Karam, S., Simeone, A., and Teti, R. (2013). Residual stress assessment in inconel 718 machining through wavelet sensor signal analysis and sensor fusion pattern recognition. *Procedia CIRP, 9*, 103–108.

Sejdić, E., Kalika, D., and Czarnek, N. (2013). An analysis of resting-state functional transcranial Doppler recordings from middle cerebral arteries. *PLoS One, 8*(2), e55405, 1–9.

Sejdić, E., Steele, C. M., and Chau, T. (2010). A procedure for denoising dual-axis swallowing accelerometry signals. *Physiological Measurement, 31*, N1–N9.

Senhadji, L., Carrault, G., Bellanger, J. J., and Passariello, G. (1995). Comparing wavelet transforms for recognizing cardiac patterns. *IEEE Engineering in Medicine and Biology Magazine, 14*(2), 167–173.

Sen, I., and Kahya, Y. P. (2005). A multi-channel device for respiratory sound data acquisition and transient detection. In *Conference Proceedings IEEE Engineering in Medicine and Biology Society, 6* (pp. 6658–6661). IEEE.

Seto, D., Clements, C. B., and Heilman, W. E. (2013). Turbulence spectra measured during fire front passage. *Agricultural and Forest Meteorology, 169*, 195–210.

Setoudeh, F., Khaki Sedigh, A., and Dousti, M. (2014). Analysis of a chaotic memristor based oscillator. *Abstract and Applied Analysis, 628169*, 1–8.

Sforza, E., Pichot, V., Cervena, K., Barthelemy, J. C., and Roche, F. (2007). *Cardiac variability and heart-rate increment as a marker of sleep fragmentation in patients with a sleep disorder: A preliminary study. Sleep, 30*(1), 43–51.

Shahbaz, M., Tiwari, A. K., and Tahir, M. I. (2015). Analyzing time-frequency relationship between oil price and exchange rate in Pakistan through wavelets. *Journal of Applied Statistics, 42*(4), 690–704.

Shandilya, S., Kurz, M. C., and Ward, K. R. (2013). Finding an optimal model for prediction of shock outcomes through machine learning. *ICCGI 2013: The Eighth International Multi-Conference on Computing in the Global Information Technology* (pp. 214–218), July 21–26, 2013. Nice, France.

Shandilya, S., Ward, K., Kurz, M., and Najarian, K. (2012). Non-linear dynamical signal characterization for prediction of defibrillation success through machine learning. *BMC Medical Informatics and Decision Making, 12*(116), 1–9.

Shankar, B. U., Meher, S. K., and Ghosh, A. (2011). Wavelet-fuzzy hybridization: Feature-extraction and land-cover classification of remote sensing images. *Applied Soft Computing, 11*, 2999–3011.

Shao, H., Shi, X., and Li, L. (2011). Power signal separation in milling process based on wavelet transform and independent component analysis. *International Journal of Machine Tools and Manufacture, 51*, 701–710.

Shaviv, N. J., Prokoph, A., and Veizer, J. (2014). Is the solar system's galactic motion imprinted in the phanerozoic climate? *Scientific Reports, 4*(6150), 1–6.

Sheybani, E., Garcia-Otero, S., Adnani, F., and Javidi, G. (2012). A fast algorithm for automated quality control in surface engineering. *Journal of Surface Engineered Materials and Advanced Technology, 2*(2), 18811, 120–126.

Shiogai, Y., Stefanovska, A., and McClintock, P. V. E. (2010). Nonlinear dynamics of cardiovascular ageing. *Physics Reports, 488*, 51–110.

Shi, S. P., Qiu, J. D., Sun, X. Y., Huang, J. H., Huang, S. Y., Suo, S. B., Liang, R. P., and Zhang, L. (2011). Identify submitochondria and subchloroplast locations with pseudo amino acid composition: Approach from the strategy of discrete wavelet transform feature extraction. *Biochimica et Biophysica Acta (BBA)-Molecular Cell Research, 1813*, 424–430.

Shokrollahi, E., and Riahi, M. A. (2013). Using continuous wavelet transform and short time Fourier transform as spectral decomposition methods to detect of stratigraphic channel in one of the Iranian south-west oil fields. *International Journal of Science and Emerging Technologies*, 5(5), 291–299.

Shugar, D. H., Kostaschuk, R., Best, J. L., Parsons, D. R., Lane, S. N., Orfeo, O., and Hardy, R. J. (2010). On the relationship between flow and suspended sediment *transport over the crest of a sand dune, Río Paraná, Argentina. Sedimentology*, 57, 252–272.

Shyu, L. Y., Wu, Y. H., and Hu, W. (2004). Using wavelet transform and fuzzy neural network for VPC detection from the Holter ECG. *IEEE Transactions on Biomedical Engineering*, 51(7), 1269–1273.

Simons, F. J., Loris, I., Nolet, G., Daubechies, I. C., Voronin, S., Judd, J. S., Vetter, P. A., Charléty, J., and Vonesch, C. (2011). Solving or resolving global tomographic models with spherical wavelets, and the scale and sparsity of seismic heterogeneity. *Geophysical Journal International*, 187, 969–988.

Singh, A., Fienberg, K., Jerolmack, D. J., Marr, J., and Foufoula-Georgiou, E. (2009). Experimental evidence for statistical scaling and intermittency in sediment transport rates. *Journal of Geophysical Research: Earth Surface*, 114, F01025, 1–16.

Singh, A., Foufoula-Georgiou, E., Porté2010Agel, F., and Wilcock, P. R. (2012). Coupled dynamics of the co-evolution of gravel bed topography, flow turbulence and sediment transport in an experimental channel. *Journal of Geophysical Research: Earth Surface*, 117, F04016, 1–20.

Singh, B. N., and Tiwari, A. K. (2006). Optimal selection of wavelet basis function applied to ECG signal denoising. *Digital Signal Processing*, 16, 275–287.

Sinou, J. J. (2010). Transient non-linear dynamic analysis of automotive disc brake squeal–on the need to consider both stability and non-linear analysis. *Mechanics Research Communications*, 37, 96–105.

Sitnikova, E., Hramov, A. E., Grubov, V., and Koronovsky, A. A. (2016). Rhythmic activity in EEG and sleep in rats with absence epilepsy. *Brain Research Bulletin*, 120, 106–116.

Sitnikova, E., Hramov, A. E., Koronovsky, A. A., and van Luijtelaar, G. (2009). Sleep spindles and spike–wave discharges in EEG: Their generic features, similarities and distinctions disclosed with Fourier transform and continuous wavelet analysis. *Journal of Neuroscience Methods*, 180, 304–316.

Sivalingam, S., and Hovd, M. (2011). Use of cross wavelet transform for diagnosis of oscillations due to multiple sources. In M. Fikar and M. Kvasnica (Eds.), *Proceedings of 18th International Conference on Process Control* (pp. 443–451), June 14–17, 2011. Bratislava: Slovak University of Technology.

Siwak, M., Rucinski, S. M., Matthews, J. M., Guenther, D. B., Kuschnig, R., Moffat, A. F., Rowe, J. F., Sasselov, D., and Weiss, W. W. (2014). A stable quasi-periodic 4.18-d oscillation and mysterious occultations in the 2011 MOST light-curve of TW Hya. *Monthly Notices of the Royal Astronomical Society*, 444(1), 327–335.

Skidmore, F., Korenkevych, D., Liu, Y., He, G., Bullmore, E., and Pardalos, P. M. (2011). Connectivity brain networks based on wavelet correlation analysis in Parkinson fMRI data. *Neuroscience Letters*, 499, 47–51.

Sleigh, J. W., Wilson, M. T., Voss, L. J., Steyn-Ross, D. A., Steyn-Ross, M. L., and Li, X. (2010). A continuum model for the dynamics of the phase transition from slow-wave sleep to REM sleep. In A. Steyn-Ross and M. Steyn-Ross (Eds.) *Modeling Phase Transitions in the Brain* (pp. 203–221). New York: Springer.

Sohrabi, M. R., and Zarkesh, M. T. (2014). Spectra resolution for simultaneous spectrophotometric determination of lamivudine and zidovudine components in pharmaceutical formulation of human immunodeficiency virus drug based on using continuous wavelet transform and derivative transform techniques. *Talanta*, 122, 223–228.

Sona, C. S., Khanwale, M. A., Mathpati, C. S., Borgohain, A., and Maheshwari, N. K. (2014). Investigation of flow and heat characteristics and structure identification of FLiNaK in pipe using CFD simulations. *Applied Thermal Engineering*, 70(1), 451–461.

Song, J. L., Hu, W., and Zhang, R. (2016). Automated detection of epileptic EEGs using a novel fusion feature and extreme learning machine. *Neurocomputing, 175*, 383–391.

Song, Y. D., Cao, Q., Du, X., and Karimi, H. R. (2013). Control strategy based on wavelet transform and neural network for hybrid power system. *Journal of Applied Mathematics, 375840*, 1–8.

Soon, W., Herrera, V. M. V., Selvaraj, K., Traversi, R., Usoskin, I., Chen, C. T. A., Lou, J. Y., et al. (2014). A review of Holocene solar-linked climatic variation on centennial to millennial timescales: Physical processes, interpretative frameworks and a new multiple cross-wavelet transform algorithm. *Earth-Science Reviews, 134*, 1–15.

Sørensen, J. S., Johannesen, L., Grove, U. S. L., Lundhus, K., Couderc, J. P., and Graff, C. (2010). A comparison of IIR and wavelet filtering for noise reduction of the ECG. In *Computing in Cardiology, 2010* (pp. 489–492). IEEE.

Sovilj, S., Van Oosterom, A., Rajsman, G., and Magjarevic, R. (2010). ECG-based prediction of atrial fibrillation development following coronary artery bypass grafting. *Physiological Measurement, 31*, 663–677.

Spence, S. M., Bernardini, E., Guo, Y., Kareem, A., and Gioffrè, M. (2014). Natural frequency coalescing and amplitude dependent damping in the wind-excited response of tall buildings. *Probabilistic Engineering Mechanics, 35*, 108–117.

Spoormaker, V. I., Schröter, M. S., Gleiser, P. M., Andrade, K. C., Dresler, M., Wehrle, R., Sämann, P. G., and Czisch, M. (2010). Development of a large-scale functional brain network during human non-rapid eye movement sleep. *Journal of Neuroscience, 30*(34), 11379–11387.

Srivastav, A., Ray, A., and Gupta, S. (2009). An information-theoretic measure for anomaly detection in complex dynamical systems. *Mechanical Systems and Signal Processing, 23*, 358–371.

Staszewski, W. J. (1997). Identification of damping in MDOF systems using time-scale decomposition. *Journal of Sound and Vibration, 203*(2), 283–305.

Staszewski, W. J. (1998a). Identification of non-linear systems using multi-scale ridges and skeletons of the wavelet transform. *Journal of Sound and Vibration, 214*(4), 639–658.

Staszewski, W. J. (1998b). Wavelet based compression and feature selection for vibration analysis. *Journal of Sound and Vibration, 211*(5), 735–760.

Staszewski, W. J., and Wallace, D. M. (2014). Wavelet-based frequency response function for time-variant systems: An exploratory study. *Mechanical Systems and Signal Processing, 47*, 35–49.

Staszewski, W. J., and Worden, K. (1999). Wavelet analysis of time-series: Coherent structures, chaos and noise. *International Journal of Bifurcation and Chaos, 9*(3), 455–471.

Steele, C. M., Sejdić, E., and Chau, T. (2013). Noninvasive detection of thin-liquid aspiration using dual-axis swallowing accelerometry. *Dysphagia, 28*, 105–112.

Stefan, W., Chen, K., Guo, H., Renaut, R. A., and Roudenko, S. (2012). Wavelet-based de-noising of positron emission tomography scans. *Journal of Scientific Computing, 50*(3), 665–677.

Stephenson, J. H., and Tinney, C. E. (2014). Extracting blade vortex interactions using continuous wavelet transforms. *American Helicopter Society 70th Annual Forum* (pp. 1–20), May 20–22, 2014. Montreal, Canada.

Stephenson, J. H., Tinney, C. E., Greenwood, E., and Watts, M. E. (2014). Time frequency analysis of sound from a maneuvering rotorcraft. *Journal of Sound and Vibration, 333*(12), 2539–2553.

Steppacher, I., Eickhoff, S., Jordanov, T., Kaps, M., Witzke, W., and Kissler, J. (2013). N400 predicts recovery from disorders of consciousness. *Annals of Neurology, 73*(5), 594–602.

Stiles, M. K., Clifton, D., Grubb, N. R., Watson, J. N., and Addison, P. S. (2004). Wavelet-based analysis of heart-rate-dependent ECG features. *Annals of Noninvasive Electrocardiology, 9*(4), 316–322.

Stirling, L. M., von Tscharner, V., Kugler, P., and Nigg, B. M. (2011). Piper rhythm in the activation of the gastrocnemius medialis during running. *Journal of Electromyography and Kinesiology, 21*, 178–183.

Stojanović, R., Karadaglić, D., Mirković, M., and Milošević, D. (2011). A FPGA system for QRS complex detection based on integer wavelet transform. *Measurement Science Review, 11*(4), 131–138.

Stojanović, R., Knežević, S., Karadaglić, D., and Devedžić, G. (2013). Optimization and implementation of the wavelet based algorithms for embedded biomedical signal processing. *Computer Science and Information Systems*, 10(1), 503–523.

Strang, G. (1989). Wavelets and dilation equations: A brief introduction. *SIAM Review*, 31(4), 614–627.

Strang, G. (1993). Wavelet transforms versus Fourier transforms. *Bulletin of the American Mathematical Society*, 28(2), 288–305.

Strang, G., and Nguyen, T. (1996). *Wavelets and Filter Banks*. Wellesley, MA: Wellesley-Cambridge Press.

Sudarshan, V. K., Mookiah, M. R. K., Acharya, U. R., Chandran, V., Molinari, F., Fujita, H., and Ng, K. H. (2016). Application of wavelet techniques for cancer diagnosis using ultrasound images: A review. *Computers in Biology and Medicine*, 69, 97–111.

Su, H., Liu, Q., and Li, J. (2011). Alleviating border effects in wavelet transforms for nonlinear time-varying signal analysis. *Advances in Electrical and Computer Engineering*, 11(3), 55–60.

Sukharev, A. L., and Aller, M. F. (2014). Wavelet analysis of variability of the radio source 3C120 in centimeter wavelength range. *Odessa Astronomical Publications*, 27(2), 146–148.

Sumathi, S., and Sanavullah, M. Y. (2009). Comparative study of QRS complex detection in ECG based on discrete wavelet transform. *International Journal of Recent Trends in Engineering*, 2(5), 273–277.

Sun, P. C., Kuo, C. D., Chi, L. Y., Lin, H. D., Wei, S. H., and Chen, C. S. (2012). Microcirculatory vasomotor changes are associated with severity of peripheral neuropathy in patients with type 2 diabetes. *Diabetes and Vascular Disease Research*, 10(3), 270–276.

Sun, Z., Hou, W., and Sun, L. (2006). Close-mode identification based on wavelet scalogram reassignment. In *24th Conference and Exposition on Structural Dynamics. IMAC-XXIV* (pp. 509–516), January 30–February 2, 2006. Bethel, Connecticut: Society for Experimental Mechanics.

Supangat, R., Grieger, J., Ertugrul, N., Soong, W. L., Gray, D. A., and Hansen, C. (2007). Detection of broken rotor bar faults and effects of loading in induction motors during rundown. In *IEEE International Electric Machines and Drives Conference, 2007. IEMDC'07* (Vol.1, pp. 196–201). IEEE.

Sweldens, W. (1996). Wavelets and the lifting scheme: A 5 minute tour. *ZAMM-Zeitschrift fur Angewandte Mathematik und Mechanik*, 76(2), 41–44.

Sweldens, W. (1998). The lifting scheme: A construction of second generation wavelets. *SIAM Journal on Mathematical Analysis*, 29(2), 511–546.

Szilagyi, J., Katul, G. G., Parlange, M. B., Albertson, J. D., and Cahill, A. T. (1996). The local effect of intermittency on the inertial subrange energy spectrum of the atmospheric surface layer. *Boundary-Layer Meteorology*, 79, 35–50.

Szilagyi, J., Parlange, M. B., Katul, G. G., and Albertson, J. D. (1999). An objective method for determining principal time scales of coherent eddy structures using orthonormal wavelets. *Advances in Water Resources*, 22(6), 561–566.

Tabatabaei, F. S., and Berkhuijsen, E. M. (2010). Relating dust, gas, and the rate of star formation in M 31. *Astronomy and Astrophysics*, 517(A77), 1–18.

Tabor, G. R., and Baba-Ahmadi, M. H. (2010). Inlet conditions for large eddy simulation: A review. *Computers and Fluids*, 39(4), 553–567.

Tafti, P. D., Delgado-Gonzalo, R., Stalder, A. F., and Unser, M. (2010). Fractal modelling and analysis of flow-field images. In *Biomedical Imaging: From Nano to Macro, 2010 IEEE International Symposium on* (pp. 49–52). IEEE.

Tafti, P. D., Van De Ville, D., and Unser, M. (2009). Invariances, Laplacian-like wavelet bases, and the whitening of fractal processes. *Image Processing, IEEE Transactions on*, 18(4), 689–702.

Tagluk, M. E., Akin, M., and Sezgin, N. (2010). Classification of sleep apnea by using wavelet transform and artificial neural networks. *Expert Systems with Applications*, 37, 1600–1607.

Talbi, M., Aouinet, A., Salhi, L., and Cherif, A. (2011). New method of R-wave detection by continuous wavelet transform. *Signal Processing: An International Journal (SPIJ), 5*(4), 165–173.

Talbi, R. B. M., Aouinet, A., and Cherif, A. (2012). ECG analysis based on wavelet transform and modulus maxima. *IJCSI International Journal of Computer Science Issues, 9*, 427–435.

Taplidou, S. A., and Hadjileontiadis, L. J. (2007). Nonlinear analysis of wheezes using wavelet bicoherence. *Computers in Biology and Medicine, 37*, 563–570.

Taplidou, S. A., and Hadjileontiadis, L. J. (2010). Analysis of wheezes using wavelet higher order spectral features. *IEEE Transactions on Biomedical Engineering, 57*(7), 1596–1610.

Tardu, S. (2011). Multiscale edge detection and imperfect phase synchronization of the wall turbulence. *Journal of Turbulence, 12*(26), 1–29.

Tarinejad, R., and Damadipour, M. (2014). Modal identification of structures by a novel approach based on FDD-wavelet method. *Journal of Sound and Vibration, 333*, 1024–1045.

Terenzi, R., and Sturani, R. (2009). Wavelet entropy filter and cross-correlation of gravitational wave data. In *Proceedings of 13th Gravitational Wave Data Analysis Workshop (GWDAW-13)* (pp. 1–10), January 19–22, 2009. San Juan, Puerto Rico.

Teti, R., Jemielniak, K., O'Donnell, G., and Dornfeld, D. (2010). Advanced monitoring of machining operations. *CIRP Annals-Manufacturing Technology, 59*, 717–739.

Thakur, G., Brevdo, E., Fučkar, N. S., and Wu, H. T. (2013). The synchrosqueezing algorithm for time-varying spectral analysis: Robustness properties and new paleoclimate applications. *Signal Processing, 93*, 1079–1094.

Thie, J., Sriram, P., Klistorner, A., and Graham, S. L. (2012). Gaussian wavelet transform and classifier to reliably estimate latency of multifocal visual evoked potentials (mfVEP). *Vision Research, 52*, 79–87.

Thomas, C., and Foken, T. (2005). Detection of long-term coherent exchange over spruce forest using wavelet analysis. *Theoretical and Applied Climatology, 80*, 91–104.

Thomas, M., Das, M. K., and Ari, S. (2015). Automatic ECG arrhythmia classification using dual tree complex wavelet based features. *AEU—International Journal of Electronics and Communications, 69*, 715–721.

Thurner, S., Feurstein, M. C., and Teich, M. C. (1998). Multiresolution wavelet analysis of heartbeat intervals discriminates healthy patients from those with cardiac pathology. *Physical Review Letters, 80*, 1544–1547.

Tian, H. J., Neyrinck, M. C., Budavári, T., and Szalay, A. S. (2011). Redshift-space enhancement of line-of-sight baryon acoustic oscillations in the sloan digital sky survey main-galaxy sample. *Astrophysical Journal, 728*(34), 1–20.

Tian, Z., Zuo, M. J., and Wu, S. (2012). Crack propagation assessment for spur gears using model-based analysis and simulation. *Journal of Intelligent Manufacturing, 23*(2), 239–253.

Timmermans, M. L., Rainville, L., Thomas, L., and Proshutinsky, A. (2010). Moored observations of bottom-intensified motions in the deep Canada Basin, Arctic Ocean. *Journal of Marine Research, 68*, 3–4.

Tiscareno, M. S., Burns, J. A., Nicholson, P. D., Hedman, M. M., and Porco, C. C. (2007). Cassini imaging of Saturn's rings: II. A wavelet technique for analysis of density waves and other radial structure in the rings. *Icarus, 189*, 14–34.

Tiscareno, M. S., Hedman, M. M., Burns, J. A., Weiss, J. W., and Porco, C. C. (2013). Probing the inner boundaries of Saturn's A ring with the Iapetus −1: 0 nodal bending wave. *Icarus, 224*(1), 201–208.

Tiwari, A. K., Dar, A. B., Bhanja, N., Arouri, M., and Teulon, F. (2015). Stock returns and inflation in Pakistan. *Economic Modelling, 47*, 23–31.

Tiwari, A. K., Dar, A. B., Bhanja, N., and Shah, A. (2013a). Stock market integration in Asian countries: Evidence from wavelet multiple correlations. *Journal of Economic Integration, 28*(3), 441–456.

Tiwari, A. K., Mutascu, M. I., and Albulescu, C. T. (2013b). The influence of the international oil prices on the real effective exchange rate in Romania in a wavelet transform framework. *Energy Economics*, 40, 714–733.

Tiwari, A. K., Oros, C., and Albulescu, C. T. (2014a). Revisiting the inflation–output gap relationship for France using a wavelet transform approach. *Economic Modelling*, 37, 464–475.

Tiwari, A. K., Suresh, K. G., Arouri, M., and Teulon, F. (2014b). Causality between consumer price and producer price: Evidence from Mexico. *Economic Modelling*, 36, 432–440.

Tjahjowidodo, T. (2012). Theoretical analysis of the dynamic behavior of presliding rolling friction via skeleton technique. *Mechanical Systems and Signal Processing*, 29, 296–309.

Tjahjowidodo, T., Al-Bender, F., and Van Brussel, H. (2007). Experimental dynamic identification of backlash using skeleton methods. *Mechanical Systems and Signal Processing*, 21, 959–972.

Tognola, G., Grandori, F., and Ravazzani, P. (1998). Wavelet analysis of click-evoked otoacoustic emissions. *IEEE Transactions on Biomedical Engineering*, 45(6), 686–697.

Tomba, E., Facco, P., Roso, M., Modesti, M., Bezzo, F., and Barolo, M. (2010). Artificial vision system for the automatic measurement of interfiber pore characteristics and fiber diameter distribution in nanofiber assemblies. *Industrial and Engineering Chemistry Research*, 49(6), 2957–2968.

Tonn, V. L., Li, H. C., and McCarthy, J. (2010). Wavelet domain correlation between the futures prices of natural gas and oil. *Quarterly Review of Economics and Finance*, 50, 408–414.

Topalova, I. (2012). Automated marble plate classification system based on different neural network input training sets and PLC implementation. *(IJARAI) International Journal of Advanced Research in Artificial Intelligence*, 1(2), 50–56.

Torrence, C., and Compo, G. P. (1998). A practical guide to wavelet analysis. *Bulletin of the American Meteorological society*, 79(1), 61–78.

Torrence, C., and Webster, P. J. (1999). Interdecadal changes in the ENSO-monsoon system. *Journal of Climate*, 12, 2679–2690.

Trabuco, M. H., Costa, M. V. C., and de Oliveira Nascimento, F. A. (2014). S-EMG signal compression based on domain transformation and spectral shape dynamic bit allocation. *Biomedical Engineering Online*, 13(22), 1–15.

Tsakiroglou, C. D., Sygouni, V., and Aggelopoulos, C. A. (2010). Using multi-level wavelets to correlate the two-phase flow characteristics of porous media withheterogeneity. *Chemical Engineering Science*, 65(24), 6452–6460.

Tsanas, A., and Clifford, G. D. (2015). Stage-independent, single lead EEG sleep spindle detection using the continuous wavelet transform and local weighted smoothing. *Frontiers in Human Neuroscience*, 9(181), 1–15.

Tsang, K. K. Y., So, R. M. C., Leung, R. C. K., and Wang, X. Q. (2008). Dynamic stall behavior from unsteady force measurements. *Journal of Fluids and Structures*, 24(1), 129–150.

Tse, N. C., and Lai, L. L. (2007). Wavelet-based algorithm for signal analysis. *EURASIP Journal on Applied Signal Processing*, 38916, 1–10.

Tse, P. W., Yang, W. X., and Tam, H. Y. (2004). Machine fault diagnosis through an effective exact wavelet analysis. *Journal of Sound and Vibration*, 277, 1005–1024.

Tsiaparas, N. N., Golemati, S., Andreadis, I., Stoitsis, J. S., Valavanis, I., and Nikita, K. S. (2011). Comparison of multiresolution features for texture classification of carotid atherosclerosis from B-mode ultrasound. *IEEE Transactions on Information Technology in Biomedicine*, 15(1), 130–137.

Tsutsumi, T., Takano, N., Matsuyama, N., Higashi, Y., Iwasawa, K., and Nakajima, T. (2011). High-frequency powers hidden within QRS complex as an additional predictor of lethal ventricular arrhythmias to ventricular late potential in post–myocardial infarction patients. *Heart Rhythm*, 8(10), 1509–1515.

Tung, W. W., Gao, J., Hu, J., and Yang, L. (2011). Detecting chaos in heavy-noise environments. *Physical Review E*, 83, 046210-1–046210-9.

Turner, B. J., and Leclerc, M. Y. (1994). Conditional sampling of coherent structures in atmospheric turbulence using the wavelet transform. *Journal of Atmospheric and Oceanic Technology*, *11*(1), 205–209.

Tuteur, F. B. (1989) Wavelet transforms in signal detection. In *C. J. M. Wavelets*, A. Grossmann, and P. Tchamitchian (Eds.) (pp. 132–138). Berlin, Heidelberg: Springer-Verlag.

Übeyli, E. D. (2009). Combined neural network model employing wavelet coefficients for EEG signals classification. *Digital Signal Processing*, *19*, 297–308.

Übeyli, E. D., Cvetkovic, D., Holland, G., and Cosic, I. (2010). Adaptive neuro-fuzzy inference system employing wavelet coefficients for detection of alterations in sleep EEG activity during hypopnoea episodes. *Digital Signal Processing*, *20*, 678–691.

Uchaipichat, N., Thanawattano, C., and Buakhamsri, A. (2013). Wavelet power spectrum analysis for PVC discrimination. In *Proceedings of the World Congress on Engineering* (Vol. 2, pp. 1316–1319), July 3–5, 2013. London, UK: WCE.

Umapathy, K., Krishnan, S., Masse, S., Hu, X., Dorian, P., and Nanthakumar, K. (2009). Optimizing cardiac resuscitation outcomes using wavelet analysis. In *31st Annual International Conference of the IEEE Engineering in Medicine and* Biology Society, *2009* (pp. 6761–6764). IEEE.

Üstündağ, M., Gökbulut, M., Şengür, A., and Ata, F. (2012). Denoising of weak ECG signals by using wavelet analysis and fuzzy thresholding. *Network Modeling Analysis in Health Informatics and Bioinformatics*, *1*(4), 135–140.

Vacha, L., and Barunik, J. (2012). Co-movement of energy commodities revisited: Evidence from wavelet coherence analysis. *Energy Economics*, *34*(1), 241–247.

Van de Ville, D., Britz, J., and Michel, C. M. (2010). EEG microstate sequences in healthy humans at rest reveal scale-free dynamics. *Proceedings of the National Academy of Sciences*, *107*(42), 18179–18184.

Van Fleet, P. (2008). *Discrete Wavelet Transformations: An Elementary Approach with Applications*. Hoboken, NJ: Wiley-Interscience.

Van Milligen, B. P., Hidalgo, C., and Sanchez, E. (1995a). Nonlinear phenomena and intermittency in plasma turbulence. *Physical Review Letters*, *74*(3), 395–398.

Van Milligen, B. P., Sanchez, E., Estrada, T., Hidalgo, C., Branas, B., Carreras, B., and Garcia, L. (1995b). Wavelet bicoherence: A new turbulence analysis tool. *Physics of Plasmas*, *2*(8), 3017–3032.

Vannozzi, G., Conforto, S., and D'Alessio, T. (2010). Automatic detection of surface EMG activation timing using a wavelet transform based method. *Journal of Electromyography and Kinesiology*, *20*, 767–772.

Van Ommen, J. R., Sasic, S., Van der Schaaf, J., Gheorghiu, S., Johnsson, F., and Coppens, M. O. (2011). Time-series analysis of pressure fluctuations in gas–solid fluidized beds: A review. *International Journal of Multiphase Flow*, *37*, 403–428.

Vaquero, J. M., Trigo, R. M., Vázquez, M., and Gallego, M. C. (2010). 155-day Periodicity in solar cycles 3 and 4. *New Astronomy*, *15*(4), 385–391.

Vázquez, L. A., Jurado, F., and Alanís, A. Y. (2015). Decentralized identification and control in real-time of a robot manipulator via recurrent wavelet first-order neural network. *Mathematical Problems in Engineering*, *451049*, 1–12.

Vázquez, R. R., Velez-Perez, H., Ranta, R., Dorr, V. L., Maquin, D., and Maillard, L. (2012). Blind source separation, wavelet denoising and discriminant analysis for EEG artefacts and noise cancelling. *Biomedical Signal Processing and Control*, *7*(4), 389–400.

Vernekar, K., Kumar, H., and Gangadharan, K. V. (2014). Gear fault detection using vibration analysis and continuous wavelet transform. *Procedia Materials Science*, *5*, 1846–1852.

Vielva, P. *(2010)*. A comprehensive overview of the cold spot. *Advances in Astronomy*, 2010, 592094, 1–20.

Vikramaditya, N. S., and Kurian, J. (2013). Amplitude and phase modulation of cavity modes in a supersonic flow. *European Journal of Mechanics-B/Fluids*, *42*, 159–168.

Vilda, P. G., Biarge, V. R., Mulas, C. M., Olalla, R. M., Fernández, L. M. M., and Marquina, A. Á. (2011). Glottal parameter estimation by wavelet transform for voice biometry. In *2011 IEEE International Carnahan Conference on Security Technology (ICCST)* (pp. 1–8). IEEE.
von Tscharner, V., Eskofier, B., and Federolf, P. (2011). Removal of the electrocardiogram signal from surface EMG recordings using non-linearly scaled wavelets. *Journal of Electromyography and Kinesiology*, 21, 683–688.
Walker, J. S. (2002). Tree-adapted wavelet shrinkage. *Advances in Imaging and Electron Physics*, 124, 343–394.
Walker, J. S. (2008). *A Primer on Wavelets and Their Scientific Applications*. 2nd Edition. Boca Raton, FL: CRC Press.
Wang, B., Zhang, H., and Wang, X. *(2013a)*. Large eddy simulation of inertial particle preferential dispersion in a turbulent flow over a backward-facing step. *Advances in Mechanical Engineering*, 2013, 493212, 1–8.
Wang, F. T., Lee, C. Y. J., Tin, H. W., Leu, S. W., Wen, C. C., and Chang, S. H. (2014). Novel Fractal-wavelet technique for denoising side-scan sonar images. *WSEAS Transactions on Signal Processing*, 10, 416–428.
Wang, G., Lewalle, J., Glauser, M., and Walczak, J. (2013c). Investigation of the benefits of unsteady blowing actuation on a 2D wind turbine blade. *Journal of Turbulence*, 14(1), 165–189.
Wang, H. A. O., Grolimund, D., Van Loon, L. R., Barmettler, K., Borca, C. N., Aeschlimann, B., and Gunther, D. (2011). Quantitative chemical imaging of element diffusion into heterogeneous media using laser ablation inductively coupled plasma mass spectrometry, synchrotron micro-X-ray fluorescence, and extended X-ray absorption fine structure spectroscopy. *Analytical Chemistry*, 83, 6259–6266.
Wang, H., Lee, S., Hassan, Y. A., and Ruggles, A. E. (2016). Laser-Doppler measurements of the turbulent mixing of two rectangular water jets impinging on a stationary pool. *International Journal of Heat and Mass Transfer*, 92, 206–227.
Wang, J., Li, X., Lu, C., Voss, L. J., Barnard, J. P., and Sleigh, J. W. (2012a). Characteristics of evoked potential multiple EEG recordings in patients with chronic pain by means of parallel factor analysis. *Computational and Mathematical Methods in Medicine*, 279560, 1–10.
Wang, J., Xu, H., Gu, J., An, T., Cui, H., Li, J., Zhang, Z., Zheng, Q., and Wu, X. P. (2010). How to identify and separate bright galaxy clusters from the low-frequency radio sky. *Astrophysical Journal*, 723, 620–633.
Wang, L., McCullough, M., and Kareem, A. (2013b). A data-driven approach for simulation of full-scale downburst wind speeds. *Journal of Wind Engineering and Industrial Aerodynamics*, 123, 171–190.
Wang, L., Wang, X., and Deng, Z. (2012b). Application of wavelet transform method for textile material feature extraction. In D. Baleanu (Ed.), *Wavelet Transforms and Their Recent Applications in Biology and Geoscience* (pp. 207–224). Rijeka, Croatia: INTECH Open Access Publisher.
Wang, L., Zhang, J., Wang, C., and Hu, S. (2003). Identification of nonlinear systems through time-frequency filtering technique. *Journal of Vibration and Acoustics*, 125, 199–204.
Wang, X., Xiang, J., Hu, H., Xie, W., and Li, X. (2015). Acoustic emission detection for mass fractions of materials based on wavelet packet technology. *Ultrasonics*, 60, 27–32.
Wan, X., Yan, K., Luo, D., and Zeng, Y. (2013). A combined algorithm for T-wave alternans qualitative detection and quantitative measurement. *Journal of Cardiothoracic Surgery*, 8, 1–7.
Watkins, L. R. (2015) Continuous wavelet transforms. In P. Rastogi and E. Hack (Eds.), *Phase Estimation in Optical Interferometry* (pp. 69–120). Boca Raton, FL: CRC Press.
Watson, J. N., and Addison, P. S. (2002). Spectral-temporal filtering of NDT data using wavelet transform modulus maxima. *Mechanics Research Communications*, 29, 99–106.
Watson, J. N., Addison, P. S., Clegg, G. R., Holzer, M., Sterz, F., and Robertson, C. E. (2000). A novel wavelet transform based analysis reveals hidden structure in ventricular fibrillation. *Resuscitation*, 43, 121–127.

Watson, J. N., Addison, P. S., Clegg, G. R., Steen, P. A., and Robertson, C. E. (2005). Wavelet transform-based prediction of the likelihood of successful defibrillation for patients exhibiting ventricular fibrillation. *Measurement Science and Technology, 16,* L1–L6.

Watson, J. N., Addison, P. S., Clegg, G. R., Steen, P. A., and Robertson, C. E. (2006a). Practical issues in the evaluation of methods for the prediction of shock outcome success in out-of-hospital cardiac arrest patients. *Resuscitation, 68,* 51–59.

Watson, J. N., Addison, P. S., Grubb, N., Clegg, G. R., Robertson, C. E., and Fox, K. A. A. (2001). Wavelet-based filtering for the clinical evaluation of atrial fibrillation. *Proceedings of the 23rd Annual International Conference of the Engineering in Medicine and Biology Society,* October 25–28, 2001. Istanbul, Turkey: IEEE.

Watson, J. N., Addison, P. S., Leonard, P., and Beatie, T. F. (2003). Patient illness classification using time frequency features derivced from the photoplethysomgram. In *Proceedings of the 25th Annual International Conference of the IEEE Engineering in Medicine and Biology Society* (Vol. 3, pp. 2978–2981), September 17–21, 2003. Cancun, Mexico: IEEE.

Watson, J. N., Addison, P. S., and Sibbald, A. (1999). The de-noising of sonic echo test data through wavelet transform reconstruction. *Shock and Vibration, 6,* 267–272.

Watson, J. N., Addison, P. S., Uchaipichat, N., Clegg, G. R., Steen, P. A., and Robertson, C. E. (2006b). Extracting heart rhythm information during CPR using wavelet transform. *MEDSIP 3rd International Conference: Advances in Medical Signals and Information Processing* (pp. 128–131), July 17–19, 2006. Stevenage, UK: IET.

Watson, J. N., Addison, P. S., Uchaipichat, N., Shah, A. S., and Grubb, N. R. (2007). Wavelet transform analysis predicts outcome of DC cardioversion for atrial fibrillation patients. *Computers in Biology and Medicine, 37,* 517–523.

Watson, J. N., Uchaipichat, N., Addison, P. S., Clegg, G. R., Robertson, C. E., Eftestol, T., and Steen, P. A. (2004). Improved prediction of defibrillation success for out-of-hospital VF cardiac arrest using wavelet transform methods. *Resuscitation, 63,* 269–275.

Watson, S. J., Xiang, B. J., Yang, W., Tavner, P. J., and Crabtree, C. J. (2010). Condition monitoring of the power output of wind turbine generators using wavelets. *IEEE Transactions on Energy Conversion, 25*(3), 715–721.

Wendt, H., Kiyono, K., Abry, P., Hayano, J., Watanabe, E., and Yamamoto, Y. (2014). MultiScale wavelet p-leader based heart rate variability analysis for survival probability assessment in CHF patients. In *2014 36th Annual International Conference of the IEEE Engineering in Medicine and Biology Society (EMBC)* (pp. 2809–2812). IEEE.

Weng, J., Guo, X. M., Chen, L. S., Yuan, Z. H., Ding, X. R., and Lei, M. (2013). Study on real-time monitoring technique for cardiac arrhythmia based on smartphone. *Journal of Medical and Biological Engineering, 33*(4), 394–399.

Wiebe, A., Sturman, A., and McGowan, H. (2011). Wavelet analysis of atmospheric turbulence over a coral reef flat. *Journal of Atmospheric and Oceanic Technology, 28*(5), 698–708.

Wiggins, M., Zhao, L., Vachtsevanos, G., and Litt, B. (2003). Non-invasive, cardiac risk stratification using wavelet coefficients. *WSEAS Transactions on Computers, 2,* 720–723.

Wiklund, U., Akay, M., and Niklasson, U. (1997). Short-term analysis of heart-rate variability of adapted wavelet transforms. *IEEE Engineering in Medicine and Biology Magazine, 16,* 113–118.

Wilczek, M., Kadoch, B., Schneider, K., Friedrich, R., and Farge, M. (2011). Wavelet analysis of the conditional vorticity budget in fully developed homogeneous isotropic turbulence. *Journal of Physics: Conference Series, 318,* 062024, 1–8.

Williams, J. R., and Amaratunga, K. (1994). An introduction to wavelets in engineering. *International Journal for Numerical Methods in Engineering, 37,* 2365–2388.

Wintoft, P. (2005). Study of the solar wind coupling to the time difference horizontal geomagnetic field. *Annales Geophysicae, 23*(5), 1949–1957.

Wu, H. T., Chan, Y. H., Lin, Y. T., and Yeh, Y. H. (2014). Using synchrosqueezing transform to discover breathing dynamics from ECG signals. *Applied and Computational Harmonic Analysis*, 36, 354–359.

Wu, H. T., Hseu, S. S., Bien, M. Y., Kou, Y. R., and Daubechies, I. (2012). Evaluating physiological dynamics via synchrosqueezing: Prediction of ventilator weaning. *IEEE Transactions on Biomedical Engineering*, 61(3), 736–744.

Wu, M. S. (2014). Genetic algorithm based on discrete wavelet transformation for fractal image compression. *Journal of Visual Communication and Image Representation*, 25, 1835–1841.

Wu, Q., Mao, J. F., Wei, C. F., Fu, S., Law, R., Ding, L., Yu, B. T., Jia, B., and Yang, C. H. (2016). Hybrid BF–PSO and fuzzy support vector machine for diagnosis of fatigue status using EMG signal features. *Neurocomputing*, 173, 483–500.

Wu, S., Shen, Y., Zhou, Z., Lin, L., Zeng, Y., and Gao, X. (2013a). Research of fetal ECG extraction using wavelet analysis and adaptive filtering. *Computers in Biology and Medicine*, 43, 1622–1627.

Wu, T. Y., and Lin, S. F. (2013). A method for extracting suspected parotid lesions in CT images using feature-based segmentation and active contours based on stationary wavelet transform. *Measurement Science Review*, 13(5), 237–247.

Wu, X., Yu, B., and Wang, Y. (2013). Wavelet analysis on turbulent structure in drag-reducing channel flow based on direct numerical simulation. *Advances in Mechanical Engineering*, 5, 514325, 1–10.

Wu, Y., and Du, R. (1996). Feature extraction and assessment using wavelet packets for monitoring of machining processes. *Mechanical Systems and Signal Processing*, 10(1), 29–53.

Wu, Y., and Noonan, J. P. (2012). Image steganography scheme using chaos and fractals with the wavelet transform. *International Journal of Innovation, Management and Technology*, 3(3), 285–289.

Wu, Y. H., and Ren, H. Y. (2014). Analysis of realistic rough surface for its globally dominant parameters using continuous wavelets. *Applied Mathematics and Mechanics*, 35, 741–748.

Wu, Y., Sun, Y., Zhan, L., and Ji, Y. (2013b). Low mismatch key agreement based on wavelet-transform trend and fuzzy vault in body area network. *International Journal of Distributed Sensor Networks*, 912893, 1–16.

Xie, L., Miao, Q., Chen, Y., Liang, W., and Pecht, M. (2012). Fan bearing fault diagnosis based on continuous wavelet transform and autocorrelation. In *IEEE Conference on Prognostics and System Health Management (PHM), 2012* (pp. 1–6). IEEE.

Xing, Y. F., Wang, Y. S., Shi, L., Guo, H., and Chen, H. (2016). Sound quality recognition using optimal wavelet-packet transform and artificial neural network methods. *Mechanical Systems and Signal Processing*, 66, 875–892.

Xu, H., Sun, S. Z., Gui, Z., and Luo, S. (2015). Detection of sub-seismic fault footprint from signal-to-noise ratio based on wavelet modulus maximum in the tight reservoir. *Journal of Applied Geophysics*, 114, 259–262.

Xue, Y. J., Cao, J. X., Tian, R. F., Du, H. K., and Shu, Y. X. (2014). Application of the empirical mode decomposition and wavelet transform to seismic reflection frequency attenuation analysis. *Journal of Petroleum Science and Engineering*, 122, 360–370.

Xue, Z., Ming, D., Song, W., Wan, B., and Jin, S. (2010). Infrared gait recognition based on wavelet transform and support vector machine. *Pattern Recognition*, 43, 2904–2910.

Xuexu, Y., Guo, R., Yinghai, G., Jifeng, Y., Yulin, S., and Yubao, S. (2013). Correlation and analysis of well-log sequence with Milankovitch cycles as rulers: A case study of coal-bearing strata of late Permian in western Guizhou. *International Journal of Mining Science and Technology*, 23, 563–568.

Yacin, S. M., Chakravarthy, V. S., and Manivannan, M. (2011). Reconstruction of gastric slow wave from finger photoplethysmographic signal using radial basis function neural network. *Medical and Biological Engineering and Computing*, 49, 1241–1247.

Yamada, M., and Ohkitani, K. (1991). An identification of energy cascade in turbulence by orthonormal wavelet analysis. *Progress of Theoretical Physics*, 86(4), 799–815.

Yang, L., and Hamori, S. (2015). Interdependence between the bond markets of CEEC-3 and Germany: A wavelet coherence analysis. *North American Journal of Economics and Finance*, 32, 124–138.

Yang, W., Tavner, P. J., Crabtree, C. J., and Wilkinson, M. (2010). Cost-effective condition monitoring for wind turbines. *IEEE Transactions on Industrial Electronics*, 57(1), 263–271.

Yang, Y., and Nagarajaiah, S. (2014). Data compression of structural seismic responses via principled independent component analysis. *Journal of Structural Engineering*, 140(7), 04014032-1–04014032-10.

Yan, J. J., Wang, Y. Q., Guo, R., Zhou, J. Z., Yan, H. X., Xia, C. M., and Shen, Y. (2012). Nonlinear analysis of auscultation signals in TCM using the combination of wavelet packet transform and sample entropy. *Evidence-Based Complementary and Alternative Medicine*, 247012, 1–9.

Yan, R., Gao, R. X., and Chen, X. (2014). Wavelets for fault diagnosis of rotary machines: A review with applications. *Signal Processing*, 96, 1–15.

Yao, C., Gao, X., and Yu, Y. *(2013)*. Wind speed forecasting by wavelet neural networks: A comparative study. *Mathematical Problems in Engineering*, 2013, 395815, 1– 7.

Yaroshenko, T. Y., Krysko, D. V., Dobriyan, V., Zhigalov, M. V., Vos, H., Vandenabeele, P., and Krysko, V. A. (2015). Wavelet modeling and prediction of the stability of states: The Roman Empire and the European Union. *Communications in Nonlinear Science and Numerical Simulation*, 26, 265–275.

Yazar, A., Keskin, F., Töreyin, B. U., and Çetin, A. E. (2013). Fall detection using single-tree complex wavelet transform. *Pattern Recognition Letters*, 34(15), 1945–1952.

Yee, E., Chan, R., Kosteniuk, P. R., Biltoft, C. A., and Bowers, J. F. (1996). Multiscaling properties of concentration fluctuations in dispersing plumes revealed using an orthonormal wavelet decomposition. *Boundary-Layer Meteorology*, 77(2), 173–207.

Yi, J., Zhang, J. W., and Li, Q. S. (2013). Dynamic characteristics and wind-induced responses of a super-tall building during typhoons. *Journal of Wind Engineering and Industrial Aerodynamics*, 121, 116–130.

Yin, J., Gao, C., and Jia, X. (2013). Wavelet packet analysis and gray model for feature extraction of hyperspectral data. *Geoscience and Remote Sensing Letters, IEEE*, 10(4), 682–686.

Yochum, M., Bakir, T., Lepers, R., and Binczak, S. (2012). Quantification of muscle fatigue with wavelet analysis based on EMG during myoelectrical stimulation. *BIODEVICES 2012*, 5(1), 53–58.

Yochum, M., Renaud, C., and Jacquir, S. (2016). Automatic detection of P, QRS and T patterns in 12 leads ECG signal based on CWT. *Biomedical Signal Processing and Control*, 25, 46–52.

Yoo, J., Lee, D. Y., Ha, T. M., Cho, Y. S., and Woo, S. B. (2010). Characteristics of abnormal large waves measured from coastal videos. *Natural Hazards and Earth System Science*, 10, 947–956.

Yoshimatsu, K., Okamoto, N., Kawahara, Y., Schneider, K., and Farge, M. (2013). Coherent vorticity and current density simulation of three-dimensional magnetohydrodynamic turbulence using orthogonal wavelets. *Geophysical and Astrophysical Fluid Dynamics*, 107(1–2), 73–92.

Yoshimatsu, K., Schneider, K., Okamoto, N., Kawahara, Y., and Farge, M. (2011). Intermittency and geometrical statistics of three-dimensional homogeneous magnetohydrodynamic turbulence: A wavelet viewpoint. *Physics of Plasmas*, 18, 092304-1–092304-9.

Young, N. J., Stappers, B. W., Weltevrede, P., Lyne, A. G., and Kramer, M. (2012). On the pulse intensity modulation of PSR B0823+ 26. *Monthly Notices of the Royal Astronomical Society*, 427(1), 114–126.

Ysusi, C. *(2009)*. Analysis of the dynamics of Mexican inflation using wavelets. Working Papers, Banco de México, (No. *2009-09)*, 1–36.

Yuenyong, S., Nishihara, A., Kongprawechnon, W., and Tungpimolrut, K. (2011). A framework for automatic heart sound analysis without segmentation. *Biomedical Engineering Online*, 10(13), 1–23.

Yun, J., Ha, D. S., Inman, D. J., and Owen, R. B. (2011). Adverse event detection (AED) system for continuously monitoring and evaluating structural health status. In M. N. Ghasemi-Nejhad (Ed.), *Proceedings of SPIE 7977, Active and Passive Smart Structures and Integrated Systems.* Conference Volume 7977, March 6, 2011. San Diego, CA. 79771R.

Zalev, J., and Kolios, M. C. (2011). Detecting abnormal vasculature from photoacoustic signals using wavelet-packet features. In *Proceedings of SPIE 7899, Photons Plus Ultrasound: Imaging and Sensing 2011,* (vol. 7899, pp. 78992M-1–78992M-15.). Bellingham, Washington, DC: International Society for Optics and Photonics.

Zaugg, S., Saporta, G., Van Loon, E., Schmaljohann, H., and Liechti, F. (2008). Automatic identification of bird targets with radar via patterns produced by wing flapping. *Journal of the Royal Society interface*, 5, 1041–1053.

Zaunseder, S., Heinke, A., Trumpp, A., and Malberg, H. (2014). Heart beat detection and analysis from videos. In *2014 IEEE 34th International Conference on Electronics and Nanotechnology (ELNANO)* (pp. 286–290). IEEE.

Zawada-Tomkiewicz, A., and Ściegienka, R. (2011). Monitoring of a micro-smoothing process with the use of machined surface images. *Metrology and Measurement Systems*, 18(3), 419–428.

Zazula, D., Đonlagić, D., and Šprager, S. (2012). Application of fibre-optic interferometry to detection of human vital signs. *Journal of the Laser and Health Academy*, 1, 27–32.

Zelelew, H., Khasawneh, M., and Abbas, A. (2014). Wavelet-based characterisation of asphalt pavement surface macro-texture. *Road Materials and Pavement Design*, 15(3), 622–641.

Zeng, K., Yan, J., Wang, Y., Sik, A., Ouyang, G., and Li, X. (2016). Automatic detection of absence seizures with compressive sensing EEG. *Neurocomputing*, 171, 497–502.

Zhai, M. Y. (2015). A new method for short-term load forecasting based on fractal interpretation and wavelet analysis. *International Journal of Electrical Power and Energy Systems*, 69, 241–245.

Zhang, B., Abbas, A., and Romagnoli, J. (2012b). Monitoring crystal growth based on image texture analysis using wavelet transformation. In *8th IFAC Symposium on Advanced Control of Chemical Processes* (Vol. 8, No. 1, pp. 33–38). Laxenburg, Austria: IFAC.

Zhang, G., Leclerc, M. Y., Duarte, H. F., Durden, D., Werth, D., Kurzeja, R., and Parker, M. (2014a). Multi-scale decomposition of turbulent fluxes above a forest canopy. *Agricultural and Forest Meteorology*, 186, 48–63.

Zhang, J., Palmer, S., and Wang, X. (2010). Identification of animal fibers with wavelet texture analysis. In *WCE 2010: Proceedings of the World Congress on Engineering* (pp. 742–747), June 30–July 2, 2010, Newswood Limited/International Association of Engineers, London, UK.

Zhang, R., McAllister, G., Scotney, B., McClean, S., and Houston, G. (2006). Combining wavelet analysis and Bayesian networks for the classification of auditory brainstem response. *IEEE Transactions on Information Technology in Biomedicine*, 10(3), 458–467.

Zhang, V. W., Zhang, Z. G., McPherson, B., Hu, Y., and Hung, Y. S. (2011b). Detection improvement for neonatal click evoked otoacoustic emissions by timefrequency filtering. *Computers in Biology and Medicine*, 41, 675–686.

Zhang, Y., Dong, Z., Wu, L., and Wang, S. (2011c). A hybrid method for MRI brain image classification. *Expert Systems with Applications*, 38(8), 10049–10053.

Zhang, Y., Li, Q., Wu, J., and Li, J. (2013b). The time-frequency characters analysis of the snoring signal. In *International Conference on Advanced Computer Science and Electronic Information. (ICACSEI* 2013) (pp. 83–86).

Zhang, Y., Xu, X., Luo, Y., Hu, H., and Zhou, H. (2012a). An improved incremental online training algorithm for reducing the influence of muscle fatigue in sEMG based HMI. In *2012 IEEE International Conference on Robotics and Biomimetics (ROBIO)* (pp. 683–688). IEEE.

Zhang, Z. G., Hung, Y. S., and Chan, S. C. (2011a). Local polynomial modeling of time-varying autoregressive models with application to time-frequency analysis of event-related EEG. *IEEE Transactions on Biomedical Engineering*, 8(3), 557–566.

Zhang, Z., Jung, T. P., Makeig, S., and Rao, B. (2013a). Compressed sensing for energy-efficient wireless telemonitoring of noninvasive fetal ECG via block sparse Bayesian learning. *IEEE Transactions on Biomedical Engineering*, 60(2), 300–309.

Zhang, Z., and Moore, J. C. (2011). Improved significance testing of wavelet power spectrum near data boundaries as applied to polar research. *Advances in Polar Science*, 22(3), 192–198.

Zhang, Z., and Moore, J. C. (2012). Comment on "Significance tests for the wavelet power and the wavelet power spectrum" by Ge (2007). *Annales Geophysicaes*, 30, 1743–1750.

Zhang, Z., Moore, J. C., and Grinsted, A. (2014b). Haar wavelet analysis of climatic time series. *International Journal of Wavelets, Multiresolution and Information Processing*, 12, 1450020-1–1450020-11.

Zhao, J., Yao, Y., Huang, W., Shi, R., Zhang, S., LeGrice, I. J., Lever, N. A., and Smaill, B. H. (2013b). Novel methods for characterization of paroxysmal atrial fibrillation in human left atria. *Open Biomedical Engineering Journal*, 7, 29–40.

Zhao, M. (2013). Flow induced vibration of two rigidly coupled circular cylinders in tandem and side-by-side arrangements at a low Reynolds number of 150. *Physics of Fluids*, 25(12), 123601-1–123601-31.

Zhao, M., Cheng, L., and An, H. (2012). Numerical investigation of vortex-induced vibration of a circular cylinder in transverse direction in oscillatory flow. *Ocean Engineering*, 41, 39–52.

Zhao, M., and Yan, G. (2013). Numerical simulation of vortex induced vibration of two cylinders of different diameters at low Reynolds numbers. *Physics of Fluids*, 25, 083601-1–083601-26.

Zhao, W., and Davis, C. E. (2009). Swarm intelligence based wavelet coefficient feature selection for mass spectral classification: An application to proteomics data. *Analytica Chimica Acta*, 651(1), 15–23.

Zhao, Y., Yuan, Z., and Chen, F. (2010). Enhancing fluid animation with adaptive, controllable and intermittent turbulence. In *Proceedings of the 2010 ACM SIGGRAPH/Eurographics symposium on computer animation* (pp. 75–84). Aire-la-Ville, Switzerland: Eurographics Association.

Zhao, Z., Shi, C., Ren, F., Zhang, L., and Luo, Y. (2013a). Adaptive analysis of diastolic murmurs for coronary artery disease based on empirical mode decomposition. In L. G. Morales (Ed.), *Adaptive Filtering: Theories and Applications* (pp. 91–120), Rijeka, Croatia: INTECH Open Access Publisher.

Zheng, F., and Liu, S. (2012). A fast compression algorithm for seismic data from non-cable seismographs. In *2012 World Congress on Information and Communication Technologies (WICT)*, (pp. 1215–1219). IEEE.

Zheng, L. M., Sone, S., Itani, Y., Wang, Q., Hanamura, K., Asakura, K., Li, F., Yang, Z. G., Wang, J. C., and Funasaka, T. (2000). Effect of CT digital image compression on detection of coronary artery calcification. *Acta Radiologica*, 41, 116–121.

Zheng, W., Liu, H., He, A., Ning, X., and Cheng, J. (2010). Single-lead fetal electrocardiogram estimation by means of combining R-peak detection, resampling and comb filter. *Medical Engineering and Physics*, 32, 708–719.

Zheng, Z., Ahn, S., Chen, D., and Laval, J. (2011). Applications of wavelet transform for analysis of freeway traffic: Bottlenecks, transient traffic, and traffic oscillations. *Transportation Research Part B: Methodological*, 45, 372–384.

Zhong, L., Wan, J., Huang, Z., Cao, G., and Xiao, B. (2013). Heart murmur recognition based on hidden markov model. *Journal of Signal and Information Processing*, 4, 140.

Zhong, R., Zong, Z., Niu, J., and Yuan, S. (2014). A damage prognosis method of girder structures based on wavelet neural networks. *Mathematical Problems in Engineering*, 130274, 1–11.

Zhou, Z., and Yang, K. (2012). Fetal electrocardiogram extraction and performance analysis. *Journal of Computers*, 7(11), 2821–2828.

Zhu, B., Ding, Y., and Hao, K. (2013). A novel automatic detection system for ECG arrhythmias using maximum margin clustering with immune evolutionary algorithm. *Computational and Mathematical Methods in Medicine*, 453402, 1–8.

Zhu, P., Zhang, J. A., and Masters, F. J. (2010). Wavelet analyses of turbulence in the hurricane surface layer during landfalls. *Journal of the Atmospheric Sciences*, *67*, 3793–3805.

Zima, M., Tichavský, P., Paul, K., and Krajča, V. (2012). Robust removal of short-duration artifacts in long neonatal EEG recordings using wavelet-enhanced ICA and adaptive combining of tentative reconstructions. *Physiological Measurement*, *33*(8), N39.

Zitto, M. E., Piotrkowski, R., Gallego, A., Sagasta, F., and Benavent-Climent, A. (2015). Damage assessed by wavelet scale bands and b-value in dynamical tests of a reinforced concrete slab monitored with acoustic emission. *Mechanical Systems and Signal Processing, 60–61*, 75–89.

Zorick, T., and Mandelkern, M. A. (2013). Multifractal detrended fluctuation analysis of human EEG: Preliminary investigation and comparison with the wavelet transform modulus maxima technique. *PLoS One*, *8*(7), e68360, 1–7.

Zuñiga, A. G., Florindo, J. B., and Bruno, O. M. (2014). Gabor wavelets combined with volumetric fractal dimension applied to texture analysis. *Pattern Recognition Letters*, *36*, 135–143.

Appendix I: Useful Books, Papers and Websites

This appendix aims to provide the reader with a shortlist of useful books, papers and websites. They have been selected by the author for their extensive content and/or clarity of presentation.

AI.1 USEFUL BOOKS

There are a large number of wavelet books in the literature, from those aimed at a mathematical audience to those which deal with a specific scientific discipline. The mathematical and statistical literature is generally weighted towards the discrete orthonormal wavelet transform and associated transforms; for example, the non-decimated discrete wavelet transform, due in part to its nice mathematical properties. However, the applied scientific literature is more balanced between the continuous and discrete wavelet transforms. There are many good texts available covering the background theory or specific applications; however, this section is restricted to three texts.

- *The World According to Wavelets*: In her book, Hubbard (1996) provides an excellent account of the history and use of the wavelet transform. As its subtitle says, the book is 'the story of a mathematical technique in the making'. Much of the text is written using no mathematics at all, and where mathematical explanations of the concepts are employed to convey some of the concepts, the treatment is minimal.

- *Ten Lectures on Wavelets*: By Ingrid Daubechies (1992), this is one of the first wavelet texts and has become a standard in the field.

- *A Wavelet Tour of Signal Processing: The Sparse Way*: The third edition of this most popular book by Mallat (2009) provides much of the useful mathematical detail underlying wavelet transform techniques that are used in practice.

AI.2 USEFUL WEBSITES

Typing 'wavelet' into a search engine should produce a very large number of sites containing wavelet material. A brief list of useful websites follows with a short note on the contents of each site. This is not a comprehensive list, but rather has been compiled to give the reader some good places to begin a search. Most of them contain a large number of hyperlinks to other useful sites. These sites were all active at the time of writing.

- https://en.wikipedia.org/wiki/Wavelet

An obvious first stop on a tour of the web. The 'Wavelet' page in Wikipedia.

- http://paos.colorado.edu/research/wavelets/

Christopher Compo and Gilbert Torrence's site. This is a very nicely presented site. It has an interactive bit where you can submit your own data for wavelet analysis. The answers to frequently asked questions and software are also available.

- http://users.rowan.edu/~polikar/WAVELETS/WTtutorial.html

Robi Polikar's site – lots of well illustrated introductory information.

- http://www-stat.stanford.edu/~wavelab/

Site containing WAVELAB software – a long list of routines for MATLAB®.

- http://www.wavelet.org/wavelet/index.html

This is the archive website for *Wavelet Digest* which contains news and views from the wavelet community over the years, including details of books and papers, theses, software, courses and conferences and not least a section devoted to questions and answers from the subscribers.

In addition to these websites, there are now many videos to be found on the web concerning wavelets, including a number of lecture courses.

Index

3-D Lissajous Figures, 308, 310

Acoustic emission signals, 174, 260
Acoustic response, 267, 321–323
Activity signals, 333
Admissibility condition, 10
Admissibility constant, 10, 12, 29
Analytic wavelets, 34, 89, 255
Analyzing wavelet, 8
Approximation coefficients, 98–99, 105–107, 109, 111–113, 115, 118, 121–122, 124, 125, 139–145, 147, 154, 161, 291
Astronomy, 371
Atmospheric processes, 211–215
Atrial fibrillation (AF), 75, 285–289

Bandpass filter, 10, 88, 243
Battle–Lemarié wavelet, 173
Bearings, 174, 249, 255–256, 259
Biorthogonal wavelets, 93, 150–151, 152, 173, 174, 213, 260, 287, 320, 330, 333, 335
Blades, 249, 255–259
Blood flow, 277, 315, 323–328, 332, 334
Blood pressure, 69, 75, 277, 316, 323–324, 326–327, 334
Borehole logging data, 350, 362
Boundary effects, 56–61
Boundary layer flow, 183, 186, 205, 211, 214–215
Bridge response, 206
B-spline wavelet, 247, 291, 297

Canopy flows, 361, 368, 369
Cantor set, 338–339, 341
Cardiac arrhythmias, 278–289
Cardio-pulmonary resuscitation (CPR), 280, 282, 285–286
Chaos, 236–238, 258
Chatter, 259, 261
Chemistry, 378
Chirp, 20, 42–43, 109–111, 121
Circular mean phase (CMP), 83–85
Cochlea, 291, 321, 323

Coefficient of variation, 181, 211, 213
Coefficient variance, 178, 181, 345–349
Coherent structures, 14–22, 177, 190, 194, 210, 211, 213–215, 224, 236
Coiflet wavelet, 148, 164, 255, 260, 273, 287, 291, 294, 330, 333, 335, 362
Compact support, 101, 105, 135, 148, 168
Complex wavelets, 10–12, 14, 29, 34, 37, 75, 89, 231, 243, 263, 275, 278, 283, 299, 305, 327, 329, 334
Compression, 1, 151, 159, 168, 231, 263–264, 267, 289, 328, 330, 349, 350, 364
Computational fluid dynamics (CFD), 200–205
Condition monitoring, 67, 88, 231, 249–259, 260–261
Continuous approximation, 98–99, 122, 126
Continuous wavelet transform, *passim*
Contourlet, 168–173
Control, 264–265
Cosmic microwave background, 376
Cricket (the game), 379
Cross-wavelet transform (CrWT), 75, 78–80, 81, 89, 90, 198, 208, 210, 213, 217, 227, 264, 265, 327, 334, 362, 371
CT scans, 364–365
Curvelet, 168–172
Cylinder flows, 181–183, 185, 198, 207–209

Damping (dynamics), 216, 232–236, 246, 252
Daubechies wavelet coefficients, 138
Daubechies wavelets, 113, 135–138, 142–143, 147–148, 162, 181, 259–260, 262–263, 270, 289, 299, 302, 312, 315, 318–319, 330–331, 333, 335, 349–350, 364, 378
Decomposition algorithm, 51, 106, 111, 117–121, 139, 144
Denoising, 26, 27, 127–135, 149, 174, 227, 238, 249, 251, 261, 271, 273, 302, 320, 328, 330, 332–335, 340, 349, 350, 355, 360, 364, 377
Derivative of Gaussian wavelets, 207, 225–226, 251, 265, 352

443

Detail coefficients, 94, 99, 107, 109, 112–113, 115, 117, 122, 124–126, 140–142, 144–145, 147, 154, 159
Dilation equation, 98, 101
Dilation parameter, 9, 11, 12, 23
Discrete approximation, 98, 124–126, 144, 147, 163
Discrete input signals, 98, 107–108, 111–117, 119, 124–127
Discrete wavelet transform *passim*
DNA, 174, 342
Doppler heart sounds, 317
Downsampled, 143
Drilling processes, 259
Dual spectrum, 182–183
Duffing oscillator, 236
Dyadic grid, 95–97, 103–104, 108, 113, 115–117, 149, 181, 268, 345, 348
Dynamics, 232–236

Earthquake excitation, 249
ECG, 267–290, 299, 333–334, 349
Ecological patterns, 378
Edge detection, 22–24, 370, 379
EEG, 290–302, 349, 353
El Niño, 218, 220, 362
Electrical systems and circuits, 265
Electromyographic signals (EMGs), 174, 299, 327, 328, 332–334
Embolism, 317, 318
Energy, scale dependent, 29, 31–33, 179, 182, 184, 347
Energy prices, 360
Energy spectrum, 9–11, 14–15, 21, 29, 31–32, 36–37, 324, 351, 352
Engines, 259, 263
Epileptic seizures, 297, 299, 349
Epileptogenic foci, 297
Event-related potentials, 290–291
Evoked potentials, 290–291, 297

Fast wavelet transform, 105–107
Fibrous materials, 263
Filter, *passim*
Filtering, 25–28, 106, 113, 140, 143–149, 164, 173, 201, 205, 226, 238, 240, 241, 243, 244, 255, 263, 264, 271, 291, 295, 312, 335, 364
Financial time series, 86, 355–357
Flatness factor, 179, 182, 183, 204, 226
Fluctuation intensity, 181–182, 211
Fluids, 177–229
Fluid-structure interaction, 75, 206–210
Foreign exchange rates, 359, 360
Fourier transform, *passim*
Fractals, 227, 236, 262, 265, 302, 315, 328, 331, 337–355, 377

Fractional Brownian motion (fBm), 342–344, 346, 347, 350, 353, 354
Frame, 93–97, 207

Gait, 333–334
Gaussian noise, 132, 134, 342
Gears, 249–255
Geophysical flows, 210–224
Geophysics, 361–371

Haar wavelet, 102–105, *passim*
Heart rate variability (HRV), 275–278, 325, 334
Heart sounds, 315–318
Heisenberg boxes, 44–51
Highpass filter, 106, 137, 139, 143–145, 164

Image registration, 370
Impulse response, 68, 232
Indexing of coefficients, 115–117
Inertial subrange (turbulence), 211, 222
Intermittency index, 185–186
Inverse discrete wavelet transform, 97
Inverse wavelet transform, 24–28, 55, 71, 240, 287

Jet flows, 194–197, 203, 204, 214, 221, 222, 225, 351, 374, 376

Laser Doppler flowmetry, 323, 325
Lena image, 161–162
Level indexing, 117
Light curve, 375, 377
Location parameter, 12, 94–95, 181
Low oscillation complex wavelets, 37, 75, 89, 243, 281, 327
Lowpass filter, 106, 135, 139, 140, 143–144, 164
Lung sounds, 318–321

Machining processes, 259–261
Magnetic resonance images (MRI), 297, 329, 331, 332
Matching pursuit, 51–55, 376
Medical images, 328–332
Mexican hat wavelet, 8–9, *passim*
  two-dimensional, 55, 89
  complex, 89, 243
Meyer wavelet, 104–105, 260, 320, 325, 335, 366
Milling machine, 259–260
Mode decoupling, 232–234
Modular approach, 90, 311
Modulus maxima, 27–28, 89, 202, 240, 243–244, 268–270, 276, 281, 285–287, 290, 299, 301–302, 325, 341, 351–352, 364, 370, 373
Moment function, 178, 185–186, 351

Moment scale correlation, 185–186
Morlet wavelet, 34–44, *passim*
    complete, 37, 243, 245
Mother wavelet, 8–9, *passim*
Multifractals, 227, 260, 276, 302, 325, 350–355, 361, 373
Multiple biosignals, 334
Multiresolution, 98–100, *passim*

Neural networks, 215, 249, 259–260
Neuroelectric waveforms, 290–302
Non-destructive testing, 238–249

Ocean drifter trajectories, 220
Ocean flows, 220
Optical images (medical), 328–330
Option prices, 360
Orthonormal basis, 95, 97
Otoacoustic emissions, 321–322

Passband centre, 11, 14, 29–30, 32, 115
Phase difference, 78–79, 83, 85, 238, 298, 326, 356, 357
Phase plot, 39–42, 44, 236, 281, 303
Phase synchronization index (PSI), 83–85
Phonocardiogram, 315, 317
Photoplethysmogram, 302–315
Piled foundations, 238–246
Plasmas, 227, 348
Posture, 333–334
Power (scale dependent), 178, 179, 184
Power spectrum, 32, 41, 114–115, 179–182, 196, 208, 211–212, 271–272, 345
Pulsar, 376
P-wave, 268, 270, 364

QRS complex, 268, 270, 272–275, 278, 285, 287–290
Quantum mechanics, 378

Rainfall, 221–224
Rapid eye movement (REM), 299–301
Reassignment, 73–75, 234, 243, 246
Reconstruction algorithm, 107, 113, 120, 126, 147
Remote sensing, 366–371
Rephasing, 69–73
Respiratory signals, 277, 299, 303–307, 312, 317, 320, 324, 330, 334, 335
Ridge/ridges, 43, 61–67, 78–79, 89, 90, 206–207, 220–222, 232–235, 269, 273, 300, 305, 308, 311, 314, 341, 370, 373, 375
    heights, 66–67
    following, 61–66
Ridgelet, 168–172

Ringdown, 240, 243
River flows, 221–224
Rossby waves, 217–218
Rotating machinery, 249–259
Running wavelet archetyping, 67–69, 90, 308, 311–314

Saturn's rings, 373–374
Scale indexing 115–116
Scaling coefficient, 101–107, *passim*
Scaling equation, 101–102
Scaling function, 98–102, *passim*
Scalogram, *passim*
Secondary wavelet feature decoupling (SWFD), 61–66, 305–307
Sediment transport, 224, 227, 352
Sequential indexing, 116
Shafts, 255–259
Shannon entropy measure, 165, 166, 191, 193–194, 216
Share price, 355
Shearlet, 168–173
Shift invariance, 149
Shockable rhythm, 285
Short-time Fourier transform (STFT), 44–51, 75, 163, 280, 283, 315
Signal detail, 99–100, 106, 110, 121, 125
Sinus rhythm, 268, 278, 285
Skewness factor, 179
Sleep apnea, 278, 299, 313, 315
Smoothing, 127–135
Snoring, 318–321
Solar flare, 371, 373
Solar phenomena, 371
Sonic echo, 238, 243
Sounds (pathological), 315–323
Spectra, *passim*
Spectrogram, 29, 31, 280, 315
Speech, 318–321
Stationary wavelet transform, 206, 268, 330, 376, 378
Statistical moments, 178
Stock market, 355–357, 359–360
Subsampled, 143
Subsurface media, 361–366
Surface characterization, 262, 263
Surface temperature, 218, 222, 227, 352
Symmlet wavelet, 148, 164, 173, 191–193, 325, 330
Synchrosqueezing, 73–75, 252–253, 312, 327, 364–366

Taps, 144
Taylor expansion, 180

Taylor's frozen flow hypothesis, 184
Threshold, 127–135, 186–191, *passim*
    universal, 131–132, 134–135, 174, 190
    scale, 123, 189, 190
Thresholding
    magnitude thresholding, 129
    hard, 129, 131–134, 187, 189, 190, 191, 264, 275, 318, 362
    soft, 129, 131, 134–135
    SURE, 135, 174
    HYBRID, 362
    Lorentz, 174, 190, 191
    Bayesian, 135
    cross validation, 135
Tight frame, 95, 207
Time-frequency atom, 48–51, 52, 54
Transform modulus differences, 76
Transform modulus ratios, 76
Transform plot, *passim*
Transient behaviour (dynamics), 206, 232, 238, 251, 263, 265
Translation invariance, 149–150
Tree crown diameter, 369
Tsunami, 217
Turbulence, 178–200, 211–215, 224, 226, 227, 228, 351, 352
Turbulent intermittency, 211
Turbulent vorticity fields, 201, 202, 203, 205
T-wave alternan (TWA), 273
Two-dimensional wavelets (continuous), 55–56
Two-dimensional wavelets (discrete), 151–163
Two-phase flows, 224–227

Ultrasonic images (medical), 330
Upsampled, 144

Vasodilator, 315, 324
Vegetation cover, 366
Venous air embolism, 317
Ventricular fibrillation (VF), 273, 278–286
Ventricular late potentials (VLPs), 273, 278
Video biosignals, 309, 311, 312, 314
Viewing the transform, 56–61
VISUSHRINK, 134, 174, 264
Vortex shedding, 30, 181, 182, 185, 187–210, 194–196, 198, 199, 205–210, 220, 226, 228

Wake flows, 198, 199
Wave slam, 216
Wavelet bicoherence (WBC), 86–88, 206, 215, 217, 227–228, 252–254, 259, 297, 302, 318–320, 360
Wavelet bispectrum (WBS), 86, 302, 319
Wavelet coefficients, *passim*
Wavelet coherence (WCOH), 85–86, *passim*
Wavelet cross-bicoherence (WcrBC), 360
Wavelet cross-bispectrum (WCrBS), 86–87
Wavelet cross-correlation (WCC), 81–83, 224, 326–327, 362
Wavelet packets, 163–168, *passim*
Wavelet ratio surface, 76, 308, 311
Wavelet transform, *passim*
Wavelet variance, 33, 211, 213, 214, 222
Waves (ocean surface), 215–221
Well logging, 361–366
Wind flow, 206, 211–215
Windowed Fourier atom, 47, 48, 297
Wraparound, 59, 108, 109, 141, 143, 145, 162, 174

Yanai waves, 211, 217

For Product Safety Concerns and Information please contact our EU representative GPSR@taylorandfrancis.com Taylor & Francis Verlag GmbH, Kaufingerstraße 24, 80331 München, Germany